The Biology of the Turbellaria

Developments in Hydrobiology 6

Series editor
H. J. Dumont

DR W. JUNK PUBLISHERS THE HAGUE–BOSTON–LONDON 1981

The Biology of the Turbellaria

Proceedings of the Third International Symposium held in Diepenbeek, Belgium

Edited by
Ernest R. Schockaert
and Ian R. Ball

Reprinted from *Hydrobiologia*, vol. 84 (1981)

DR W. JUNK PUBLISHERS THE HAGUE-BOSTON-LONDON 1981

Distributors:

for the United States and Canada

Kluwer Boston, Inc.
190 Old Derby Street
Hingham, MA 02043
U.S.A.

for all other countries

Kluwer Academic Publishers Group
Distribution Center
P.O. Box 322
3300 AH Dordrecht
The Netherlands

This volume is listed in the Library of Congress Cataloging in Publication Data

ISBN-13: 978-94-009-8670-1 e-ISBN-13: 978-94-009-8668-8
DOI: 10.1007/978-94-009-8668-8

Cover design: Max Velthuijs

Preface

The Third International Symposium, 'The Biology of the Turbellaria', was held at the Limburgs Universitair Centrum, Diepenbeek, Belgium from 11 to 15 August 1980. The first (Chicago, 1970) and second (Helsinki, 1977) symposia were held in memory of Libbie Hyman and Alex Luther respectively. This third symposium, however, was held in honour of a living legend within the field of turbellarian biology, Tor G. Karling. The co-organizers of the symposium took special pride in this, not only for its acknowledgement of Tor Karling's scientific contributions, but also because they could themselves acknowledge the kindnesses they have received from him, as well as this interest in, and encouragement of, their own work. That many others have been so influenced, directly or indirectly, is indicated by the fact that some 57 participants representing 17 nations attended or contributed. An especially rewarding feature was the preponderance of younger 'new faces'. We are sure that it will please Tor to know that interest in Turbellaria continues, even grows, among the following generation of zoologists.

During the course of the symposium some 30 formal presentations were given. These ranged from critical review papers to poster presentations of specialized research topics. The diversity of the presentations contributed in no small way to the success of the symposium. Not all the presentations, however, were submitted for publication and some that were not given, because of illness of the invited speakers, have been included. We believe that these proceedings give a good overview of the current status of research into the biology of the Turbellaria.

The papers are not published here in precisely the same order as they were given at the symposium. What we have done is to group the more technical papers by subject matter and to preface each group where possible by one or two of the invited review papers covering the same field. We hoped in this way to make the published proceedings more unified and more useful as a general overview of turbellarian biology.

The success of the symposium was in no small way due to the splendid facilities which were made available by the Limburgs Universitair Centrum, and in acknowledging this help we must make special mention of the Rector, Professor Dr. L. Verhaegen and the Vice Rector, Professor Dr. M. van Poucke. The costs of the symposium were met by a subvention from the National Fonds voor Wetenschappelijk Onderzoek of Belgium, the Ministerie voor Nationale Opvoeding and the Limburgs Universitair Centrum. We also owe our gratitude to Dr. A. Grijpdonck, Head of Administration at the Limburgs Universitair Centrum, whose staff gave us their considerable help, particularly Miss T. de Backere and Miss G. Vlieghe, who both functioned as symposium secretaries. We are also indebted to the Governor of the Province of Limburg and the Mayor of Hasselt for their help with the social events, i.e. the reception at the provincial Open Air Museum of Bokrijk and the concert at the Hasselt Music School, respectively. On behalf of all the participants of the symposium we thank them all most heartily.

Tor, we sincerely hope that you have derived as much personal pleasure and satisfaction from this symposium as we have ourselves, and that you feel further rewarded by the dedication of this volume to yourself.

Ian R. Ball

Ernest R. Schockaert

Contents

ECOLOGY & FAUNISTICS

NUTRITION & REPRODUCTION

REGENERATION & DIFFERENTIATION

ULTRASTRUCTURE

1. Alan Pike
2. Elisabeth De Vries
3. Nicole Gourbault
4. Julian Smith
5. Marianne Wikgren
6. Reinhard Rieger
7. Irmeli Palmberg
8. George Shinn
9. Andrzej Czubaj
10. Alberto Lanfranchi
11. Michael Burt
12. Andrew Dorey
13. Peter Schwank
14. Jerzy Kolasa
15. Jan Hendelberg
16. Leigh Winsor
17. Jean-Pierre L'Hardy
18. Paul Martens
19. Jan Van der Land
20. Beate Sopott-Ehlers
21. Isabelle Martelly
22. Ulrich Ehlers
23. Jerzy Moraczewski
24. Maria Reuter
25. Nathan Riser
26. Barbara MacKinnon
27. Angela MacFarlane
28. Lodovico Galleni
29. Iwashiro Oki
30. Els Martens
31. Dirani Moreira
32. Vittorio Gremigni
33. Patrick Boaden
34. Aage Christensen
35. Waturu Teshirogi
36. Jaime Baguñà
37. Jacinto Gamo
38. Rafael Romero
39. Ian R. Ball
40. Eveline Dubois Reymond-Marcus
41. Jozef Hauser
42. Anne-Lise Karling
43. Tor Karling
44. Tom Reynoldson
45. Ernest Schockaert

List of participants and contributors

Jaime Baguñà, Departamento de Genetica, Fac. de Biologia, Universidad de Barcelona, Plaça Universidat, Barcelona 7, Spain

Ian R. Ball, Instituut voor Taxonomische Zoologie, Universiteit van Amsterdam, P.O. Box 20125, NL - 1000 HC Amsterdam, The Netherlands

Guiseppina Benazzi-Lentati, Istituto di Zoologia e Anatomia comparata, Universita di Pisa; Private address: Via dell'Ordine di S. Stefano 11, Marina di Pisa, Italy

Mario Benazzi, Istituto di Zoologia e Anatomia comparata, Universita di Pisa; Private address: Via dell'Ordine di S. Stefano 11, Marina di Pisa, Italy

Rob Biersma, Instituut voor Taxonomische Zoologie, Universiteit van Amsterdam, P.O. Box 20125, NL - 1000 HC Amsterdam, The Netherlands

Patrick J. S. Boaden, Marine Biology Station, Queen's University of Belfast, Portaferry BT 22 1PF, Northern Ireland

Michael D. B. Burt, Department of Biology, University of New Brunswick, P.O. Box 4400, Fredericton, E3B 5A3, NB, Canada

Aage M. Christensen, Marine Biological Laboratory, University of Copenhagen, DK - 3000 Helsingør, Denmark

Andrzej Czubaj, Department of Cytology, Institute of Zoology, University of Warsaw, Krakowskie Przedmiescie, 26/28, P - 00 - 927/1, Warsaw, Poland

Elisabeth J. De Vries, Instituut voor Taxonomische Zoologie, Universiteit van Amsterdam, P.O. Box 20125, NL - 1000 HC, Amsterdam, The Netherlands

Andrew E. Dorey, Department of Zoology, University of Bristol, Woodland Road, Bristol BS8 1UG, England

Eveline Dubois Reymond-Marcus, Department of Zoology, University of São Paulo, P.O. Box 6994, 01000 São Paulo, Brazil

Ulrich Ehlers, II. Zoologisches Institut der Universität, Berliner Strasse 28, D - 3400 Göttingen, F.R.G.

Lesley C. Fleming, Department of Biology, University of New Brunswick, P.O. Box 4400, Fredericton E3B 5A3, NB, Canada

Murina Galena, Institute of Biology of the Southern Seas of the Academy of Sciences, Nakhimow 2, Sevastopol 335000, U.S.S.R.

Lodovico Galleni, Istituto di Zoologia e Anatomia comparata, Universita di Pisa, Via Volta 4, I - 56100 Pisa, Italy

Jacinto Gamo, Departamento di Zoologia, Universidad de Alcalá, Alcalá de Henares (Madrid), Spain

Nicole Gourbault, Laboratoire des Vers, Museum national d'Histoire naturelle, 43 Rue Cuvier, F - 75231 Paris Cedex 05, France

Vittorio Gremigni, Istituto di Zoologia e Anatomia comparata, Universita di Pisa, Via Volta 4, I - 56100 Pisa, Italy

Jozef Hauser, Departamento de Biologia, Centro de Ciencias Biomedicas, Universidade do Vale do Rio dos Sinos, P.O. Box 275, 93000 São Leopoldo, R.S., Brazil

Jan Hendelberg, Department of Zoology, University of Göteborg, S - 40033 Göteborg, Sweden

Joseph B. Jennings, Department of Pure and Applied Zoology, University of Leeds, Baines Wing, Leeds LS2 9JI, England

Tor G. Karling, Section of Invertebrate Zoology, Swedish Museum of Natural History, S - 10405 Stockholm, Sweden

Jerzy Kolasa, Department of Animal Morphology, Institute of Biology, University of Poznan, nl. Szamarzewskiego 91, Poznan, Poland

Alberto Lanfranchi, Istituto di Biologia Generale, Universita di Pisa, Via Volta 6, I - 56100 Pisa, Italy

Jean-Pierre L'Hardy, Laboratoire de Biologie Animale, Faculté des Sciences, Université du Maine, Route de Laval, F - 72017 Le Mans Cedex, France

Angela E. MacFarlane, Department of Zoology, University of Bristol, Woodland Road, Bristol BS8 1UG, England

B. M. MacKinnon, Department of Zoology, University of Aberdeen, Tillydrone Ave., Aberdeen, Scotland

Isabelle Martelly, Laboratoire de Biologie Animale, Université de Paris, Val de Marne, Avenue Général De Gaulle, F - 94010 Créteil Cedex, France

Els E. Martens, Departement SBM, Limburgs Universitair Centrum, B - 3610 Diepenbeek, Belgium

Paul M. Martens, Departement SBM, Limburgs Universitair Centrum, B - 3610 Diepenbeek, Belgium

Jerzy Moraczewski, Department of Cytology, Institute of Zoology, University of Warsaw, Krakowskie Przedmiescie, 26/28, P - 00 - 927/1 Warsaw, Poland

Dirani Moreira, Departamento de Biologia, Centro de Ciencias Biomedicas, Universidade do Vale do Rio dos Sinos, P.O. Box 275, 93000 São Leopoldo, R.S., Brazil

Iwashiro Oki, Ôsaka Prefectural Institute of Public Health, Nakamichi 1 chôme 3–39, Kigashinari-ku, Ôsaka 537, Japan

Irmeli Palmberg, Institutionen för Biologi, Åbo Akademi, Porthansgatan 3, SF – 20500 Åbo 50, Finland

A. W. Pike, Department of Zoology, University of Aberdeen, Tillydrone Ave, Aberdeen, Scotland

Maria Reuter, Institutionen för Biologi, Åbo Akademi, Porthansgatan 3, SF – 20500 Åbo 50, Finland

Tom B. Reynoldson, Department of Zoology, University College of N. Wales, Bangor, Gwynedd LL57 2UW, Wales

Reinhard M. Rieger, Department of Zoology, University of North Carolina, 135 Wilson Hall, Chapel Hill, NC 27514, U.S.A.

Nathan W. Riser, Marine Science Institute, Northeastern University, East Point, Nahant, MA 01909, U.S.A.

Rafael Romero, Departamento de Genetica, Fac. de Biologia, Universidad de Barcelona, Plaça Universidat, Barcelona 7, Spain

Ernest Schockaert, Departement SBM, Limburgs Universitair Centrum, B – 3610 Diepenbeek, Belgium

Peter Schwank, Limnologische Flussstation des Max-Planck-Instituts für Limnologie, P.O. Box 260, D – 6407 Schlitz, F.R.G.

George Schinn, Department of Zoology, University of Washington, NJ-15 Seattle, WA 98195, U.S.A.

Julian P. S. Smith III, Department of Zoology, University of North Carolina, 202 Wilson Hall, Chapel Hill, NC 27514, U.S.A.

Beate Sopott-Ehlers, II. Zoologisches Institut der Universität, Berliner Strasse 28, D – 3400 Göttingen, F.R.G.

Waturu Teshirogi, Department of Biology, Faculty of Science, Hirosaki University, 3 Bunkyo-cho, Hirosaki, Aomori-ken 036, Japan

Seth Tyler, Department of Zoology, University of Maine at Orono, Murray Hall, Orono, ME 04473, U.S.A.

Jan Van der Land, Rijksmuseum van Natuurlijke Historie, P.O. Box 9517, NL – 2300 RA Leiden, The Netherlands

Marianne Wikgren, Institutionen för Biologi, Åbo Akademi, Porthansgatan 3, SF – 20500 Åbo 50, Finland

Leigh Winsor, Department of Zoology, James Cook University of North Queensland, Townsville, Queensland 4811, Australia

J. O. Young, Department of Zoology, University of Liverpool, P.O. Box 147, Liverpool L69 3BX, England

This volume is dedicated to Tor Karling

50 years of turbellarian research

A brief biography of Tor G. Karling

Jan Hendelberg
Dept. of Zoology, University of Göteborg, Box 25059, S-400 31 Göteborg, Sweden

Keywords: Turbellaria, T. G. Karling, biography

This third international symposium on the biology of the Turbellaria is held in honour of Professor Tor G. Karling. He is a very experienced person within the field, having now worked for half a century with morphological, taxonomical, systematic and other problems of the Turbellaria, and we are happy to find him here, taking part in the symposium.

Who is Tor Karling? When I was asked to give a biography of him at this symposium I felt honoured, but also hesitant. Is it possible in a short symposium paper to account for scientific results published under more than 80 titles? Even if I restrict myself to those dealing with turbellarians, there are still 66. And Karling is not only important as a scientist – it has been said that he is just as important as a man. As you can understand, it is not possible to give a full account of why this symposium is given in his honour, but, hopefully, this brief biography will give some explanation.

Karling was born in Hangö, Finland, in 1909. He retired from the chair of invertebrate zoology at the Swedish Museum of Natural History in 1975. Some of you may think that his rich scientific production is the result of a straight academic career – but this is not so. However, it began at a university, in Helsinki, where he was a student of Alex. Luther, who introduced him into the field of turbellarian morphology and systematics.

In 1929 Karling started his field studies of turbellarians at the Tvärminne Zoological Station, situated on the south coast of Finland, not far from Hangö. He took an interest in the small forms, about 1 mm long or somewhat more, many of them living in the interstices between the sand grains of the sea bottom. His first scientific paper, a report on *Bresslauilla relicta*, a dalyellioid turbellarian, was published in 1930.

A turbellarian group to which Karling has devoted much of his interest is the Kalyptorhynchia, a group characterized by having a proboscis for capturing prey, an organ in the anterior end of the animal, separate from the mouth which is ventral. In 1931 Karling published his first descriptions of turbellarian species new to science, three new kalyptorhynchs. Already in this early paper we find some of the skill in making the drawings which characterizes his publications. We also find that the descriptions are based on thorough morphological and anatomical studies which make them valuable not only as descriptions of species, but also as a basis for studying relationships of the Turbellaria as a whole and the evolution of the group.

In 1934 Karling was awarded a stipendium which enabled him to collect turbellarians at Herdla, on the west coast of Norway. In 1936 he worked for a period at the Swedish Museum of Natural History, in Stockholm. Here he met two other turbellarian scientists, Sixten Bock and Einar Westblad.

Karling wrote his thesis 'Zur Morphologie und Systematik der Alloeocoela Cumulata und Rhabdocoela Lecithophora (Turbellaria)' before the war. It was printed in 1940, and he defended it during the so-called Moscow Peace in an icy cellar in Helsinki with Alex. Luther as the opponent. The dissertation is a very thorough morphological study of a number of species, and also a thorough study of the international literature with very precise references. New ideas of the evolution of organs were put forward, and based on this, a new turbellarian

Hydrobiologia 84, 1–5 (1981). 0018-8158/81/0841-0001/$01.00.

system was erected. Just before this was published, similar ideas were put forward by Meixner, a scientist who had already been an inspiration to Karling earlier. In the new system suggested by Karling, the old groups Alloeocoela and Rhabdocoela were divided. The Macrostomida were no longer placed in the same group as many of the other former rhabdocoels, which were now called the Neorhabdocoela. Instead they were suggested to be a subgroup of a new order, the Archoophora, in which all the small turbellarians with the seemingly primitive type of eggs, the entolecithal eggs, were collected.

Karling spent some of the war years in uniform, and he was even wounded. For 28 years he was a teacher in different secondary schools, first in Finland where he was also a headmaster. After a five-year period at the Swedish Museum of Natural History in Stockholm 1951–1956 he returned to the teaching profession, now in Sweden, undertaking most of his turbellarian studies in his spare time. However, these studies were so successful that in 1967 he was appointed a professor in invertebrate zoology at the Swedish Museum of Natural History. He held this chair until his retirement in 1975.

One of the secrets behind Karling's successful studies is the very thorough approach he always adopts. This already starts in the field. When he collects material, he always makes very careful notes about the habitat and about the bottom material from which the turbellarians are later extracted. He is also interested in collecting devices. Among others, he constructed a special sand scraper for small animals (Karling 1937).

Karling collected much of his material in the Baltic Sea along its Finnish and Swedish coasts. In 1974 he published a very valuable synopsis of the turbellarian fauna of this brackish water sea. Based on this synopsis, I have compiled a diagram (Fig. 1) showing the number of turbellarians described during different decades. As can be seen most of them were described during the last five decades, and about a quarter of the total number of new descriptions were made by Karling.

In the synopsis published in 1974, there are keys to the identification of all the Baltic turbellarians known up to then, as well as figures showing the characteristic features of them all, for instance the hard structures of the genital organs which, when occurring, are species-specific. Besides a systematic list, there is a general informative list of species, in alphabetical order, with references to the numbers in the systematic list. The idea of using alphabetic order for systematic material at first I found somewhat strange. However, I have since found it very useful. It has other advantages than just to put the first species Karling described, *Acrorhynchides robustus*, at the top of the list. The synopsis also provides much information about ecology and biogeography of the Baltic species, a result, among others, of the careful notes Karling made during the collection of material.

Karling's work was not limited to the Scandinavian countries. In 1960 he visited two marine stations on the Pacific coast of North America where he collected material. Apart form the polyclads studied by Hyman and others, only four free-

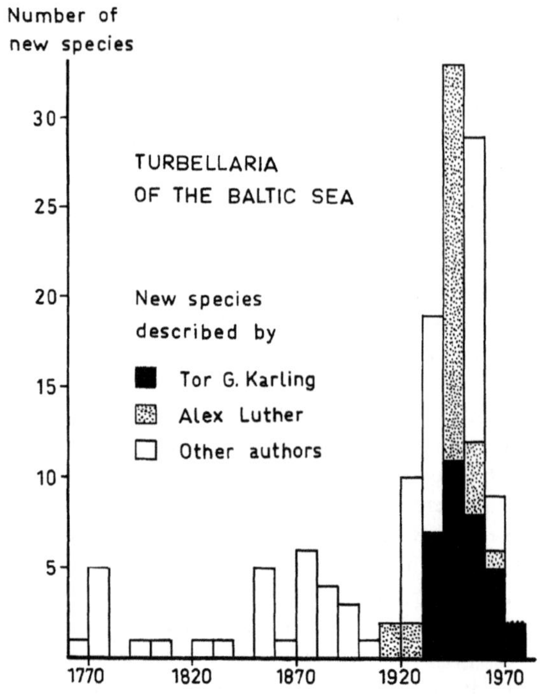

Fig. 1. Turbellaria of the Baltic: number of species described in different decades. As can be seen, most of the species were described in the last five decades (1921–1970) preceding Karling's (1974b) synopsis, from which the data were extracted. The majority of species were described by one or other of three authors: Tor G. Karling, Alex Luther and Peter Ax.

living turbellarian species were known at that time from this coast. In a short period Karling collected about 80 species, most of them new to science, and started to describe them in a series of publications (1962a, b, 1964a, 1966c). Since then he has visited many other parts of the world, and found and described other new turbellarians, always with the same careful investigation and concise interpretation of structural details.

One of the new species, found in Bermuda, was called by Karling (1978) *Annalisella bermudensis.* Thus, the new genus was dedicated to his wife, Anna-Lisa, his support in life and not least during the many years in which he was using much of his spare time for studying the Turbellaria. Everyone who has met Anna-Lisa knows how much this support must have meant to Tor.

Among all the theoretical discussions and ideas in Tor Karlings papers I have chosen a few to give you examples of this side of his work. One example is his paper (Karling 1961a) dealing with the bifurcated proboscis found in many Kalyptorhynchia. Besides giving information about the morphology of this structure in different genera, Karling gives detailed information about its function and also discusses its evolution.

One of Karling's ideas, now generally accepted, I think, was that the evolution from a primitive to a more complex structure often follows the same way in different turbellarian organ systems. Thus, for example, the copulatory apparatus, the pharynx structure of the mouth, and the adhesive organs-proboscis, all characterized by functioning with muscular contraction and secretion, were found by Karling (1963b) to show similar trends of evolution, starting with a simple folding and continuing with more complicated alterations. From these observations Karling concluded that the spontaneous variability affording the basis of selection, is restricted by the structural properties of turbellarian tissues.

In one of his papers Karling (1967a) discussed the systematic values of the categories Archoophora and Neoophora, categories characterized by entolecithal and ectolecithal eggs, respectively. This paper was published in German, a language used by Karling in many of his earlier papers. Karling was born in a Swedish-speaking part of Finland, but he had to learn Finnish, too. He started to publish in German, but from 1961 and onwards most of his papers are written in English. But now and then, especially when dealing with theoretical problems, he has returned to German.

A further development of the systematics of the Turbellaria was made by Karling in 1970 at the Hyman Memorial Symposium in Chicago. Adopting Hennig's ideas of analysis of relationships, based on the sister-group principle, Karling then discussed the affinities of the turbellarian subgroups. He also reconstructed a theoretical turbellarian archetype. His detailed discussion, published in the Hyman Memorial Volume (Karling 1974a), will be of great value for coming generations of turbellarian scientists.

Karling's collections of material from different geographical regions have made it possible for him to compare how structural details of closely related species have evolved in different parts of the world. An example of this is his study (Karling 1977a) of the genus *Austrorhynchus,* species of which were collected in the Mediterranean, the Antarctic regions, different parts of the Pacific, and the Indian Ocean. On the basis of the structural studies, the phylogeny and biogeography of the genus were discussed.

Science is international, and few persons have shown this better than Tor Karling. In his search for material he has worked in laboratories in different parts of the world. He has helped and encouraged many of us younger students from different countries. He has led international courses in the study of the Turbellaria. He has taken part in many symposia in Europe and America, and he has taken part in the organization of such symposia, such as the Symposium on Turbellaria held at Tvärminne, in 1977.

In Tvärminne this limerick was dedicated to Tor Karling by Pat Boaden:

There once was a fellow called Tor
Who said 'I won't write any more'
But with exhortation
From friends of each nation
He wrote twice as much as before.

I do not know if it is possible for Karling to write twice as much as before, but I know that most days he works for a number of hours in his emeritus room at the Swedish Museum of Natural History. And I really hope that he will spare some time for publishing a check-list, or perhaps a full synopsis, of the Kalyptorhynchia of the world, that is, of the group he knows better than any other person. Anyway, what is sure is that he has made friends wherever he has been, and this, of course, is one of the reasons why we are happy to have him here with us to-day. We now thank him for what he has done and wish him many more prosperous years of turbellarian research.

References

Tor G. Karling's papers on Turbellaria and other aquatic worms are listed in this reference list.

Karling, T. G., 1930. Bresslauilla relicta Reisinger (Turbellaria, Rhabdocoela) zum ersten Male in Finnland angetroffen. Mem. Soc. Fauna Flora fenn. 6: 128–130.

Karling, T. G., 1931. Untersuchungen über Kalyptorhynchia (Turbellaria Rhabdocoela) aus dem Brackwasser des Finnischen Meerbusens. Acta zool. fenn. 11: 1–67.

Karling, T. G., 1934a. Ein Beitrag zur Kenntnis der Nemertinen des Finnischen Meerbusens. Mem. Soc. Fauna Flora fenn. 10: 76–90.

Karling, T. G., 1934b. Haplobranchus balticus n.sp. Eine neue sedentäre Polychäte aus dem Finnischen Meerbusen. Mem. Soc. Fauna Flora fenn. 10: 242–244.

Karling, T. G., 1935. Mitteilungen über Turbellarien aus dem Finnischen Meerbusen: 1. Dalyellia nigrifrons n.sp. 2. Promesostoma cochlearis n.sp. Mem. Soc. Fauna Flora fenn. 10: 388–395.

Karling, T. G., 1937. Ein Apparat zum Auffangen von Kleintieren des Meeressandes. Acta Soc. Fauna Flora fenn. 60: 387–391.

Karling, T. G., 1940. Zur Morphologie und Systematik der Alloeocoela Cumulata und Rhabdocoela Lecithophora (Turbellaria). Acta zool. fenn. 26: 1–260.

Karling, T. G., 1943. Studien an Halammovortex nigrifrons (Karling) (Turbellaria Neorhabdocoela). Acta zool. fenn. 37: 1–23.

Karling, T. G., 1947. Studien über Kalyptorhynchien (Turbellaria). I. Die Familien Placorhynchidae und Gnathorhynchidae. Acta zool. fenn. 50: 1–64.

Karling, T. G., 1949. Studien über Kalyptorhynchien (Turbellaria). II. Die Familien Karkinorhynchidae und Diascorhynchidae. Acta zool. fenn. 58: 1–42.

Karling, T. G., 1950. Studien über Kalyptorhynchien (Turbellaria). III. Die Familie Schizorhynchidae. Acta zool. fenn. 59: 1–33.

Karling, T. G., 1952a. Kalyptorhynchia (Turbellaria). Further zool. Results Swed. Antarct. Exped. 1901–03 4, 9: 1–50.

Karling, T. G., 1952b. Studien über Kalyptorhynchien (Turbellaria). IV. Einige Eukalyptorhynchia. Acta zool. fenn. 69: 1–49.

Karling, T. G., 1953a. Cytocystis clitellatus n.gen., n.sp., ein neuer Eukalyptorhynchien-Typus (Turbellaria). Ark. Zool. (2) 4: 493–504.

Karling, T. G., 1953b. Zur Kenntnis der Gattung Rogneda Ulianin (Turbellaria, Kalyptorhynchia). Ark. Zool. (2) 5: 349–368.

Karling, T. G., 1954a. Einige marine Vertreter der Kalyptorhynchien-Familie Koinocystididae. Ark. Zool. (2) 7: 165–183.

Karling, T. G., 1954b. Echinoderes levanderi n.sp. (Kinorhyncha) aus der Ostsee. Ark. Zool. (2) 7: 189–192.

Karling, T. G., 1954c. Über einige Kleintiere des Meeressandes des Nordsee-Ostsee-Gebietes. Ark. Zool. (2) 7: 241–249.

Karling, T. G., 1955. Studien über Kalyptorhynchien (Turbellaria). V. Der Verwandtschaftskreis von Gyratrix Ehrenberg. Acta Zool. fenn. 88: 1–39.

Karling, T. G., 1956a. Morphologisch-histologische Untersuchungen an den männlichen Atrialorganen der Kalyptorhynchia (Turbellaria). Ark. Zool. (2) 9: 187–279.

Karling, T. G., 1956b. Zur Kenntnis einiger Gnathorhynchiden nebst Beschreibung einer neuen Gattung. Ark. Zool. (2) 9: 343–353.

Karling, T. G., 1956c. Alexlutheria acrosiphoniae n.gen., n.sp., ein bemerkenswerter mariner Vertreter der Familie Dalyelliidae (Turbellaria). Ark. Zool. (2) 10: 331–345.

Karling, T. G., 1957. Drei neue Turbellaria Neorhabdocoela aus dem Grundwasser der schwedischen Ostseeküste. K. Fysiogr. Sällsk. Lund Förh. 27: 25–33.

Karling, T. G., 1958a. Zur Kenntnis von Stygocapitella subterranea Knöller und Parergodrilus heideri Reisinger (Annelida). Ark. Zool. (2) 11: 307–342.

Karling, T. G., 1958b. Zur Kenntnis der Gattung Coelogynopora Steinböck (Turbellaria Proseriata). Ark. Zool. (2) 11: 559–568.

Karling, T. G., 1961a. Zur Morphologie, Entstehungsweise und Funktion des Spaltrüssels der Turbellaria Schizorhynchia. Ark. Zool. (2) 13: 253–286.

Karling, T. G., 1961b. On a species of the genus Multipeniata Nasonov (Turbellaria) from Burma. Ark. Zool. (2) 15: 105–111.

Karling, T. G., 1962a. Marine Turbellaria from the Pacific Coast of North America. I. Plagiostomidae. Ark. Zool. (2) 15: 113–141.

Karling, T. G., 1962b. Marine Turbellaria from the Pacific Coast of North America. II. Pseudostomidae and Cylindrostomidae. Ark. Zool. (2) 15: 181–209.

Karling, T. G., 1962c. Gnathostomula paradoxa Ax, 1956 (Gnathostomulida) from the vicinity of Bergen, Norway. Sarsia 7: 23–24.

Karling, T. G., 1963a. Ulianinia mollissima Levinsen, 1879, re-discovered (Turbellaria Prolecithophora). Vidensk. Medd. dansk naturh. Foren. 125: 496–508.

Karling, T. G., 1963b. Some evolutionary trends in turbellarian morphology. In: Ed. Dougherty, E. C. The Lower Metazoa, Univ. Calif. Press. Berkeley and Los Angeles, pp. 225–233.

Karling, T. G., 1963c. Die Turbellarien Ostfennoskandiens. V. Neorhabdocoela 3. Kalyptorhynchia. Fauna fenn. 17: 1–59.

Karling, T. G., 1964. Ueber einige neue und ungenügend bekannte Turbellaria Eukalyptorhynchia. Zool. Anz. 172: 159–183.

Karling, T. G., 1965a. Marine Turbellaria from the Pacific Coast of North America. III. Otoplanidae. Ark. Zool. (2) 16: 527–541.

Karling, T. G., 1965b. Haplopharynx rostratus Meixner (Turbellaria) mit den Nemertinen vergleichen. Z. Zool. Syst. Evolut.-Forsch. 3: 1–18.

Karling, T. G., 1966a. On nematocysts and similar structures in turbellarians. Acta zool. fenn. 116: 1–28.

Karling, T. G., 1966b. Rhodope veranyi (Gastropoda, Euthyneura) von der norwegischen Küste. Sarsia 24: 33–35.

Karling, T. G., 1966c. Marine Turbellaria from the Pacific Coast of North America. IV. Coelogynoporidae and Monocelididae. Ark. Zool. (2) 18: 493–528.

Karling, T. G., 1966d. On the defecation apparatus in the genus Archimonocelis (Turbellaria, Monocelididae). Sarsia 24: 37–44.

Karling, T. G., 1967a. Zur Frage von dem systematischen Wert der Kategorien Archoophora und Neoophora (Turbellaria). Commentat. biol. 30: 1–11.

Karling, T. G., 1967b. On the genus Promesostoma (Turbellaria), with descriptions of four new species from Scandinavia and California. Sarsia 29: 257–268.

Karling, T. G., 1968. On the genus Gnosonesima Reisinger (Turbellaria). Sarsia 33: 81–108.

Karling, T. G., 1969. Ein überzähliger Genitalapparat bei einem rhabdocoelen Turbellar. Z. Morph. Tiere 65: 202–208.

Karling, T. G., 1970a. On Pterastericola fedotovi (Turbellaria), commensal in sea stars. Z. Morph. Tiere 67: 29–39.

Karling, T. G., 1970b. Kinorhyncha. In: Ed. Gray, P., The Encyclopedia of the Biological Sciences. New York: Reinhold Co, pp. 470–471.

Karling, T. G., 1970c. Bothriomolus balticus Meixner 1938 (Turbellaria) in dem schwedischen Binnensee Vättern. Zool. Anz. 184: 120–121.

Karling, T. G., 1970d. Aspects on the anatomy and affinities of the turbellarian groups. Am. Zool. 10: 525.

Karling, T. G., 1973a. Anatomy and taxonomy of a new otoplanid (Turbellaria, Proseriata) from South Georgia. Mikrofauna Meeresboden 16: 361–369.

Karling, T. G., 1973b. Ist Hypoblepharina Böhmig (Turbellaria) eine kommensalistische Gattung? Mit Beschreibung einer neuen Art aus der Amphipode Cheirimedon femoratus. Z. Morph. Tiere 75: 51–58.

Karling, T. G., 1974a. On the anatomy and affinities of the turbellarian orders. In: Eds. Riser, N. W. & Morse, M. P. Biology of the Turbellaria. New York: McGraw-Hill, pp. 1–16.

Karling, T. G., 1974b. Turbellarian fauna of the Baltic proper. Identification, ecology and biogeography. Fauna fenn. 27: 1–101.

Karling, T. G., 1977a. Taxonomy, phylogeny and biogeography of the genus Austrorhynchus Karling (Turbellaria, Polycystididae). In: Ed. Sterrer, W. & Ax, P., The meiofauna species in time and space. Mikrofauna Meeresboden 61, pp. 153–165.

Karling, T. G., 1977b. On Alex Luther's contribution to our knowledge of the Turbellaria. In: Eds. Karling, T. G. & Meinander, M., The Alex Luther Centennial Symposium on Turbellaria. Acta zool. fenn. 154, pp. 4–9.

Karling, T. G., 1978. Anatomy and systematics of marine Turbellaria from Bermuda. Zool. Scripta 7: 225–248.

Karling, T. G., 1980. Revision of Koinocystididae (Turbellaria). Zool. Scripta 9: 241–269.

Karling, T. G. & Kinnander, H., 1953. Några virvelmaskar från Östersjön, Svensk faun. Revy 3: 73–79.

Karling, T. G., Mack-Fira, V. & Dörjes, J., 1972. First report on marine microturbellaria from Hawaii. Zool. Scripta 1: 251–269.

Karling, T. G. & Mack-Fira, V., 1973. Zur Morphologie und Systematik der Gattung Paramesostoma Attems (Turbellaria Typhloplanoida). Sarsia 52: 155–170.

Karling, T. G. & Meinander, M. (eds.), 1977. The Alex Luther Centennial Symposium on Turbellaria. Proceedings of a symposium at Tvärminne Zoological Station, Finland, 1977. Acta zool. fenn. 154.

Karling, T. G. & Nilsson, A.-M., 1974. Further studies on the genus Hypoblepharina Böhmig (Turbellaria, Dalyellioida) with description of H. serrifera sp.n. Zool. Scripta 3: 59–63.

Karling, T. G. & Schockaert, E. R., 1977. Anatomy and systematics of some Polycystididae (Turbellaria, Kalyptorhynchia) from the Pacific and S. Atlantic. Zool. Scripta 6: 5–19.

Bashiruddin, M. & Karling, T. G., 1970. A new ectocommensal turbellarian (fam. Pterastericolidae) from the sea star Astropecten irregularis. Z. Morph. Tiere 67: 16–28.

Dörjes, J. & Karling, T. G., 1975. Species of Turbellaria Acoela in the Swedish Museum of Natural History, with remarks on their anatomy, taxonomy and distribution. Zool. Scripta 4: 175–189.

Holmquist, C. & Karling, T. G., 1972. Two new species of interstitial marine triclads from the North American Pacific coast, with comments on evolutionary trends and systematics in Tricladida (Turbellaria). Zool. Scripta 1: 175–184.

Schockaert, E. & Karling, T. G., 1970. Three new anatomically remarcable Turbellaria Eukalyptorhynchia from the North American Pacific coast. Ark. Zool. (2) 23: 237–253.

Schockaert, E. R. & Karling, T. G., 1975. Anatomy and taxonomy of some species of Polycystididae (Turbellaria, Kalyptorhynchia) from N. Atlantic and Mediterranean coastal areas. Zool. Scripta 4: 133–143.

The phyletic status of the Paludicola

Ian R. Ball
Institute of Taxonomic Zoology, University of Amsterdam, Plantage Middenlaan 53, Amsterdam, The Netherlands

Keywords: Turbellaria, Tricladida, Paludicola

Abstract

Is the assumption of the monophyly of the Paludicola (freshwater triclads) justified? There seems little doubt on morphological grounds of the close relationship between the Planariidae and the Dendrocoelidae but it is difficult to connect, phylogenetically, the Dugesiidae with these groups. A connection between the former and some of the latter has been unknowingly implied by the author in previous papers, thus raising an anomaly since this would mean that the Dugesiidae are not monophyletic.

Is it possible that the Dugesiidae and the Planariidae + Dendrocoelidae have arisen independently from marine ancestors? The implications of such a speculation are examined.

Introduction

The purpose of this paper is to draw attention to an anomaly in previously developed theories of the origin and distribution of freshwater triclads (Ball 1974a, 1977a). I have failed to resolve this anomaly while retaining the assumption of the monophyly of the Paludicola, but the anomaly disappears if that assumption is dropped. Nonetheless, a proposal that the Paludicola may be di- or even polyphyletic has other implications that need exploring.

The original hypothesis

In its strict formulation (Ball 1975) the original hypothesis has been summarised by Pielou (1980) and by Lewis (1980). Pielao has misunderstood some of the arguments and so her discussion is invalid. In brief, I regarded the Dugesiidae as the most primitive family of freshwater triclads and proposed that they had originated south of the present day equator, in Gondwanaland, prior to continental break-up. The northwards dispersal of elements of this original fauna coincided with the early disruption of Gondwanaland and after separation was well underway the subgenus *Dugesia* arose in Africa and after closure of the Tethys sea dispersed northwards and throughout Palaearctis (after the opening of the North Atlantic Ocean). The subgenus *Schmidtea* arose later in western Palaearctis. The higher freshwater triclads, the Planariidae and Dendrocoelidae, together constituting the sister group of the Dugesiidae, arose in Palaearctis also after the closure of the Tethys sea and the elevation of the Himalayan chain.

In a later paper (Ball 1977a) it was argued that *Schmidtea* probably was not 'derived' but rather represented the modern descendents of an originally widespread (Pangaean) proto-dugesiid which later differentiated vicariantly, so much of the dispersal element of the earlier formulation could be abandoned. Ultrastructural studies by Gremigni (1979) support this alternative as does continuing histological work by E.J. de Vries and myself.

For the purposes of the discussion which now

Hydrobiologia 84, 7–12 (1981). 0018-8158/81/0841-0007/$01.20.

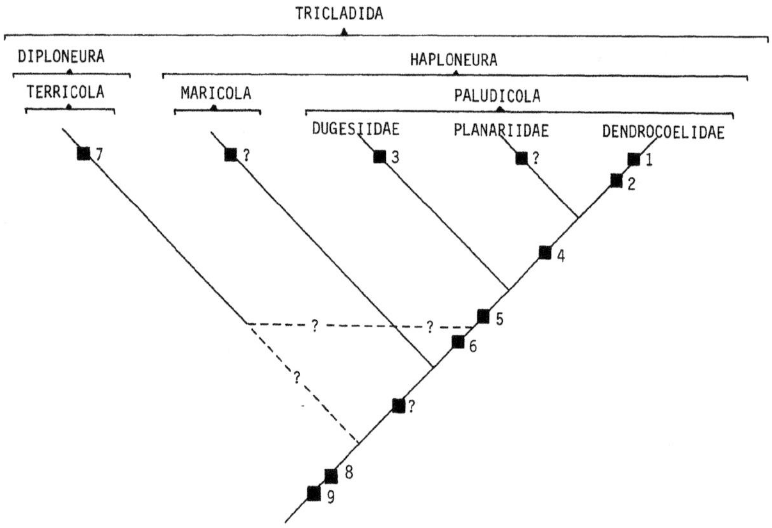

Fig. 1. Phylogenetic relationships of the Tricladida showing the defining autapomorphies for each phyletic line (black squares). The characters are: (1) inner pharyngeal muscles intermingled, (2) anterior adhesive organ present, (3) multicellular eye-cup, (4) common oviduct opens into atrium, (5) probursal, (6) reduced cephalic duct (7) diploneuran nervous system, (8) tricladoid intestine, (9) germaria anterior in female gonads.

Table 1. Classification of the Tricladida.

1. DIPLONEURA
1.1. Terricola
2. HAPLONEURA
2.1. Maricola
2.2. Paludicola
2.2.1. Dugesioidea
2.2.1.1. Dugesiidae
2.2.2. Planarioidea
2.2.2.1. Planariidae
2.2.2.2. Dendrocoelidae

follows the classification and sistergroup relationships that the original hypothesis require are shown in Table 1 and Fig. 1. I shall refer to the Planariidae plus Dendrocoelidae as the superfamily Planarioidea co-ordinate with the Dugesioidea, the latter containing the single family Dugesiidae.

The anomaly

In strict phylogenetic classifications, which are essential for biogeographical analysis, the members of each phyletic line must be united by uniquely derived shared features, or synapomorphies. The characters defining precisely each line are then referred to as autapomorphies for the group in question.

The systems outlined in Table 1 and Fig. 1 require that there be autapomorphies defining the Haploneura, the Maricola, and the Planariidae, but these remain to be discovered, if indeed they exist. More important for the present discussion is the anomalous state of the Paludicola for if the Planarioidea arose in the Northern Hemisphere after the closure of Tethys then they must have closer phylogenetic relationships with some (i.e. northern) Dugesioidea than with others (i.e. southern) (Fig. 2), thus implying that there are one or more synapomorphies uniting some of the Dugesiidae with the Planarioidea. In other words the Dugesiidae *sensu* Ball 1974a are non-monophyletic, and yet they seem to be well defined by the autapomorphy of their peculiar eye structure (Ball 1974b). If Fig. 2 is correct then there has been secondary loss of the 'dugesiid eye' and a return to the primitive condition in the Planariidae, and evolution in a new direction in the Dendrocoelidae. It could be claimed that the proto-dugesiid (eg. pre *Cura* – *Schmidtea*) had the primitive 'maricolan eye', leading direct to the Planariidae, but then it

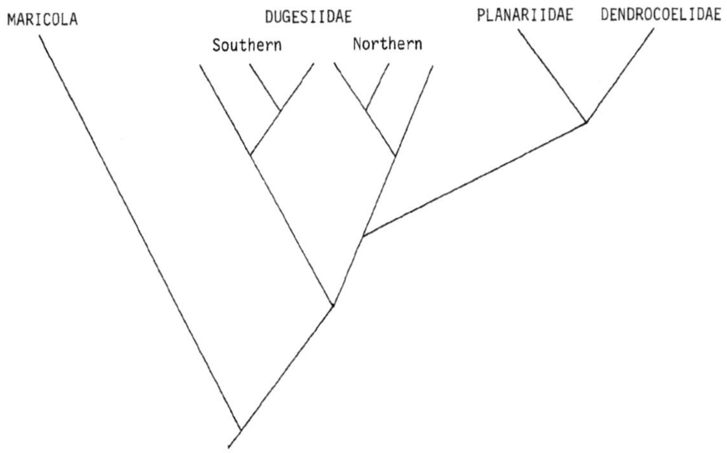

Fig. 2. A possible, stylised, phylogeny of the Haploneura as implied in earlier publications. For further discussion see text.

would be necessary, under the vicariance hypothesis (Ball 1977a), to assume a multiple origin of the complex 'dugesiid eye'.

A possible resolution of the anomaly

The problems of the status of the paludicolan groups can be resolved by assuming that the freshwater triclads are diphyletic. The phyletic relationships of the turbellarian groups, and their defining characters, have been discussed by Karling (1974) and by Ball (1974a) for the Tricladida. Within the latter the integrity of the Dugesiidae and the Planarioidea seem well assured by autapomorphous characters (Fig. 1) but it is difficult to link the two, wholly or in part. The supposed autapomorphies of the Paludicola are two (Fig. 1, characters 5,6); their probursal condition and the reduced cephalic duct, lacking precerebral diverticula, of the intestine. These are positional characters rather than complex structural characters and by generally accepted criteria of homologous relationships (see Riedl 1978) must be considered as weak defining characters. If one ignores them (but see below) an alternative phylogeny of the Tricladida could be based on Fig. 3. This assumes a diphyletic origin of the freshwater

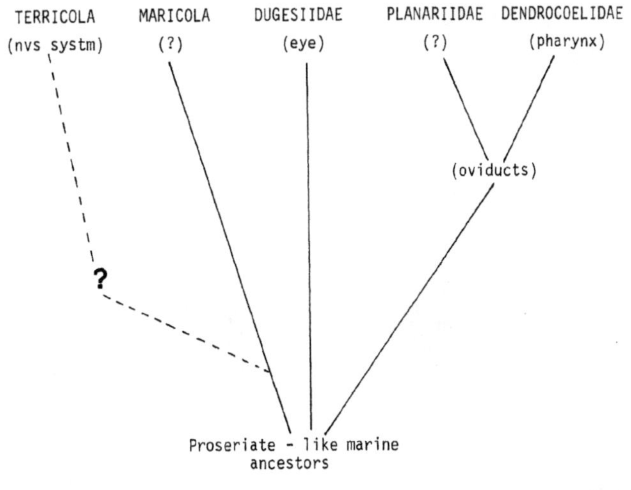

Fig. 3. A proposed open phylogeny of the Tricladida in which the Paludicola are diphyletic. See text.

triclads, and an independent origin of the Maricola, from marine proseriatelike ancestors. This admittedly speculative view, which expresses the degree of our ignorance concerning the Tricladida rather than the depth of our knowledge, has many implications.

Are the Paludicola diphyletic?

If the proposal of the non-monophyly of the Paludicola, as indicated in Fig. 3, is accepted then this implies that there should be autapomorphies defining (a) the Dugesiidae, (b) the Planarioidea, (c) the Planariidae, (d) the Dendrocoelidae, and not for (e) the Dugesiidae plus any or all of the Planariidae and/or Dendrocoelidae. Conditions (a), (b) and (d) are met (Fig. 1, characters 3,4,1 respectively); (c) is not met but then nor is it in the alternative schemes, wherein this same criterion is required. Condition (e) is violated only by the two weak characters (Fig. 1, characters 5,6) already discussed, and these need further consideration.

With respect to the probursal condition we may note that this is generally accepted as having arisen independently and severally in some marine triclads of the genera *Probursa* (Hyman 1944), *Pacificides* (Holmquist & Karling 1972) and *Bdellasimilis* (Ball 1976), and with a tendency to this condition in *Tryssosoma* (Ball 1977b). There are also retrobursal planarians occurring in freshwaters, *Rhodax* (Marcus 1946), *Opisthobursa* (Benazzi & Giannini 1973), *Eviella* (Ball 1977c) and *Balliania* (Gourbault 1978) which have been variously assigned to the Paludicola and the Maricola (for discussions see Ball 1977a; Gourbault 1978). It seems that there are no compelling reasons for assuming the probursal condition of the Paludicola to be uniquely derived.

Similar arguments can be advanced concerning the cephalic duct. The fact that the anterior ramus of the intestine does not form precerebral diverticula in the Paludicola, whereas it does in the Maricola, has been used as a defining or diagnostic character in the appropriate group (Meixner 1928; Luther 1960). Yet in the maricolan species of the family Bdellouridae this duct is reduced, whereas in some Tasmanian freshwater triclads of the genus *Spathula* there are marked precerebral or preocellar intestinal diverticula (Ball & Tran 1979), and

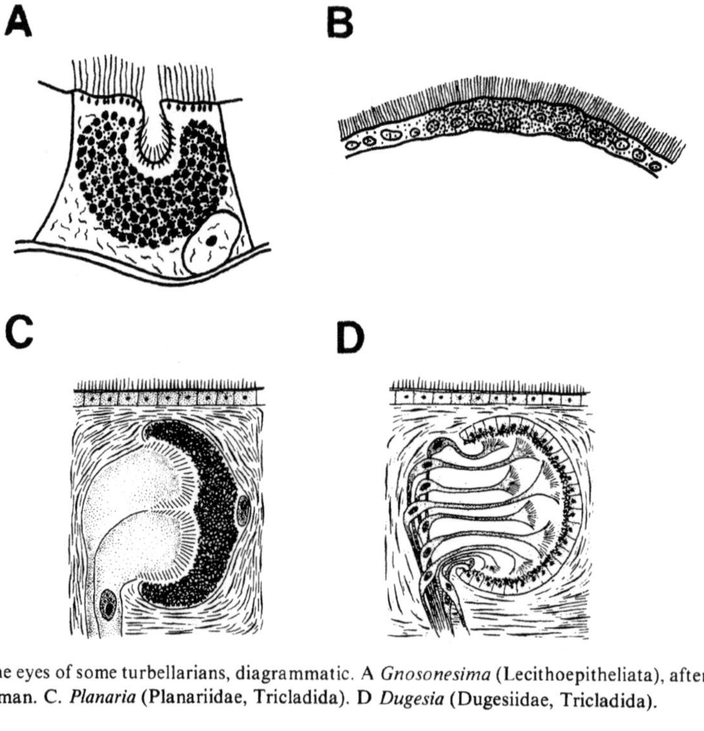

Fig. 4. Sections through the eyes of some turbellarians, diagrammatic. A *Gnosonesima* (Lecithoepitheliata), after Karling. B. *Alaurina* (Microstomidae) after Hyman. C. *Planaria* (Planariidae, Tricladida). D *Dugesia* (Dugesiidae, Tricladida).

some Australian specimens of *Cura pinguis* exhibit the same phenomenon. The features of the cephalic duct in the Paludicola, i.e. reduction, are neither necessary nor sufficient for membership in the group, nor are they confined to the group.

The ecological criteria separating the Maricola and the Paludicola have also fallen and indeed such criteria are usually of low phylogenetic value. It may be noted that on several southern hemisphere islands, such as Tierra del Fuego, Gough Island, St. Helena, Posession Island and Kerguelen Island, maricolan planarians are known from freshwaters, and this is true also of one Australian marine triclad which is known only from freshwaters (see Ball 1977b).

One important implications remains to be discussed and this concerns the types of eyes found in the Dugesiidae and the Planarioidea, and their respective origins, for under the scheme of Fig. 3 they must have had separate origins.

The eye of the Dugesiidae comprises a multicellular pigment cup embracing numerous retinal cells. In the Maricola and the Planariidae the pigment cup comprises one cell containing one to three visual cells; the Dendrocoelidae are similar but have numerous visual cells. An evolutionary transformation from the planariid to the dendrocoelid condition is readily evisaged but not so for the planariid to dugesiid condition. But if the Paludicola are diphyletic then the latter has never occurred. The single celled eye cup of the Planariidae (Fig. 4c) perhaps could be derived from one like that of some of the lower Turbellaria (Fig. 4a) and the eye of the Dugesiidae (Fig. 4d) from one like the multicellular epidermal light sensitive structure of some Monocelididae (Fig. 4b). In fact a condition intermediate between that of Fig. 4b and 4d is known from *Spathula camara,* an australian freshwater triclad lacking normal eyes but possessing deeply invaginated epidermal canals which are surrounded by pigment granules (Ball 1977d). Elucidation of the possibilities suggested here would perhaps best come from ultrastructural work on a wider range of turbellarians.

From the information just discussed I suggest that there are no good reasons, other than tradition, for believing the Paludicola to be monophyletic. This does not mean that they are not, but only that their monophyly has not been demonstrated. Here is a weakness in our knowledge. The suggestion that they might be diphyletic (Fig. 3) not only solves certain problems but it also opens up new avenues of enquiry if the hypothesis is to be properly tested.

Implications for the Tricladida as a whole

The open scheme presented as Fig. 3 requires, under the assumption that the diploneuran nervous system is autapomorphous for the Terricola, that the Haploneura be para- or polyphyletic. Indeed no good defining character(s) for the group has been found. Moreover, the Maricola may be monophyletic or not, but hitherto no autapomorphy for the infraorder has been discovered. The position of the Terricola causes many problems if only because we know so little of phylogenetic relationships within the land planarians. In Fig. 1 a number of alternatives are shown. Marcus (1953) considers them to be close to the Maricola but obviously if they have an ancestor uniquely in common with the latter then the integrity of the Haploneura must fall, and likewise if they originated at a point between characters 5 and 6 in the same Figure. The exact position of the Terricola is an unsolved problem.

References

Ball, I. R., 1974a. A contribution to the phylogeny and biogeography of the freshwater triclads (Platyhelminthes: Turbellaria). In: Eds. Riser, N. W. & Morse, M. P. The Biology of the Turbellaria. McGraw-Hill, New York, pp. 339–401.

Ball, I. R., 1974b. A new genus and species of freshwater planarian from Australia (Platyhelminthes, Turbellaria). J. Zool., Lond. 174: 149–158.

Ball, I. R., 1975. Nature and formulation of biogeographical hypotheses. Syst. Zool. 24: 407–430.

Ball, I. R., 1976. Observations on Bdellasimilis barwicki, a marine triclad from Australian freshwaters (Platyhelminthes: Turbellaria). Aust. Zool. 19: 95–101.

Ball, I. R., 1977a. On the phylogenetic classification of aquatic planarians. Acta zool. fenn. 154: 21–35.

Ball, I. R., 1977b. La Faune Terrestre de l'Ile de Sainte-Hélène IV-2. Turbellaria. Ann. Mus. roy. Afr. centr. Zool. 220: 492–511.

Ball, I. R., 1977c. A new and primitive retrobursal planarian from Australian freshwaters (Platyhelminthes, Turbellaria, Tricladida). Bijdr. Dierkunde 47: 149–155.

Ball, I. R., 1977d. A monograph of the genus Spathula (Platyhelminthes, Turbellaria, Tricladida). Aust. J. Zool. Suppl. 47: 1–47.

Ball, I. R. & Tran, T. V. H., 1979. New freshwater triclads from Tasmania (Platyhelminthes, Turbellaria). Bijdr. Dierkunde 49: 153–161.

Benazzi, M. & Giannini, E., 1973. A remarkable cave planarian: Opisthobursa mexicana Benazzi, 1972. Acc. Naz. Lincei, Quaderno 171: 47–54.

Gourbault, N., 1978. Une nouvelle planaire primitive d'origine marine, Turbellarié Triclade de Polynésie. Cah. Biol. mar. 19: 23–36.

Gremigni, V., 1979. An ultrastructural approach to planarian taxonomy. Syst. Zool. 28: 345–355.

Holmquist, C. & Karling, T. G., 1972. Two new species of interstitial marine triclads from the North American Pacific coast, with comments on evolutionary trends and systematics in Tricladida (Turbellaria). Zool. Scr. 1: 157–184.

Hyman, L. H., 1944. Marine Turbellaria from the Atlantic coast of North America. Am. Mus. Novit. 1266: 1–15.

Karling, T. G., 1974. On the anatomy and affinities of the turbellarian orders. In: Eds. Riser, N. W. & Morse, M. P., The Biology of the Turbellaria. McGraw-Hill, New York, pp. 1–16.

Lewis, R. W., 1980. Evolution: a system of theories. Perspectives Biol. Med. 1980: 551–572.

Luther, A., 1961. Die Tubellarienfauna Ostfennoskandiens II. Tricladida. Fauna Fennica 11: 1–42.

Marcus, E., 1946. Sobre Turbellaria brasileiros. Bol. Fac. Cienc. Letr. Zool. S. Paulo 11: 5–254.

Marcus, E., 1953. Turbellaria Tricladida. Inst. Parcs Nat. Congo Belge, Expl. Parc Nat. Upemba, Mission G. F. de Witte, 21: 1–62.

Meixner, J., 1928. Der Genitalapparaat der tricladen unde seine Beziehungen zu ihrer allgemeinen Morphologie, Phylogenie, Okologie und Verbreitung. Zeitschr. Morph. Okol. Tiere. 11: 570–612.

Pielou, E. C., 1980. Biogeography. John Wiley & Sons, New York. pp. ix + 351.

Riedl, R., 1978. Order in Living Organisms. John Wiley & Sons, Chichester. pp. xii + 313.

The geographical and bathymetrical distribution of the Fecampiidae (Turbellaria, Rhabdocoela)

Aage Møller Christensen
Marine Biological Laboratory, DK-3000 Helsingør, Denmark

Keywords: Turbellaria, Rhabdocoela, Fecampiidae, parasites, crustaceans, distribution

Abstract

All members of the Fecampiidae are endoparasites. Since 1964 when only four species were known, four new species and 11 cocoon types, made by as many unknown species, have been described.
The Fecampiidae are distributed in all major oceans from shallow waters to more than 5000 m depth.

Introduction

The dalyelloid rhabdocoele family Fecampiidae is the only family of turbellarians in which all members lack mouth, pharynx and intestine, at least in the postlarval stages. Hence they are compelled to absorb food through the epidermis, and all of them do this by parasitizing marine invertebrates, notably crustaceans, in which they in nearly all cases occupy part of the body cavity.

In order to map the distribution of the Fecampiidae it does not suffice to record finds of specimens within their hosts. Upon reaching sexual maturity all fecampids leave their hosts to enclose themselves in a cocoon made of secretory material from the many large subepidermal gland cells, which are so characteristic for them. Shortly after they start to deposit numerous egg capsules in their cocoons (see e.g. Christensen & Kanneworff 1965).

In the case of the hermaphroditic genus *Fecampia* the cocoon is generally attached to a hard substratum (Caullery & Mesnil 1903; Christensen & Hurley 1977), whereas in the dioecious genus *Kronborgia* the cocoon may be (1) anchored in the bottom (Kanneworff & Christensen 1966), (2) attached to a host tube (Christensen & Kanneworff 1965), (3) attached to shells inhabited by hermit crab hosts (Christensen & Kanneworff 1967) or (4) attached to the host itself (Baylis 1949; Christensen 1976). In many cases, however, they are apparently just deposited on the bottom or possibly among macroalgae. The cocoon of the monotypic genus *Glanduloderma* is unknown.

The tubiform cocoons of the *Kronborgia* species may reach a length between 8 and 110 cm, depending on species and size of female producing them. Since they are also very durable, it is no wonder that they often constitute the only material at hand to prove that an unknown species occur in a locality where a particular cocoon type has been found. Although the worm stays in the cocoon at least to the end of spawning, it has only in one case been possible to describe a new species on the basis of a worm found in its cocoon (Christensen 1981). The procedure is further complicated by the fact that great changes in internal anatomy take place in connection with cocoon formation and subsequent spawning.

Generally the fecampids have been considered to be rare animals, at least with regard to number of species. This paper aims to show that this point of view is no longer tenable.

Distribution

The first described species, *Fecampia erythro-*

Hydrobiologia 84, 13-16 (1981). 0018-8158/81/0841-0013/$00.80.

cephala Giard, was found in the crabs *Carcinus maenas* and *Cancer pagurus* and the hermit crab *Pagurus bernhardus* on the north coast of France (Giard 1886). Later Caullery & Mesnil (1903) described another species *F. xanthocephala* from the same area, but from a different host, namely the isopod *Idotea neglecta*. Giard's species also occurs in shallow water along the southern and western coasts of England (Southward 1951). The genus is furthermore represented by one or more species in the Mediterranean (Brun 1967).

A third species, *Glanduloderma myzostomatis* Jägersten, was described from finds in the body cavity of two species of the polychaete genus *Myzostoma* collected in south Japanese waters (Jägersten 1940).

When a fourth species, *Kronborgia* (*Fecampia*) *spiralis* (Baylis) was found in the antarctic isopod *Serolis schythei* and described by Baylis (1949) it became next to certain that such a distribution (Europe, Japan, Falkland Islands) could not be accounted for unless one assumed that the Fecampiidae had other unknown species which, so to speak, in their distribution linked the above areas. That supposition has proved to be correct.

Since 1949 one new genus and four new species have been described. They are *Kronborgia amphipodicola* found in the tube-building amphipods *Haploops tenuis* (but not *H. tubicola*) and *Ampelisca* spp. in inner Danish waters (Christensen & Kanneworff 1964), *K. caridicola* found in the shrimps *Lebbeus polaris, Eualis machilenta* and *Paciphaea tarda* in Greenland waters (Kanneworff & Christensen 1966), *Fecampia balanicola* from the mantle cavity of the barnacles *Balanus pacificus, B. trigonus, Membranobalanus orcutti. Megabalanus californicus* and *Gonopea galeata* in Californian waters (Christensen & Hurley 1977), and *Fecampia abyssicola*, the host of which is unknown, from deep waters (1500–4730 m) in the Atlantic off Africa, the Indian Ocean north of Madagascar, and the Kermadec trench northeast of New Zealand (Christensen 1981). The latter species is no doubt well distributed in the Atlantic and Indo-West Pacific areas between the 40° latitudes.

In addition, 11 cocoon types, which definitely belong to as many unknown species, have been described by Christensen & Kanneworff (1967, types A-F incl.) and Christensen (1980, types G-K incl.). A 12th type (type L., unpubl.) was found in shallow, sheltered waters in southern Chile by the Lund University Chile Expedition 1948–49 and kindly placed at my disposal by professor Hans Brattström. It is typical for cocoons of the *Kronborgia* type (Fig. 1). Details will be published elsewhere.

Finally, a still undescribed species, most probably to be placed in the genus *Kronborgia*, parasitizes one or more shrimp species in Puget Sound on the northwest coast of the USA (unpublished finds by drs Paul Illg & Claus Nielsen).

All the mentioned finds of fecampid parasites and cocoons, except for finds belonging to one and the same species or cocoon type close to one another, are plotted in Fig. 2. Also plotted is the find of an unknown worm in an *Ampelisca* sp. published and figured by Bacesgu & Mayer (1960), as I am fairly convinced that it is a fecampid worm.

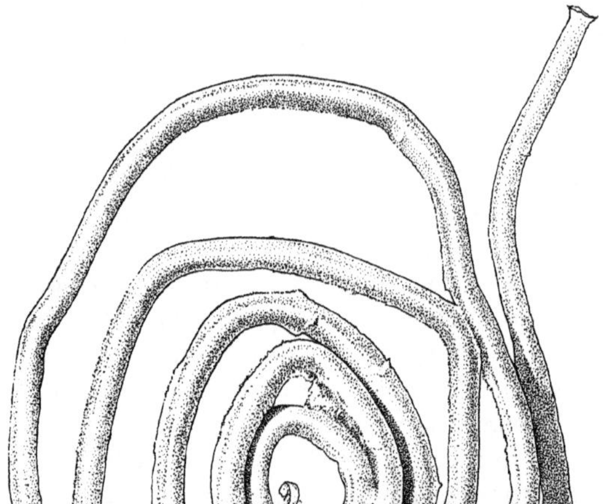

Fig. 1. Cocoon type L. Collected by the Lund University Chile Expedition, Canal Moraledo, St. M74, 5–7 m depth, 22 Feb. 1949.

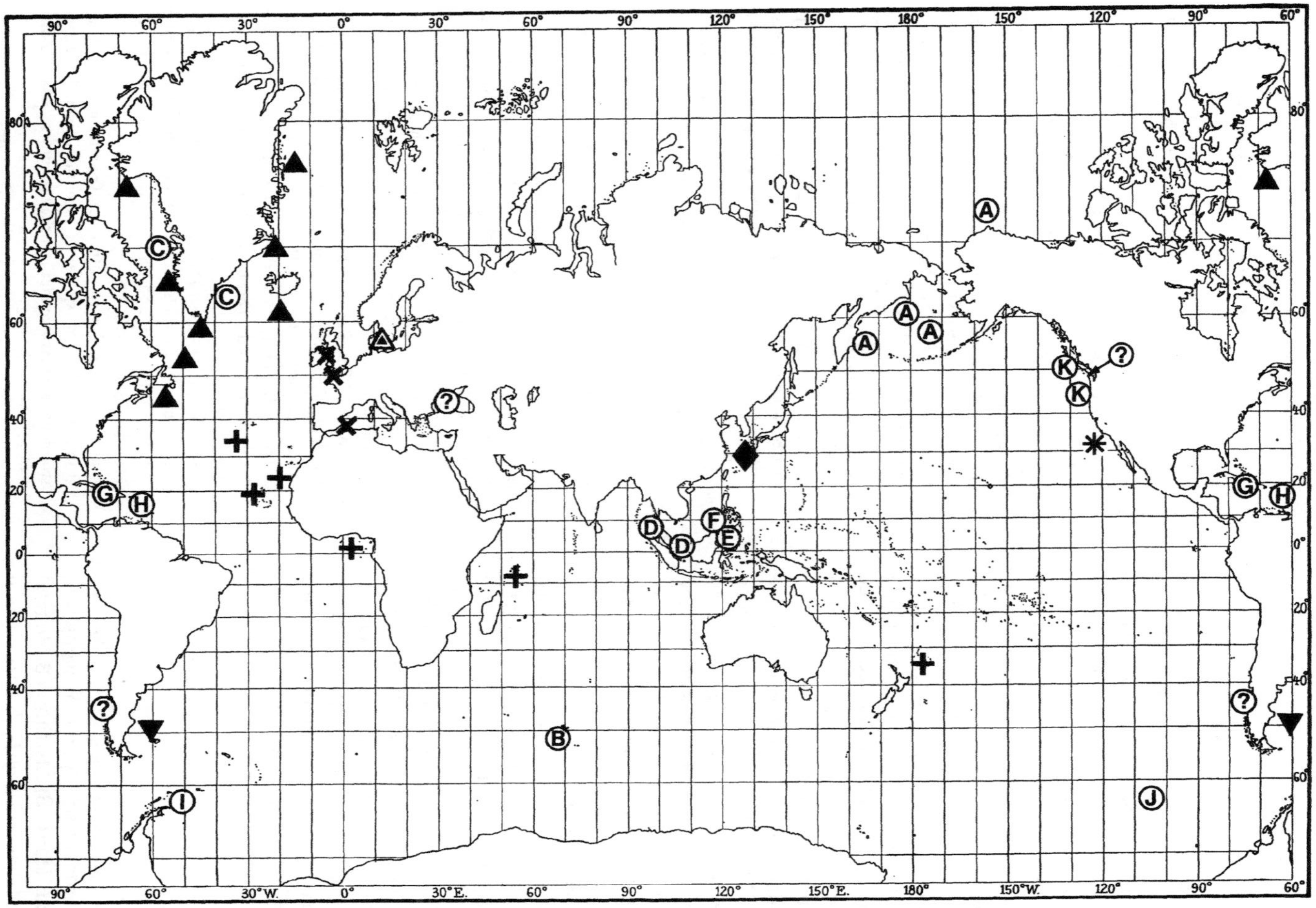

× { *Fecampia erythrocephala* Giard / *Fecampia xanthocephala* Caullery & Mesnil }

✚ *Fecampia abyssicola* Christensen

✱ *Fecampia balanicola* Christensen & Hurley

◆ *Glanduloderma myzostomatis* Jägersten

◬ *Kronborgia amphipodicola* Christensen & Kanneworff

▲ *Kronborgia caridicola* Kanneworff & Christensen

▼ *Kronborgia spiralis* (Baylis)

○ Cocoons, type indicated by capital letter

(?) Undescribed species or cocoon

Fig. 2. Geographical distribution of known species and cocoon types of Fecampiidae. Included is unpublished information based on collections of cocoons of *Kronborgia caridicola* and of type *A* loaned from the Smithsonian Institutuion.

Discussion

As the data plotted in Fig. 2 show, the Fecampiidae are represented in all major oceans on both hemispheres from arctic to antarctic waters, although it should be noted that no species of the genus *Fecampia* has so far been recorded in the last mentioned two cold-water areas. Vertically the Fecampiidae occur from very shallow water to abyssal depths, and this is true for both the hermaphroditic *Fecampia* species and the dioecious *Kronborgia* species. Thus, *F. abyssicola* occurs to a depth of 4730 m, and two cocoon types (G & H), which are reminiscent of *Fecampia* cocoons, were found at depths of 1760 and 3852 m, respectively (Christensen 1981). The same paper records the findings of a *Kronborgia* type cocoon (Type J) at a depth of 5–42 m in the Antarctic Sea.

The available data show that the Fecampiidae comprises at least 22 species, of which 8 have been described. However, I am convinced that many more than 14 species remain to be discovered and described. Certain other turbellarian families also contain species that subsist as parasites. They are listed by Jennings (1971) and here too there is no doubt that more species remain to be found. Thus, the same year Jespersen & Lützen (1971) described a new species, *Triloborrhyncus psilastericola,* which quite clearly feeds on tissue of its asteroid host.

On this background and considering also that e.g. the genus *Kronborgia* with its dwarf males, mode of larval entrance into the host etc. seems to show as intesting a life history as that of many trematodes, it seems pertinent to wonder why one rarely see turbellarian parasites mentioned in general text books on parasitology.

Acknowledments

Grateful thanks are extended to drs Marian Pettybone (Smithsonian Institution, Washington, D.C., USA) and Hans Brattström (Biological Station, Blomsterdalen, Norway) for loan of cocoon materials.

References

Bacesgu, M. & Mayer, R., 1960. Nouveaux cas de commensalisme (Colomastix et Tritaeta) et de parasitisme (Rhizorhina) pour la Mer Noire et quelques observations sur l'Ampelisca des eaux Prébosphoriques. Trav. Mus. Hist. nat. Gr. Antipa 2: 87–96.

Baylis, H. A., 1949. Fecampia spiralis, a cocoon-forming parasite of the antarctic isopod Serolis schythei. Proc. Linn. Soc. Lond. 161: 64–71, 2 plates.

Brun, B., 1967. Sur la présence de turbellariés du genre Fecampia en Méditerranée. Bull. Mus. Hist. nat. Marseille 32: 141–145.

Caullery, M. & Mesnil, F., 1903. Recherches sur les Fecampia Giard, Turbellariés Rhabdocèles, parasites internes des Crustacés. Ann. Fac. Sci. Marseille 13: 131–167, 2 plates.

Christensen, A. M., 1976. On the morphology and biology of Kronborgia spiralis (Baylis 1949) (Turbellaria, Neorhabdocoela), with a note on its systematic positions. Ophelia 15: 77–97.

Christensen, A. M., 1981. Fecampia abyssicola n. sp. (Turbellaria: Rhabdocoela) and five cocoon types of undescribed species of Fecampiidae from the deep sea. Galathea Rep. 19: (in press).

Christensen, A. M. & Hurley, A. C., 1977. Fecampia balanicola sp.nov. (Turbellaria Rhabdocoela), a parasite of Californian barnacles. In: Eds., Karling, T. G. & Meinander, M., The Alex. Luther Centennial Symposium on Turbellaria. Acta zool. fenn. 154: 119–128.

Christensen, A. M. & Kanneworff, B., 1964. Kronborgia amphipodicola gen. et sp. nov., a dioecious turbellarian parasitizing ampeliscid amphipods. Ophelia 1: 147–166.

Christensen, A. M. & Kanneworff, B., 1965. Life history and biology of Kronborgia amphipodicola Christensen & Kanneworff (Turbellaria, Neorhabdocoela). Ophelia 2: 237–251.

Christensen, A. M. & Kanneworff, B., 1967. On some cocoons belonging to undescribed species of endoparasitic turbellarians Ophelia 4: 29–42.

Giard, M. A., 1886. Sur un Rhabdocèle nouveau, parasite et nidulant (Fecampia erythrocyphala). C. r. hebd. Séanc. Acad. Sci. Paris 103: 499–501.

Jennings, J. B., 1971. Parasitism and commensalism in the Turbellaria. In Ed., Dawes, B. Advances in Parasitology, Acad. Press, pp. 1–32.

Jespersen, Å. & Lützen, J., 1971. Triloborhynchus psilastericola n. sp., a parasitic turbellarian (Fam. Pterastericolidae) from the starfish Psilaster andromeda (Müller and Troschel). Z. Morph. Tiere 71: 290–298.

Jägersten, G., 1940. Zur Kenntnis von Glanduloderma myzostomatis n. gen., n. sp., einer eigentümlichen, in Myzostomiden schmarotzenden Turbellarienform. Ark. Zool. 33A (3): 24 pp. 2 tables.

Kanneworff, B. & Christensen, A. M., 1966. Kronborgia caridicola sp. nov., an endoparasitic turbellarian from North Atlantic shrimps. Ophelia 3: 65–80.

Southward, A. J., 1951. On the occurrence in the Isle of Man of Fecampia erythrocephala Giard, a platyhelminth parasite of crabs. Rep. mar. biol. Stn Port Erin (1950): 27–30.

The taxonomy, zoogeography and biology of Bipalium kewense Moseley, 1878 (Tricladida, Terricola)

L. Winsor
Zoology Dept., James Cook University of North Queensland, Douglas 4811, Queensland, Australia

Keywords: Turbellaria, Terricola, taxonomy, zoogeography

Abstract

Knowledge of this cosmopolitan land planarian is briefly reviewed. The external morphology and anatomy of the copulatory organs are considered with emphasis on the variation of taxonomic characters. This species can be distinguished from similar planarians by differences in markings, particularly those of the head and neck; positions of body apertures and structure of the copulatory organs.

Karyological studies on neoblasts of *B. kewense* reveal a diploid chromosome complement of 16; one subtelocentric, three submetacentric and four metacentric pairs of chromosomes.

The present distribution of the species is mapped. Its natural range extends through Vietnam to Kampuchea. Elsewhere it has been introduced by man. Outside its normal range *Bipalium* is always confined to habitats in man-modified environments. Its natural habitat is probably tropical rainforest in the uplands of Indochina. Temperatures in these regions are more representative of a subtropical than tropical-humid type climate. This may explain the occurence of the species in low temperature hothouses and its tolerance to low outdoor temperatures.

Man has apparently passively dispersed *B. kewense* together with rooted plants. A sequential spread of the species from country to country worldwide cannot be ascertained. In Australia and North America *B. kewense* shows an immature disjunct distribution typical of an introduced species. Generally, older records in each centre are from hothouses in coastal cities; more recent records from outdoors further inland.

Asexual reproduction by architomy in *B. kewense* appears to be largely influenced by environmental temperature and the availability of earthworms, its sole food-source. Sexual specimens of the species have been found mainly in regions with tropical or subtropical climates. Sexual reproduction may be seasonal. Sexual *Bipalium* can reproduce asexually by architomy. This probably does not occur during periods of cocoon production. Factors influencing sexual reproduction are not clear.

Outdoors the species is usually prevalent in spring and autumn, less common in summer and winter. Mortality and 'aestivation' may account for reduced numbers of specimens observed during these periods. *B. kewense* also exhibits seasonality in hothouses where it reaches maximum numbers in winter months and is scarce in summer.

B. kewense is of marginal medical and veterinary importance because of its implication in cases of pseudoparasitism in humans and domestic animals. It is of minor commercial importance as a pest on earthworm farms in the southern United States.

Hydrobiologia 84, 17 (1981). 0018-8158/81/0841-0017/$00.20.

A new species of freshwater turbellarian from Africa, predatory on mosquitoes: Mesostoma zariae n. sp. (Typhloplanoida)

Jerzy Kolasa[1] and Alan P. Mead[2]
[1] *Dept. of Animal Morphology, A. Mickielwicz University, Szamarzewskiego 91, 60-569 P, Poland*
[2] *University of Technology, Lae, Papua New Guinea*

Keywords: Turbellaria, Typhloplandoe, *Mesostoma zariae*

Abstract

A new species of the genus *Mesostoma* Ehrenberg 1935, *M. zariae* n.sp. is described and its relationship discussed. The new species belongs to the *M. lingua* species-group and is characterised by the presence of two kinds of prostate secretions and a pear-shaped penis papilla. It occurs in small standing or slowly running waters at Zaria, Nigeria. It has previously been proven (Mead 1978) that this species is predatory on the aquatic stages of mosquitoes.

Introduction

Turbellarian fauna of Africa has been poorly studied. Young (1976) reported only eighty valid freshwater species for Africa, a number contrasting with the over three hundred species known from Europe. Until now about eight valid species of *Mesostoma* are found in various parts of Africa, and recently Mead (1978) mentioned a new form of *Mesostoma* from Nigeria. Its biological significance results from its exceptional potentialities to reduce mosquito populations in small standing water bodies. This species is described in the present paper.

Material and methods

Specimens of the new species were collected in the basin and gully areas, including ponds, of the River Kubani near Ahmadu Bello University, Zaria, Nigeria (Typelocality). These animals were cultured and a large number of their offspring studied alive on squash preparation. Several individuals (narcotised by adding a few drops of alcohol to a small petri dish for an hour or two) were fixed is Stieve's fluid*, serially sectioned (at 8 μm) and stained in Delafields haematoxylin and eosin.

Field samples were collected using a plancton mesh net swept along the water surface, through and beside immersed vegetation and through grasses and sedges along the edge. The net sweepings, including vegetation and filamentous algae were washed into a suitable jar. A sample of bottom mud was added and the bottle or jar topped up with water. In the laboratory the whole sample is tipped into clear glass or perspex tanks and left to settle, shaded. After some hours, or overnight, the tank was illuminated and examined. The animals emerged from the mud or the vegetation were then taken with a pasteur pipette and put into small petri dishes for further examination. Like most *Mesostoma* species, this species can be cultured in clean water, or boiled tap water, using cladocera, chironomid larvae or mosquito larvae as food. Oligochaete worms can be fed as well but population growth seems slower than when arthropods are used as food.

* Stieve's fluid: 19 cc $HgCl_2$ saturated, 5 cc formaldehyde 1 cc acetic acid.

Hydrobiologia 84, 19–22 (1981). 0018-8158/81/0841-0019/$00.80.

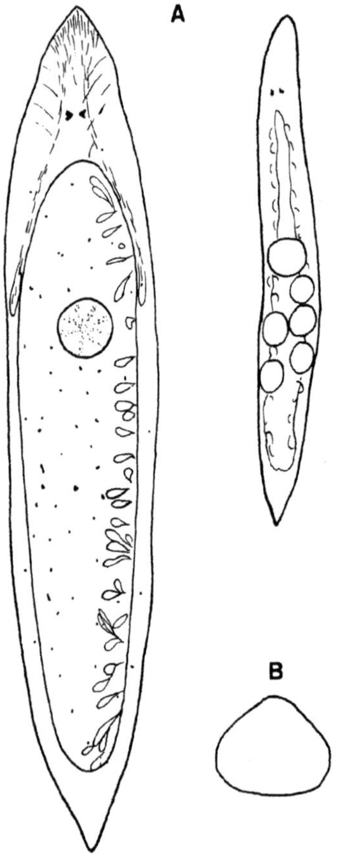

Fig. 1. Mesostoma zariae sp. nov. A. General view of living animals specimens. B. Optical body cross-section.

Description of the new species

The animals become sexually mature at a length of 4 mm. The longest worms reached 7 mm. They are irregularly spindle like; with a weak constriction at the level of the eyes and a slightly outlined tail region (Fig. 1A). The body is ovoid in cross section (Fig. 1B). Thus, the shape is almost identical with that of the other species of *M. lingua* species-group except for the anterior body extremity which seems more pointed. They are generally colourless with some dots of dark-brown pigment. Some animals appear brown or whitish due to the intestine content or the masses of brown dormant or white subitaneous eggs occupying major parts of the body.

The two black eyes (about 45 × 64 μm) are closer to each other than to the body margin. They do not anastomose. Dermal and adenal rhabdites occur (resp. 10 μm and 20–25 μm long), the latter forming light refracting tracts in the anterior of the body and on both sides of the intestine. Within the glands, adenal rhabdites are about 14 μm. The epidermis is about 12 μm high and covered with 6 μm long locomotory cilia.

The intestine occupies about two thirds of the whole body length. The pharynx, 330–380 μm in diameter, lies a little before the body middle. Its height in fixed specimens is half its width. It has a ciliated outer margin.

In the material studied the male gonads were very weakly developed and were not observed in either living or fixed animals. They seem to be distributed above the intestine as indicated by indistinct bundles of sperm cells seen in various parts of body. Two deferent ducts carry sperm cells from the testicles towards the copulatory organ. In the proximity of that organ they run parallel and finally enter it together apically. The shape of the copulatory organ (Fig. 2) may vary depending on the amount of sperm as well as on the compression exerted by neighbouring tissues. It consists of a seminal vesicle, a prostate vesicle and a penis papilla. The border between the seminal and the prostate vesicle is oriented obliquely with the entrance of the prostate glands in about the middle of the copulatory bulb. Unlike in other *Mesostoma* species two kinds of prostate secretions were observed, the coarser ones closest to the seminal vesicle. The penis papilla is a slender process, large at its base. The ejaculatory duct is lined with a thin cuticular membrane. The penis papilla is lying in the male atrium, a diverticle of the common genital atrium.

The stalk of the copulatory bursa opens into the common genital atrium through its dorsal wall (Fig. 3). This bursa stalk is longer than the bursa diameter. The entrance of the bursa is guarded by a sphincter, though the dominant direction of the muscles around the stalk is longitudinal. The single ovary and its oviduct together may be more than 380 μm long. The margins of the oviduct are undulated near the seminal vesicle as in *M. africanum* Kolasa, 1978. The female duct enters the atrium from the right side. It is joined by the two yolk ducts near the atrium. The yolk glands are of

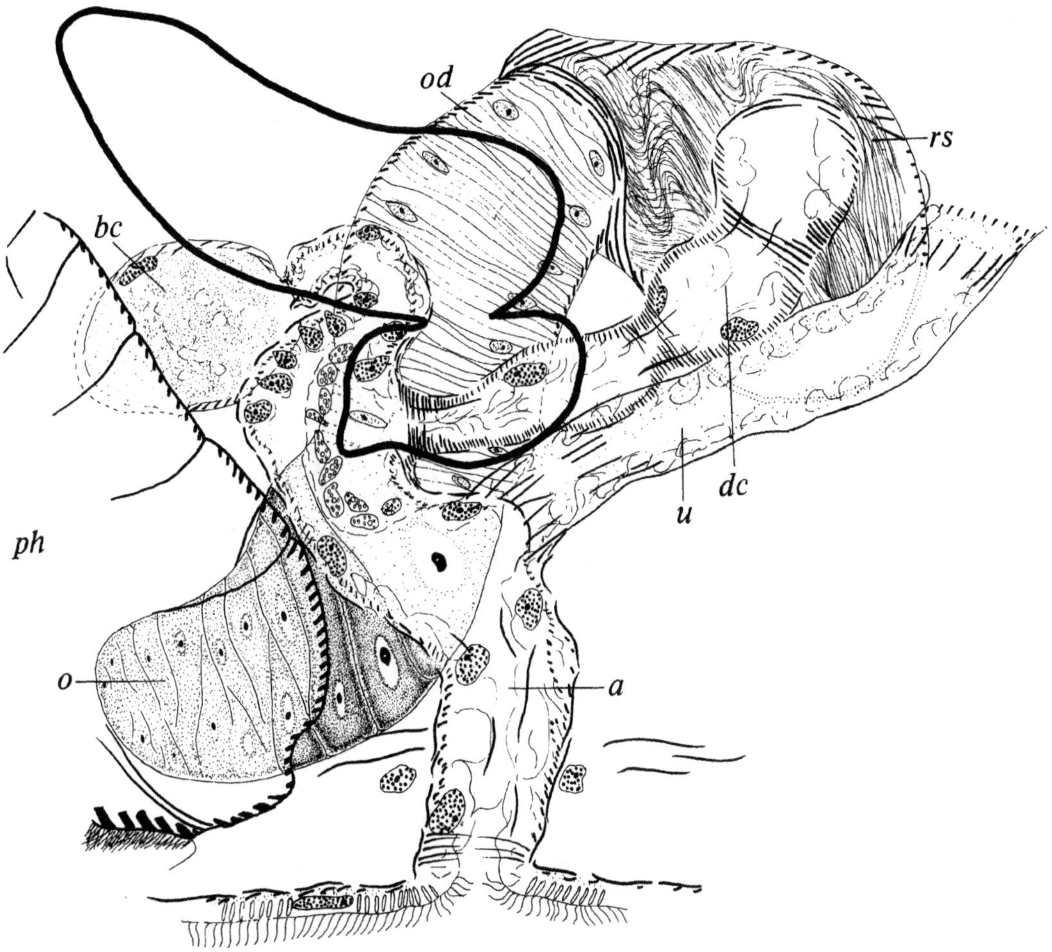

Fig. 2. Mesostoma zariae sp. nov. Sagittal reconstruction of the reproductive organs from serial sections.

the follicular type and are distributed all over the body, being particularly numerous laterally, and absent at the extremities.

From the distribution of the eggs in the body it can be concluded that the uteri are T-shaped. The dormant eggs are brown and they number up to ten. Their size is about 280 × 190 μm. The number of whitish subitaneous eggs vary from seven in a young specimen up to 50 in larger specimens. The common gential atrium is elongated, lined by a loose epithelium with some nuclei and surrounded by internal longitudinal and external circular muscles. The genital pore is separated from the mouth. The genital organs described, except the gonads, vitellaries and uteri, are situated behind the pharynx.

Discussion and differential diagnosis

The new species belongs to a group of species related to *Mesostoma lingua* (Albigaard 1789). It can easily be distinguished from its relatives by the presence of two kinds of secretions in the prostate vesicle and the large penis papilla. Such a large penis papilla is also found in *M. sibollae* Kolasa, 1976, but this species is coloured green, while *M. zariae* n.sp. is colourless.

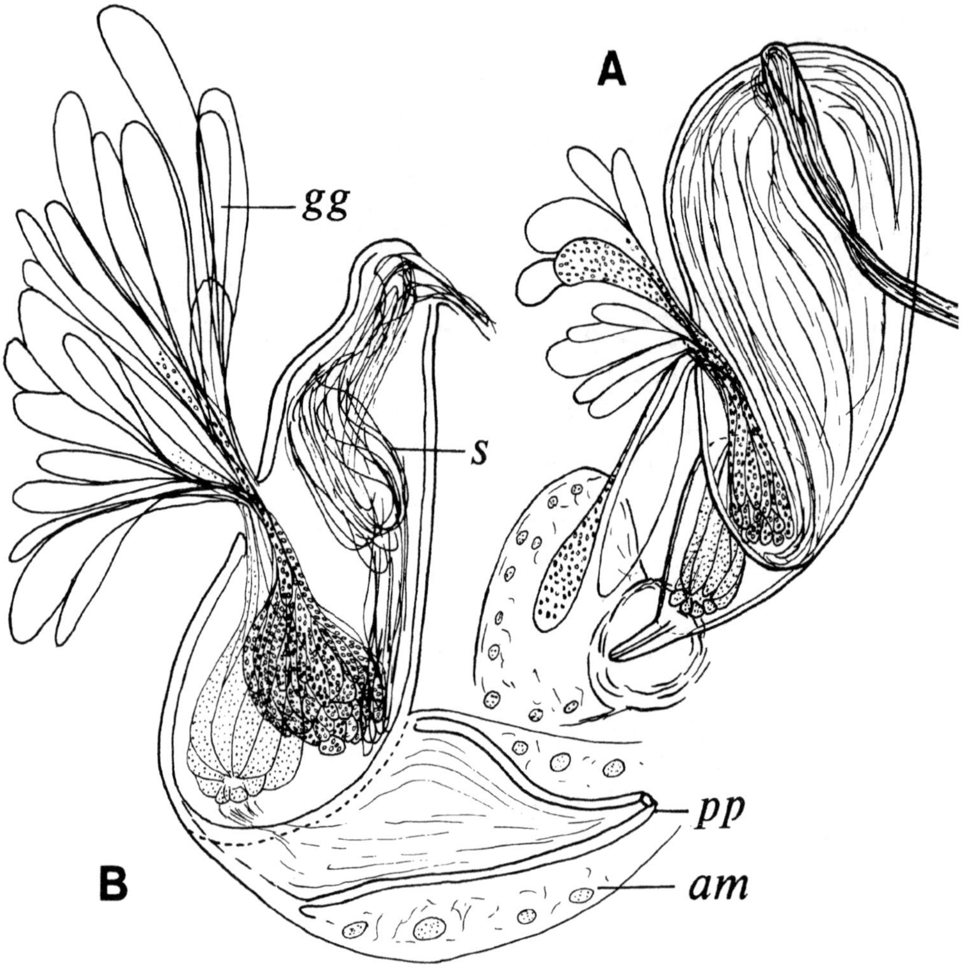

Fig. 3. Mesostoma zariae sp. nov. A. copulatory organ filled up entirely with sperm as seen in a slightly squashed individual. B. The same from another specimen, weakly filled with sperm, strongly squashed.

Acknowledgements

The authors are indebted to Dr. Johnstone O. Young (Liverpool) for sectioning several specimens of *M. zariae* sp. nov.

Abbreviations

a – common genital atrium
am – male atrium
bc – copulatory bursa
dc – female duct
gg – prostate glands
o – ovary
od – oviduct
ph – pharynx
pp – penis papilla
rs – seminal receptacle
s – sperm cells
u – uterus

References

Kolasa, J., 1976. Systematical studies on Typhloplanidae (Turbellaria) I. Mesostoma africanum spec. nov., M. viareginum spec. nov., and M. sibollae spec. nov. Boll. Zool. 43: 273–291.

Mead, A. P., 1978. A rhabdocoele turbellarian predator on the aquatic stages of mosquitoes. Ann. Trop. Med. Parasit. 72: 591–594.

Young, J. O., 1976. The freshwater Turbellaria of the African continent. Zool. Anz. 197: 419–432.

Paravortex karlingi sp. nov. from Cerastoderma edule L., in Britain

Alan W. Pike[1] & Michael D. B. Burt[2]
[1] *Dept. of Zoology, Univ. of Aberdeen, Tillydrone Avenue, Aberdeen AB9 2TN, Scotland.*
[2] *Dept. of Biology, Univ. of New Brunswick, Canada.*

Keywords: Turbellaria, Neorhabdocoela, *Paravortex*, parasites

Abstract

Paravortex karlingi sp. nov. collected from the intestine of the bivalve mollusc *Cerastorderma edule* from the Ythan Estuary, N. E. Scotland, and elsewhere, is distinguished from a closely related species, *P. cardii*, also occurring in this host, on the basis of differences in habitat occupied by the two species as well as behavioural and morphological differences. *P. karlingi* is smaller, has fewer embryos in the gravid adult and shows a different behaviour pattern when released from the host intestine. It is also negatively phototactic whereas *P. cardii* is initially positively phototactic, only later becoming negatively phototactic. The occurrence of both species in Britain is briefly described.

Introduction

The three species of *Paravortex* Wahl that have been described inhabit three different species of bivalve molluscs. *P. scrobiculariae* is found, at different times, in the digestive gland tubules or intestine of *Scrobicularia plana* (Freeman 1957; Jennings & Phillips 1978). *P. cardii* lives in the digestive gland tubules of *Cerastoderma edule*, contrary to Hallez's (1909) original description of the habitat as the stomach (Jennings & Phillips 1978). *P. gemellipara* is the exceptional species of the three in that its habitat is said to be the gills of *Modiolus plicatulus* (Linton 1910; Ball 1916); although Patterson (1912) argued that the correct habitat of the worm is the lumen of the kidney.

All three species have been recorded from British intertidal molluscs but the one record, by Leigh-Sharpe (1933), of *P. gemellipara* from *Cerastoderma edule* in Plymouth was most probably not this species but *P. cardii*, as already suggested by Atkins (1934). There are only two previously published records of *P. cardii* from the British Isles, by Atkins (1934) and Jennings & Phillips (1978), although both Dr. H. H. Williams (personal communication) and Professor T. B. Reynoldson (personal communication) have observed this parasite in *Cerastoderma edule* in South and North Wales respectively.

During the present study, worms recovered exclusively from the intestine of *C. edule* collected from the Ythan Estuary, Grampian, N.E. Scotland, were carefully compared with individuals of *P. cardii* collected from habitats north and south of the Ythan Estuary. The morphological and behavioural differences between specimens of *Paravortex* recovered from the digestive gland of *C. edule* (i.e. *P. cardii)* and those collected from the intestine of *C. edule* from the Ythan Estuary, and elsewhere, are sufficient, we believe, to necessitate the designation of a new species. We propose the name *P. karlingi* after Professor Tor Karling, in whose honour this conference was convened.

Because the two species parasitize the same host species and because of uncertainties about their distribution in the British Isles a small survey was conducted, the results of which will be published in more detail elsewhere.

Hydrobiologia 84, 23–30 (1981). 0018-8158/81/0841-0023/$01.60.

Materials and methods

Regular samples of *Cerastoderma edule* were obtained from the same site near the mouth of Tarty Burn on the Ythan Estuary by raking through a sand/mud substrate. Other samples from different localities in Britain were sent by rail or air freight by their collectors. All samples were examined as soon as possible after collection, usually within 2–3 days. Those not examined immediately were maintained in running sea water on a sand/mud substrate that was agitated daily to provide a food source for the bivalves.

Each individual was examined microscopically after the valves were opened by incision of the adductor muscles. To avoid premature release of *P. cardii* the visceral mass was removed with extreme care to avoid rupturing the dorsal side of the alimentary tract. On removing this mass to clean filtered seawater the digestive gland was first opened and thoroughly searched for worms. The wall of the foot was then incised and the edges pulled apart to reveal the intestinal loops adjacent to the stomach. These were removed intact, by severing their connection with the stomach and the anus, and then placed in fresh filtered seawater. In this way the digestive gland and intestine could be examined without the possibility of intermixing of worms from the two habitats. The intestine was opened along its entire length and any worms present were transferred to seawater at 4 °C.

Worms intended for wholemounts were fixed in precisely the same way to achieve uniformity for measurement. They were dropped into boiling 10% formalin which produced good symmetrical worms unless the fixative had inadvertently cooled slightly in which case more contracted specimens were obtained. The latter were not used for measurement. Histological material was fixed in seawater Bouins. Whole worms were stained with acid alum carmine and sections with haematoxylin and eosin or Hubschmanns stain.

The phototactic behaviour of the worm was investigated using a small perspex chamber (Fig. 1) the internal dimensions of which were 2 × 1 × 1 cm. The outside surfaces were painted matt black with the exception of the base and one end, through which light was directed. A fibre-optic light source providing 80 lumens at the surface of the chamber was used throughout. Worms to be tested were quickly removed from their hosts and placed in seawater at room temperature in a darkened container. Samples of six or ten worms were used for each experiment according to the availability of

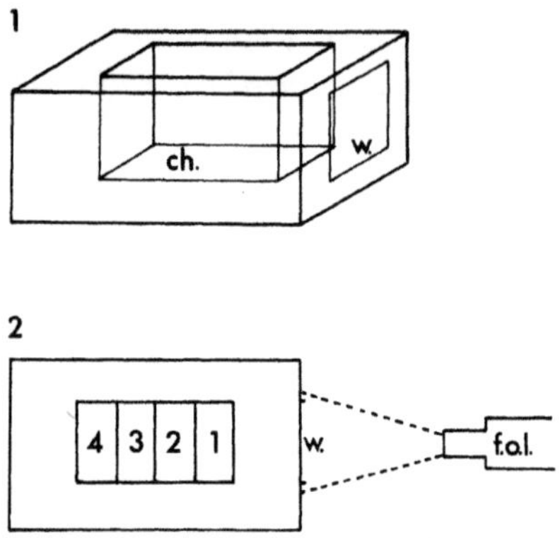

Figs. 1 & 2. Side and plan views of perspex chamber for experiments on phototactic responses of *P. karlingi* and *P. cardii.*

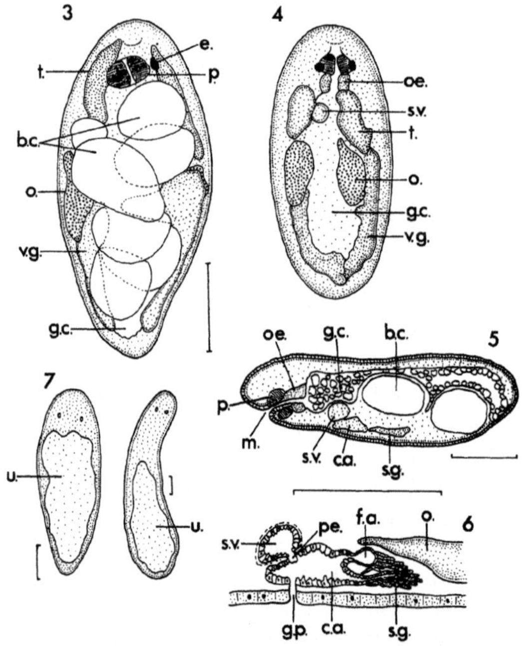

Figs. 3–7. Fig. 3. Holotype, gravid *P. karlingi* in dorsal view. *Fig. 4.* Mature *P. karlingi* dorsal view. *Fig. 5.* Longitudinal section of gravid *P. karlingi* showing alimentary canal and part of reproductive system. *Fig. 6.* Reconstruction of copulatory apparatus of a gravid *P. karlingi. Fig. 7.* Diagrams depicting live body shape and typical area occupied by the uterus in *P. karlingi* (left) and *P. cardii* (right). All scales are 0.1 mm.

material. They were introduced into the test chamber in a darkroom and allowed to acclimate for 10 min before the experiment was run. The initial distribution of worms in the chamber was recorded using the four divisions marked on the base of the chamber (Fig. 2), the count being made in light shining through the bottom of the chamber. The bottom light was extinguished and the side light turned on at the same time that a stopwatch was started. The distribution of worms in the four sections of the chamber was recorded at 60 sec intervals for 4 min, after which the light was extinguished and the worms were left in total darkness for 5 min. The distribution was again noted in bottom illumination. Fresh worms were used for each experiment except for occasions when subsequent responses of the test animals were being assessed. In these instances the animals were returned to a holding dish after the initial test and kept in total darkness until used again.

A 2×2 Contingency test was applied to the data using summations of divisions one and two, and three and four. The comparison was made between the initial distribution and the distribution after 4 min exposure to light.

Observations and description

The habitat

Paravortex karlingi sp. nov. of all ages live only in the intestine of the host. The host intestine, which is a long coiled tube, is usually filled with a core of indigestible particles mixed with mucus (Fig. 8). This core is transported along the intestine by the cilia of the intestinal epithelial cells. The worms live between this core and the epithelial surface so that their habitat is a very restricted one; individuals normally have their ventral surfaces applied to the epithelial surface. Distribution of the worms was not recorded since they seemed to occupy most of the gut length and also because their mobility, even within a confined space was such that the results obtained may not have given a true reflection of any preferred microhabitat within the intestine. It was often convenient to starve the hosts for a few hours, before examining them, to reduce the amount of food passing through the intestine and thereby lessen the possibility of losing worms.

Behaviour of worms within the host gut and after release

In the intact, partly-filled intestine worms glide over the epithelium with their ventral surfaces in contact with it. The movement is slow but regular and the worms occasionally turn around and move back over the same area. Movement is even slower and more irregular in well fed hosts, due to the large core of material in the gut lumen.

The worms become very active on being released from the host. Their swimming speed and rate of turning increase dramatically but will diminish again if the worm comes into contact with a host surface. Worms isolated from the host swim constantly, but with fewer changes of direction after the first minute or two of isolation.

Adult and gravid *P. karlingi* are negatively phototactic. This behaviour is immediately apparent when the worms are removed from the host and it persists thereafter. Table 1 shows the results of experiments using the perspex test chamber. These results are based on mean figures obtained from four experiments each using six worms. Comparative data were also collected for *P. cardii.* This species exhibits an initial positive phototaxis that subsequently reverses after a period of indeterminate behaviour. The data for this species, included in Table 1, are the mean figures from five experiments each one using ten worms. The worms were left in total darkness for 15 to 20 hours and

Table 1. Phototactic responses of gravid adult Paravortex karlingi and P. cardii

	Dark 4	3	2	Light 1	χ^2
A. P. karlingi Numbers are totals for four experiments each using six worms					
Control (bottom light)	8	5	5	6	$p < 0.001$
Test (directional light)	21.4	1.15	1.05	0.4	
B. *P. cardii.* Numbers are totals for five experiments each using ten worms					
Control	11	17	10	12	$p < 0.001$
Test	7.4	4.4	7.2	31	
P. cardii. Experiments repeated after 15–20 h. Totals are from three experiments using 10, 10 and 9 worms each					
Control	7	6	11	5	$p < 0.1$ N.S.
Test	16.8	2.6	2.2	7.4	

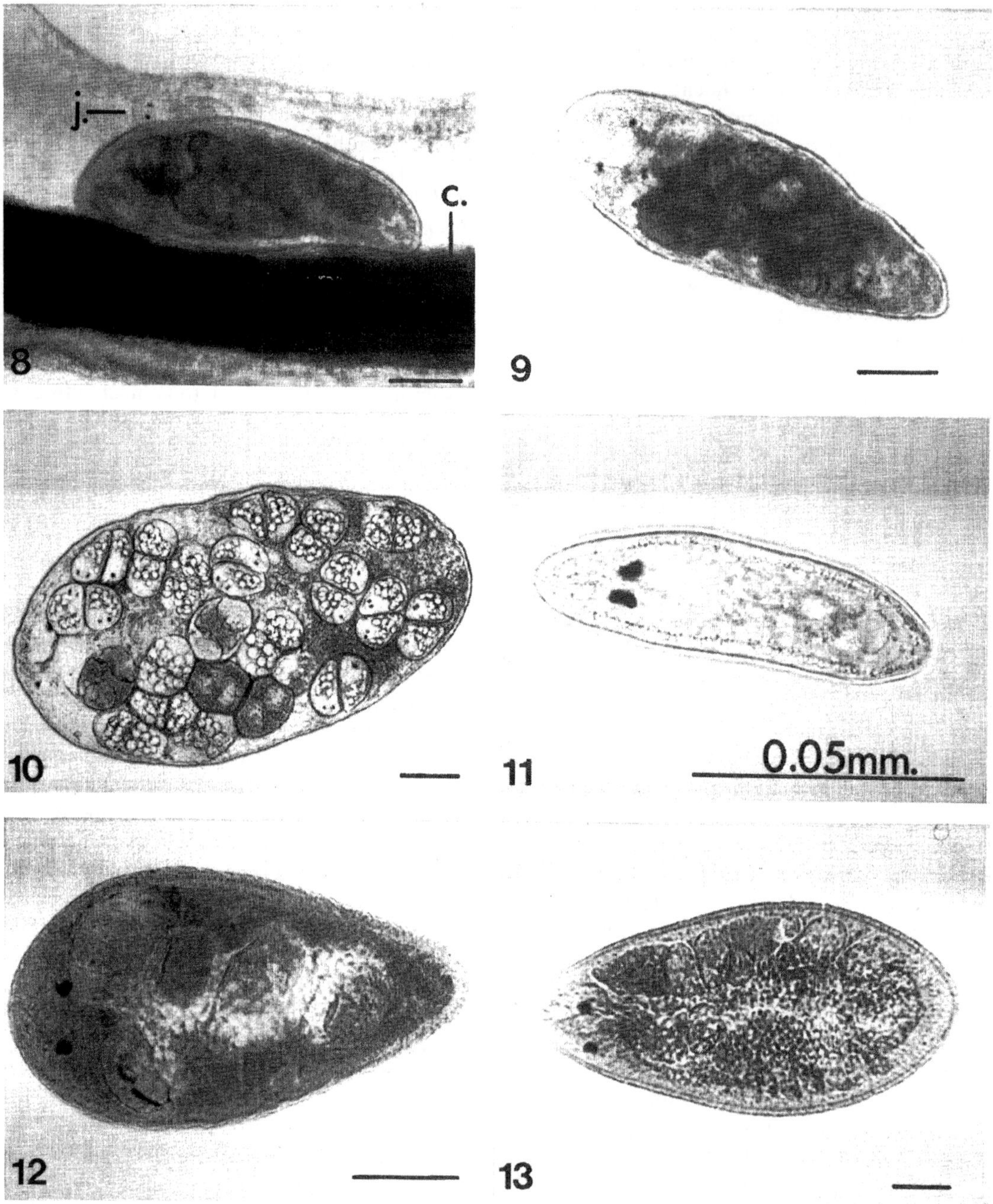

Figs. 8–13. *Fig. 8.* Gravid and juvenile *P. karlingi* in situ in *C. edule* intestine. *Fig. 9.* A living unflattened gravid *P. karlingi* showing typical body shape. *Fig. 10.* A flattened gravid *P. karlingi* showing brood capsules. *Fig. 11.* An unflattened living new born juvenile *P. karlingi*. *Fig. 12.* A fixed and stained but unflattened gravid *P. karlingi*. *Fig. 13.* A fixed, stained but unflattened *P. cardii*. All scales 0.1 mm except where indicated.

Table 2. Measurements of *Paravortex karlingi* in mm.

		Range (No. measured)	Mean ± 1 S.D.
A. Fixed worms, not stained or mounted			
1. Juvenile	Body length	0.10–0.36(9)	0.21 ± 0.09
	Body breadth	0.08–0.18(9)	0.12 ± 0.04
2. Mature	Body length	0.30–0.57(28)	0.39 ± 0.07
	Body breadth	0.10–0.23(28)	0.18 ± 0.03
3. Gravid	Body length	0.30–0.58(27)	0.47 ± 0.07
	Body breadth	0.18–0.32(27)	0.23 ± 0.03
B. Fixed, stained and mounted worms			
Gravid	Body length	0.33–0.51(20)	0.42 ± 0.06
	Body breadth	0.15–0.27(20)	0.21 ± 0.03
	Pharynx length	0.03–0.05(19)	0.04 ± 0.004
	Pharynx breadth	0.05–0.06(19)	0.05 ± 0.01
	Seminal vesicle diameter	0.04–0.06(14)	0.04 ± 0.01

tested again. The data show that the phototaxis is now reversed.

Morphology of adult worms

Gravid worms contain numerous brood capsules that obscure and distort the internal organs to such an extent that the normal morphology is not visible (Fig. 3). The following description is based on several mature, but not gravid worms. Table 2 gives measurements of all stages.

Living worms are bluntly rounded anteriorly with a tapering tail region which is usually slightly pointed (Fig. 9). This body shape is retained in fixed worms (Fig. 10). A pair of crescentic pigmented eyespots is present anteriorly and lateral to the pharynx. There does not appear to be a lens present in the concavity of the eyespot.

A simple alimentary canal is present (Fig. 5) opening through the mouth which is ventral and generally very inconspicuous. This mouth opens into an oval muscular pharynx behind which is a short oesophagus that immediately opens into the large, saccate intestinal caecum. The latter has a bubbly appearance due to the structure of the epithelial layer.

The reproductive system is similar to that described for other species of *Paravortex* (Fig. 4). The paired testes are antero-laterally situated, often reaching the lateral margins of the pharynx in fixed specimens. Their shape is usually cylindrical, sometimes irregularly so. The vasa deferentia were not seen in whole mounts or sections but the globular seminal vesicle was always visible in the ventral median plane. In longitudinal sections the small plug like penis (Fig. 6) could be seen opening into the common atrium from the base of the seminal vesicle.

The ovaries, lying immediately behind the testes, are similar in size to the testes but more irregularly shaped, often appearing roughly triangular. Oviducts were not seen but a small atrium opening into the posterior end of the common atrium could be seen in longitudinal sections of the worm (Fig. 6). Numerous conspicuous shell gland cells lie behind, and lateral to, this atrium. The vitellaria are large, elongate, only slightly lobed organs that lie mostly behind the ovaries and lateral to the gut caecum.

Gravid worms are only slightly longer than mature worms (Fig. 3) but the organs, even in unflattened specimens, are pushed to the margins of the body by the numerous brood capsules present in the parenchyma. Up to 20 brood capsules (Fig. 10 have been recorded (range 1–20 in 24 specimens, mean 4.7) in worms from the Ythan Estuary. Each capsule contains only two developing embryos. Only a few new born juveniles were obtained and these were elongate, cylindrical organisms (Fig. 11) with prominent eyespots and well developed alimentary canal. They are active swimmers but no experiments were done to identify their phototactic behaviour.

Distribution of P. karlingi and P. cardii

It was possible, through the generous co-operation of many individuals in various parts of

Table 3. Distribution of *Paravortex cardii* and *P. karlingi* in Britain.

Locality	*P. cardii*	*P. karlingi*
Loch Ewe, W. Scotland	–	+
Isle of Cumbrae, W. Scotland	+	+
Dornoch Firth, E. Scotland	+	+
Ythan Estuary, E. Scotland	–	+
Eden Estuary, E. Scotland	+	+
Strangford Lough, N. Ireland	+	+
Conway Estuary, N. Wales	+	+
Bury Estuary, S. Wales	+	+
Tamar Estuary, S. W. England	+	–
Snettisham, Kings Lynn, E. England	+	+

the country, to examine small samples of *Cerastoderma edule* for the presence of the two species of *Paravortex*. The data are presented in Table 3 and indicate the wide distribution of the two species.

Material deposited

The holotype, gravid specimen of *P. karlingi* has been deposited with the British Museum (Natural History). A second slide, bearing worms of various ages, has also been deposited. Other whole mounts and sections are in the authors' collections.

Discussion

Although *P. cardii* and *P. karlingi* are very similar morphologically, and share the same host species, we believe that the differences we have found are quite clear and sufficient to distinguish them. However, there are several points that must be clarified concerning the two species.

The first relates to the distribution of *P. cardii*. Hallez (1909) stated that the gravid adult worms are to be found in the stomach of *Cerastoderma edule*, and that newly emerged young leaving the parent pass into the intestine where they remain for several days. During their stay in the intestine Hallez said that the worms grew in size and copulated before passing out of the host to become temporarily free living. Recently Jennings & Phillips (1978) have given the habitat of *P. cardii* as the digestive gland tubules and not the stomach. From our own investigations on *P. cardii* we would agree with their view that *P. cardii* is a parasite of the digestive gland tubules.

It is reasonable to believe that the progeny of *P. cardii* pass into the host's stomach and on into the intestine before being discharged from the anus. It is possible, however, that the worms Hallez (1909) described from the intestine were in reality *P. karlingi* and not developing *P. cardii*. He has also referred to the occasional presence of gravid worms in the intestine of *C. edule*. Alternately it may be argued (i) that *P. cardii* can occur in both habitats, but develops differently in each or (ii) that the stages we have found in the intestine are *P. cardii* developing in the intestine before migrating out of the body to invade another host, or will migrate forward in the same host to establish themselves permanently in the digestive gland tubules.

It is not possible to know whether Hallez (1909) was working with one or two species since he gave few details about the instestinal stages. The possibility that we are dealing with one species that differs according to the habitat in which it grows has some precedents in the other groups of parasitic platyhelminths. Differences in such features as size could be realistically referred to such an occurrence but the behaviour of the two species seems to us to be too fundamentally different to be accounted for by such an occurrence. Furthermore the differences in epidermal structure, referred to by MacKinnon *et al.* (1981), confirm that this possibility is very unlikely.

The possibility that the worms found in the intestine are developing *P. cardii* is also unlikely in view of the fact that no digestive gland forms have ever been found in any of more than 1200 *C. edule* from the Ythan Estuary. The worms that have been found in the intestine have included all stages from new born juveniles to gravid adults containing up to 20 embryos. Furthermore both *P. cardii* in the digestive gland tubules and *P. karlingi* from the intestine occur concurrently in *C. edule* collected from sites to the north and south of the Ythan Estuary. It is unlikely in this case that the host population has any bearing on the nature or site of infection, or that geographical situation has any significance.

The morphological comparisons that we have made between *P. cardii* and *P. karlingi* have been made not only on the basis of Hallez's (1909) description but also on specimens of *P. cardii* that have been collected from many parts of Britain. All our comparisons of measurements, body shape and behaviour have been made on our own material. This is especially important as far as measurements are concerned because previous authors have not specified the conditions under which their measurements were made. All of ours were taken from specimens that were fixed unflattened and then measured on uncovered slides or after mounting in Canada balsam. From our observations gravid *P. cardii* are indeed larger than *P. karlingi,* the former having a mean length of 0.79 ± 0.07 mm (range of ten specimens 0.7–0.9 mm) after fixation staining and mounting. The mean breadth is 0.33 ± 0.03 mm (range of ten specimens 0.28–0.36 mm). Few other measurements could be reliably made, of which mean pharynx dimensions were 0.06 ± 0.01 × 0.07 ± 0.01 mm (range of ten specimens 0.05–0.07 × 0.06–0.08 mm). The ratio of mean pharynx length to mean body length in *P. cardii* was 1:13.2. The seminal vesicle diameter in *P. cardii* was 0.06 ± 0.01 mm (range of nine measurements 0.05–0.08 mm). These measurements show that fixed *P. cardii* are 1.9 × the length of *P. karlingi,* that they have a marginally bigger pharynx and seminal vesicle and that the ratio of pharynx length to body length is different.

The differences in body shape observed for these two species in life are also seen in fixed specimens (Fig. 12, 13). These differences may be due, in part, to the distribution of brood capsules which in *P. cardii* appears to be posterior, but more anterior in *P. karlingi* (Fig. 7). The number of brood capsules in *P. cardii* is also much larger. Atkins (1934) reported up to 60 capsules for *P. cardii* at Plymouth which is three times the maximum found in *P. karlingi.*

The behaviour of *P. cardii* on release from the host was emphasized by Hallez (1909). This behaviour was also noted in specimens we recovered and differed considerably from that of *P. karlingi.* Furthermore the behaviour of the two species in response to directional sources of light is different. It would be interesting to compare the behaviour of the other two species in the same way, especially since there is some dispute concerning *P. gemellipara.* Linton (1910) stated very clearly that the worms were negatively phototactic whereas Ball (1916) claimed that this was not the case, but that the behaviour of adult worms was neutral. Perhaps *P. gemellipara* behaves in a way similar to *P. cardii* which would explain the discrepancy.

It is clear from our study that a lot more work should be done with the members of this genus, both with regard to the taxonomic relationships of the species that have been described thus far and also to their biology, which is virtually unknown apart from the studies of Jennings & Phillips (1978), Jennings & LeFlore (1979) & Phillips (1978).

Acknowledgements

We wish to thank Drs/Messrs E. Bowers, J. R. Burt, W. Finlayson, A. Franklin, J. Green, D. Halton, B. Hepper, G. Pickett, D. Saward, J. W. Smith and H. H. Williams for their generous and much appreciated assistance in obtaining mollusc samples from around the coast of the British Isles. We are also very grateful to Mrs. Huggins of the Eastern Sea Fisheries Committee at Kings Lynn for her willing assistance in obtaining a further sample from that area. Mr. W. Edwards gave us considerable technical assistance for which we are most appreciative. Finally we wish to acknowledge the photographic assistance of Mr. R. Smith.

We also wish to acknowledge with grateful thanks financial assistance to M.D.B.B. by NATO through a Sen. Sci. grant No. B(109) and by NSERC of Canada through operating grant No. A-2358, and to A.W.P. by Royal Society through a travel grant.

References

Atkins, D., 1934. Two parasites of the common cockle Cardium edule; a rhabdocoel Paravortex cardii Hallez and a copepod Paranthessius rostratus (Canu). J. Mar. Biol. Assoc. U.K. 19: 669–676.

Ball, S. J., 1916. Development of Paravortex gemellipara. J. Morphol. 27: 453–557.

Freeman, R. F. H., 1957. Paravortex scrobiculariae (Graff) in Great Britain. Nature. 180: 1213–1214.

Hallez, P., 1909. Biologie, organisation, histologie et embryologie d'un rhabdocoele parasite du Cardium edule L., Paravortex cardii n.sp. Arch. Zool. Exp. Gén. Sér. 4, 9: 429–544.

Jennings, J. B. & LeFlore, W. B., 1979. Occurrence and possible adaptive significance of some histochemically demonstrable dehydrogenases in two entosymbiotic rhabdocoels (Platyhelminthes: Turbellaria). Comp. Biochem. Physiol. 62B: 301–304.

Jennings, J. B. & Phillips, J. I., 1978. Feeding and digestion in three entosymbiotic graffillid rhabdocoels from bivalve and gastropod molluscs. Biol. Bull. 155: 542–562.

Leigh-Sharpe, W. H., 1933. Note on the occurrence of Graffilla gemellipara Linton (Turbellaria) at Plymouth. Parasitology 25: 108.

Linton, E., 1910. On a new rhabdocoele commensal with Modiolus plicatulus. J. Exp. Zool. 9: 371–386.

MacKinnon, B. M., Burt, M. D. B. & Pike, A. W., 1981. Ultrastructure of the epidermis of adult and embryonic Paravortex spp. (Turbellaria, Eulecithophora). In: Schockaert, E. R. & Ball, I. R. Turbellaria. Proc. Third Int. Symp. Hydrobiologia (this volume).

Patterson, J. T., 1912. Early development of Graffilla gemellipara – a supposed case of polyembryony. Biol. Bull. Mar. Biol. Lab., Woods Hole. 22: 173–204.

Phillips, J. I., 1978. The occurrence and distribution of haemoglobin in the entosymbiotic rhabdocoel Paravortex scrobiculariae (Graff) (Platyhelminthes: Turbellaria). Comp. Biochem. Physiol. 61A: 679–683.

Abbreviations

bc – brood capsules
c – core of indigestible particles
ca – common atrium
ch – chamber
e – eye
fa – female atrium
fol – fibre optic light
gc – gut caecum
gp – genital pore
j – juvenile worm
m – mouth
o – ovary
oe – oesophagus
p – pharynx
pe – penis
sg – shell gland
sv – seminal vesicle
t – testis
u – uterus
vg – vitelline gland
w – window

Karyological observations on Polyclads

Lodovico Galleni & Ileana Puccinelli
Istituto di Zoologia e Anatomia comparata, Univ. di Pisa Via A. Volta 4, I 56100 Pisa, Italy

Keywords: Turbellaria, Polycladida, karyology

Abstract

A karyological analysis of eight species of Polyclads has been carried out.

A chromosome number of n = 10 for *Stylochus alexandrinus* and for *Stylochus mediterraneus* was ascertained on germinal lines.

A chromosome number of 2n = 20 was ascertained from metaphasic plates from regenerative blastemas of *Notoplana alcinoi* and *Leptoplana tremellaris*. A karyometric analysis of this last species permitted the reconstruction of an idiogram.

An idiogram was also prepared from metaphasic plates of *Stylostomum ellipse* obtained from early embryonic mitoses. The chromosome number of this species also was 2n = 20.

Chromosome numbers of n = 8 and 2n = 16 were ascertained on germinal lines and metaphasic plates of blastemas in *Echinoplana celerrima* and on germinal lines and early embryonic mitoses of *Stylochoplana maculata*.

Finally the diploid number 2n = 18 was ascertained on metaphasic plates from regenerative blastema of *Yungia aurantiaca*.

Introduction

Our knowledge of the karyology of Polyclads is still unsatisfactory. At present the chromosome numbers of only 29 species are known, and about 14 of these numbers almost certainly are in need of revision (Galleni & Puccinelli 1975). These data we consider to be reliable, in that they have been confirmed by more than one author, or are based on modern karyological studies, are shown in Table 1.

This paucity of informations has many causes. In most species it is difficult to collect enough specimens suitable for a detailed karyological analysis. Moreover it is not possible to obtaine regenerative blastema from all the species analysed. Finally many metaphasic plates exhibit chromosomes of such small dimensions that a careful karyometric analysis is exceedingly difficult to carry out.

At present our efforts are mainly devoted to determining the chromosome numbers of as many species as possible and to preparing idiograms suitable for comparative analysis.

Material and methods

Most of the species examined were collected on the rocky shores of the tyrrhenian coasts of Italy. *Leptoplana tremellaris* and *Stylostomum ellipse*, however, were collected in British waters and studied during a six month stay by one of us (L.G.) at the Laboratory of the Marine Biological Association of U.K. in Plymouth (Devon).

The haploid sets were studied from the male and female meioses and, when possible, the mean chiasma frequency was calculated.

The diploid sets were studied in metaphasic plates from regenerative blastemas or from early

Hydrobiologia 84, 31–44 (1981). 0018-8158/81/0841-0031/$02.80.

Table 1. Chromosome numbers in Polyclads.
♀ I, II = oocyte I, II; ♂ I, II = spermatocyte I, II; sperm. = spermatogonium; embr. = embryonal mitosis; neobl. = neoblast; som. = somatic mitosis.

Species	n	2n	Authors
Suborder Acotylea			
fam. Stylochidae			
Stylochus (Stylochus) alexandrinus Steinböck	10 ♂ I,II	–	present paper
Stylochus (Imogine) ellipticus (Girard)	10 ♀ I,II	20 embr.	Van Name (1899)
	–	20 embr.	Bahr and Hillmann (1966)
Stylochus (Imogine) mediterraneus Galleni	10 ♂ I, ♀ I	–	present paper
fam. Leptoplanidae			
subfam. Leptoplaninae			
Leptoplana tremellaris (O. F. Müller)	8 ♂ I, ♀ I	16 embr.	present paper
Stylochoplana maculata (Quatrefages)	–	20 neobl.	present paper
subfam. Notoplaninae			
Notoplana alcinoi (O. Schmidt)	10 ♂ I, ♀ I	20 neobl.	present paper
Notoplana humilis (Stimpson)	10 ♀ I,II	–	Kato (1940)
Notoplana igiliensis Galleni	10 ♀ I, ♂ I,II	20 embr. sperm. neobl.	Galleni & Puccinelli (1975)
subfam. Euplaninae			
Euplana gracilis (Girard)	9 ♀ I,II	18 embr.	Christensen (1971)
subfam. Hoploplaninae			
Hoploplana inquilina (Wheeler)	9-10(?) ♀ I	–	Wheeler 1894
	10 ♀ I,II	–	Patterson & Wieman (1912)
fam. Planoceridae			
Echinoplana celerrima Haswell	8 ♂ I, ♀I	16 neobl.	present paper
Suborder Cotylea			
fam. Pseudoceridae			
Thysanozoon brocchii (Risso)	9 ♀ I,II, ♂ I	18 embr.	Van der Stricht (1897, 1898)
	9 ♀ I,II	18 embr.	Schockaert (1902, 1905)
	9 ♂ I,II	18 som.	Bertani (1947)
	–	18 neobl.	Galleni & Puccinelli (1977)
Yungia aurantiaca (Delle Chiaje)	–	18 neobl.	present paper
fam. Euryleptidae			
Prostheceraeus vittatus (Montagu)	6 ♀ I,II	–	Francotte (1897, 1898)
	6 ♀ I,II	12 embr.	Von Klinckowström (1897)
	6 ♀ I	–	Gerard (1901)
Stylostomum ellipse (Dalyell)	–	20 embr.	present paper

embryonic mitoses or from the germinal lines.

Regenerative power was not homogenous in all species. Most of the Acotyleans produced blastemas with numerous mitoses within two or three days after cutting the part in front of the brain. In species of the suborder Cotylea mitoses from blastemas were obtained three or more days after the transection of a large part of the caudal end of the animal. We failed to obtain good regeneration with enough mitoses in a few species as reported in the text.

All the specimens with regenerative blastema or early embryos were put in a 0.15% colchicine SIGMA solution in sea water for 3 to 4 h. Then the portion with blastema or the egg-plates or body fragments were stained for 30 min with acetic orcein or lactic acetic orcein and finally squashed between slide and coverglass.

Relative length (r.l.) was calculated as length of chromosome × 100/total length of haploid genome. Centromeric index (c.i.) was calculated as length of the short arm × 100/length of the whole chromosome.

Chromosome classification is according to Levan *et al.* (1964).

In some plates the r.l. values seemed inverted in respect of their mean values. This fact is easily explained considering that the differences among chromosomes are often so slight that their order can be changed by small variations in the level of spiralization. In these cases we always arranged the chromosomes according to the position of the centromere.

The germinal lines have been examined with the same squashing methods using body fragments containing tests or ovaries and from laid unfertilized oocytes.

Results and discussion

Suborder ACOTYLEA

Fam. Stylochidae

Stylochus (Stylochus) alexandrinus Steinböck. Karyological investigations on this species have been carried out on specimens collected in autumn along the coasts of Livorno among mussels and barnacles. The species showed a low regenerative power; therefore we failed to obtain from blastemas metaphasic plates suitable for any karyological investigation.

As regards the germinal lines, our specimens presented many young oocytes, not suitable for karyological investigation, while, on the contrary, the male germinal line was fully mature with numerous spermatocytes I and II. The haploid number was n = 10.

The 10 bivalents of spermatocytes I are large or medium in size, except for the last two, which appear very small.

Most of male bivalents are rod-shaped with one terminal or intercalar chiasma. Only a few bivalents (from 1 to 4 for cell) are ring shaped with two chiasmata. Spermatocytes with all bivalents with one chiasma are often present. Therefore the chiasma frequency is rather low: about 11.0/cell (Plate I, Fig. 1).

The secondary spermatocyte chromosomes were useful for a rough examination of the chromosome morphology (Plate I, Fig. 2).

The haploid set presents eight chromosomes large or medium in size, and gradually decreasing, and two smaller ones.

Seven chromosomes are meta- or submeta-centric while three chromosomes seem to be subtelocentric.

Stylochus (Imogine) mediterraneus Galleni. Numerous specimens of this worm were collected through all the year on the mussel beds near Pisa and Livorno: the species is, in fact, predator on mussels (see Galleni *et al.* 1980).

The haploid number n = 10 was obtained in the germinal lines both from spermatocytes I and freshly laid unfertilized oocytes I (Plate I, Figs. 3 and 4).

General remarks. As reported in Table 1 a chromosome number of n = 10 has also been found in *Stylochus (Imogine) ellipticus* (Van Name 1899). It is noteworthy that the three species of *Stylochus* have the same chromosome number in spite of the fact that they belong to different subgenera.

Fam. Leptoplanidae

Notoplana alcinoi (O. Schmidt). Numerous specimens of *N. alcinoi* were collected along the Tuscan coasts, mainly in the mussel beds.

A haploid number of n = 10 was ascertained in spermatocytes I (Plate II, Figs. 3 and 4) and in uterine oocytes or in freshly laid unfertilized oocytes I (Plate II, Figs. 1 and 2).

A diploid number of 2n = 20 was ascertained in mitoses from blastemas.

The analysis of germ cells was carried out on a great number of plates: male bivalents show mainly one terminal chiasma: spermatocytes I with all bivalents with one chiasma are often observed (Plate II, Fig. 4). A few ring shaped bivalents, (from 1 to 3) may be present (Plate II, Fig. 3). Mean chiasma frequency is low (11.0).

In the female line the chiasma frequency is higher: many bivalents showing two chiasmata are present (Plate II, Figs. 1 & 2). Uterine oocytes show a higher chiasma frequency (18.0) than the laid ones (16.0). This fact is probably due to the separation of chiasmata in aging unfertilized oocytes. Analysis of the diploid complement was possible because of the good regenerative power of the species. The karyometric analysis is still in progress, but chromosome measurements accomplished on one plate suggest the presence of three subtelocentric and three submetacentric chromosomes and of four metacentric ones (Plate III, Figs. 1 & 2).

Stylochoplana maculata (Quatrefages). Some specimens of *Stylochoplana maculata* collected in the winter in the Orbetello lagoon (Grosseto-

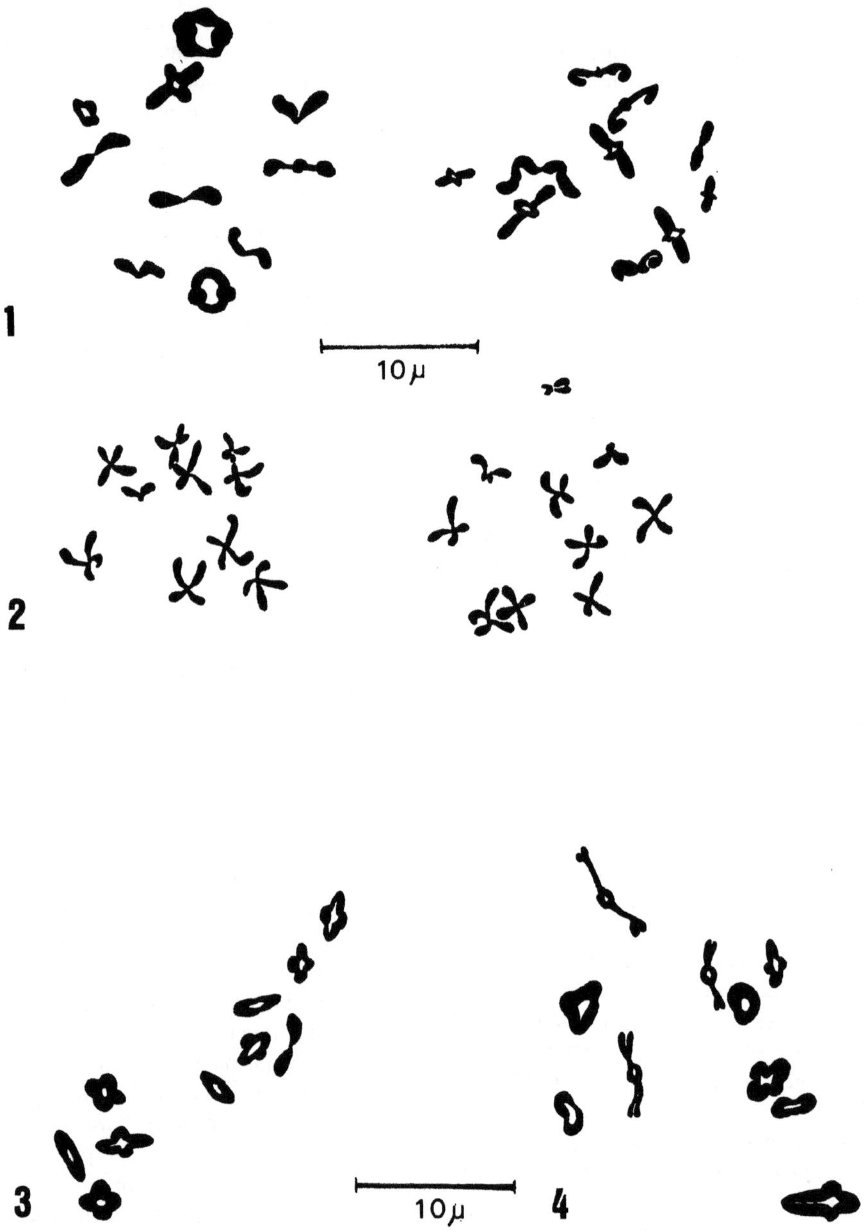

Plate I. (1) *Stylochus alexandrinus* - Camera lucida drawings of metaphasic bivalents from spermatocytes I.
(2) *Stylochus alexandrinus* - Camera lucida drawings of chromosomes from two metaphasic plates of spermatocytes II.
(3) *Stylochus mediterraneus* - Camera lucida drawing of metaphasic bivalents from spermatocyte I.
(4) *Stylochus mediterraneus* - Camera lucida drawing of metaphasic bivalents from oocyte I.

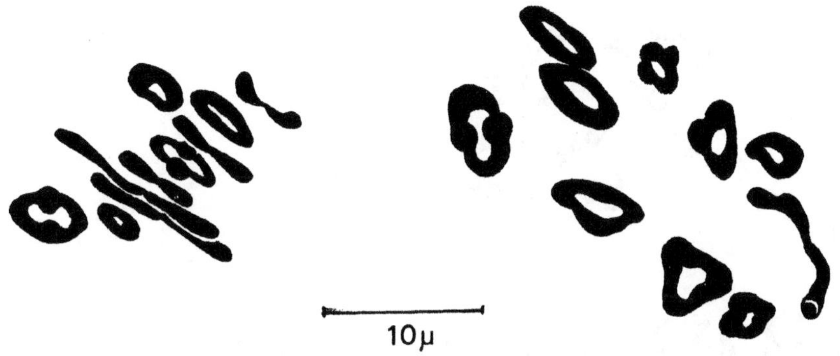

1 2

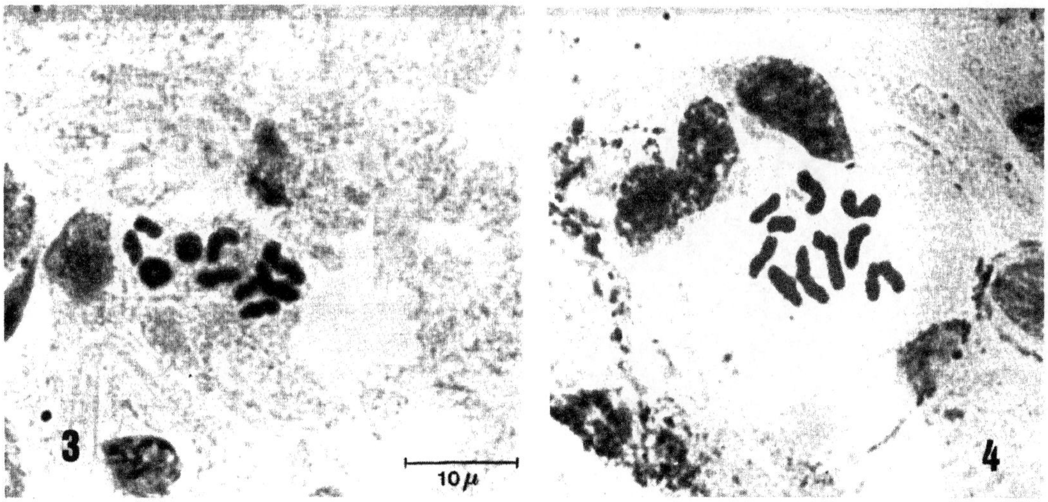

Plate II. (1) *Notoplana alcinoi* – Camera lucida drawing of metaphasic bivalents from laid unfertilized oocyte I.
(2) *Notoplana alcinoi* – Camera lucida drawing of metaphasic bivalents from uterine-fertilized oocyte I.
(3) *Notoplana alcinoi* – Metaphasic bivalents from spermatocyte I.
(4) *Notoplana alcinoi* – Metaphasic bivalents from spermatocyte I (enlargement as in Fig. 3).

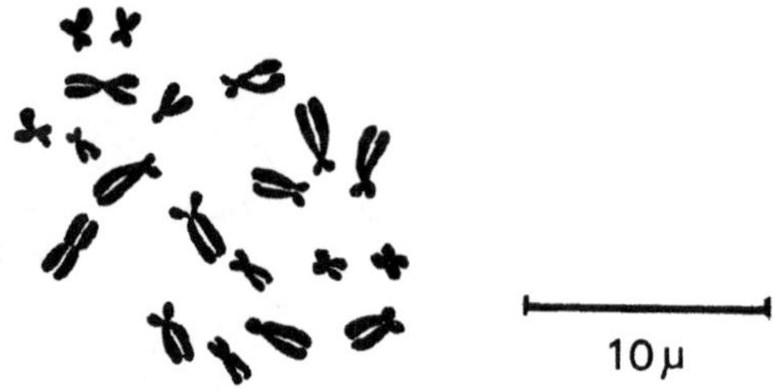

1

2

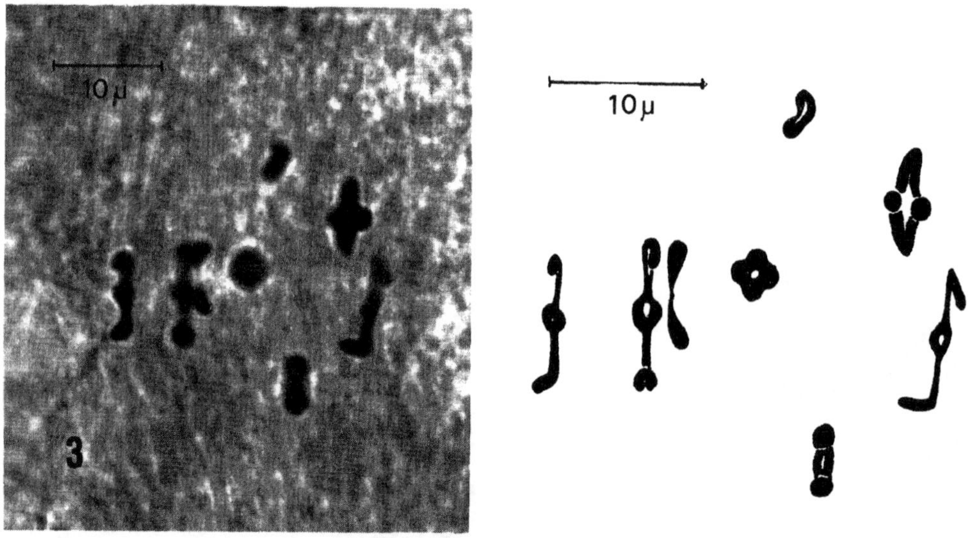

Plate III. (1) *Notoplana alcinoi* – Camera lucida drawing of a metaphasic plate from a neoblast.
(2) *Notoplana alcinoi* – Derived karyogram from the plate of Fig. 1.
(3) *Stylochoplana maculata* – Metaphasic bivalents from oocyte I and related camera lucida drawing.

Tuscany) allowed us to furnish the chromosome number of the species.

The haploid number n = 8 was observed in spermatocytes I (Plate IV, Fig. 2) and in laid unfertilized oocytes I (Plate III, Fig. 3). The eight male bivalents are large or medium in size, except two small bivalents, easily distinguishable. They are mainly with one terminal chiasma; ring shaped bivalents with two chiasmata may be present in various number (from 1 to 3) and may be formed by either medium large or small chromosomes. Mean chiasma frequency is 11.0.

Also in laid oocytes a low chiasma frequency is present (mean chiasma frequency 10.5) in contrast to the observations made in other Polyclads. This fact may be explained with the aging of the oocytes. Many bivalents, in fact, are stretched along spindle fibres.

A diploid number of 2n = 16 was observed in early embryonic mitoses (Plate IV, Fig. 1), but the chromosomes were not suitable for a careful analysis of the karyotype.

Leptoplana tremellaris (O. F. Müller). The idiogram of *Leptoplana tremellaris* was prepared on cephalic blastemas of some specimens collected among weeds in Drake Island (Plymouth) in autumn 1976. The idiogram was prepared through the analysis of five metaphasic plates (Table 2) and shows ten pairs of homologous chromosomes (Plate V, Figs. 1, 2 & 3). Their relative lengths decrease from 12.86 to 6.72 and it is quite impossible to divide the chromosomes into groups. With respect to centromeric position chromosomes 4, 7 and 10 are submetacentric, chromosomes 1, 5, 8 and 9 are metacentric and chromosomes 2, 3 and 6 are subtelocentric.

General remarks. The similarities between the karyometric data of *Notoplana alcinoi* and *Notoplana igiliensis* (see Table 3) are evidence of the close relationship between these species which are distinguishable only from a few characteristics of the reproductive system.

Some problems are posed by the chromosome number of *Leptoplana tremellaris*. Our data are different from those obtained by Francotte (1897, 1898) who found the number n = 8, 2n = 16. In our opinion the most probable explanation of this discrepancy is that the specimens examined by Francotte could belong to a different species. Many Polyclads have been classified, only on the basis of external characters, as *Leptoplana tremellaris* because the external morphology of Leptoplanids is quite similar. To avoid any taxonomic confusion the posterior half of the specimen which furnished us with most of the karyological preparations was sectioned and the copulatory apparatus observed.

It is interesting that *Leptoplana tremellaris* shows an idiogram similar, in some aspects, with that of *Notoplana igiliensis* reported by Galleni & Puccinelli (1975). In both idiograms ten pairs of homologous chromosomes are present four pairs being metacentric. In contrast, *Stylochoplana maculata*, which, according to Marcus & Marcus (1966) belongs to the same subfamily as *Leptoplana tremellaris*, presents a different chromosome number. It is to be noted that the family Leptoplanidae comprises many species and its subdivisions are quite artificial. Cytotaxonomical investigations may be expected to help in preparing a more natural division of this family.

Fam. Planoceridae

Echinoplana celerrima Haswell. Karyological investigation was carried out on some specimens collected near Naples and obtained from the 'Stazione Zoologica' or collected along the coast of Livorno (see for a taxonomic discussion: Galleni 1978).

A haploid number of n = 8 was ascertained on spermatocytes I (Plate VI, Fig. 1) and in freshly laid oocytes I (Plate VI, Fig. 2).

In the male line bivalents show a mean chiasma frequency of 12.6 while in the female line most of the bivalents were ringshaped and the mean chiasma frequency higher (15.0). A few metaphasic plates from blastemas were obtained from which a diploid number of 2n = 16 was observed (Plate VI, Fig. 3).

General remarks. Echinoplana celerrima is the only species among the Planoceridae whose chromosome number has been investigated. Its chromosome number, n = 8, 2n = 16, is the lowest among the Acotylea. It is to be noted that, according to the way of insemination (Haswell 1907, see also Karling 1963) this species seems to be one of the most primitive of the Acotylea.

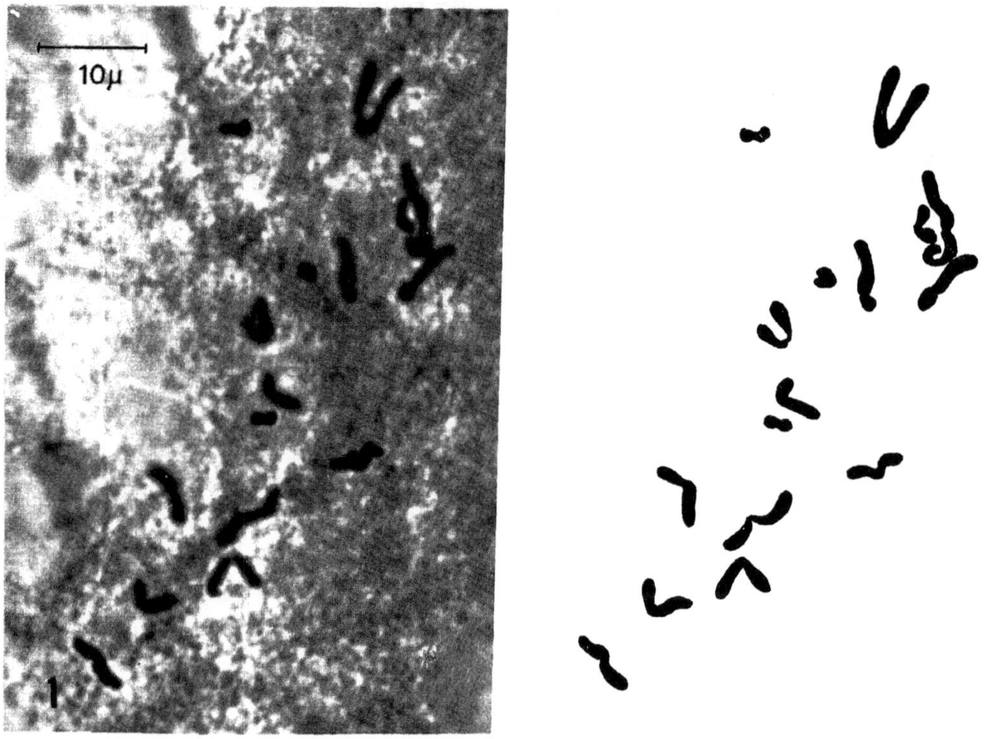

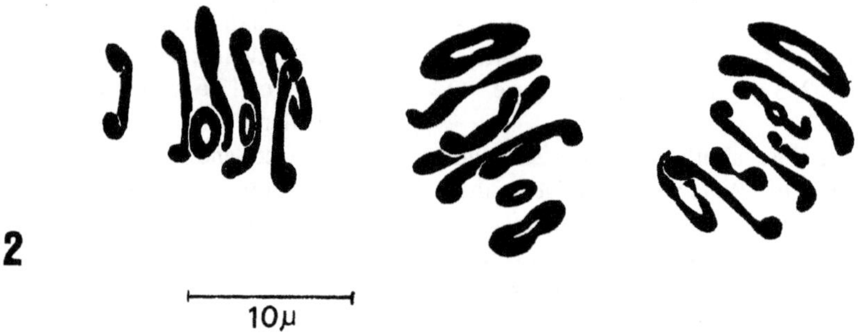

Plate IV. (1) *Stylochoplana maculata* – Metaphasic plate from an early embryonic mitosis and related camera lucida drawing.

(2) *Stylochoplana maculata* – Camera lucida drawings of metaphasic bivalents from spermatocytes I.

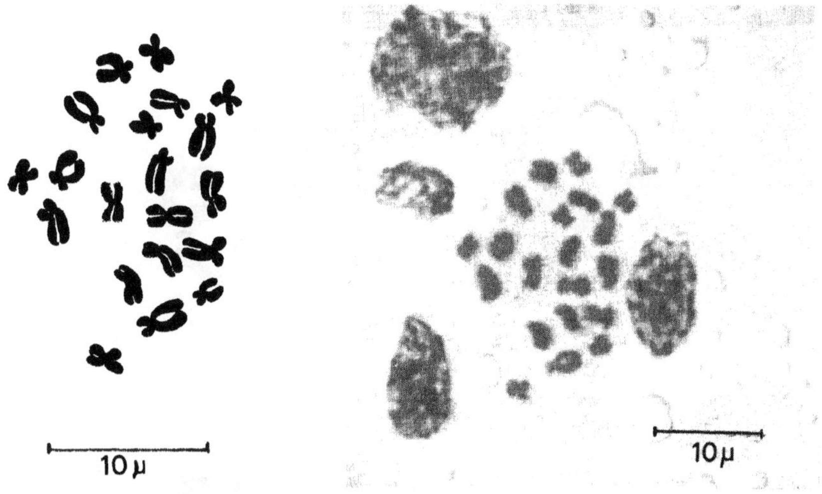

1

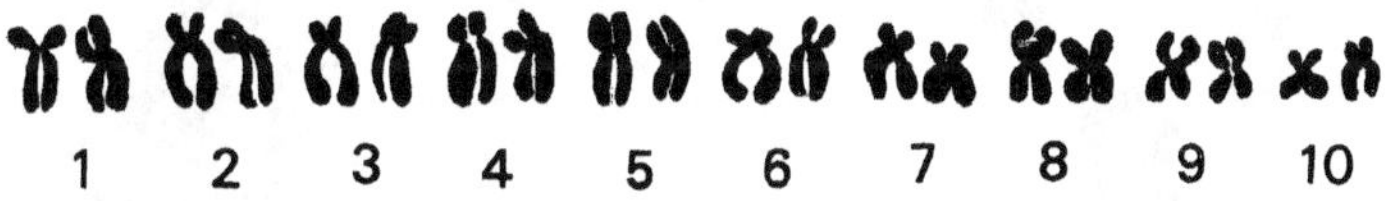

2

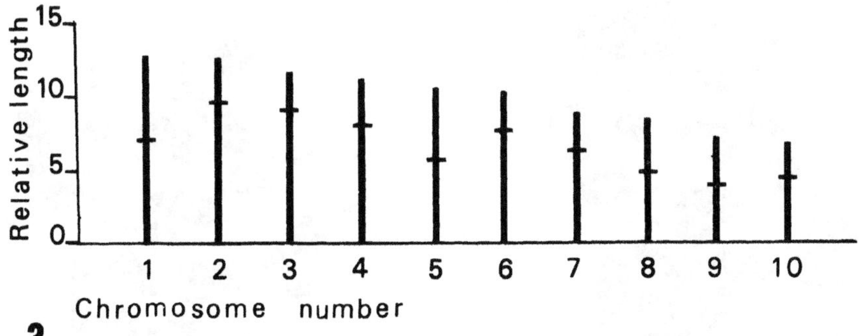

3

Plate V. (1) *Leptoplana tremellaris* – Metaphasic plate from neoblast and related camera lucida drawing.

(2) *Leptoplana tremellaris* – Derived karyogram from the plate of Fig. 1.

(3) *Leptoplana tremellaris* – Idiogram based on the mean values reported in Table 2.

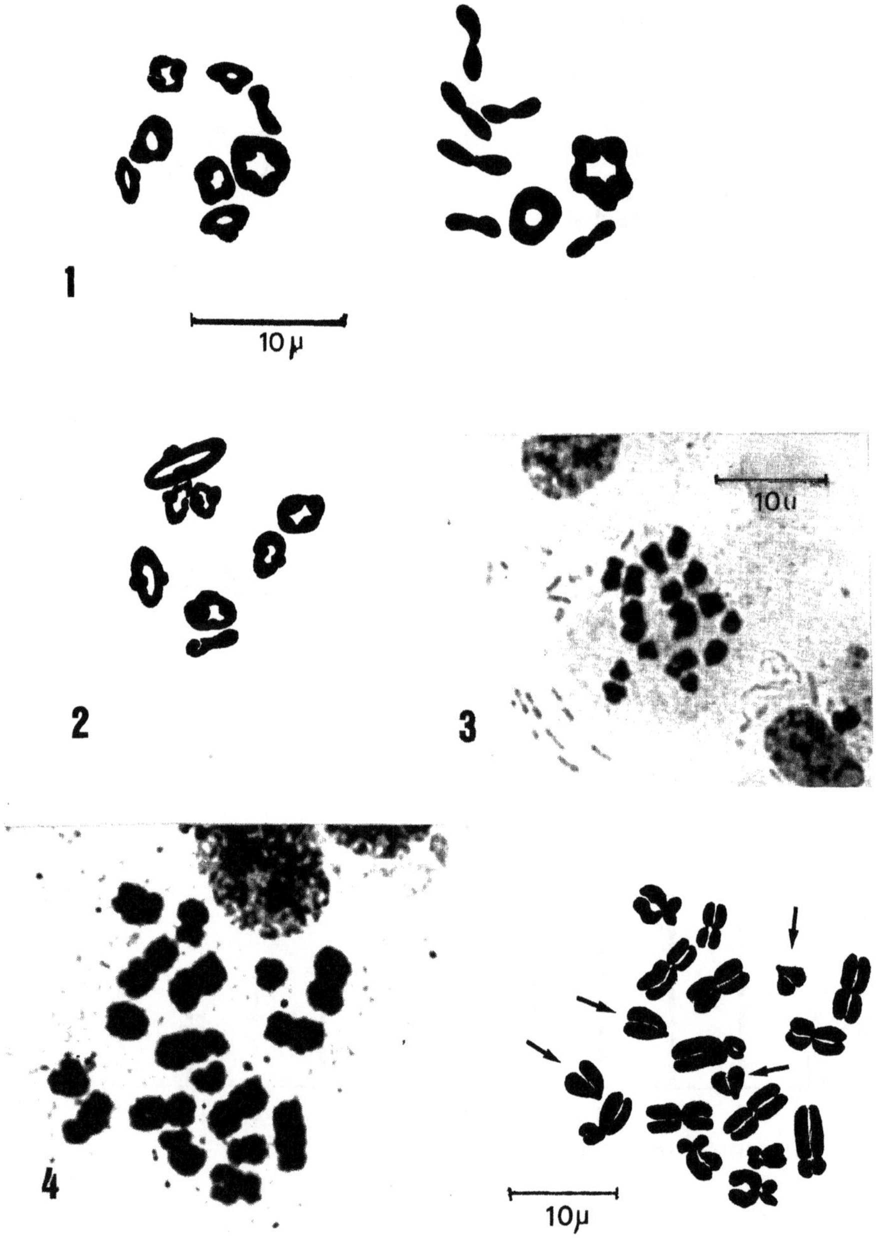

Plate VI. (1) *Echinoplana celerrima* - Camera lucida drawings of metaphasic bivalents from spermatocytes I.
(2) *Echinoplana celerrima* - Camera lucida drawing of metaphasic bivalents from oocyte I.
(3) *Echinoplana celerrima* - Metaphasic plate from neoblast.
(4) *Yungia aurantiaca* - Metaphasic plate from neoblast and related camera lucida drawing. Arrows show the telocentric chromosomes.

Table 2. Chromosome measurements in five cells of *Leptoplana tremellaris.*

Chromosome no.		Cells 1	2	3	4	5	Means s.d.
1	r.l.	12.95	14.76	12.37	12.42	11.93	12.89 ± 1.10
	c.i.	40.45	46.25	49.32	49.02	38.83	44.78 ± 4.88
2	r.l.	12.95	13.47	12.37	12.41	12.37	12.71 ± 0.49
	c.i.	21.28	27.40	24.68	25.38	23.16	24.38 ± 2.31
3	r.l.	11.29	11.88	11.89	12.27	11.18	11.70 ± 0.46
	c.i.	19.61	24.03	22.97	20.22	22.32	21.83 ± 1.86
4	r.l.	10.74	10.87	11.08	12.39	10.83	11.18 ± 0.69
	c.i.	28.28	28.81	27.54	25.58	29.64	27.97 ± 1.54
5	r.l.	10.06	10.33	10.44	10.86	10.79	10.50 ± 0.33
	c.i.	42.49	46.43	43.08	46.59	47.33	45.18 ± 2.23
6	r.l.	10.61	9.59	9.96	10.02	10.28	10.09 ± 0.38
	c.i.	22.09	30.77	20.97	26.68	23.33	24.77 ± 3.98
7	r.l.	9.64	8.30	9.23	8.42	8.61	8.84 ± 0.57
	c.i.	25.71	26.67	25.24	36.30	29.37	28.66 ± 4.56
8	r.l.	7.71	8.21	8.84	8.15	9.48	8.49 ± 0.69
	c.i.	46.71	44.94	42.27	43.34	45.44	44.54 ± 1.75
9	r.l.	7.02	5.54	6.98	6.72	8.87	7.02 ± 1.19
	c.i.	47.22	46.67	47.15	46.87	44.74	46.53 ± 1.02
10	r.l.	7.02	7.01	6.82	6.34	6.63	6.76 ± 0.29
	c.i.	31.48	26.32	38.82	39.00	33.33	33.79 ± 5.33

Table 3. (a) Mean values and standard deviations of chromosome measurements in five cells of *Notoplana igiliensis* (from Galleni & Puccinelli 1975).
(b) Chromosome measurements of the metaphasic plate of *Notoplana alcinoi* reported in Plate 2.

		Chromosome no. 1	2	3	4	5	6	7	8	9	10
a	r.l.	13.37	12.69	11.81	11.71	10.87	9.45	9.04	7.59	6.83	6.07
	s.d.	0.53	0.31	0.80	0.40	0.61	0.72	0.84	0.64	0.21	0.28
	c.i.	17.79	16.90	46.92	18.60	18.59	22.86	44.59	22.87	46.26	44.91
	s.d.	3.47	3.50	1.41	1.23	2.02	4.79	3.20	4.65	2.09	1.96
b	r.l.	13.39	12.85	12.46	12.23	10.76	9.52	8.67	7.43	6.66	6.04
	c.i.	17.32	48.17	21.73	25.32	28.77	22.64	38.75	30.94	41.77	47.62

Suborder COTYLEA

Fam. Pseudoceridae

For this family we have already published the idiogram of *Thysanozoon brocchii* (Galleni & Puccinelli 1977). At present we are working on a karyometric analysis of the related species *Yungia aurantiaca.* A diploid chromosome number of 2n = 18 was observed from metaphasic plates of caudal blastemas (Plate VI, Fig. 4). From the analysis of the karyotype (the idiogram is still in progress) the presence of two pairs of telocentric chromosomes is to be underlined. This fact is particularly noteworthy because they are known within the order only from the two species here discussed.

Table 4. Chromosome measurements in five cells of *Stylostomum ellipse.*

Chromosome no.		Cells 1	2	3	4	5	Means s.d.
1	r.l.	13.36	15.20	12.41	12.50	13.31	13.36 ± 1.12
	c.i.	47.33	45.53	47.81	49.08	49.09	47.75 ± 1.46
2	r.l.	11.86	12.81	12.71	12.20	11.68	12.52 ± 0.50
	c.i.	42.31	42.80	45.50	45.46	38.43	42.90 ± 2.90
3	r.l.	11.73	11.15	12.17	12.39	10.45	11.58 ± 0.79
	c.i.	49.50	45.62	49.73	47.74	48.75	48.27 ± 1.67
4	r.l.	11.14	10.88	11.08	11.04	10.68	10.96 ± 1.86
	c.i.	46.21	41.15	46.39	39.67	44.05	43.49 ± 3.01
5	r.l.	10.52	10.24	9.64	9.90	9.72	10.04 ± 0.37
	c.i.	44.62	41.07	46.56	45.65	44.40	44.46 ± 2.08
6	r.l.	8.81	9.25	9.59	9.59	9.03	9.25 ± 0.34
	c.i.	48.96	45.01	48.81	49.37	43.60	47.15 ± 2.56
7	r.l.	8.48	9.36	8.83	9.67	9.29	9.13 ± 0.47
	c.i.	38.57	34.06	36.68	41.93	39.64	38.17 ± 2.98
8	r.l.	8.65	8.68	9.03	8.50	9.18	8.81 ± 0.28
	c.i.	26.75	31.25	20.33	24.16	25.61	25.62 ± 3.97
9	r.l.	8.45	6.58	7.63	7.67	8.44	7.75 ± 0.77
	c.i.	39.81	34.03	25.32	31.09	38.48	33.75 ± 5.86
10	r.l.	6.97	5.83	6.91	6.60	8.20	6.90 ± 0.86
	c.i.	46.75	41.86	35.04	42.53	48.22	42.88 ± 5.15

Fam. Euryleptidae

Stylostomum ellipse (Dalyell). Some specimens of *Stylostomum ellipse* collected on the mouth of the River Yealm (Plymouth) in autumn 1976 allowed us to reconstruct the idiogram of this species from early embryonic mitoses.

The idiogram was prepared through the analysis of five metaphasic plates (Table 4) and shows ten pairs of homologous chromosomes (Plate VII, Figs. 1, 2 & 3) whose relative lengths decrease from 13.36 to 6.90. With respect to the centromere position, chromosomes 1, 2, 3, 4, 5, 6, 7 and 10 are metacentric while chromosomes 8 and 9 are submetacentric.

At present the chromosome number of *Stylostomum ellipse* 2n = 20 is the highest found in the suborder Cotylea.

Conclusions

Our present data are too few to trace a general view on karyological evolution in Polyclads. Meanwhile we wish to stress that pairs of telocentric chromosomes have been found in only two species of Polyclads: one pair in the complement of *Thysanozoon brocchii* (Galleni & Puccinelli 1977) and two pairs in that of *Yungia aurantiaca.* Both these species belong to the family Pseudoceridae and to the most primitive suborder Cotylea (For a discussion about the primitivity of Cotylea based on spermatozoa structures: see Hendelberg 1969).

Moreover this suborder shows species with chromosome numbers of 2n = 12, 2n = 18 and 2n = 20 while the suborder Acotylea shows species with chromosome numbers of 2n = 16 and 2n = 20. So far the most primitive suborder presents telocentric chromosomes and the lowest chromosome number and the suborder Acotylea presents seven of the ten studied species with chromosome numbers of 2n = 20.

Finally, there seems to be some uniformity of chromosome numbers in the same family: 2n = 18 in Pseudoceridae, 2n = 20 in Stylochidae and 2n = 20 in five of the seven studied species in the family Leptoplanidae.

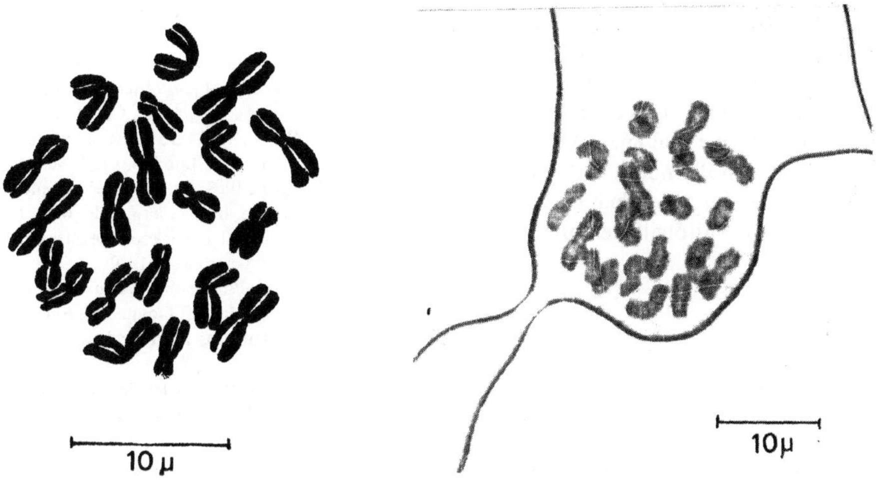

1

2

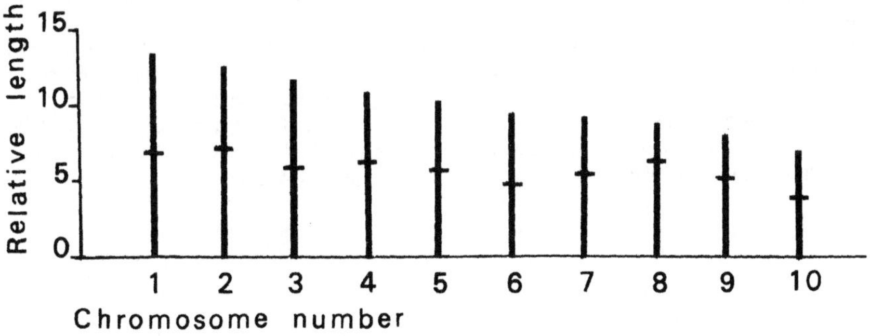

3

Plate VII. (1) *Stylostomum ellipse* – Metaphasic plate from early embryonic mitosis and related camera lucida drawing.
(2) *Stylostomum ellipse* – Derived karyogram from the plate of Fig. 1.
(3) *Stylostomum ellipse* – Idiogram based on the mean values reported in Table 4.

Acknowledgements

The present research was supported by the Italian Ministry of Education (Ricerca Scientifica, capitolo 6562, art. 47, impegno 78/8721).

Some part of the work was accomplished at the Laboratory of the Marine Biological Association of the United Kingdom in Plymouth with the support of a NATO scholarship awarded to L. Galleni through the Italian National Research Council (CNR).

References

Bahr, L. M. jr. & Hillman, R. E., 1966. Chromosome number of Stylochus ellipticus (Girard). Trans. Am. microsc. Soc. 85: 323-324.

Bertani, G., 1947. Distribuzione dei chiasmi nella meiosi maschile di Thysanozoon brocchii. Pubbl. Staz. zool. Napoli 21: 110-116.

Christensen, D. J., 1971. Early development and chromosome number of the Polyclad Flatworm Euplana gracilis. Trans. Am. microsc. Soc. 90: 457-463.

Francotte, P., 1897. Recherches sur la maturation, la fécondation et la segmentation chez les Polyclades. Mem. cour. Mem. Sav. étr. Acad. r. Sci. Bel. 55: 5-65.

Francotte, P., 1898. Recherches sur la maturation, la fécondation et la segmentation chez les Polyclades, Archs. Zool. exp. gén. 3s. 6: 189-298.

Galleni, L., 1978. Policladi delle coste toscane - III - Echinoplana celerrima Haswell, Planoceride nuovo per il Mediterraneo e note sul genere Echinoplana. Atti Soc. Tosc. Sci. Nat. Mem. Serie B 85: 139-148.

Galleni, L. & Puccinelli, I., 1975. Karyology of Notoplana igiliensis Galleni (Polycladida - Acotylea). Caryologia 28: 357-387.

Galleni, L. & Puccinelli, I., 1977. Karyometric analysis in Thysanozoon brocchii (Risso) (Turbellaria - Polycladida). Acc. Lincei Rendic. Cl. Sci., Fis., Mat. Nat. 63: 436-439.

Galleni, L., Tongiorgi, P., Ferrero, E. & Salghetti, U., 1980. Stylochus mediterraneus (Turbellaria-Polycladida), predator on the mussel Mytilus galloprovincialis. Mar. Biol. 55: 317-326.

Gerard, O., 1901. L'ovocyte de prémier ordre du Prostheceraeus vittatus avec quelques observations relative à la maturation de trois autres Polyclades. Cellule 18: 138-248.

Haswell, W. A., 1907. Observation on australasian Polyclads. Trans. Linn. Soc. London, s.II 9: 465-485.

Hendelberg, J., 1969. On the development of different types of spermatozoa from spermatids with two flagella in the Turbellaria with remarks on the ultrastructure of the flagella. Zool. Bidr. Uppsala 38: 1-49.

Karling, G. T., 1963. Some evolutionary trends in turbellarian morphology. In: Ed. Dougherty, E. C., The lower Metazoa, Berkeley and Los Angeles: Univ. of California Press, pp. 225-233.

Kato, K., 1940. On the development of some Japanese Polyclads. Jap. J. Zool. 8: 537-573.

Klinckowström, A. von, 1897. Beiträge zur Kenntniss der Eireifung und Befruchtung bei Prostheceraeus vittatus. Arch. Microsc. Anat. 48: 587-605.

Levan, A., Fredga, K. & Sandberg, A. A., 1964. Nomenclature for centromeric position on chromosomes. Hereditas 52: 201-220.

Marcus, E. & Marcus, E., 1966. Systematische Übersicht der Polycladen. Zool. Beitr. 12: 319-343.

Patterson, J. T. & Wieman, H. L., 1912. The uterine spindle of the Polyclad Planocera inquilina. Biol. Bull. 23: 271-293.

Schockaert, R., 1902. L'ovogènese chez le Thysanozoon brocchii. Deuxieme partie. Cellule 20: 103-175.

Schockaert, R., 1905. La Fécondation et la Segmentation chez le Thysanozoon brocchii. Cellule 22: 7-37.

Van Name, W. O., 1899. The maturation, fertilization and development of the planarians. Trans. Conn. Acad. Arts. Sci. 10: 262-295.

Van der Stricht, O., 1897. Les ovocentres et les spermatocentres de l'ovule de Thysanozoon brocchii. Verh. anat. Ges. Jena 11: 92-99.

Van der Stricht, O., 1898. La formation des deux globules polaires et l'apparition des spermatocentres dans l'oeuf de Thysanozoon brocchii. Archs. Biol. 15: 367-461.

Wheeler, W. H., 1894. Planocera inquilina, a Polyclad inhabiting the branchial chamber of Sycotypus canaliculatus Gill. J. Morph. 9: 195-201.

The karyotypes of Dugesia species from Spain (Turbellaria, Tricladida)

Nicole Gourbault
Laboratoire des Vers, associé au C.N.R.S., Muséum national d'Histoire naturelle, 43 rue Cuvier, F 75231 Paris Cedex 05, France

Keywords: Turbellaria, Tricladida, *Dugesia,* karyology, Spain

Abstract

Some species of Planarians, new to Spain, are recorded. *Dugesia polychroa, D. sicula, D. iberica* and *D. gonocephala s. l.* have been investigated karyologically. The former possesses a diploid complement characteristic of the biotype A (2n = 8); the second is diploid with 2n = 18; diploidy and triploidy were found in sexual populations of *D. iberica* with n = 8. Triploidy occurred in all the asexual strains of the *D. gonocephala* group with a basic number of either 8 or 9. In this latter case B-chromosomes were occasionally found.

Introduction

During a field-trip to continental Spain and the island of Mallorca to collect hypogeous fauna from the low permeability interstitial sediments in river underflows and water tables, several specimens of common epigeous planarians were collected; the majority of these are referred to the genus *Dugesia.*

Hitherto few records deal with the Spanish Turbellarian fauna. *Crenobia alpina* (Dana) was reported in the Pyrenees by Borelli (1905), and *Dendrocoelum lacteum* (Müller) from the vicinity of Madrid was studied by Bordas (1921). Arndt (1926) identified *Polycelis felina* (Dalyell) from numerous localities between Gerona and Huesca, *Phagocata vitta* (Dugès) in a well from Tarragona province and *Dugesia lugubris* (Schmidt) from Lerida. Recently, Benazzi *et al.* (1975) described a new species of the *'lugubris-polychroa'* group, i.e. *Dugesia mediterranea*; asexual populations were found in Barcelona whereas sexual individuals were recorded on certain Mediterranean islands (Corsica, Sardinia, Sicily). Among the subterranean forms, *Dendrocoelopsis brementi* (de Beauchamp) is known from a Pyrenean cave and *Atrioplanaria racovitzai* (de Beauchamp) from the water table of the river Henares in Guadalajara (Gourbault 1971).

The material examined herein* is used to increase the chorological data concerning European Triclads and to confirm the identity of species through a karyological approach.

Materials and methods

The specimens studied were collected as follows:

– *Dugesia polychroa* (Schmidt, 1861). Rio Henares, Chiloeches, Alcalá de Henares; coll. 15 Sept. 1976, with numerous cocoons.

– *Dugesia sicula* Lepori, 1948. Ull de la Font outflow, Pollença, Mallorca; coll. 23 May 1976. Large population (cf. Gourbault & Lescher-Moutoué 1979).

– *Dugesia iberica* Gourbault & Benazzi, 1979. Outflow and brooklet close to the Cova dets Estudiants, Soller, Mallorca; coll. 22 and 26 May

* Field-work of Nicole Gourbault, Françoise Lescher-Moutoué and Raymond Rouch. Collecting trips supported by MNHN, Paris and Laboratoire souterrain du CNRS, Moulis, France.

1976. Rio Júcar, Villelva de la Sierra, Cuenca; coll. 17 Sept. 1976.

- *Dugesia* aff. *iberica.* Small brooklet feeding a water through under Calobra Road, Mallorca; coll. 26 May 1976. Large dark specimens.

- Many asexual strains of *Dugesia* sp. of the '*D. gonocephala* group': (1) Small tributary of the Rio Eresma, Trescasas, Segovia; coll. 12 Sept. 1976. (2) Rio Rambla de Mal Burgo, Fortunate, Teruel; coll. 20 Sept. 1976. (3) Rio Guadalete, Algodonales near Ronda; coll. 18 Sept. 1977. (4) Rio Turon, El Burgo, Ronda; coll. 18 Sept. 1977. (5) Rio Guadalhorce, Alora, Malaga; coll. 19 Sept. 1977. (6) Rio Salia, La Viñuela, Malaga; coll. 20 Sept. 1977. Very numerous specimens. (7) Rio Guadalfeo, Vêlez de Benaudalla, Granada; coll. 21 Sept. 1977. (8) Deifontes, vauclusian spring near Granada; coll. 22 Sept. 1977. (9) Subterranean water, Font de Can Salas, Pollença, Mallorca; coll. 23 and 25 May 1976.

About five specimens of each of these populations were studied both morphologically and karyologically. All the karyotypes were obtained from enlarged camera lucida drawings of mitotic metaphase stages in regenerative blastemas immersed in colchicine, stained in lactic-aceto-orcein and squashed. The chromosomes are classified according to the nomenclature of Levan *et al.* 1964.

It is of interest to note here the collection of the following additional species:

- *Dugesia tigrina* (Girard) collected with *D. polychroa* in the river Henares; this may be the first record of this american species in Spain.

- *Polycelis felina,* Rio Lobos Springs (right tributary of the river Duero), El Burgo de Osma, Soria; coll. 13 Sept. 1976; Rio Cuervo Springs (left tributary of the river Tajo), Tragacete, Cuenca; coll. 17 Sept. 1976; Rio Frio outflow, Granada; coll. 23 Sept. 1977.

- *Phagocata* sp. from a well in Sale del Rio Seco, Nerja (paper in progress), coll. 21 Sept. 1977.

- Indetermined blind and unpigmented species: from a well in Huértor-Tájar, Granada; coll. 23 Sept. 1977 and from a brooklet underflow, Rio Burró, left R. Fluvia, Besalu, Figueras; coll. 18 Sept. 1978.

Results

- *Dugesia polychroa*

The diploid complement of *Dugesia polychroa* consists of 8 elements. They are easy to pair, a

Table 1. Relative length and centromeric index of the four chromosomes pairs of *Dugesia polychroa.* Means and standard deviations from ten mitotic metaphase plates, compared with the values recalculated from the idiogramm of biotype A established by Benazzi *et al.* 1970.

N° chromosome	Relative length	Centromeric index
1	34.86 ± 1.85	46.85 ± 1.58
2	27.05 ± 1.47	16.23 ± 1.70
3	21.43 ± 1.39	19.04 ± 1.56
4	16.52 ± 1.02	22.83 ± 1.18
	Biotype A	
1	34.30	46
2	25.50	11
3	22.20	16
4	17.80	20

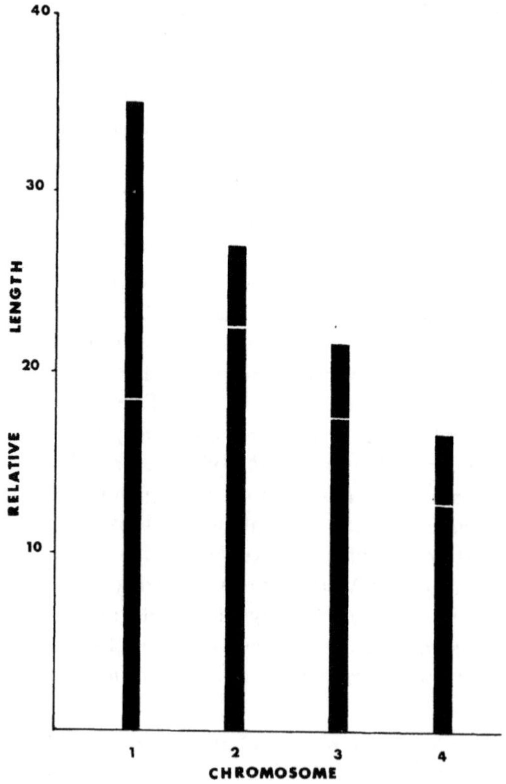

Fig. 1. Idiogram of the karyotype of *Dugesia polychroa*

Table 2. Relative length and centromeric index of the nine chromosome pairs of *Dugesia sicula*. Means and standard deviations from ten mitotic metaphase plates.

N° chromosome	Relative length	Centromeric index
1	14.13 ± 0.65	42.40 ± 2.20
2	12.77 ± 0.49	45.68 ± 2.54
3	11.83 ± 0.30	41.17 ± 2.64
4	11.67 ± 0.41	44.84 ± 2.41
5	10.98 ± 0.37	42.47 ± 3.90
6	10.55 ± 0.52	43.73 ± 4.06
7	10.15 ± 0.31	45.66 ± 2.08
8	9.59 ± 0.59	46.16 ± 1.46
9	8.91 ± 0.62	42.02 ± 1.73

condition that is perfectly supported by measurements of the relative lengths and centromeric indices. The means and standard deviations calculated for ten metaphase plates are given in Table 1; the karyotype is represented by the idiogram of Figure 1. Four chromosomes characterize the haploid set: one large metacentric and three smaller acrocentrics of different lengths, the second being the most heterobrachial, (similar to findings of Benazzi & Puccinelli 1961, 1973).

– *Dugesia sicula*

Karyologically this species is 'poorly known; it seems to be a diploid but 9 bivalents are found during oogenesis' (Benazzi & Benazzi Lentati 1976, p. 17).

The Mallorcan strain possesses a diploid complement of 2n = 18. This number was observed in more than thirty plates, two excepted (with 2n = 16).

The chromosomes are easily paired. Karyometric data (Table 2) form the basis of Figure 2. All chromosomes are metacentric and decrease very gradually in size.

– *Dugesia iberica*

Soller populations: A mitotic chromosome com-

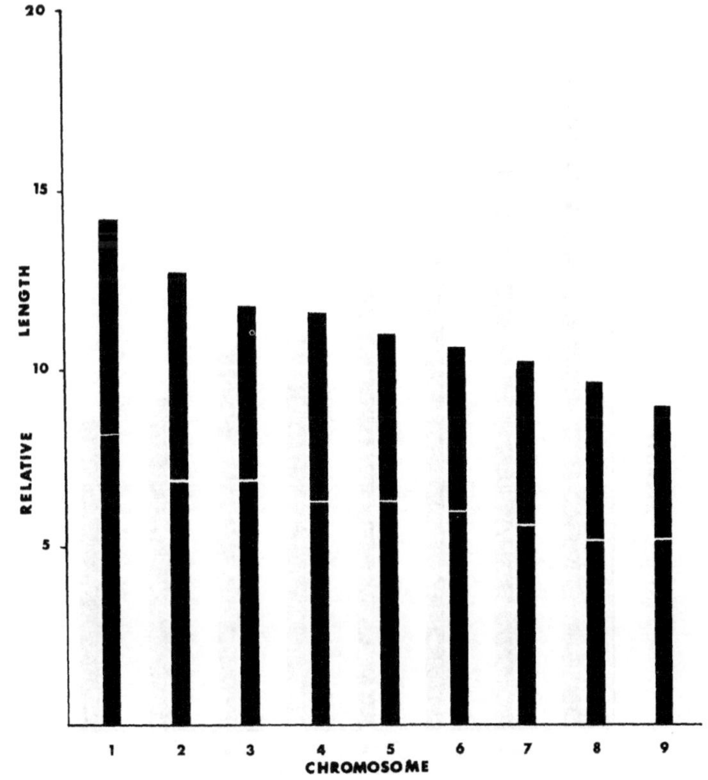

Fig. 2. Idiogram of the karyotype of *Dugesia sicula*.

Table 3. Relative length and centromeric index of the eight chromosomes pairs of *Dugesia iberica* (Soller population). Means and standard deviations from ten mitotic metaphase plates.

N° chromosome	Relative length	Centromeric index
1	18.96 ± 1.14	48.25 ± 0.57
2	15.56 ± 0.65	47.66 ± 0.68
3	13.45 ± 0.40	34.50 ± 2.15
4	12.48 ± 0.97	39.24 ± 1.37
5	10.90 ± 0.52	46.69 ± 1.88
6	10.62 ± 0.68	37.46 ± 1.84
7	9.28 ± 0.50	47.53 ± 0.39
8	8.66 ± 0.40	38.97 ± 1.49

plement of 16 was observed in all metaphase plates. The diploid condition is well confirmed by the measurements. Means and standard deviations were calculated for ten plates (Table 3). The idiogram of figure 3 illustrates the karyotype related to the description given by Benazzi & Benazzi Lentati (1976) for the haploid set of the Mediterranean *D. gonocephala* complex.

Most of the chromosomes are metacentric, although some are intermediate between meta and submetacentric, and the third is definitively submetacentric. Chromosome 8 is about half the size of chromosome 1.

Rio Júcar population: the complement of 24 found in all metaphases might be referable to a triploid condition. This viewpoint is significantly supported by the karyometrical studies as eight groups of three homologous chromosomes were noted for each series.

The basikaryotype of this triploid *Dugesia iberica* is represented by the idiogram of Figure 4, derived from Table 4. Most of the chromosomes are metacentric but differences in centromeric indices occur between the diploid and triploid strains, three elements being submetacentric in the latter.

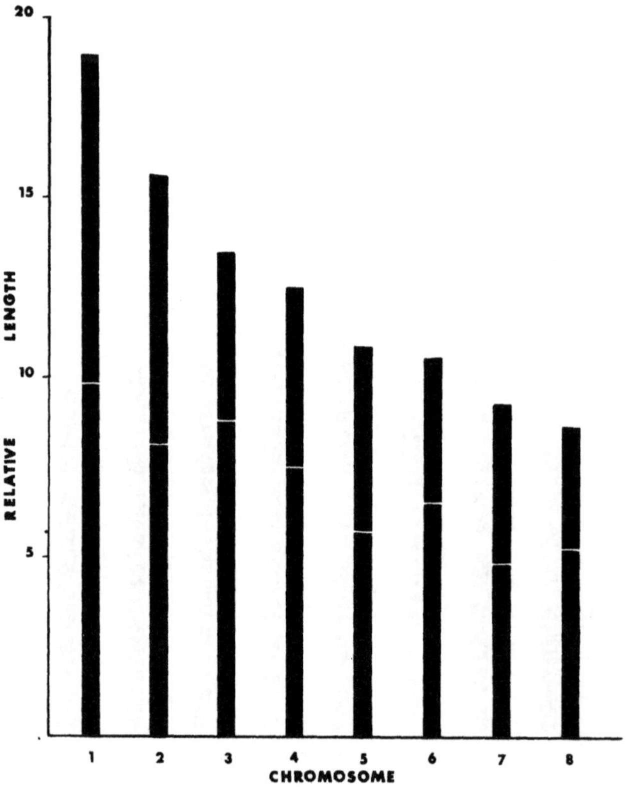

Fig. 3. Idiogram of the karyotype of the Soller population of *Dugesia iberica*.

Table 4. Relative length and centromeric index of the eight chromosomes of the haploid set of *Dugesia iberica* (Rio Júcar population with 3 n = 24). Means and standard deviations from ten mitotic metaphase plates.

N° chromosome	Relative length	Centromeric index
1	17.37 ± 0.73	46.56 ± 0.71
2	15.20 ± 0.88	34.80 ± 2.72
3	13.02 ± 0.35	36.56 ± 2.30
4	12.34 ± 0.69	47.57 ± 1.25
5	11.34 ± 0.43	39.42 ± 1.94
6	10.63 ± 0.28	39.06 ± 2.02
7	10.27 ± 0.40	35.81 ± 1.02
8	9.66 ± 0.35	46.14 ± 0.82

– *Dugesia* aff. *iberica*

In spite of similarities in the copulatory apparatus with the previous species, these populations exhibit different karyotypes. The diploid condition is ascertained by the measurements of the 18 chromosomes. The length of the nine elements decreases gradually from the first to the eight; while the ninth is markedly reduced to a third of the size of the first one. All of them are clearly metacentric (Table 5 and Figure 5).

– *Asexual populations of the* Dugesia gonocephala *group*

As already well established, the haploid chromosome number of the *D. gonocephala* complex is either eight or nine; so the complement of 24, or 27 found in the Spanish strains may be referred to a triploid condition. This triploidy is confirmed by the analysis of the metaphase plates; the chromosomes decrease gradually in size and can be arranged in groups of three metacentric or submetacentric elements.

For instance, a triploid set of 24 elements was found in the populations from the stations n° 1, 2 and 8. Their karyotype is similar to the one de-

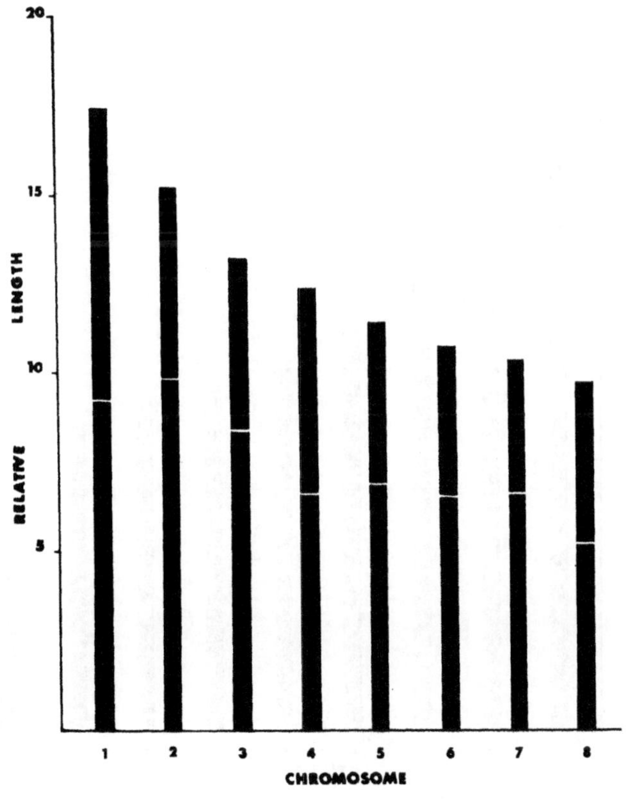

Fig. 4. Idiogram of the karyotype of the Rio Júcar population of *Dugesia iberica.*

Table 5. Relative length and centromeric index of the nine chromosome pairs of *Dugesia* aff. *iberica* (La Calobra population). Means and standard deviations from ten mitotic metaphase plates.

N° chromosome	Relative length	Centromeric index
1	17.18 ± 0.84	48.18 ± 0.87
2	15.01 ± 1.07	46.89 ± 1.08
3	12.95 ± 1.05	41.54 ± 1.04
4	11.63 ± 0.69	48.88 ± 0.84
5	10.26 ± 0.44	46.91 ± 1.18
6	9.57 ± 0.36	47.97 ± 1.01
7	9.14 ± 0.49	42.26 ± 1.36
8	8.50 ± 0.50	43.43 ± 1.63
9	5.92 ± 1.09	48.74 ± 0.52

scribed for *D. iberica* from Rio Júcar, and exhibits three pairs of submetacentric chromosomes.

On the contrary in stations 3, 4, 5, 6, 7 and 9, the haploid number is 9. the measurements of some complements show that all the elements are metacentric and decrease very gradually in size; the smallest being half the size of the first one. Moreover, in most of the metaphases, B-chromosomes are present: one or two are found in a few plates of the *Dugesia* from the Rio Guadalete, and Turon, and in most of the specimens from the Rio Guadalhorce, Salia and Guadalfeo. Two and even three are frequent in the population from the Font de Can Salas. All of them are metacentric.

(All the data are summarised in Table 6).

Discussion

The status of the *Dugesia polychroa* – *lugubris* complex was recently subjected to intensive researches (summarized in Benazzi & Benazzi Lentati 1976). The most recent study concerns the karyotype of a Greek population of *D. polychroa* from Corfu (Ball 1979). The same species, in central Spain, possesses the diploid complement of biotype A (after nomenclature of Benazzi 1957), which characterizes both the topotype and the italian

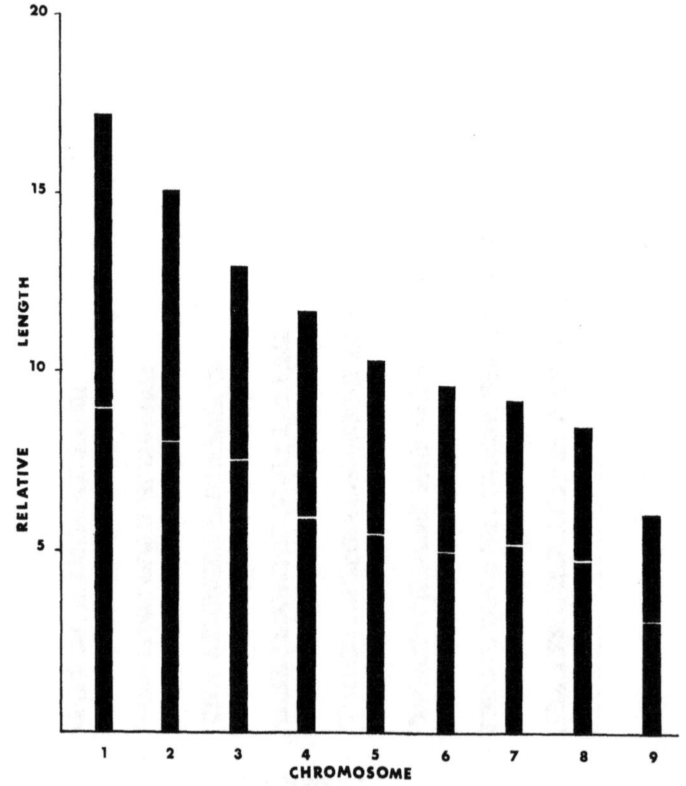

Fig. 5. Idiogram of the karyotype of *Dugesia* aff. *iberica* (La Calobra population).

Table 6. Chromosome numbers in Dugesiidae from Spain.

Species	Localities	Chromosome nb. n	2n	3n
D. polychroa	Rio Henares Chiloeches	4	8	
D. sicula	Ull de la Font Pollença, Mallorca	9	18	
D. iberica	Cova dets Estudiants Soller, Mallorca	8	16	
	Rio Jucar Villelva de la Sierra			24
D. aff. iberica	Brooklet La Calobra, Mallorca	9	18	
D. gonocephala group (asexual population)	(1) Eresma tributary Trescasas			24
	(2) Rio Rambla de Mal Burgo, Fortunate			24
	(3) Rio Guadalete Algodonales			27*
	(4) Rio Turon El Burgo			27*
	(5) Rio Guadalhorce Alora			27*
	(6) Rio Salia la Vinuela			27*
	(7) Rio Guadalfeo Velez de Benaudalla			27*
	(8) Vauclusian spring Deifontes			24
	(9) Font de Can Salas Pollença, Mallorca			27*

* Presence of B-chromosomes.

specimens from Pisa, Elba and Toblino lake. It is the one described as *D. lugubris* by Dutrillaux & Lenicque (1971) from the vicinity of Paris. This same biotype A is also present in specimens from some localities in Corsica and Sardinia where the biotype G of *D. mediterranea* is observed. The haploid number of biotype G is also four but the chromosome morphology differs from that of biotype A. One fissiparous race of *D. mediterranea* is known from Barcelona, and this peculiar distribution was explained by Benazzi *et al.* (1975) in terms of late Oligocene microplate dispersal resulting in isolation of these western Mediterranean islands. Thus it is of interest to note here the presence also of *D. sicula* in Mallorca. This species was first recorded from Sicily and Elba and the present finding is in perfect agreement with the biogeographical explanation of Benazzi *et al.* (1975).

The karyometric data of the chromosomes of *D. iberica* are similar to those of the basic set of the *D. gonocephala* group, in spite of some differences in centromere positions. However centromeric indices may vary as already observed in the allied species of this group (Benazzi & Benazzi Lentati 1976). *D. iberica* possesses at least three pairs of heterobrachial elements, but in the population from La Calobra, in *D. sicula* and in most of the fissiparous Southern strains from continental Spain, all the chromosomes are metacentric. Similarly, all the chromosomes of *D. gonocephala* from Corfu (Ball 1979) are also metacentric. This constant morphology of the elements within the same species can help to clarify the relationships of asexual populations. For instance, it might be possible to assign to *D. iberica* all the triploid strains with 24 meta- and submetacentric chromosomes which occur more commonly in the Northern part of Spain. On the contrary, the triploid fissiparous strains of the southern part, and of Mallorca, which exhibit metacentric chromosomes are not referable to the above species and assignment to another taxon is not possible until some sexual specimens have been found.

The presence of B-chromosomes in these last strains is of interest. This type of tiny supernumerary elements was first noticed by Melander (1950) in *Polycelis tenuis* Ijima, and by Benazzi (1960, 1969) in specimens of the *D. gonocephala* group. Bromley (1974) found one to five of them in the triploid populations of *D. biblica* Benazzi & Banchetti from Israel, and Kawakatsu *et al.* (1976) noticed heterobrachial supernumerary chromosomes in *D. japonica* Ichikawa & Kawakatsu. These chromosomes were studied in detail by Deri (1971, 1975) for some Corsican populations of *D. benazzii*. In a culture of polyploid strains of this same species, spontaneous appearance of B-chromosomes was observed by Benazzi Lentati & Deri (1977). A similar phenomenon was reported by Deri (1980) for a eudiploid population kept for a prolonged period in laboratory culture. This occurred with the transformation of the diploid complement into the hyperdiploid state. Specimens were fissiparous at this time.

The cytological observations of numerous planarian strains allowed Benazzi Lentati & Deri (1980) to conclude that, although, the occurrence of heterogenous chromosome sets often seemed to be related to fission as seems to be the case in the spanish populations, they were not genetically correlated, but, could exist concomitantly.

References

Arndt, W., 1926. Spongilliden und Turbellarien, aus dem nördlichen und östlichen Spanien, gesammelt von Dr. F. Haas in den Jahren 1914–1919. Senckenbergiana 8: 22–30.

Ball, I. R., 1979. The karyotypes of two Dugesia species from Corfu, Greece (Platyhelminthes, Turbellaria). Bijdr. Dierk. 48: 187–190.

Benazzi, M., 1957. Cariologia di Dugesia lugubris (O. Schmidt) (Tricladida Paludicola). Caryologia 10: 276–303.

Benazzi, M., 1960. Evoluzione cromosomica e differenziamento razziale e specifico nei Tricladi. Acad. nac. Lincei, Evoluzone e Genetica, 47: 273–297.

Benazzi, M., 1969. Annotazioni citosistematische sui Tricladi di alcune isole tirreniche. Accad. naz. Lincei, Rend. Cl. Sci. fis. mat. nar., sér. 8, 46: 605–609.

Benazzi, M., Baguna, J., Ballester, R., Puccinelli, I. & Del Papa, R., 1975. Further contribution to the taxonomy of the 'Dugesia lugubris-polychroa group' with description of Dugesia mediterranea n. sp. (Tricladida, Paludicola). Boll. Zool. Ital., 42: 81–89.

Benazzi, M. & Benazzi Lentati, G., 1976. The cytogenetics of Platyhelminthes. Stuttgart: Gebrüder Borntraeger, 182 pp.

Benazzi, M. & Puccinelli, I., 1961. Analisi comparativa del cariogramma dei biotipi di Dugesia lugubris (Tricladida Paludicola). Atti. Assoc. Genet. Ital. 6: 419–426.

Benazzi, M. & Puccinelli, I., 1973. A Robertsonian translocation in the freshwater Triclad Dugesia lugubris: karyometric analysis and evolutionary inferences. Chromosoma 40: 193–198.

Benazzi Lentati, G. & Deri, P., 1977. Insorgenza di B-cromosomi in individui poliploidi di Dugesia benazzii (triclade paludicolo) allevati in laboratorio. Accad. Naz. Lincei, R. Cl. Sc. Fis. Mat. Nat., Sér. 8, 62: 847–851.

Benazzi Lentati, G. & Deri, P., 1980. On the origin of heterogeneous chromosome sets in some fissiparous planarians. Acc. Naz. Lincei, R. Cl. Sci. Fis. Mat. Nat., ser. 8 (in press).

Bordas, P., 1924. La profase de reduccion en la ovogénesis de Dendrocoelum lacteum. Trab. Mus. Nac. Madrid, ser. Zool. 44: 5–101 (1921).

Borelli, A., 1905. Sulla presenza della Planaria alpina e della Polycelis cornuta nei Pirenei. Bull. Zool. Anat. Comp. Torino 20, 483: 1–4.

Bromley, H. J., 1974. Morpho-karyological Types of Dugesia (Turbellaria, Tricladida) in Israel and their distribution patterns. Zool. Scripta 3: 239–242.

Deri, P., 1971. Ricerche cariologiche e di biologia della riproduzione in popolazioni di Dugesia benazii della Corsica. Atti Soc. tosc. Sci. Nat. Pisa, B. Ital. 78: 55–74.

Deri, P., 1975. B-cromosomi in popolazioni polisomiche di Dugesia benazzii (Tricladida Paludicola) della Corsica. Atti Soc. Tosc. Sci. Nat. Mem. 82: 25–38.

Deri, P., 1980. Incremento del numero cromosomico e comparsa di B-cromosomi durante l'allevamento in laboratorio di una popolazione diploide di Dugesia benazzii (triclade paludicolo). Acc. Naz. Lincei, R. Cl. Sci. Fis. mat. Nat., ser. 8 (in press).

Dutrillaux, B. & Lenicque, P., 1971. Analyse du caryotype de cinq espèces de Planaires par la méthode du choc hypotonique. Acta Zool., Suède, 52: 241–248.

Gourbault, N., 1971. turbellariés Triclades des eaux souterraines d'Europe occidentale: Nouvelles données géonémiques. Ann Spéléol. 26: 181–188.

Gourbault, N. & Benazzi, M., 1979. Une nouvelle espèce ibérique du 'groupe Dugesia gonocephala' (Turbellariés, Triclades). Bull. Mus. natn. Hist. nat., Paris, 4e sér. 1 A (2): 329–337.

Gourbault, N. & Lescher-Moutoué, F., 1979. Faune des eaux souterraines de Majorque. Endins, Mallorca, 5–6: 43–54.

Kawakatsu, M., Oki, I., Tamura, S. & Sugino, H., 1976. Studies on the morphology, karyology and taxonomy of the Japanese freshwater Planarian Dugesia japonica Ichikawa et Kawakatsu, with a description of a new subspecies, Dugesia japonica ryukyuensis subspec. nov. Bull. Fuji Women's College 14 ser. II: 81–126.

Lepori, N. G., 1948. Descrizione di Dugesia sicula, nuova sp. di Triclade d'acqua dolce dei dintorni di Catania. Arch. Zool. Ital. 33: 461–472.

Levan, A., Fredga, K. & Sandberg, A. A., 1964. Nomenclature for centromeric position on chromosomes. Hereditas 52: 201–220.

Melander, Y., 1950. Accessory chromosomes in animals, especially in Polycelis tenuis. Hereditas 36: 19–38.

Schmidt, O., 1861. Untersuchungen über Turbellarien von Corfu und Cephalonia, nebst Nachträgen zu früeheren Arbeiten. Z. wiss. Zool. 11: 1–30.

Karyological and taxonomic studies of Dugesia japonica Ichikawa et Kawakatsu in the Far East

Iwashiro Oki, Sachiko Tamura, Takao Yamayoshi & Masaharu Kawakatsu
Ôsaka Prefectural Institute of Public Health, Nakamichi-1-chôme 3-69, Higashinari-ku, Ôsaka 537, Japan; Biological Laboratory, Fuji Women's College, Kita-16, Nishi-2, Kita-ku, Sapporo (Hokkaidô) 001, Japan

Keywords: Turbellaria, Tricladida, *Dugesia*, karyology, taxonomy, Japan

Abstract

A review of previous studies on the taxonomy, karyology and chorology of a polymorphic species *Dugesia japonica* from the Far East is presented. Two subspecies are now known: *D. j. japonica* (n = 8, 2x = 16, 3x = 24) and *D. j. ryukyuensis* (n = 7, 2x = 14, 3x = 21). An attempt has also been made to determine the definition of the B-chromosome as LB and SB and the variation of the karyotypes of both subspecies is described. Every known karyotype of *D. japonica* is classified into six groups (see Table 2). *D. japonica* from many localities has a diploid karyotype (2x), a triploid karyotype (3x) and an orthoploidic mixoploid karyotype of 2x & 3x. The origin and the karyological significance of these karyotypes are discussed.

1. Introductory review of the previous studies

Dugesia japonica Ichikawa et Kawakatsu, 1964, is the commonest freshwater planarian of the Far East including the Japanese Islands and the Ryûkyû Islands. This widely distributed species is found everywhere both in running and standing waters, including shallow underground habitats. It is a polymorphic species; the animals from different localities show a considerable degree of variation in its morphological, anatomical and physiological characters (Kawakatsu, 1965, 1967, 1971, 1974). Taxonomically, the species now exists in two subspecies, i.e., *Dugesia japonica japonica* Ichikawa et Kawakatsu, 1964 and *Dugesia japonica ryukyuensis* Kawakatsu, 1976 (in Kawakatsu, Oki, Tamura & Sugino 1976b).

In 1887, Ijima mentioned the Japanese freshwater triclad with a head of the triangular shape as '*Planaria gonocephala* Dugès'. The first published morphological description of this common Japanese species was that of Ijima & Kaburaki (1916). Later, Kaburaki (1922) gave a more detailed account of the species. The species referred to as *Dugesia* (= *Planaria;* = *Euplanaria*) *gonocephala* has been recorded by many turbellarian specialists from the large geographical areas of the Japanese Islands, the Ryûkyû (Loo Choo, Lew-Chew) Islands, Taiwan or Formosa, and the northeastern part of the Eurasian Continent (i.e., Middle and North China including Manchuria and Kwantung and Korea). The literature is listed in Kawakatsu (1968).

As to the taxonomic position of the Japanese species referred to as *D. gonocephala,* the first reasonable doubt was offered by Marcus (1953). He clearly expressed his opinion, based on morphological data, that the so-called '*D. gonocephala*' from Japan (and the Far East) differs from the European and African *D. gonocephala* in anatomical details of the copulatory apparatus and in the histology of the pharynx. Later, Dahm (1963) who studied the karyotype of a sexually reproducing population from Matsumoto City in Honshû, Central Japan, demonstrated that the chromosome number of the sexual specimens was found to be 2n = 16 and during meiosis 8 completely paired bivalents appeared. He also mentioned the differences of the

Hydrobiologia 84, 53-68 (1981). 0018-8158/81/0841-0053/\$03.20.

karyotypes between the Japanese *Dugesia* species and European *D. gonocephala.* Dahm's chromosomal observations was confirmed by Teshirogi & Itagaki (1965) who studied the karyotypes of the Japanese *Dugesia* species from four localities in North, Central and South Japan.

Kawakatsu re-examined the *Dugesia* species from various localities in Japan and its neighbouring countries. For comparison, he also examined European specimens of *D. gonocephala* s.s. As a result of these comparative studies, the Japanese *Dugesia* species was at last described as a new species *Dugesia japonica* Ichikawa et Kawakatsu, 1964. Since the life-history of the species had been thoroughly studied by Okugawa & Kawakatsu (1954–1958), in specimens collected from a spring-fed stream at Genchi in Matsumoto City, Nagano Pref., Honshû, Central Japan, these were employed as the type-series. Dahm's (1963) material mentioned above was collected from this type-locality. Externally, both *D. gonocephala* and *D. japonica* can easily be separated by the differences of the head shape and the position of eyes in living specimens (Kawakatsu, Teshirogi, Ôgawara & Tarui 1965).

Since *D. japonica* was originally described, the morphological variation of the genital anatomy of this polymorphic species from numerous localities in the Japanese Islands including the Southwest Islands of Japan (i.e., the Satsunan Islands, the Ryûkyû Islands and the Yaeyama Islands) and the neighbouring countries (i.e., Taiwan, China, Hong Kong, and Korea) has been studied by Kawakatsu and his co-researchers. They also studied the distributional ecology and the determination of the geographical distribution range. These studies were discussed and reviewed by Kawakatsu (1965, 1967, 1970, 1974), and listed in Kawakatsu (1968–1980).

The life-history and the reproductive biology of *D. japonica* have been studied by Okugawa & Kawakatsu (1954–1958) (developed along the line of the studies by Curtis (1902), Vandel (1922) and Kenk (1935, 1937, 1940, 1941)) (see also Kawakatsu 1965, 1967, 1971, 1974; Okugawa 1955, 1957). They divided *D. japonica* into two main groups or so-called physiological races according to the difference of the mode of reproduction, sexual or asexual. The former includes many populations each having a different frequency of formation of the copulatory organ. The frequency of fission is also different in each population belonging to the asexual race. Tanaka (1965) studied the reproductive biology of *D. japonica* from Okinawa Island. Sugino (1962, 1969), who made mutual transplantation experiments on *D. japonica* from various localities of Japan, concluded that the animals can be classified into three ecological types: a type inhabiting clean and cold mountain streams, a type inhabiting nonpolluted rivers and a type inhabiting polluted waters in the plains. Both sexual and asexual populations are included in each of Sugino's types. The ecological classifications of *D. japonica* mentioned here at least show the physiological polymorphism of this species.

The chromosomal characters of *D. japonica* from several localities in Central Japan have been studied by Sugino and his co-researchers (Hirose, Katô & Sugino 1974; Sugino, Hirose & Katô 1973) and Oki & Tamura (1974, 1975a, b). According to them, the chromosome number of the animals from almost all of the populations is $2x = 16$, but variation in chromosome numbers occurs in some populations. In many specimens examined from several localities there are two types of somatic cells in a single individual, viz.: $2x = 16$ and $3x = 24$. Moreover, it has been found out that some specimens have a chromosome number of 25.

In the comparative genital anatomy of *D. japonica* from vairous localities of Japan and its neighbouring countries, the most noticiable variation has been found in the shape of the penis papilla and the degree of development of the vagina. The animals from the type-locality have a large and highly asymmetrical penis papilla and a well-developed vagina surrounded by thick muscle layers. On the other hand, a well-developed constriction or the valve surrounding the basal part of the penis papilla is found in the specimens from the Ryûkyû Islands in Southwest Japan. This character in the penial anatomy is also observed in the specimens from Hangchow in Middle China. When *D. japonica* was originally described as a new species, the animals from Okinawa (Ryûkyû) and Middle China were also identified as the same species after some hesitation (Ichikawa & Kawakatsu 1964, pp. 186–187, footnote 2). Ball (1970), who studied the taxonomy of several *Dugesia* species from the Oriental Region, expressed the opinion that *D.*

japonica with the 'atrial fold' (i.e., the valve surrounding the basal part of the penis papilla) should be given separate status, at least at the level of subspecies. His proposal was not accepted by Kwakatsu because a transitional form (with a considerably large valve at the basal part of the dorsal lip of the penis papilla) was found in Kagoshima, the southernmost part of Kyûshû in South Japan (Kawakatsu & Iwaki 1967; see also Kawakatsu 1971, 1975; Kawakatsu, Horikoshi & Akama 1972; Kawakatsu & Mack-Firā 1975; Kawakatsu & Tanaka 1971, 1976).

The recent chromosomal findings of *D. japonica* already mentioned led Kawakatsu, and Oki and co-workers to reconsider the taxonomy of this species based upon morphological, anatomical and karyological data. Kawakatsu reexamined the slides of *D. japonica* from various localities including the type-series. The chromosomal observations of the animals from several localities of Japan including Okinawa Island and Korea in comparison with the karyotype of specimens recollected in the type-locality were carried out. The chromosome number of the specimens from a stream at Nakamayama-gawara in Urasoe City, Okinawa, was found to be $2x = 14$. During meiosis seven bivalents were found. A triploid form ($3x = 21$) and specimens that had cells with few B-chromosomes were also found (Kawakatsu, Oki, Tamura & Sugino 1976a; Oki, Tamura, Kawakatsu & Sugino 1976).

Finally, Kawakatsu, Oki, Tamura & Sugino (1976b), divided *D. japonica* into two subspecies. A redescription of the nominate subspecies *(D. j. japonica)* and a description of the new subspecies *(D. j. ryukyuensis),* as well as descriptions of animals from several localities in Japan, Korea and China, were given. They also discussed the possible dispersal routes of the proto-species of *D. japonica* in relation to its speciation. Some of the results of further studies on *D. japonica* from Japan have been published as separate papers (Kawakatsu, Oki, Tamura, Yamayoshi & Takahashi 1980; Tamura, Yamayoshi, Oki & Kawakatsu 1979; Tamura, Yamayoshi, Oki, Murayama & Kawakatsu 1978; Yamayoshi, Tamura, Oki & Kawakatsu 1980), and *D. japonica* from Taiwan was recently studied from the taxonomic and karyological viewpoints (Kawakatsu, Oki, Tamura, Yamayoshi, Lue & Hagiya 1979).

The purpose of the present paper is to give a review of the authors' continuing studies on the karyology and taxonomy of *D. japonica,* together with new descriptions and viewpoints about this widely distributed, polymorphic species.

2. Karyological aspects

For chromosomal studies, a squash method was employed (Kawakatsu, Oki, Tamura & Sugino 1976b, pp. 83–84).

For the description of the karyotypes, the authors classify the excess chromosomes or B-chromosomes in two groups, large B-chromosomes (LB) and a small B-chromosomes (SB). Both types are defined as follows:

Large B-chromosome: with a length longer than the long arm of the smallest chromosome in the basic karyotype. Usually, no variation of the number occurs.

Small B-chromosome: with a length shorter than the long arm of the smallest chromosome in the basic set, so shorter than LB. The number of SB in one animal varies widely, and thus, SB is eliminated from the chromosome number.

Dugesia japonica japonica Ichikawa et Kawakatsu, 1964

This nominate subspecies is widely distributed in the Japanese Islands (except the Amami Islands and the Ryûkyû Islands), Taiwain, Korea, Northeast China, and Hong Kong (Fig. 1).

Hitherto, the authors have examined the chromosomes of the animals from 45 localities in the Far East (37 in Japan, 6 in Taiwan and 2 in Korea). The geographical areas where these stocks were collected and the names of each stock are as follows:

Japan. Hokkaidô (Sapporo-Heiwa stock, Sapporo-Higashi-misumai stock, Hokkaidô-Nîkappu stock); Kantô Region in Honshû (Nikkô-Chûgûji stock, Tochigi-Daichûji stock, Gunma-Akagi stock, Gunma-Haruna stock, Mitsukaidô-Sugao stock, Mitsukaidô-Yakkôin stock, Chôshi-Kasuga stock, Chiba-Unakami stock, Fuchû-Tôkyô stock, Yokohama-Hiyoshi stock, Hatano-Minoge stock); Chûbu Region in Honshû (Kashiwazaki-Kôchi stock, Kashiwazaki-Tan'ne stock, Matsumoto-Genchi stock, Kanazawa-Yutanihara stock, Gifu-

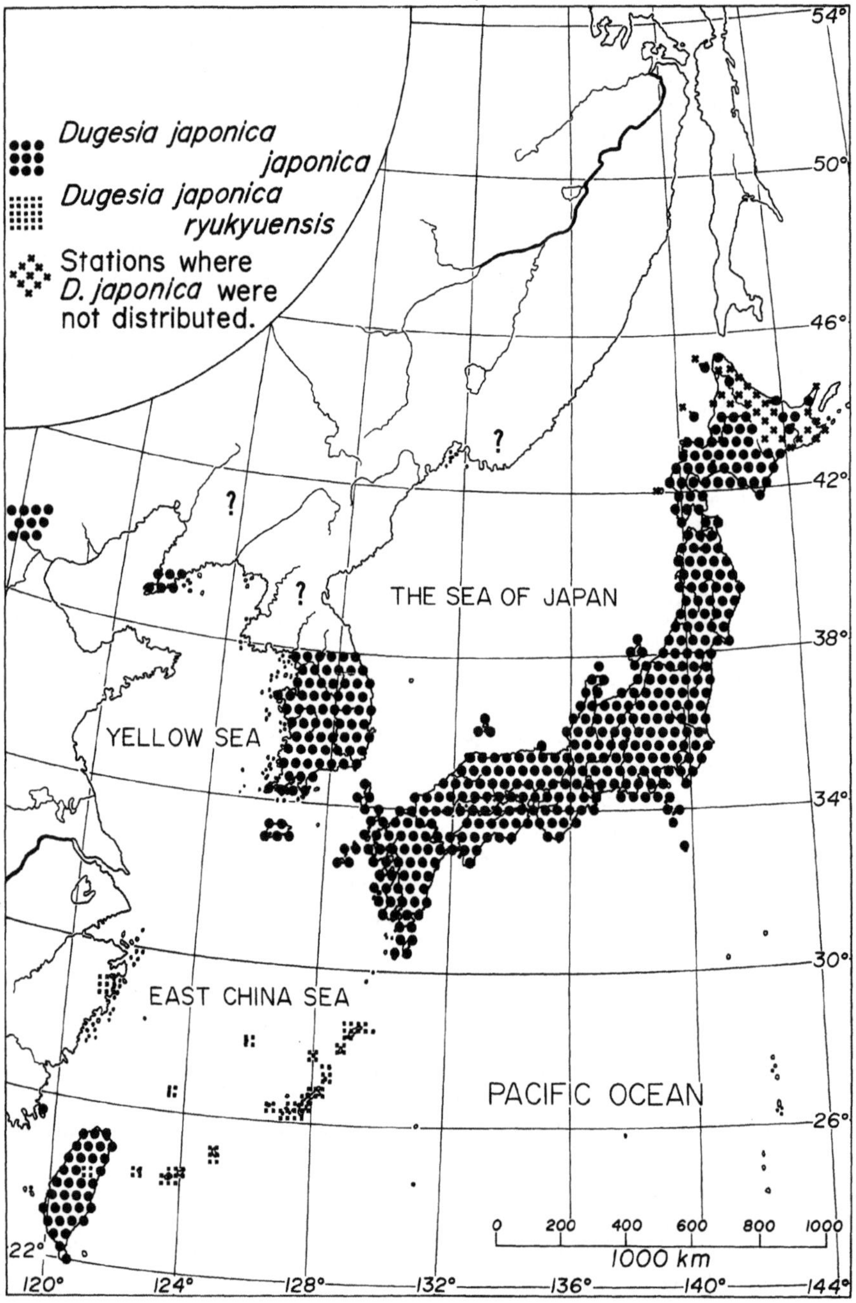

Fig. 1. Map of the Japanese Islands and adjacent countries in the Far East, showing the geographical distribution of *Dugesia japonica*. Areas dotted with solid circles: *D. j. japonica*. Areas dotted with small solid squares: *D. j. ryukyuensis*. X: Districts where the species do not occur.

Table 1. Variation of the chromosome numbers of *Dugesia japonica (D. j. japonica* and *D. j. ryukyuensis).*

Chromosome numbers	Karyotypes	*No. of localities (no. of the specimens examined in parentheses)* JAPAN: HOKKAIDŌ	HONSHŪ	SHIKOKU	KYŪSHŪ	RYŪKYŪ	TAIWAN	KOREA	TOTAL*
Dugesia japonica japonica									
16	2×	–	18 (83)	3 (9)	3 (12)	–	3 (16)	1 (1)	28 (121)
17	2× + 1LB	–	1 (2)	–	–	–	–	–	1 (2)
24	3×	–	2 (3)	–	1 (1)	–	–	1 (3)	4 (7)
25	3× + 1LB	–	4 (24)	–	2 (5)	–	–	–	6 (29)
24	(3×-1) + 1LB	–	–	–	–	–	1 (4)	–	1 (4)
25	(3×-1) + 2LB	–	–	–	2 (3)	–	2 (7)	–	4 (10)
16 & 24	2× & 3×	3 (20)	17 (115)	–	1 (1)	–	–	2 (6)	23 (142)
16 & 32	2× & 4×	–	1 (3)	–	–	–	–	–	1 (3)
16 & 24 & 32	2× & 3× & 4×	–	1 (3)	–	–	–	–	–	1 (3)
25	3× + 1LB & (3×-1) + 2LB	–	1 (1)	–	–	–	–	–	1 (1)
25	3× + 1LB & (3×-1-1) + 3LB	–	1 (1)	–	–	–	–	–	1 (1)
25	3× + 1LB & (3×-1) + 2LB & (3×-1-1) + 3LB	–	1 (2)	–	–	–	–	–	1 (2)
24 & 25	(3×-1) + 1LB & (3×-1) + 2LB	–	–	–	–	–	1 (1)	–	1 (1)
25 & 26	3× + 1LB & 3× + 2LB	–	–	–	1 (1)	–	–	–	1 (1)
25 & 26	3× + 1LB & (3×-1) + 3LB	–	–	–	–	–	–	1 (2)	1 (2)
25 & 26	3× + 1LB & 3× + 2LB & (3×-1) + 3LB	–	1 (3)	–	–	–	–	–	1 (3)
25 & 26	(3×-1-1) + 3LB & (3×-1) + 3LB	–	1 (2)	–	–	–	–	–	1 (2)
25 & 26	3× + 1LB & (3×-1) + 2LB & 3× + 2LB & (3×-1) + 3LB	–	1 (4)	–	–	–	–	–	1 (4)
25 & 26	3× + 1LB & (3×-1) + 2LB & (3×-1-1) + 3LB & 3× + 2LB & (3×-1) + 3LB	–	1 (1)	–	–	–	–	–	1 (1)
25 & 34	3× + 1LB & 4× + 2LB	–	–	–	1 (1)	–	–	–	1 (1)
26 & 27	Unidentified karyotype	–	1 (1)	–	–	–	–	–	1 (1)
TOTAL*		3 (20)	52 (248)	3 (9)	11 (24)	0	7 (28)	5 (12)	81 (341)
Dugesia japonica ryukyuensis									
14	2×	–	–	–	–	2 (8)	–	–	2 (8)
14	(2×-1) + 1LB	–	–	–	–	1 (1)	–	–	1 (1)
21	3×	–	–	–	–	3 (8)	–	–	3 (8)
21	(3×-1) + 1LB	–	–	–	–	2 (2)	–	–	2 (2)
22	(3×-1) + 2LB	–	–	–	–	1 (1)	–	–	1 (1)
14 & 21	2× & 3×	–	–	–	–	1 (2)	–	–	1 (2)
21 & 22	(3×-1) + 1LB & (3×-1) + 2LB	–	–	–	–	2 (2)	–	–	2 (2)
21 & 22	(3×-3-1) + 4LB & (3×-2-1) + 4LB	–	–	–	–	1 (1)	–	–	1 (1)
22 & 23	(3×-2) + 3LB & (3×-2) + 4LB	–	–	–	–	1 (1)	–	–	1 (1)
21 & 22 & 23	(3×-2) + 2LB & (3×-2) + 3LB & (3×-3-1) + 5LB & (3×-3-1) + 6LB	–	–	–	–	1 (1)	–	–	1 (1)
TOTAL*		0	0	0	0	15 (27)	0	0	15 (27)

* The actual number of the localities in each area is as follows: Hokkaidō (3), Honshū (27), Shikoku (3), Kyūshū including the Tsushima Islands (4), Ryūkyū (6), Taiwan (6), and Korea (2). Sometimes the animals having the different karyotypes were intermingled in one locality.

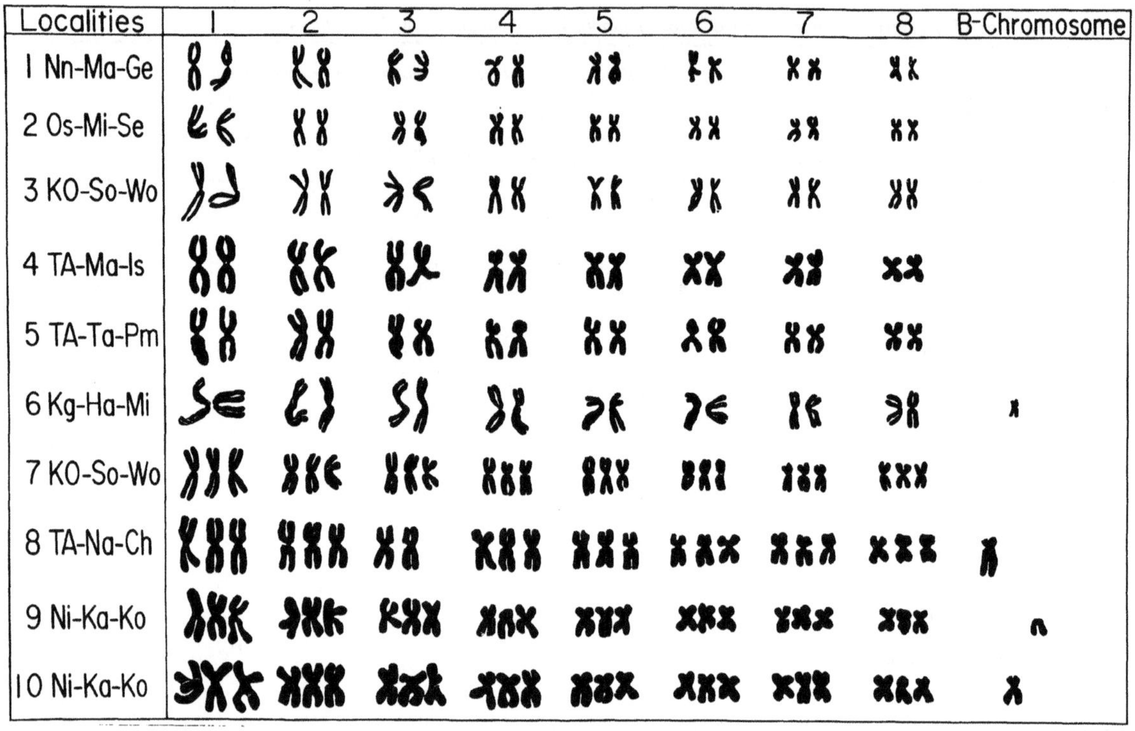

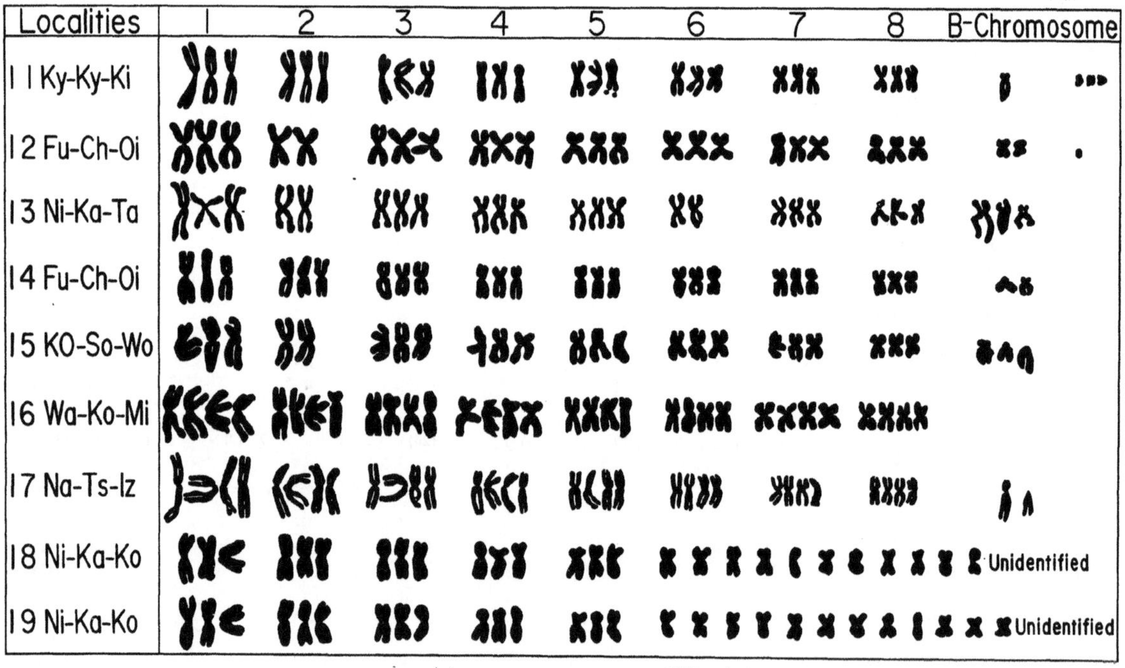

Figs 2 (top) and 3 (bottom). Idiograms of *Dugesia japonica japonica* from 13 localities in Japan, Taiwan and Korea. 1, Nn-Ma-Ge: Matsumoto-Genchi stock, Japan (2x; the typelocality of *D. j. japonica*). 2, Os-Mi-Se: Ôsaka-Mino'o stock, Japan (2x). 3, KO-So-Wo: Seoul-Wooyi-dong stock, Korea (2x). 4, TA-Ma-Is: Matsu-tao Island stock, Taiwan (2x). 5, TA-Ta-Pm: Taipei-National Palace Museum stock, Taiwan (2x). 6, Kg-Ha-Mi: Hastano-Minoge stock, Japan (2x + 1LB). 7, KO-So-Wo: (3x). 8, TA-Na-Ch: Nantou-Chunghsing-hsintsun stock, Taiwan [(3x-1) + 1LB]. 9, Ni-Ka-Ko: Kashiwazaki-Kôchi stock, Japan (3x + 1LB). 10, Ni-Ka-Ko: (3x + 1LB). 11, Ky-Ky-Ki: Kyôto-Kibune stock, Japan (3x + 1LB). 12, Fu-Ch-Oi: Chikushino-Ôishi stock, Japan [(3x-1) + 2LB]. 13, Ni-Ka-Ta: Kashiwazaki-Tan'ne stock, Japan [(3x-1-1) + 3LB]. 14, Fu-Ch-Oi: (3x + 2LB). 15, KO-So-Wo: [(3x-1) + 3LB]. 16, Wa-Ka-Mi: Wakayama-Misato stock, Japan (4x). 17, Na-Ts-Iz: Tsushima-Izuhara stock, Japan (4x + 2LB). 18, Ni-Ka-Ko: Unidentified karyotype. 19, Ni-Ka-Ko: Unidentified karyotype.

Nagara stock, Seto-Jôkôji stock); Kinki Region in Honshû (Mie-Seibôji stock, Mie-Kamori stock, Suzuka-Tsubakijinja stock, Kyôto-Kibune stock, Ôsaka-Mino'o stock, Sen'nan-Kinyûji stock, Kôbe-Suma stock, Wakayama-Misato stock, Wakayama-Hioki stock); Chûgoku Region in Honshû (Kurashiki-Kojima stock); Shikoku (Iyomishima-Nakanoshô stock, Iyomishima-Nakasone stock, Ehime-Masaki stock); Kyûshû (Chikushino-Ôishi stock, Ôita-Kujû stock); the Tsushima Islands near Kyûshû (Tsushima-Mitake stock, Tsushima-Izuhara stock).

Taiwan. Matsu-tao Island stock, Taipei-Museum stock, Taipei-Teinmoo stock, Taipei-Uulei stock, Taoyuan-Tsouho stock, Nantou-Chunghsing-hsintsun stock.

Korea. Seoul-Wooyi-dong stock, Seoul-Gwanac stock.

The variation of the chromosome numbers and karyotypes of *D. j. japonica* from Japan, Taiwan and Korea is shown in Table 1. The idiograms of the animals from 13 localities are also shown in Figs. 2 & 3.

The chromosome number of the diploid specimens of the nominate subspecies of *D. japonica* is 2x = 16. The karyotypes consist of eight pairs of meta- or submetacentric chromosomes in descending order of the size. *D. j. japonica* is, however, a polymorphic form not only from its morphological and anatomical characters, but also from the karyotypes. As will be seen in Table 1, animals from some localities have a triploid complement (3x = 24). Animals that have cells with the diploid and the triploid chromosomes in one body (2x & 3x) are rather common. Animals having cells of the other types of the orthoploidic mixoploid chromosomes (2x & 4x, 2x & 3x & 4x) in one body were recently found in Central Japan. Moreover, animals having cells with B-chromosomes (including LB and SB chromosomes) in addition to the diploid and/or the triploid chromosomes have also occur.

Dugesia japonica ryukyuensis Kawakatsu, 1976

This subspecies is distributed in the Amami Islands and the Ryûkyû Islands in Southwest Japan, the Mt. Alishan area in Taiwan, and the vicinity of Hangchow in Middle China (Fig. 1).

Since this subspecies was described from Okinawa Island (type-loc.: a stream at Nakamayamagawara, Urasoe City), the authors have examined the chromosomes of the animals from the other 5 localities in Okinawa in addition to those from the type-locality. The names of the stocks

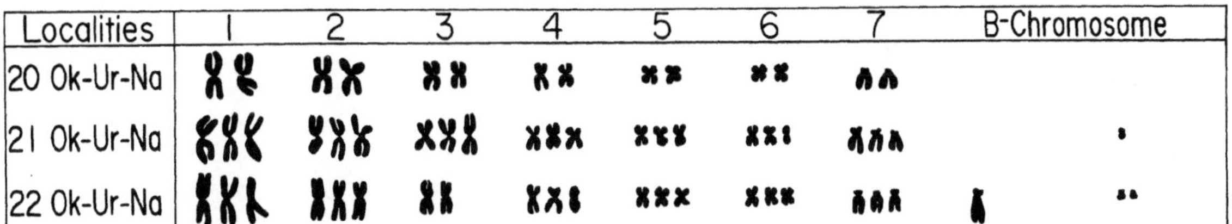

Fig. 4. Idiograms of *Dugesia japonica ryukyuensis* from two localities in Okinawa, Japan. 20, On-Ur-Na: Urasoe-Nakamayamagawara stock, Okinawa, Japan (2x; the type-locality of *D. j. ryukyuensis*). 21, On-Ur-Na: (3x). 22, On-Ur-Na: [(3x–1) + 1LB]. 23, On-Ur-At: Urasoe-Atogâ stock, Okinawa, Japan [(3x–1) + 2LB]. 24, On-Ur-At: [(3x–2) + 2LB]. 25, On-Ur-At: [(3x–2) + 3LB]. 26, On-Ur-At: [(3x–2) + 4LB]. 27, On-Ur-At: [(3x–3–1) + 4LB]. 28, On-Ur-At: [(3x–3–1) + 6LB].

examined cytologically are as follows:

Okinawa. Urasoe-Nakamayamagawara stock, Urasoe-Atogâ stock, Naha-Shuri stock, Okinawa-Kinson stock, Okinawa-Nakijin stock, Okinawa-Gushiken stock.

The variation of the chromosome numbers and karyotypes of *D. j. ryukyuensis* from Okinawa are shown in Table 1. The idiograms of the animals from 2 localities are shown in Fig. 4.

The diploid specimens of *D. j. ryukyuensis* have the chromosome number of 2x = 14. The karyotype consists of 6 pairs of metacentric chromosomes and one pair of acrocentric chromosomes. Animals have triploid complement (3x = 21) and cells with the diploid and the triploid chromosomes (2x & 3x) may also occur in the same individuals. A few number of large B-chromosomes is found in some of the diploid and triploid specimens. Moreover, an irregular number of small B-chromosomes (1–16 in number) is found in the karyotypes of the diploid and the triploid specimens.

3. Discussion

The geographical distribution area of *D. japonica* in the Far East of the Asiatic Continent has not yet been determined. Primorsk, North Korea,

Table 2. Karyotypes of *Dugesia japonica (D. j. japonica* and *D. j. ryukyuensis).*

Ploidy	Karyotypes*	*D. j. japonica* n = 8, 2X = 16, 3X = 24, 4X = 32	*D. j. ryukyuensis* n = 7, 2X = 14, 3X = 21
Diploid	2X	+	+
Diploidy	2X + 1LB	+	-
Diploidic Aneuploid	(2X-1) + 1LB	-	+
Triploid	3X	+	+
Triploidy	3X + 1LB	+	-
Triploidic Aneuploid	(3X-1) + 1LB	+	+
	(3X-1) + 2LB	+	+
Orthoploidic Mixoploid	2X & 3X	+	+
	2X & 4X	+	-
	2X & 3X & 4X	+	-
Triploidic Mixoploid; Triploidic & Tetraploidic Mixoploid	3X + 1LB & 3X + 2LB	+	-
	3X + 1LB & (3X-1) + 2LB	+	-
	3X + 1LB & (3X-1) + 3LB	+	-
	3X + 1LB & (3X-1-1) + 3LB	+	-
	3X + 1LB & 4X + 2LB	+	-
	(3X-1) + 1LB & (3X-1) + 2LB	+	+
	(3X-2) + 3LB & (3X-2) + 4LB	-	+
	(3X-1-1) + 3LB & (3X-1) + 3LB	+	-
	(3X-3-1) + 4LB & (3X-2-1) + 4LB	-	+
	3X + 1LB & 3X + 2LB & (3X-1) + 3LB	+	-
	3X + 1LB & (3X-1) + 2LB & (3X-1-1) + 3LB	+	-
	3X + 1LB & (3X-1) + 2LB & 3X + 2LB & (3X-1) + 3LB	+	-
	(3X-2) + 2LB & (3X-2) + 3LB & (3X-3-1) + 5LB & (3X-3-1) + 6LB	-	+
	3X + 1LB & (3X-1) + 2LB & (3X-1-1) + 3LB & 3X + 2LB & (3X-1) + 3LB	+	-
	Unidentified karyotype	+	-

* Small B-chromosomes were excluded from this table.

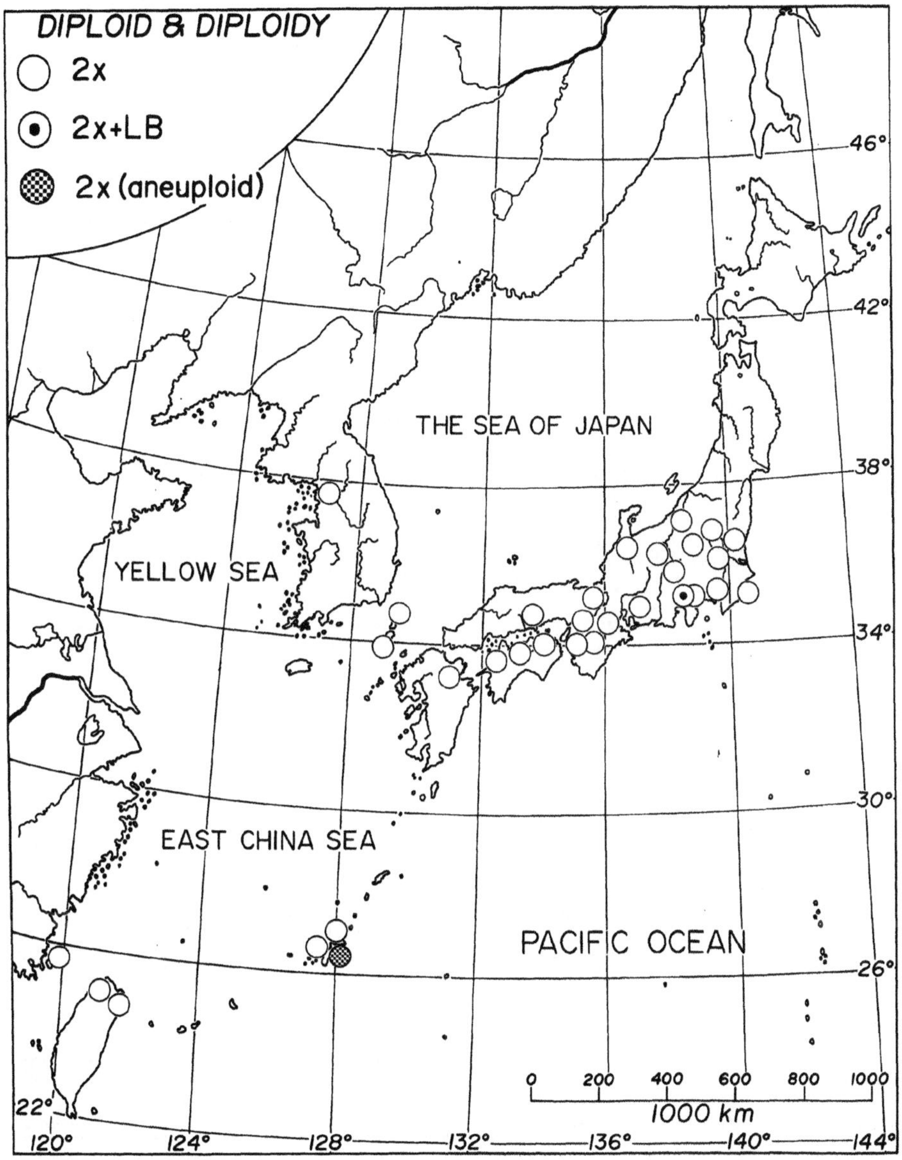

Fig. 5. Map of the Japanese Islands and adjacent countries in the Far East, showing the geographical distribution of the karyotypes of diploid and diploidy in *Dugesia japonica (D. j. japonica* and *D. j. ryukyuensis).*

China including Hai-Nan Island, and the Indo-China Peninsula are poorly known as regards their planarian fauna. Judging from the present distributional data of *D. japonica* and the topography of the Far East, the previous geographical distribution map of this species (Kawakatsu 1974, p. 296, fig. 3) has been revised in Fig. 1.

Considerations about the possible explanation of the relationship between the two karyotypes of *D. j. japonica* (2x = 16) and *D. j. ryukyuensis* (2x = 14) and their subspeciations were discussed in a previous paper (Kawakatsu, Oki, Tamura & Sugino 1976b). Only a further discussion about the karyotypes of *D. japonica* is given here.

Listed in Table 2 are every known variation of the karyotypes of *D. japonica* based upon the authors' latest data of the chromosomal analysis. The karyotypes can be classified into six main groups,

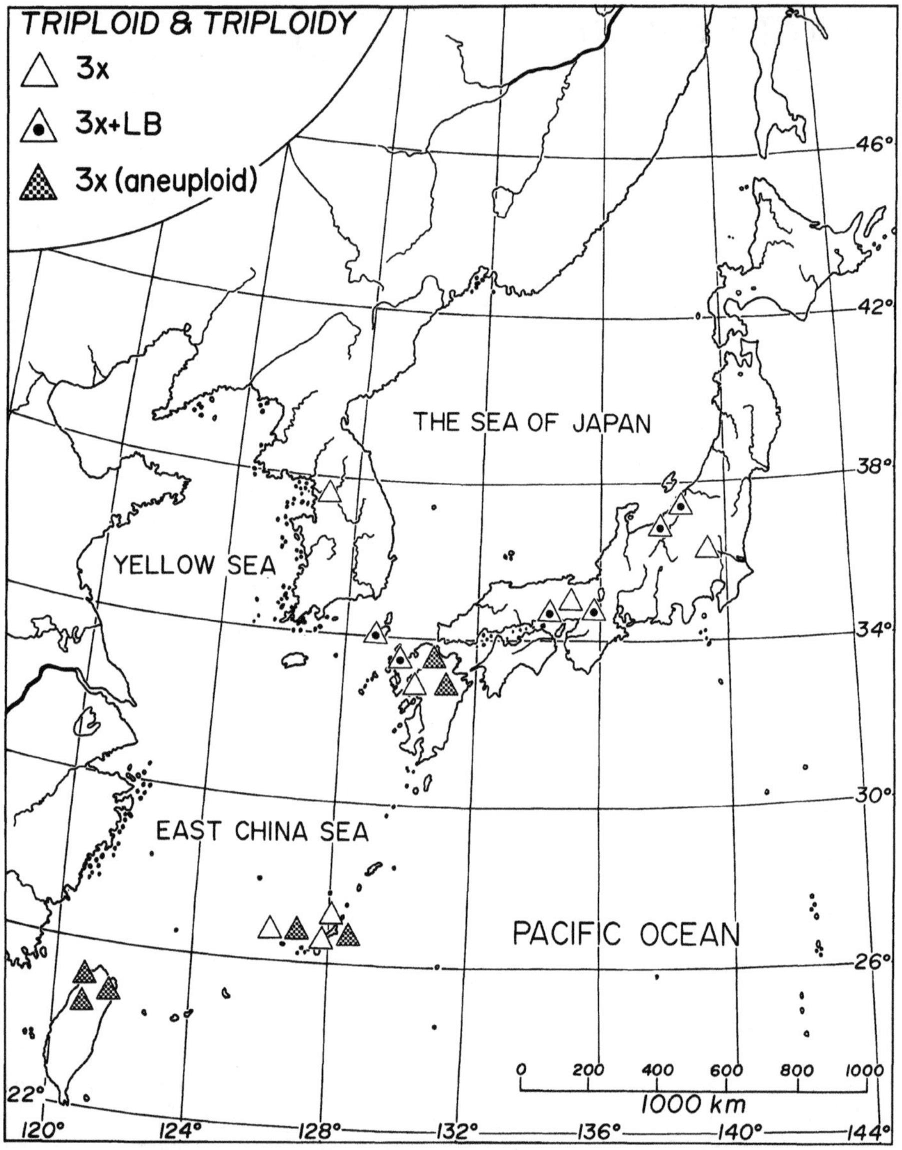

Fig. 6. Map of the Japanese Islands and adjacent countries in the Far East, showing the geographical distribution of the karyotypes of triploid and triploidy in *Dugesia japonica (D. j. japonica* and *D. j. ryukyuensis).*

namely: (1) diploid and diploidy; (2) diploidic aneuploid; (3) triploid and triploidy; (4) triploidic aneuploidy; (5) orthoploidic mixoploid; and (6) triploidic mixoploid and triploidic & tetraploidic mixoploid. The geographical distribution of these karyotypes is shown in Figs. 5, 6 & 7. In some of the localities, two or three or four karyotypes were recorded. Fig. 8 shows all of the records on a single map.

In *D. j. japonica,* the diploid form (2x) is distributed widely in Honshû, Shikoku and Kyûshû in Japan, and Taiwan and Korea. The diploidy form (2x + LB) is, however, only found in one locality in Honshû (Figs. 5 & 8). The localities of the orthoploidic triploid form (3x) are limited to Honshû and Kyûshû in Japan and Korea. The triploidy form (3x + LB) is distributed in Honshû and Kyûshû in Japan. The triploidic aneuploid

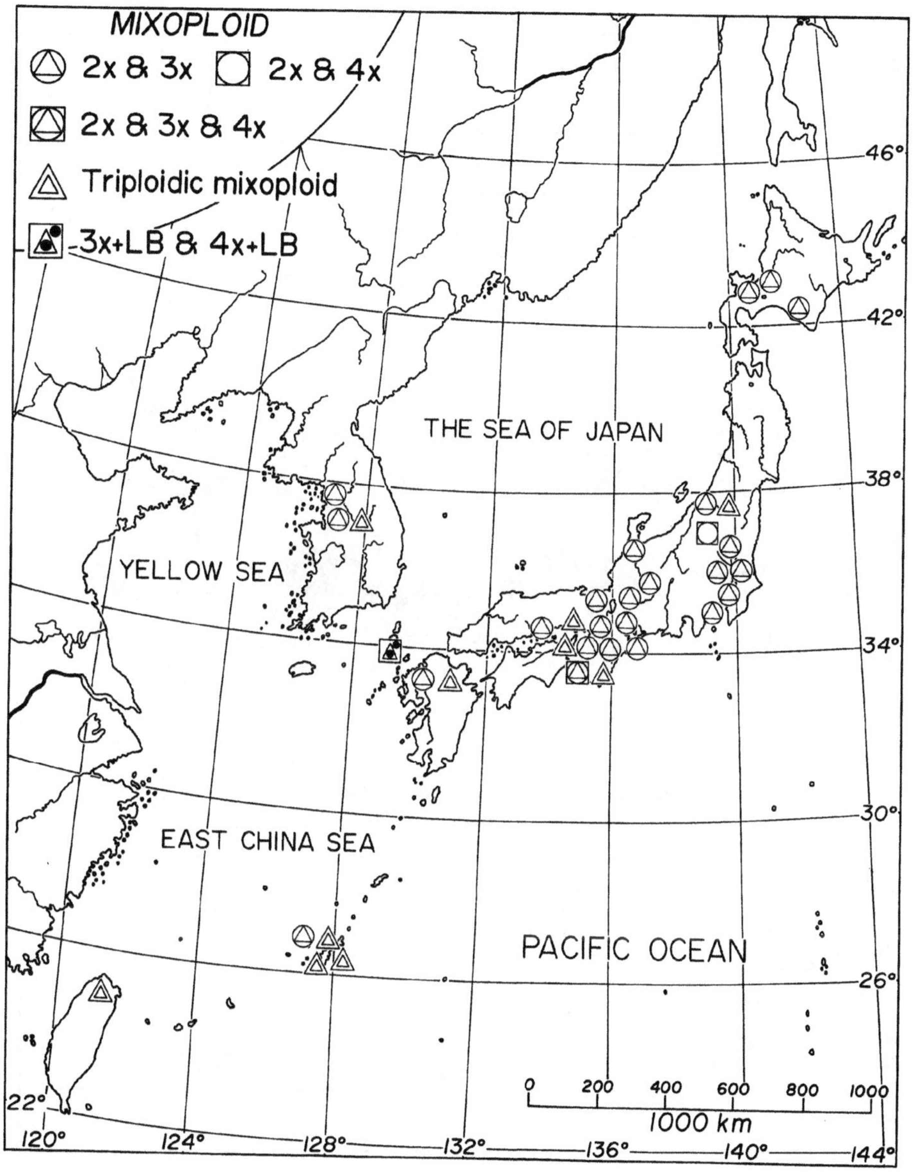

Fig. 7. Map of the Japanese Islands and adjacent countries in the Far East, showing the geographical distribution of the karyotypes of mixoploid in *Dugesia japonica (D. j. japonica* and *D. j. ryukyuensis).*

forms [(3x–1) + LB] with one or two B-chromosomes are distributed in Kyûshû in Japan and Taiwan (Figs. 6 & 8).

Among the orthoploidic mixoploid forms, the animals having the karyotype of 2x & 3x are distributed rather commonly in Hokkaidô, Honshû and Kyûshû in Japan and Korea. The distribution of both the orthoploidic karyotypes of 2x & 4x and 2x & 3x & 4x is restricted to Honshû in Japan. The triploidic mixoploid karyotypes show a high degree of variation in their chromosome numbers. Although these karyotypes are distributed in Honshû and Kyûshû in Japan and Taiwan and Korea, the number of localities of each karyotype is very limited. The triploidic & tetraploidic mixoploid karyotype is only found in Tsushima Island (Figs. 7 & 8).

For *D. j. ryukyuensis* the karyological data are

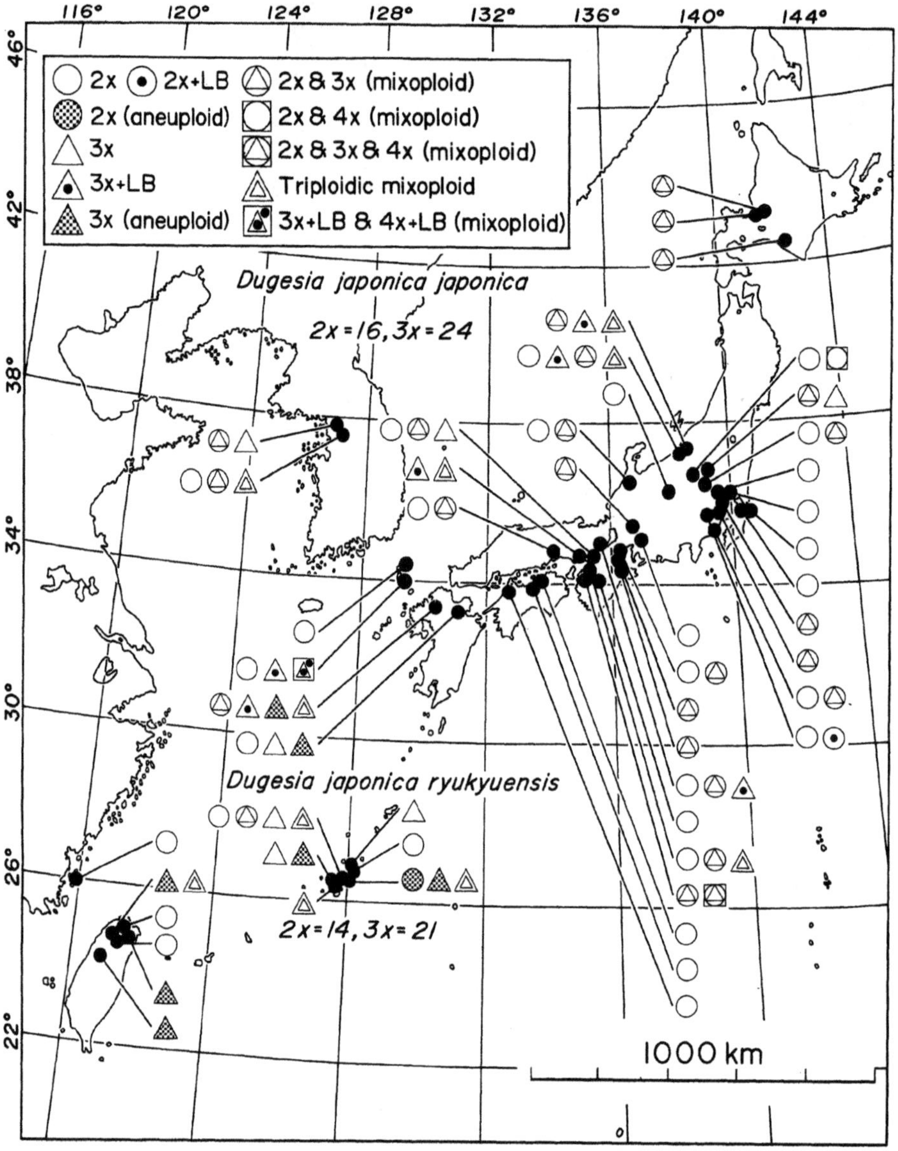

Fig. 8. Map of the Japanese Islands and adjacent countries in the Far East, showing the geographical distribution of every known karyotype in *Dugesia japonica (D. j. japonica* and *D. j. ryukyuensis).* Explanation in the text.

restricted to Okinawa Island. Animals having the karyotypes of diploid (2x), diploidic aneuploid [(2x–1) + 1LB], triploid (3x), triploidic aneuploid [(3x–1) + LB], orthoploidic mixoploid (2x & 3x only) and of several other triploidic mixoploid are found (Figs. 5, 6, 7 & 8).

As will be seen in Table 1 and Fig. 7, the number of localities of *D. j. japonica* having the karyotype of 2x & 3x (orthoploidic mixoploid) is nearly equal to that of the diploid form (2x). They are now recorded from the various areas of the Far East except Shikoku in Japan and Taiwan. This orthoploidic mixoploid karyotype is also found in *D. j. ryukyuensis.*

The occurrence of the orthoploidic mixoploid karyotype of 2x & 3x in some European and North

American dugesiid planarians was already reported by Benazzi and his team and by Dahm (Benazzi 1963; Dahm 1958; see also Benazzi & Benazzi-Lentati 1976). According to the authors' own data (mostly unpublished), the karyotype of 2x & 3x has been found in several species in addition to *D. japonica*. These species are: *Dugesia siamana* Kawakatsu, 1980, from Thailand (2x = 16 & 3x = 24) (Kawakatsu, Oki, Tamura & Yamayoshi 1980), *Dugesia schubarti* (Marcus 1946) from the vicinity of São Leopoldo in South Brazil (2x = 8 & 3x = 12, 2x = 8 & 3x = 12 & 4x = 16) (Kawakatsu, Hauser & Friedrich 1976; Kawakatsu, Oki, Tamura, Yamayoshi, Hauser & Friedrich 1980; Oki, Tamura, Yamayoshi, Kawakatsu, Hauser & Friedrich 1980) and *Polycelis auriculata* Ijima et Kaburaki, 1916, from Nikkô in Honshû in Central Japan [2x = 6 & (3x-1) + 2LB = 10]*.

It is true from these findings that the presence of an orthoploidic mixoploid karyotype is very common among many fissioning species of freshwater planarians. At present the authors have not sufficient data for discussing the mechanism of its formation in *D. japonica,* but they now strongly incline to the view that the ancestral orthoploidic mixoploid form may have differentiated from the diploid proto-species of *D. japonica* by the mitotic non-disjunction of the blastmeres in an early stage of cleavage. Since *D. japonica* is a fissioning planarian, the animals having a mixoploid karyotype can survive by way of asexual reproduction without any influence of the chromosomal sterility. The authors found that sometimes sexual maturity occurs in some of the mixoploidic animals of 2x & 3x from the Ôsaka-Mino'o locality. In the meiosis of the male line of these sexual specimens, only the haploid complement (n = 8) was found. The juveniles originating from the sexual reproduction of these mixoploid sexual specimens have only the diploid karyotype (2x = 16).

In *D. j. japonica* the other two orthoploidic mixoploid karyotypes (2x & 4x, 2x & 3x & 4x) and a triploidic & tetraploidic mixoploid karyotype (3x + 1LB & 4x + 2LB) are known. All these mixoploidic karyotypes, which probably differentiated from the diploid ancestral form, can be maintained by asexual reproduction.

In theory, three hypotheses are possible for the explanation of the formation of the orthoploidic triploid karyotype of 3x in *D. japonica*. These are: (1) the natural hibridization of the diploid (2x) and the tetraploidic (4x) animals, (2) the disappearance of the diploid cells from the animals having the orthoploidic mixoploid karyotype of 2x & 3x, and (3) the fertilization of the abnormal diploid gametes (2x) and the normal haploid gamete (n) formed in the diploid animals.

In the first case, a population of the animals having only the tetraploidic karyotype (4x) has not yet been found in *D. japonica*. Since mixoploidic karyotypes are present only in lower animal groups such as planarians and Japanese liver flukes (*Fasciola* sp.; cf. Moriyama, Tsuji & Seto 1979), the second hypothesis has but limited possibilities. From these considerations, the third hypothesis seems to be the most probable explanation of the formation of the orthoploidic triploid karyotype (3x) in *D. japonica*. This karyotype can be maintained by both sexual and asexual reproduction.

It is worthy of special note that a few B-chromosomes are frequently to be found in the various karyotypes in *D. japonica* (Tables 1 & 2, Figs. 2 & 3). What is the origin and the karyological significance of these B-chromosome? Many triploidic mixoploid karyotypes have large and/or small B-chromosomes. In the case of the hypoploidic triploid karyotype found in some populations [(3x-1) + 1LB] (see Fig. 2, TA-Na-Ch; Fig. 5, Ok-Ur-Na), it may safely be explained that one large B-chromosome in this karyotype originated from the trisomic triploidy by a separation of one chromosome in a certain set of chromosomes. On the other hand, a hyperploidic triploid karyotype is found in some populations (3x + LB) (see Fig. 2, Ni-Ka-Ko; Fig. 3, Ky-Ky-Ki, etc.). Moreover, the animals from many localities have an unstable number of small B-chromosomes in their karyotypes. As will be seen in Fig. 4, the number of small B-chromosomes varies from zero to 16 in animals from the Urasoe-Atogâ stock of *D. j. ryukyuensis*. An unidentified karyotype found in one animal from the Kashiwazaki-Kôchi stock of *D. j. japonica* that has chromosomes of 26 and 27 in one body remains an enigma. As was already discussed preliminarily in a previous paper (Tamura, Yamayoshi, Oki, Murayama & Kawakatsu 1978), B-chromosomes of *D. japonica* may have origi-

* Teshirogi & Ishida (this volume) have analysed this species in detail.

The authors may also consider the possibility that the formation of various kinds of polyploidic karyotypes in *D. japonica* may be the effect of extremely low temperatures during the past glacial age. Polyploidic karyotypes of *D. japonica* may perhaps be induced by several environmental factors, such as the chemical and physical conditions today. More light could be thrown on this problem by a further taxonomic-karyological analysis of widely distributed, polymorphic planarian species, including *D. japonica.*

References

Ball, I. R., 1970. Freshwater triclads (Turbellaria, Tricladida) from the Oriental Region. J. Linn. Soc., 49: 271–294 (+ pls. 1–2).

Benazzi, M., 1963. Genetics of reproductive mechanisms and chromosome behavior in some fresh-water triclads. In: Ed. Dougherty, E. C., The Lower Metazoa – Comparative Biology and Phylogeny. Berkeley: Univ. California Press, pp. 405–422.

Benazzi, M. & Benazzi-Lentati, G., 1976. Platyhelminthes. In: Ed. John, B., Animal Cytogenetics. Berlin: Gebrüder Borntraeger, (i-vi) + pp. 1–182.

Curtis, W. C., 1902. The life history, the normal fission, and the reproductive organs of Planaria maculata. Proc. Boston Soc. Nat. Hist., 30: 515–583 (+ pls. 9–19).

Dahm, A. G., 1958. Taxonomy and Ecology of Five Species Groups in the Family Planariidae (Turbellaria, Tricladida, Paludicola). Nya Litografen, Malmö, pp. 1–241 (+ pls. I–V).

Dahm, A. G., 1963. The karyotypes of some freshwater triclads from Europe and Japan (Turbellaria, Tricladida, Paludicola). Arkiv Zool., ser. 2, 16: 41–67.

Hirose, E., Katô, F. & Sugino, H., 1974. Chromosomes of freshwater planarian, Dugesia japonica, II. Zool. Mag. (Tôkyô), 83: 442. (In Japanese)

Ichikawa, A. & Kawakatsu, M., 1964. A new freshwater planarian, Dugesia japonica, commonly but erroneously known as Dugesia gonocephala (Dugès). Annot. Zool. Japon, 37: 185–194.

Ijima, I., 1887. Über einige Tricladen Europas. J. Coll. Sci. Imp. Univ. Tokyo, 1: 337–358 (+ pl. XXV).

Ijima, I. & Kaburaki, T., 1916. Preliminary descriptions of some Japanese triclads. Annot. Zool. Japon., 9: 153–171.

Kaburaki, T., 1922. On some Japanese freshwater triclads; with a note on the parallelism in their distribution in Europe and Japan. J. Coll. Sci. Imp. Univ. Tokyo, 44: 1–71 (+ pl. I).

Kawakatsu, M., 1965. On the ecology and distribution of freshwater planarians in the Japanese Islands, with special reference to their vertical distribution. Hydrobiologia, 26: 349–408.

Kawakatsu, M., 1967. On the ecology and distribution of freshwater planarians in the Japanese Islands, with special reference to their vertical distribution (Rev. Ed.). Bull. Fuji Women's Coll., no. 5: 117–177.

Kawakatsu, M., 1970. Further studies on the vertical distribution of freshwater planarians in the Japanese Islands. Am. Zool., 10: 546.

Kawakatsu, M., 1971. Problems on the morphological variation and the physiological races of a Japanese freshwater planarian, Dugesia japonica Ichikawa et Kawakatsu. In: Eds. Kawakatsu, M. & Iba, M., Comm. Compil. Sci. Papers Publ. Occ. Retir. Prof. Hisao Sugino at the Age of Sixty-Five, ser. Turbellarians, Biol. Lab. Ôsaka Kyôiku Univ., Ôsaka, pp. 43–52. (In Japanese with English summary)

Kawakatsu, M., 1974. Further studies on the vertical distribution of freshwater planarians in the Japanese Islands. In: Eds. Riser, N. W. & Morse, M. P., The Hyman Memorial Volume – Biology of the Turbellaria. New York: McGraw-Hill. pp. 291–338.

Kawakatsu, M., 1975. Problems on the morphological variation and the taxonomic position of a Japanese freshwater planarian, Dugesia japonica Ichikawa et Kawakatsu. Zool. Mag. (Tôkyô), 84: 444. (In Japanese)

Kawakatsu, M., 1968–1980. A list of publications on Japanese Turbellarians (1855–1967) – Including titles of publications on foreign Turbellarians written by the Japanese authors. Bull. Fuji women's Coll., no. 6: 71–127. Do. (1968). Ibid., no. 7: ser. II: 23–43 (+ pls. I–V). Do. (1969). Ibid., no. 8, ser. II: 107–114. Do. (1970). Ibid., no. 9, ser. II: 29–40 + 1 folder. Do. (1971). Ibid., no. 10, ser. II: 45–55. Do. (1972). Ibid., no. 11, ser. II: 57–65. Do. (1973). Ibid., no. 12, ser. II: 57–67. Do. (1974). Ibid., no. 13, ser. II: 49–58. Do. (1975). Ibid., no. 14, ser. II: 73–79. Do. (1976). Ibid., no. 15, Ser. II: 57–68. Do. (1977). Ibid., no. 16, ser. II: 67–82. Do. (1978). Ibid., no. 17, ser. II: 45–58. Do. (1979). Ibid., no. 18, ser. II: 117–127.

Kawakatsu, M., Hauser, J. & Friedrich, S. M. G., 1976. The freshwater planaria from South Brazil. Bull. Natn. Sci. Mus. Tokyo, ser. A (Zool.), 2: 205–223.

Kawakatsu, M., Horikoshi, I. & Akama, H., 1972. Report on freshwater planarians from the Izu Peninsula and the Izu Islands in Japan. Zool. Mag. (Tôkyô), 81: 119–126. (In Japanese with English summary)

Kawakatsu, M. & Iwaki, S., 1967. Report on freshwater planaria from the Satsunan Islands and Kagoshima (Kyûshû) in South Japan. Bull. Fuji Women's Coll., no. 5: 179–185.

Kawakatsu, M. & Mack-Firā, V., 1975. The fauna of the insular lava caves in West Japan. XI. Proseriata et Tricladida (Turbellaria). Bull. Natn. Sci. Mus. Tôkyô, ser. A (Zool.), 1: 77–84 (+ pl. 1).

Kawakatsu, M., Oki, I., Tamura, S. & Sugino, H., 1976a. Morphological and karyological reexamination of the taxonomy of the freshwater planarian Dugesia japonica, II. Considerations about the subdivision of the species into two subspecies, with special reference to their subspeciations and phylogenetical problems. Zool. Mag. (Tôkyô), 85: 508. (In Japanese)

Kawakatsu, M., Oki, I., Tamura, S. & Sugino, H., 1976b. Studies on the morphology, karyology and taxonomy of the Japanese freshwater planarian Dugesia japonica Ichikawa et Kawakatsu, with a description of a new subspecies, Dugesia japonica ryukyuensis subspec. nov. Bull. Fuji Women's Coll., no. 14, ser. II: 81–126.

Kawakatsu, M., Oki, I., Tamura, S. & Yamayoshi, T., 1980. The freshwater planarians from Thailand and South India. Annot. Zool. Japon., 53: 254–268.
Kawakatsu, M., Oki, I., Tamura, S., Yamayoshi, T., Hauser, J. & Friedrich, S. M. G., 1980. Karyological and taxonomic studies of freshwater planarians in Brazil. I. Review of the previous studies, with special reference to the taxonomy of Dugesia species from South Brazil. Zool. Mag. (Tôkyô), 89: 628. (In Japanese)
Kawakatsu, M., Oki, I., Tamura, S., Yamayoshi, T., Lue, K.-Y. & Hagiya, M, 1979. Additional report on freshwater planarians from Taiwan. Bull. Fuji Women's Coll., no. 17, ser. II: 59–91.
Kawakatsu, M., Oki, I., Tamura, S., Yamayoshi, T. & Takahashi, N., 1980. Morphological, karyological and taxonomic studies of Dugesia japonica Ichikawa et Kawakatsu from the Tsushima Islands. Proc. Jap. Soc. Syst. Zool., no. 19: 1–10 (+ pls. 1–2).
Kawakatsu, M. & Tanaka, I., 1971. Additional report on freshwater planaria from the Southwest Islands of Japan. Biol. Mag. Okinawa (Naha), no. 8: 46–52 (+ pls. I–II). (In Japanese with English summary)
Kawakatsu, M. & Tanaka, I., 1976. Additional report on freshwater planaria from the Southwest Islands of Japan, II. Zool. Mag. (Tôkyô), 85: 73–77.
Kawakatsu, M., Teshirogi, W., Ôgawara, G. & Tarui, Y., 1965. Photographic gleanings of planarians. I. Dugesia japonica Ichikawa et Kawakatsu and Dugesia gonocephala (Dugès). Collect. & Breed. (Tôkyô), 27: 330–335. (In Japanese)
Kenk, R., 1935. A morphological proof of the existence of zooids in Euplanaria dorotocephala. Physiol. Zoöl., 8: 442–456.
Kenk, R., 1937. Sexual and asexual reproduction in Euplanaria tigrina (Girard). Biol. Bull., 73: 280–294.
Kenk, R., 1940. The reproduction of Dugesia tigrina (Girard). Am. Nat., 74: 471–475.
Kenk, R., 1941. Induction of sexuality in the asexual form of Dugesia tigrina (Girard). J. Exp. Zool., 87: 55–69.
Marcus, E., 1953. Turbellaria Tricladida. Inst. Parcs Nat. Congo Belge, Explor. Parc Nat. de l'Upemba. Miss. G. F. de Witte (1946–1949), no. 21: 1–62.
Moriyama, N., Tsuji, M. & Seto, T., 1979. Three karyotypes and their phenotypes of Japanese liver flukes (Fasciola sp.). Jap. J. Parasitol., 28: 23–33. (In Japanese with English summary)
Oki, I. & Tamura, S., 1974. Studies on the chromosome of freshwater planaria lived in the Mino River. Zool. Mag. (Tôkyô), 83: 457. (In Japanese)
Oki, I. & Tamura, S., 1975a. A freshwater planarian with heterogeneous cell populations of two different chromosome numbers. Ibid., 84: 61–63. (In Japanese with English summary)
Oki, I. & Tamura, S., 1975b. A freshwater planarian with heterogeneous cell populations of two different chromosome numbers. Ibid., 84: 444. (In Japanese)
Oki, I., Tamura, S., Kawakatsu, M. & Sugino, H., 1976. Morphological and karyological reexamination of the taxonomy of the freshwater planarian Dugesia japonica, I. Chromosomal analysis of the animals from different localities in Japan and Korea. Ibid., 85: 507. (In Japanese)
Oki, I., Tamura, S., Yamayoshi, T., Kawakatsu, M., Hauser, J. & Friedrich, S. M. G., 1980. Karyological and taxonomic studies of freshwater planarians in Brazil. II. Chromosomes of Dugesia species from South Brazil. Ibid., 89: 628. (In Japanese)
Okugawa, K. I., 1955. On the supernumerary sexual organs of Dugesia gonocephala (Dugès), induced by the low temperature. Bull. Kyoto Gakugei Univ., ser. B, no. 6: 1–14 (+ pls. I–II).
Okugawa, K. I., 1957. An experimental study of sexual induction in the asexual form of Japanese freshwater planarian, Dugesia gonocephala (Dugès). Ibid., ser. B, no. 11: 8–27 (+ pls. 1–6).
Okugawa, K. I. & Kawakatsu, M., 1954a. Studies on the fission of Japanese freshwater planaria, Dugesia gonocephala (Dugès). I. Comparative studies on fission rates and frequencies of sexual and asexual races influenced by temperatures, starvation and distilled water. Ibid., ser. B, no. 4: 25–34. (In Japanese with English summary)
Okugawa, K. I. & Kawakatsu, M., 1954b. Studies, II. On the fission plane and fission pieces. Ibid., ser. B, no. 4: 35–40 (+ 1 folder). (In Japanese with English summary)
Okugawa, K. I. & Kawakatsu, M., 1954c. Studies, III. Comparative studies on breeding and fission frequencies of sexual and assumed asexual races which had been collected in ten localities in Japan, with an additional study on the fission plane. Ibid., ser. B, no. 5: 42–52. (In Japanese with English summary)
Okugawa, K. I. & Kawakatsu, M., 1956a. Studies, IV. Comparative studies on breeding and fission frequencies of sexual and assumed asexual races which were observed in laboratory cultures and natural habitats. Ibid., ser. B, no. 8: 23–42. (In Japanese with English summary)
Okugawa, K. I. & Kawakatsu, M., 1956b. Studies, V. On the influence of fission frequencies of the animals of sexual and asexual races by means of head removed operation. Ibid., ser. B, no. 8: 43–59. (In Japanese with English summary)
Okugawa, K. I. & Kawakatsu, M., 1957. Studies, VI. Comparative studies on breeding and fission frequencies of sexual and asexual races which were observed under by different food cultures. Ibid., ser. B, no. 10: 18–37. (In Japanese with English summary)
Okugawa, K. I. & Kawakatsu, M., 1958. Studies, VII. Comparative studies on breeding and fission frequencies of sexual and asexual races influenced by various concentrations of Ringer solutions and hydrogen-ion concentrations. Ibid., ser. B, no. 12: 19–44. (In Japanese with English summary)
Sugino, H., 1962. Mutual tansplantation in planarians collected from various localities. Mem. Ôsaka Univ. Liberal Arts & Educat., ser. B, no. 11: 123–127.
Sugino, H., 1969. Collecting, breeding and experiments of a common Japanese freshwater planarian, Dugesia japonica (with annotations and an appendix written by M. Kawakatsu: On the physiological races of *D. japonica* as a polytypic species). Collect. & Breed. (Tôkyô), 31: 222–229, 294–303. (In Japanese)

This article is transcribed in Comm. Compil. Sci. Papers Publ. Occ. Retir. Prof. Hisao Sugino at the Age of Sixty-five, ser. Science Education and Other Subjects, pp. 1–16; Ibid., ser. Turbellarians, pp. 1–16.

Sugino, H., Hirose, E. & Katô, F., 1973. The chromosomes of a Japanese freshwater planarian, Dugesia japonica Ichikawa et Kawakatsu. Nature Study (Ôsaka), 19, no. 4: 41–43. (In Japanese)

Tamura, T., Yamayoshi, T., Oki, I. & Kawakatsu, M., 1979. Karyological and taxonomic studies of Dugesia japonica Ichikawa et Kawakatsu. II. Chromosomes of Dugesia japonica japonica collected from eighteen localities in Japan. Proc. Jap. Soc. Syst. Zool., no. 17: 1–14 (+ pls. 1–2). (In Japanese with English summary)

Tamura, S., Yamayoshi, T., Oki, I., Murayama, H. & Kawakatsu, M., 1978. Karyological and taxonomic studies of Dugesia japonica Ichikawa et Kawakatsu. I. Chromosomes of the animals of Dugesia japonica japonica collected from five localities in the central part of Honshû and Shikoku, Japan. Ibid., no. 15: 8–18 (+ pls. 1–2). (In Japanese with English summary)

Tanaka, I., 1965. Observation on the breeding of Dugesia japonica Ichikawa et Kawakatsu from Okinawa (with an appendix written by M. Kawakatsu). Collect. & Breed. (Tôkyô), 27: 458–459. (In Japanese)

Teshirogi, W. & Itagaki, G., 1965. The chromosomes of Dugesia species, a Japanese freshwater planarian known as Dugesia gonocephala. Zool. Mag. (Tôkyô), 74: 38–45. (In Japanese with English summary)

Vandel, A., 1922. Recherches expérimentales sur les modes de reproduction des planaires Triclades Paludicoles. Bull. Biol. France Belg., 55: 343–518.

Yamayoshi, T., Tamura, S., Oki, I. & Kawakatsu, M., 1980. A survey of the physico-chemical analysis of river water in relation to the chromosomal variation of freshwater planarians in Ôsaka Prefecture. Proc. Ôsaka Prefect. Inst. Public Health, Ed. Publ. Health, no. 18: 147–158. (In Japanese)

Studies on the speciation of Japanese freshwater planarian Polycelis auriculata based on the analysis of its karyotypes and constitutive proteins

Wataru Teshirogi & Sachiko Ishida
Dept. of Biology, Faculty of Science, Hirosaki Univ. Hirosaki 036, Japan

Keywords: Turbellaria, karyology, mixoploidy, electrophoresis

Abstract

Various kinds of chromosomal polymorphisms or karyotypic variations are found in the Japanese freshwater planarian *Polycelis auriculata*. Within this species, there are found worms whose chromosome numbers are 2n = 6, 10, 11, 12 and others, and 3x = 6 and 9. There are some which have cells with triploidy and tetraploidy complements (3x = 6 & 4x = 8), and others which have cells with triploidy and hexaploidy complements (3x = 6 & 6x = 12). These worms with such varied karyotypes are usually found in separate habitats, though occasionally they occur together. Electrophoretic analysis of the proteins extracted from the karyotypically different worms which belong to three different local populations shows some dissimilarity in the constitutive proteins according to their karyotypic differences. The results obtained suggest that this species is still in the process of speciation or chromosomal evolution.

Introduction

Planarians of the genus *Polycelis* found so far in Japan comprise four species, *P. sapporo, P. akkeshi, P. schmidti,* and *P. auriculata,* and their chromosome numbers are 2n = 42, 42, 32 and 6 respectively (Teshirogi *et al.* 1978, 1979; Nîmura 1978). In *P. auriculata,* however, there are also found various kinds of chromosomal polymorphisms or karyotypic variations. Therefore, these karyotypic variations in this species were considered from the viewpoint of chromosomal evolution or speciation. Furthermore electrophoretic patterns of the proteins extracted from the worms belonging to several populations with different karyotypes were compared.

Materials and methods

Materials used in this investigation were *Polycelis auriculata* Ijima & Kaburaki, collected from streams of various localities in Japan as shown in Fig. 12. This species is distributed in almost all parts of Hokkaido except its northern part, and north of the Chubu (Central) Districts of the Main Island (Honshū) of Japan (Kawakatsu 1967).

Chromosomal studies. Adult planarians, 14–25 mm in length, were used. After the worms were starved for a week, they were cut into transverse pieces and cultivated in dechlorinated tap water for three to four days. The mitotic division was observed in the neoblasts of the regenerating pieces and the meiotic division was observed in the primary spermatocytes and oocytes by air drying method (cf. Imai *et al.* 1977; Teshirogi *et al.* 1980). The dried specimens were stained with Giemsa solution for 10–15 min. G-band staining was performed using urea (Kato & Yoshida 1972). C-banded chromosomes were obtained by a more prolonged period of urea treatment (at 37 °C for 30 min).

Hydrobiologia 84, 69–77 (1981). 0018-8158/81/0841-0069/$02.20.

Polyacrylamide gel electrophoresis. Four groups of worms, which were starved for 1, 2, 3 and 4 weeks after feeding on chicken liver, were used for the experiment. One fresh whole worm or twenty pharynges were homogenized in cold phosphate buffer (pH 7.4) at 2 000 rpm for 1 to 2 min. These homogenates were centrifuged at 15 000 rpm for 10–15 minutes, and 50–100 μl of the supernatants were used as samples. By a modification of the method of Davis (1964), electrophoresis was carried out at 4 mA per gel for about one hour in a glass tube measuring 5 mm (diameter) × 70 or 100 mm. The gels were fixed in 12.5% TCA for 30 min. and stained in a mixture of 1 part of 1% Commassie brilliant blue and 15 parts of 12.5% TCA for one hour and then destained in 7% acetic acid. The optical density of the stained bands was traced at 640 nm using an OZUMOR SD-92 densitometer.

Results

A. Morphological features of Polycelis auriculata from various localities

This species exhibits polymorphism in many features such as the shape of the head, body length, body color and the number of eyespots (Fig. 1). The eyespot number among the adults of the populations of Sakunami (Miyagi Pref.) Zatoishi (Aomori Pref.) and Tashirotai (Aomori Pref.) were found to range between 91–184, 79–145 and 50–94, respectively.

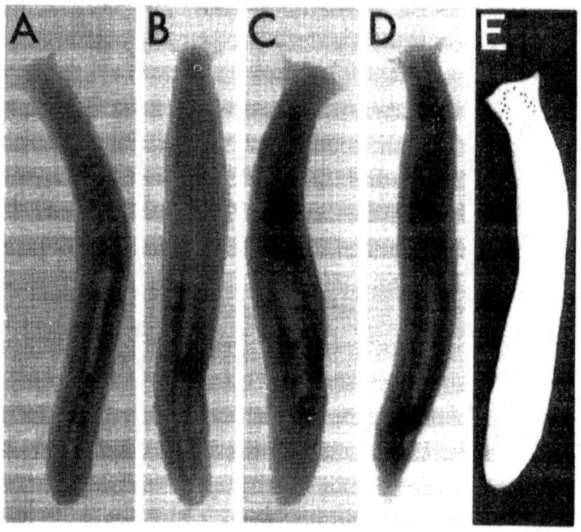

Fig. 1. Polycelis auriculata from various localities. × ca. 4. A: from Sakunami, Miyagi Prefecture; 2n = 6. B: from Zatoishi, Aomori Pref.; 2n = 12. C: from Tashirotai, Aomori Pref., 3x = 6. D: from Mt. Iwaki, Aomori Pref.; 3x = 9. E: from Mt. Iwaki; 2n = 12. Unpigmented except eyespots.

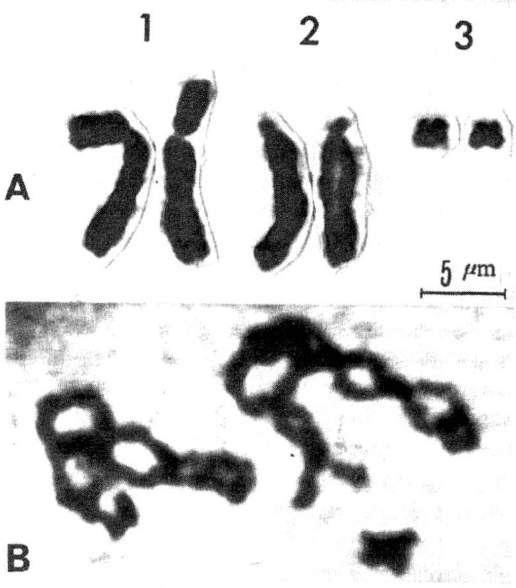

Fig. 2. Polycelis auriculata from Sakunami. A: karyotype, 2n = 6. B: three bivalents in oocyte.

B. Karyotypic variations in Polycelis auriculata

1. Specimens from Sakunami.

The chromosome number of this population was 2n = 6 and n = 3. The karyotype consisted of one pair of submetacentric chromosomes, large in size, one pair of acrocentric, medium, and one pair of metacentric, small (Fig. 2). Worms which have the same karyotype as the Sakunami population have been collected from several other localities (Fig. 12).

2. Specimens from Zatoishi.

The chromosome number of this population was 2n = 12 and n = 6. The karyotype consisted of two pairs of submetacentric chromosomes, large in size, one pair of acrocentric, medium, one pair of telocentric, medium, and two pairs of metacentric, small. In this population some of the worms were found with 2 B-chromosomes (Fig. 3). Two pairs of the large chromosomes showed very similar banding patterns both in G- and C-band staining, and

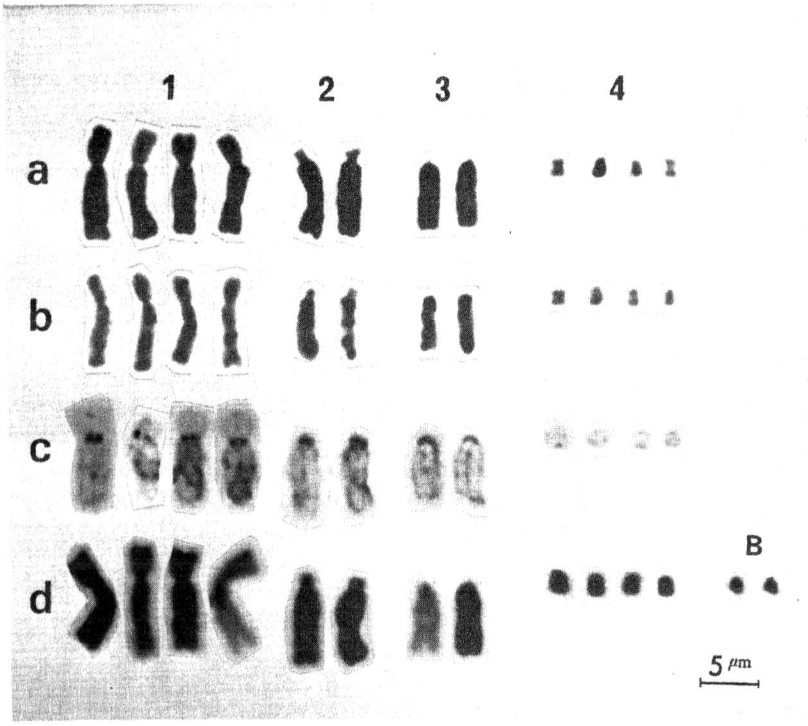

Fig. 3. Polycelis auriculata from Zatoishi. a–d: karyotypes of the metaphase plates from four different specimens. Chromosomes are arranged according to their length. a and d: Giemsa staining. b: G-banded chromosomes. c: C-banded chromosomes. B; B-chromosome.

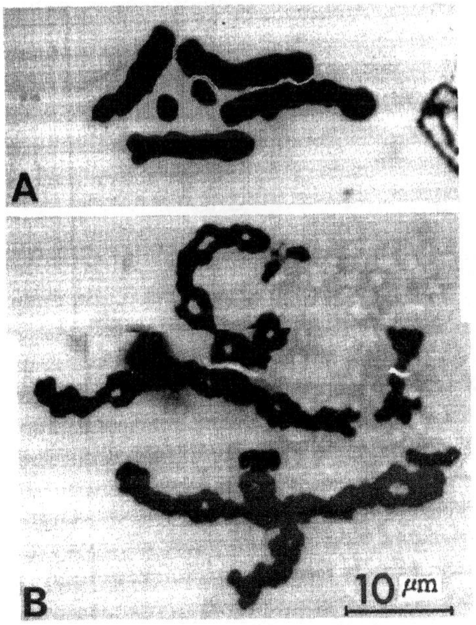

Fig. 4. Chromosomes in germ cells of *Polycelis auriculata* from Zatoishi, Giemsa staining. A: six bivalents at metaphase of primary spermatocyte. B: six bivalents at diakinesis stage of primary oocyte.

two pairs of the small chromosomes also showed similar patterns with both staining techniques (Fig. 3, b, c). Six bivalents were observed in the primary spermatocytes and primary oocytes (Fig. 4). Worms which have the same karyotypes as the Zatoishi population were also found at Mt. Iwaki, the Sasaya Pass and Lake Hibara (Fig. 12).

3. Specimens at Mt. Iwaki.

Fig. 5 shows five different types of chromosomal polymorphism in *P. auriculata* collected at Mt. Iwaki in Aomori Prefecture. In Fig. 6, these chromosomes are arranged according to their length. Fig. 6, 'a' shows the chromosomes of $3x = 9$, which consist of three submetacentric, acrocentric and metacentric chromosomes. The chromosomes shown in 'b', $2n = 10$; when compared with 'a', have two telocentric chromosomes, medium and small in size, instead of lacking one submetacentric chromosome. The chromosomes shown in 'c' ($2n = 10$), when compared with 'a', have two pairs of telocentric chromosomes, one medium, the other small in size, instead of lacking two submetacentric chromosomes, and lack one small

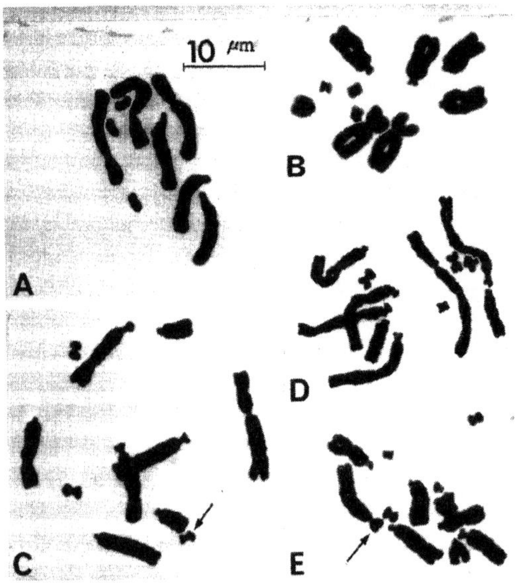

Fig. 5. Five chromosomal polymorphisms in metaphase plates of neoblasts in *P. auriculata* from Mt. Iwaki. A: 3x = 9. B: 2n = 10. C: 2n = 10 + 1 B-chromosome. Arrow indicates B-chromosome. D: 2n = 11 + 1 B-chromosome. E: 2n = 11 + 1 B-chromosome. Arrow indicates deletion of short arms of chromosome 1.

metacentric chromosome, but have 1 B-chromosome. In 'd', the karyotype (2n = 11) is very similar to 'c', but it has three small chromosomes. The karyotype in 'e' is very similar to that in 'd', but in 'e' deletion of the anterior part of the short arms of chromosome No. 1 can be recognized. The worms with this deletion lacked eyes.

4. Specimens from Tashirotai.

Most of the worms from this population, have a chromosome number of 3x = 6 (somatic cell) which consists of three submetacentric chromosomes, large in size, and three acrocentric, medium. G-banding also showed that they are triploidy (Fig. 7). In this population sexual worms were also found, but their frequency was about 3% throughout the year. In the sexual worms, there were found some which have both cells with the chromosome number 3x = 6 and 4x = 8, and also some which have both cells with the chromosome number 3x = 6 and 6x = 12. In most cases the bivalent chromosome (n) of spermatocyte was n = 2, but in some cases n = 3 or 4 was also found in addition to n = 2 in the same individuals (Fig. 8). The spermatogenesis of the Tashirotai population is still under investigation. No detailed observation has been made as yet about the chromosomes of the oocytes.

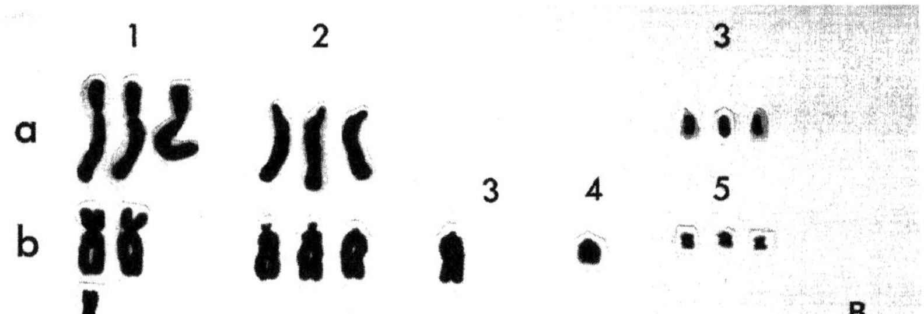

Fig. 6. Comparative analysis of various chromosomal polymorphisms in Fig. 5. Chromosomes are arranged according to length. a–e: correspond to Fig. 5, A–E, respectively. B; B-chromosome.

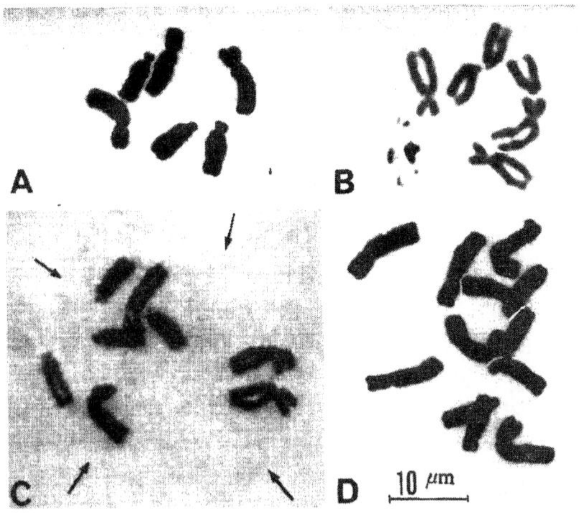

Fig. 7. Metaphase plates in neoblasts of *P. auriculata* from Tashirotai. A, C and D: Giemsa staining. B: G-banded chromosomes. A and B: 3x = 6. C: 4x = 8. Arrows indicate the border line of cytoplasm. D: 6x = 12.

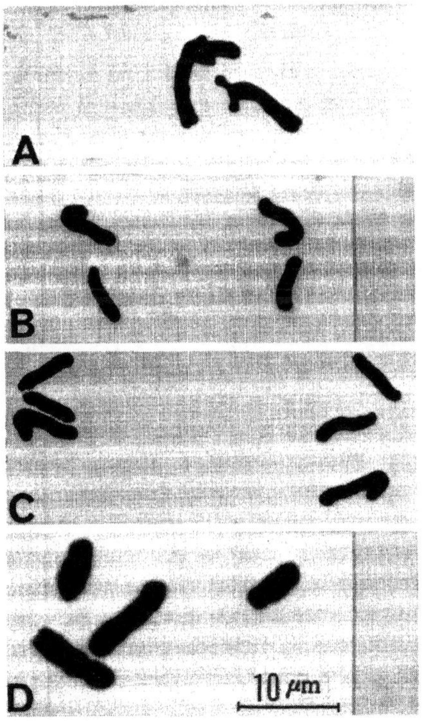

Fig. 8. Chromosomes in spermatocytes of *P. auriculata* from Tashirotai. Giemsa staining. A: two bivalent chromosomes at the metaphase stage of primary spermatocyte. B: four univalents at the anaphase stage of secondary spermatocyte. C: anaphase stage of secondary spermatocyte with six univalents, moving to both poles. D: four bivalents at the metaphase stage of primary spermatocyte.

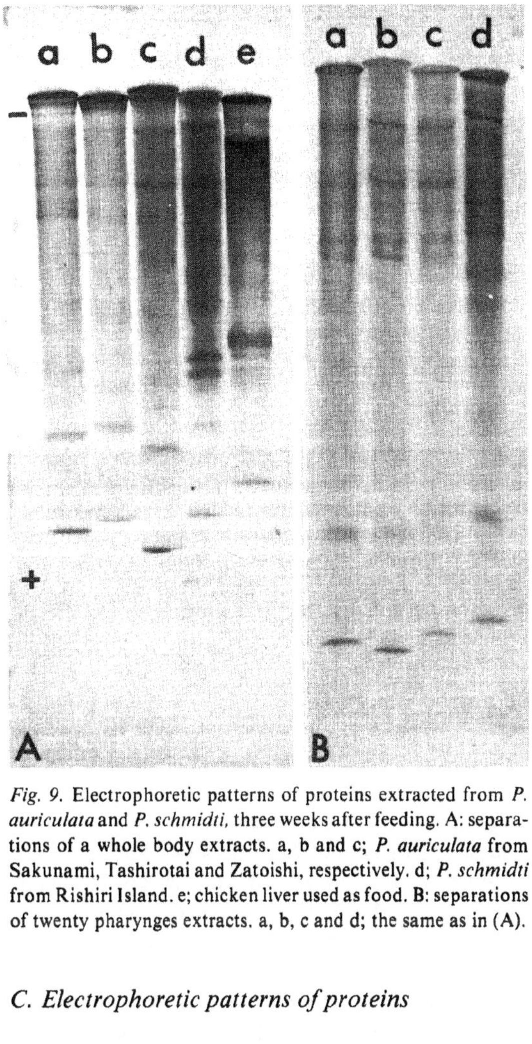

Fig. 9. Electrophoretic patterns of proteins extracted from *P. auriculata* and *P. schmidti*, three weeks after feeding. A: separations of a whole body extracts. a, b and c; *P. auriculata* from Sakunami, Tashirotai and Zatoishi, respectively. d; *P. schmidti* from Rishiri Island. e; chicken liver used as food. B: separations of twenty pharynges extracts. a, b, c and d; the same as in (A).

C. Electrophoretic patterns of proteins

Electrophoretic patterns of the proteins extracted from the karyotypically different worms which belong to the three different local populations (Sakunami, Tashirotai & Zatoishi) were observed. Fig. 9, A shows photographs of the gel electrophoretic patterns of the samples obtained from one whole worm starved for three weeks after feeding. The band patterns of the specimens from the three localities show some dissimilarities with one another. On the contrary, the band patterns of both *P. schmidti* and the chicken liver given as their food (used for comparison) were remarkably different from those of *P. auriculata* (see Fig. 10). Fig. 9, B shows the band patterns of the proteins of the pharynges from 20 worms. In this case more similarity was found among the band patterns of *P.*

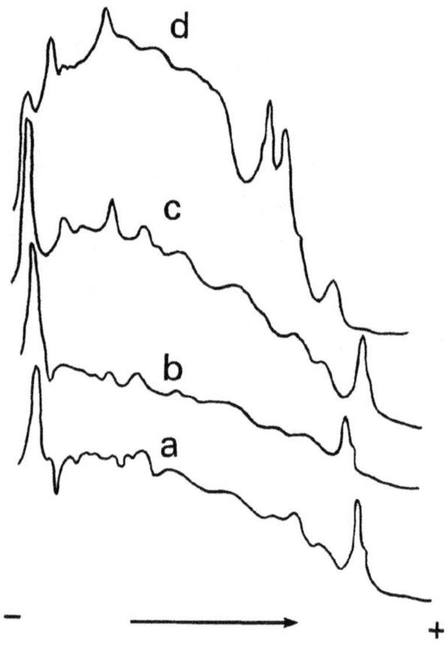

Fig. 10. Densitometric tracings of the gels shown in Fig. 9, A. a, b and c: *P. auriculata* from Sakunami, Tashirotai and Zatoishi, respectively. d: *P. schmidti.*

auriculata from the three different localities than in the case of 'A', but the band patterns of *P. schmidti* were distinguishable from those of *P. auriculata.*

Table 1 shows the electrophoretic profiles of asexual worms of *P. auriculata* and *P. schmidti.* The band patterns of *P. auriculata* from the three localities and *P. schmidti* from the Rishiri Island were compared as to their presence (1) and absence (0). The relative mobility of each band was calculated and numbered in decending order. Bands 13 and 36 were present only in the Sakunami population, but absent in the other two populations of *P. auriculata,* while bands 12 and 20 were present only in the Tashirotai population. Thus, some differences in their constitutive proteins were found among the three populations of *P. auriculata,* but a remarkable difference in the band patterns was found between *P. auriculata* and *P. schmidti.* When the similarity coefficients based on the presence or absence data of the bands were calculated by the formula of Jaccard (1908), they were 0.56 between Sakunami and Tashirotai populations, 0.64 between Tashirotai and Zatoishi, and 0.54 between Zatoishi and Sakunami. The Sakunami population showed a lower similarity value than the other two populations.

Table 1. Presence (1) or absence (0) of protein bands in homogenates of 4 populations of *Polycelis auriculata* and *P. schmidti.*

Band	Rm value	Population			
		Polycelis auriculata			*P. schmidti*
		Sakunami	Tashirotai	Zatoishi	Rishiri Island
1	0.80	1	1	1	1
2	0.71	1	1	1	1
3	0.67	0	0	0	1
4	0.64	1	1	1	1
5	0.62	0	1	1	0
6	0.60	0	0	0	1
7	0.56	1	1	0	0
8	0.55	0	0	0	1
9	0.54	1	1	1	0
10	0.51	1	1	1	1
11	0.48	0	0	0	1
12	0.44	0	1	0	1
13	0.41	1	0	0	0
14	0.39	1	1	1	0
15	0.38	0	0	0	1
16	0.32	0	1	1	0
17	0.30	1	0	1	0
18	0.29	1	1	0	0
19	0.28	0	0	0	1
20	0.26	0	1	0	0
21	0.24	1	1	0	1
22	0.23	1	0	1	0
23	0.21	1	1	1	1
24	0.20	0	1	1	0
25	0.18	1	1	1	1
26	0.16	1	1	0	1
27	0.15	0	0	0	1
28	0.14	1	1	1	0
29	0.13	0	0	0	1
30	0.12	1	0	1	0
31	0.11	1	1	1	0
32	0.10	0	0	0	1
33	0.09	0	1	1	0
34	0.07	1	1	1	0
35	0.05	0	1	1	1
36	0.04	1	0	0	0

Rm value = Relative mobility value.

Discussion

In *Polycelis auriculata,* various kinds of chromosomal polymorphisms or karyotypic variations are observed. These karyotypes can be classified into eleven categories as shown in Fig. 11, a–k. Karyotype a is $2n = 6$, karyotype b $3x = 9$, karyotype c is

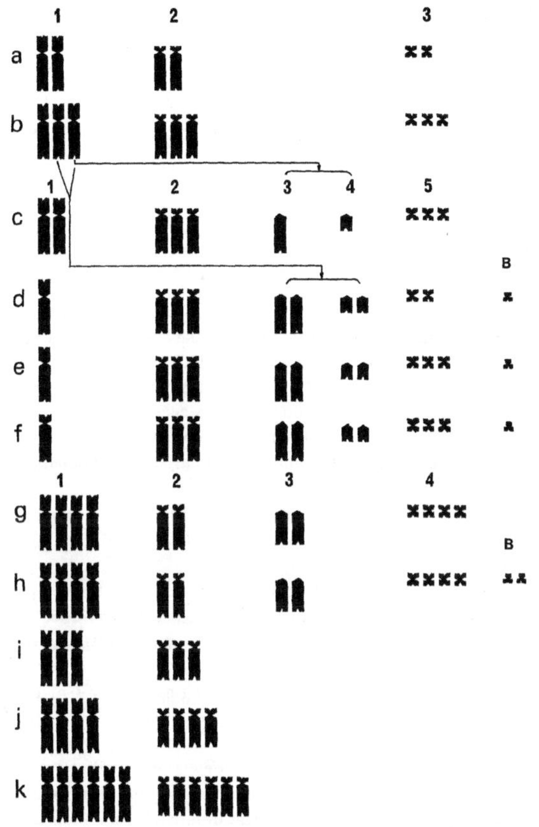

Fig. 11. Karyograms of neoblasts of *Polycelis auriculata* from various localities. a: 2n = 6. Specimens from Mt. Soranuma-dake, Mt. Hakkoda-ōdake, Sakunami and other localities (see Fig. 12). b–f: specimens from Mt. Iwaki. b: 3x = 9. c: 2n = 10. d: 2n = 10 + 1 B-chromosome, e: 2n = 11 + 1 B-chromosome. f: 2n = 11 + 1 B-chromosome. Notice the deletion of short arms of chromosome No. 1. g and h: specimens from Zatoishi, Mt. Iwaki and other localities (see Fig. 12). g: 2n = 12. h: 2n = 12 + 2 B-chromosomes. 1–k: specimens from Tashirotai. i: 3x = 6. j: 4x = 8. k: 6x = 12. B; B-chromosome.

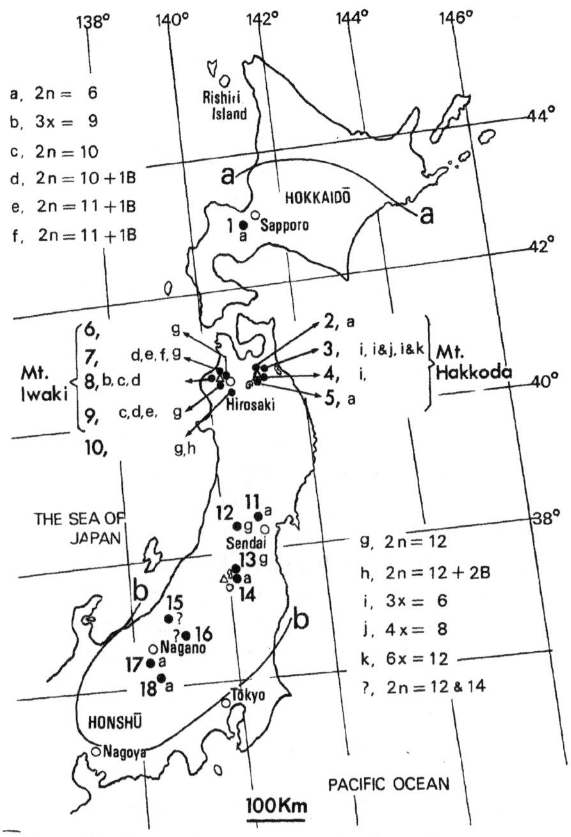

Fig. 12. Sketch map of Hokkaido and the northern half of the Main Island of Japan, showing the geographical distribution of the karyotypes of *P. auriculata* at 18 localities (1–18). 1, Mt. Soranuma-dake. 2–10, Aomori Pref. 2, Odake. 3, Tashirotai. 4, Tsutanuma. 5, Oirase-keiryu. 6, Yayoi. 7, Shakujyo-shimizu Spring. 8, half way up Mt. Iwaki. 9, Iwaki Shrine. 10, Zatoishi. 11 and 12, Miyagi Pref. 11, Sakunami. 12, Sasaya Pass. 13 and 14, Lake Hibara (Fukushima Pref.). 15–18, Nagano Pref. 15, Nozawa Spring. 16, Shiga Heights. 17, Hijiriyama. 18, Utsukushigahara. a–k, correspond to a–k in Fig. 11. ?, unknown karyotype. a–a; northern – limit of distribution of *P. auriculata*. b–b; southern – limit of distribution of *P. auriculata*. 15–18; from the data of Nîmura (1978, 1979).

2n = 10, which is considered to have been caused by centric fission of one chromosome 1 of 'b'. In the same way, karyotype d is considered to have been caused by centric fission of two of the chromosomes 1 in 'b'. In this case, one of small the chromosomes (chromosome 5) is lacking. Karyotype e is almost the same as 'd', but it has three small chromosomes. Karyotype f, though very similar to 'e' shows deletion of the anterior part of the short arms of chromosome 1. The worms with this deletion are eyeless planarians in phenotype which may be caused by the chromosome aberration. Both karyotypes g and h are 2n = 12. Arguing from both the number of chromosomes 1 and 4 (four of each) and the fact that the worms of karyotypes g and h are distributed in almost all the ranges of this species, it is assumed that they originated from tetraploidy of 'a'. In the karyotype g population some of the worms are milk-white in body color (Fig. 1, E). This seem to have been caused by gene mutation. Karyotype i, which is 3x = 6 (most of the worms from the Tashirotai population), had no small chromosome (such as chromosome 3 in 'a' and 'b', chromosome 5 in 'c'–'f' and chromosome 4 in 'g' and 'h'). Since sexual worms are also found in this population, it is considered that these small chromosomes have no relation with sexual differentiation. Both karyotype j ($4x = 8$) and k ($6x = 12$), observed in the neoblasts of the sexual worms in the Tashirotai population, are mixed with 'i' ($3x = 6$). That is, some of the sexual worms in this population are mixoploid. The chromosome numbers of germ cells in this species also differ according to the different habitats i.e., $n = 2, 3, 4$ and 6. It is not yet clear, however, whether the basic chromosome number of this species is $x = 2$ or 3, or both $x = 2$ and $x = 3$.

The distribution of this species with various karyotypes is shown in Fig. 12. The populations with karyotypes i, j and k are found only from Mt. Hakkoda. Worms with various karyotypes are also obtained from Mt. Iwaki. At Mt. Iwaki it has been found that the number of worms with chromosomal polymorphism increases according to the height of the mountain. However, relationships between the polymorphism or karyotypic variation and its geographical distribution remain to be studied further in the future. In populations with wide karyotypic variations from different localities, Kawakatsu and the present authors (1980) have found no discernible difference in the morphology of their copulatory apparatus. Nevertheless, in the populations with various karyotypes mentioned above, several differences are recognized in their morphological features and also in their electrophoretic patterns of proteins. These results suggest that this species is still in the progress of speciation or chromosomal evolution.

Karyological studies of European species of the genus *Polycelis* have been carried out on *P. nigra*, *P. tenuis* and *P. felina*. In these species, various types of polyploidy and heteroploidy, and chromosome number variations from different localities have been reported (Schleip 1907; Rappeport 1915; Melander 1950; Lepori 1949, 1955; Dahm 1958; Le Moigne 1962; Benazzi 1963; also see Benazzi & Lentati 1976). The karyotypes of *P. auriculata* show little similarity with those of the above-mentioned European species or with those of the other species of the genus *Polycelis* found in Japan (Teshirogi *et al.* 1978, 1979; Nîmura 1978, 1979).

Acknowledgment

The present study was partly supported by a Grant-in-Aid for Scientific Research from the Japanese Ministry of Education, Science and Culture (1979).

References

Benazzi, M., 1963. Il problema sistematico delle Polycelis del gruppo nigra-tenuis alla luce di ricerche citologiche e genetiche. Monit. Zool. Ital. 70–71: 288–300.

Benazzi, M. & Lentati, G. B., 1976. 'Animal Cytogenetics' Vol. 1: Platyhelminthes. Gebrüder, Borntraeger, Berlin.

Dahm, A. G., 1958. Taxonomy and ecology of five species groups in the family Planariidae. (Turbellaria Tricladida Paludicola). Nya Litografen Malmö.

Davis, B. J., 1964. Disc electrophoresis II. Method and application to human serum proteins. Ann. N.Y. Acad. Sci. 121: 404–427.

Imai, H. T., Crozien, R. H. & Taylor, R. N., 1977. Karyotype evolution in Australian ants. Chromosoma 59: 341–393.

Jaccard, P., 1908. Nouvelles recherches sur la distribution florale. Bull. Soc. Vaud. Sci. Mat. 44: 223–270.

Kato, H. & Yoshida, T. H., 1972. Banding patterns of Chinese hamster chromosomes revealed by new techniques. Chromosoma 36: 272–280.

Kawakatsu, M., 1967. On the ecology and distribution of freshwater planarians in the Japanese islands, with special reference to their vertical distribution. Bull. Fuji Women's College 5: 117–177.

Kawakatsu, M., Teshirogi, W. & Ishida, S., 1980. Morphological variation of the copulatory apparatus of a Japanese freshwater planarian, Polycelis (Seidlia) auriculata IJIMA et KABURAKI, collected from five localities in the Tôhoku Region, Honshu, Japan. Sci. Rep. Hirosaki Univ. 27: 130–146.

Le Moigne, A., 1962. Etude de formules chromosomiques de quelques Polycelis (Turbellariés Triclades) de la région parisienne. Bull. Soc. Zool. Fr. 87: 259–270.

Lepori, N. G., 1949. Ricerche sulla ovogenesi e sulla fecondazione nella planaria Polycelis nigra Ehrenberg con particolare riguardo all'ufficio del nucleo spermatico. Caryologia 1: 280–295.

Lepori, N. G., 1955. La differeziazione specifica di Polycelis nigra Ehrenberg e Polycelis tenuis Ijima (Tricladida Paludicola) e la loro distribuzione geografica in Europe. Atti Soc. Tosc. Sc. Nat. Ser. B 62: 50–71.

Melander, Y., 1950. Accessory chromosomes in animals, especially in Polycelis tenuis. Hereditas 36: 19–38.

Nîmura, F., 1978. Karyotype of a Japanese freshwater planarian, Polycelis auriculata. Shinanoji 35: 1–4. (in Japanese)

Nîmura, F., 1979. Chromosome of a Japanese freshwater planarian, Polycelis auriculata. Shinanoji 36: 4–6. (in Japanese)

Rappeport, T. von, 1915. Zur Spermatogenese der Süsswasser-Tricladen. Arch. Zellforsch. 14: 1–25.

Schleip, W., 1907. Die Samenreifung beiden Planarien. Zool. Jb. Anat. Ontog. Tiere 24: 129–174.

Teshirogi, W., Nîmura, F., Ishida, S. & Hasebe, K., 1978. Karyotypes of some freshwater planarians, and chromosomal variations in neoblasts of regenerating pieces. Zool. Mag., Tokyo, 87: 535 (in Japanese)

Teshirogi, W., Hasebe, K. & Ishida, S., 1979. Karyological studies of the freshwater planarian genus Polycelis. Zool. Mag., Tokyo, 88: 664. (in Japanese)

Teshirogi, W., Ishida, S. & Hasebe, K., 1980. Karyological studies of a Japanese freshwater planarian, Dendrocoelopsis lactea. Zool. Mag., Tokyo, 89: 41–48.

Studies on the speciation of the European freshwater planarians Polycelis nigra and Polycelis tenuis based on the analysis of enzyme variation by means of iso-electric focusing

Rob Biersma[1] and Henri J. W. Wijsman[2]
[1] *Institute of Taxonomic Zoology, and* [2] *Institute of Genetics, University of Amsterdam, The Netherlands*

Keywords: Turbellaria, Tricladida, *Polycelis*, speciation, Europe, enzyme variation, iso-electric focusing

Abstract

Polycelis nigra Ehrenberg and *Polycelis tenuis* Ijima differ morphologically and karyologically. No difference, however, was found in the isozyme pattern of malate dehydrogenase and tetrazolium oxidase, indicating a close relationship. Most sibling species differ at half of the loci.

It could be deduced that the reproductive behaviour of a single population of *Polycelis nigra* in a Dutch pond was not panmictic. Two genetically different strains retained their identities during two years of observation. If pseudogamy occurs in this diploid planarian, the presence of heterozygous specimens indicates the absence of a true meiosis.

The iso-electrofocusing technique by which these population-genetical studies were carried out, also lends itself to a comparison of overall protein banding patterns. The membrane proteins especially are conservative. The sodium dodecyl sulphate extracted proteins of *Polycelis nigra-tenuis, Planaria torva* and *Phagocata vitta* were very similar, while their water soluble proteins were not. This technique may be of great help in taxonomic studies of the higher taxa.

Introduction

Estimates of genetic divergence between closely related species support the notion of major genic changes during speciation. Even species that appear closely related by other evidence, such as morphological similarity, are often completely different in allelic composition at a great many of their loci. Populations continue to accumulate genic differences during and following speciation. Therefore, the genetic distance between semi-species, sibling species and other groups of uncertain status is of particular interest for the taxonomist.

There is an increasing number of studies on genetic similarity in vertebrates and invertebrates. In a recent comprehensive study of Thorpe (1979) the intraspecific similarity between populations typically is in the order of 0.8–1.0; the interspecific similarity between congeneric species is in the order of 0.4–0.7. Sibling species of the *Drosophila* group show a similarity in the order of 0.5–0.7 (Ayala, Tracey, Hedgecock & Richmond 1974).

The taxonomic status of *Polycelis nigra* and *P. tenuis* based on morphological, karyological and chorological studies, has not always been clear (see Benazzi 1963 for a review). Biochemical data, resulting in an estimate of the genetic distance between them, might be expected to shine some light on this question.

Several authors have reported pseudogamy as occurring in the *Polycelis nigra-tenuis* group (Benazzi 1963; Benazzi & Benazzi-Lentati 1976), based on cytological observations of the gamatogenesis of the sperm and the oocytes in polyploid populations of both species. The outcome of genetic crossing experiments by Benazzi often was obscured by the possibility of autofertilization. In some instances he observed maternal inheritance of

Hydrobiologia 84, 79–86 (1981). 0018-8158/81/0841-0079/$01.60.

the body colour. The question arises as to what extent pseudogamy or autofertilization is the rule in natural population of these species.

An alternative approach to establishing taxonomic relationships is the comparison of overall protein banding patterns. Until recently it seemed reasonable to believe that comparison of these patterns was of limited value for taxonomic studies of higher taxa. Even within a population the electrophoretically detectable variation of enzymes is in the order of 20%; interspecific variation is much higher. There was no reason to assume that this variation did not exist among other proteins, for instance the structural proteins such as membrane proteins (Johnson 1976). Indeed, comparison of the banding pattern of the water soluble proteins of two species, related on the genus level, often was not very satisfactory, owing to the difficulty of deciding exactly which bands represented homologues.

To some extent this problem has been overcome by the finding that structural proteins are less variable than enzymes (McConkey, Taylor & Phan 1979; Racine & Langley 1980). Using a strong detergent such as sodium dodecyl sulphate (SDS) it is possible to extract these structural proteins, which normally are not water soluble. Comparison of the banding patterns of this fraction with a suitable technique may prove of great help for elucidating the taxanomic relationships of higher taxa.

Materials and methods

Iso-electric focusing

A semi-micro technique depending on iso-electric focusing was developed. Iso-electric focusing (Arbuthnott & Beeley 1975) is a steady state technique in which proteins seek their iso-electric point in a linear pH-gradient under influence of an electric current. The pH-gradient is generated by a mixture of hundreds of biologically almost inert synthetic acids of different pK, the so-called Ampholines. The upper and lower limits are formed by a strong base and a strong acid, which function as electrode buffers. The supporting material of the gel is polyacrylamide.

The gel tubes here used were 130 mm long and had an inner diameter of 4 mm. Gels were prepared as follows. A solution of 5% Ampholines range 3.5–10 (LKB), 7.5% acrylamide and 0.15% bis-acrylamide (Bio-Rad), and 10% glycerol was thoroughly mixed, but with avoidance of air bubbles. The glass tubes were sealed at one side with Parafilm. Just before filling (by aid of a syringe fitted with a narrow polyethylene tube) polymerization was brought about by adding a small drop (10 microliter pro 10 ml acrylamide) of 10% ammonium persulphate (Bio-Rad) and a similar drop of N,N,N′,N′.tetraethylenediamine (TEMED). The amount of ammonium persulphate may be doubled, if the filling procedure is carried out quickly and in a cold room. The tubes were filled up to 10 mm from the top. The polymerization reaction was complete after one hour, if the gels were placed in an oven at 40 °C; at room temperature it took longer.

After polymerization the Para-film was removed and the glass tubes were placed in an appropriate pipe gel electrophoresis apparatus. The upper buffer was 0.2 M NaOH, the lower 0.1 M H_3PO_4. The positive pole is connected to the acid, the negative pole to the base. Since gels may slide out of the glass pipes, it is important to place the glass tubes very near the bottom of the apparatus so as to restrict the gels, or to block the glass tubes with some permeable inert material, such as fine nylon mesh.

Sources of specimens

Polycelis nigra was collected from Ambleteuse (near Calais, France) at three different localities: in lowland ditches, in hill streams and in dune ponds. In the hill streams they were found together with *Crenobia alpina* and *Polycelis felina.* In Amsterdam (The Netherlands) the specimens were collected from a small pond in the University Garden (Hortus Botanicus). This pond was visited regularly and supplied most of the Dutch specimens. In addition some ditches in the Watergraafsmeer and at Durgerdam (both near Amsterdam) produced some worms. In these Amsterdam localities *Polycelis nigra* was found associated with *Dendrocoelum lacteum* and *Planaria torva.* Other specimens were collected in a mountain stream in the vicinity of Rome (Italy), and in a mountain pond near Bangor, Wales (U.K.) and from Cherry

Hinton near Cambridge (U.K.).

Polycelis tenuis was found in the Hortus pond already mentioned. By December 1980, two years after these findings, the species seems to have been completely replaced by *Polycelis nigra*. A few *P. tenuis* specimens were collected in Ambleteuse (France). The third source was a laboratory stock, kept alive for over five years, the original worms having come from a small river north of Paris.

Polycelis felina and *Crenobia alpina* were collected in Ambleteuse (France), *Planaria torva* and *Dendrocoelum lactum* in the Kinselmeer and IJsselmeer near Durgerdam (Amsterdam) and *Phagocata vitta* in Sherkin Island near Baltimore (Ireland). *Dugesia lugubris* was collected in Diemen (Amsterdam) and the *Dugesia gonocephala* came from Montpellier (France). The last seven species were collected in small numbers. They were used for comparison in preliminary experiments.

Preparation of samples and enzyme staining

The weight of the specimens varied from 1 to 10 mg. A single individual was ground in four volumes of chemically purified quartz sand (Merck) in a 2 ml polyethylene 'nose' vial (Finn) with a drop of grinding buffer (0.1 ml 10% glycerol, 0.1 M Tris/HCl pH 8.5). For efficient grinding it was necessary to ensure that the rod by means of which the worm was ground fitted exactly into the conical vial. After centrifugation with a table centrifuge the supernatant was applied on top of the gel. The sample was overlayered with pure water.

Separation was complete after 6 h at 300 V but overnight operation at this voltage was also possible. Adequate cooling was accomplished by placing the focusing apparatus in an ordinary refrigerator.

About 15 enzyme stainings were tested. Malate dehydrogenase (MDH), esterases, glucose-phosphate isomerase (GPI), tetrazolium oxidase (TO) and glutamate-oxaloacetate transaminase (GOT) were particularly successful. Esterases show much variation in most animals. In the *Polycelis* species 8–12 bands appeared within five minutes, and more after prolonged staining, more or less overcrowding the gel. GOT showed little or no variation in one single band. GPI probably consists of two different isozymes, both of which are variable in the *Polycelis* species.

For practical reasons the population genetic study on *Polycelis nigra* and *P. tenuis* was centered upon MDH and TO, because these enzymes can be stained on the same gel and with the same staining procedure. This was as follows. The gel rods were carefully squeezed out of the glass tubes by aid of a syringe filled with water. They were soaked in ten volumes of 0.5 M Tris/HCl pH 8.0. After one or two minutes two volumes of 1 M malate pH 7.5 were added in which was freshly dissolved: 12 mg nicotinamide adenine dinucleotide (NAD), 15 mg nitro blue tetrazolium (NBT) and 1 mg phenazine metasulphate (PMS) pro 10 ml. All chemicals were obtained from Sigma.

The solution, as well as the gels, were shielded from the light. At room temperature a MDH banding pattern, mostly consisting of 3–6 sharp bands, appeared within half an hour. Prolonged incubation could give rise to diffuse additional bands. Illumination turned the whole gel blue, except for two bands, which represent the two TO isozymes.

Overall protein staining

For overall protein staining the procedure was adjusted in the following way. Flatworms were ground in three volumes sand and two volumes urea (Merck). Only a sall amount of water was added, just enough to dissolve all the urea. After centrifugation the supernatant, containing the water soluble proteins, was applied on a 7.5% polyacrylamide gel. The gel contained 8 M urea, 5% Ampholines, 2% Triton X-100 or NP-40 (Imperial Shell).

For extracting the membrane proteins the pellet of the last procedure was washed with 0.5 ml water and again centrifuged. The new pellet was ground with 0.1 ml 10% SDS (Sigma). After centrifugation the supernatant, containing the water insoluble proteins, was applied on top of the gel. The samples were overlayered with water.

After iso-electric focusing (300 V, 8–16 h) the proteins were fixed and the Ampholines washed out by soaking the gels in ten volumes 5% tricloroacetic acid (TCA) (Sigma) for three hours at room temperature. Ampholines behave more or less as small proteins with respect to chemical staining. The dye Coomassie Blue G-250 (BDH) turned out to be the best staining agent. Three to six hours at room temperature was enough in a 0.5% G-250 Coomassie Blue solution in 40% methanol. The dye Coomassie Brillant Blue R-250 (BDH), which is normally used for protein staining, gave bad results. Remainders of Ampholines heavily stain with this sensitive dye, resulting in overstained gels.

Destaining is accomplished by repeated incubations in 40% methanol at room temperature. Alternative methods that will be of value are described in 'LKB application note 250'.

Results

Cells of higher animals contain at least two malate dehydrogenase isozymes: one in the mitochondria and the other in the cytosol. They have a function in the basic metabolic pathway in the cell. Both isozymes are variable in *Polycelis nigra* and *P. tenuis*. Also in other flatworm species some variation was found.

The first isozyme, MDH-1, with an isoelectric point around pH 7.0 normally showed up as a strong band and a faint one; in some specimens four bands were observed, two strong ones and two faint.

The second isozyme, MDH-2, with an isoelectric point around pH 5.5 normally showed one band; in some specimens three bands were observed (Fig. 1).

In all animals of the *Polycelis nigra-tenuis* group studied there were two tetrazolium oxidase bands in the acidic region, which did not vary. TO presumably is a protein with another natural function than to oxidize tetrazolium; the artificial dye nitro blue tetrazolium, used to stain dehydrogenases, is oxidized under *in vitro* conditions. As a result the TO bands appear white against a blue background after illumination of the gel (Fig. 2). Also in other planarians two TO bands were found

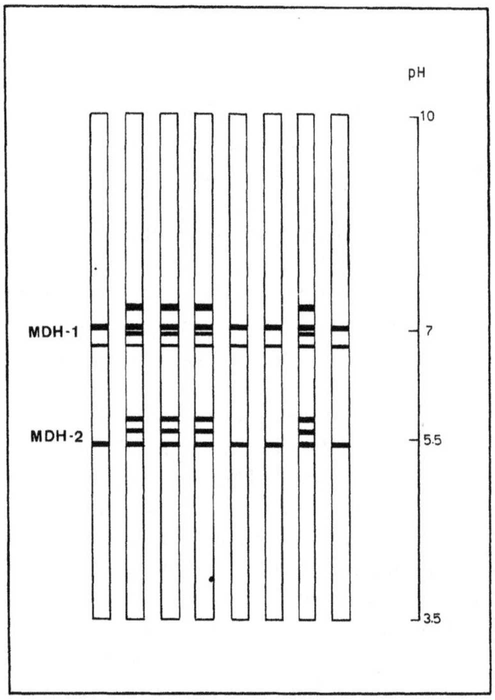

Fig. 1. Schematic drawing of iso-electric focused pipe gels of *Polycelis nigra* from an Amsterdam pond. The gels were stained for malate dehydrogenase and show the banding pattern for two isozymes, MDH-1 and MDH-2. Each gel contains the extracts of a single worm.

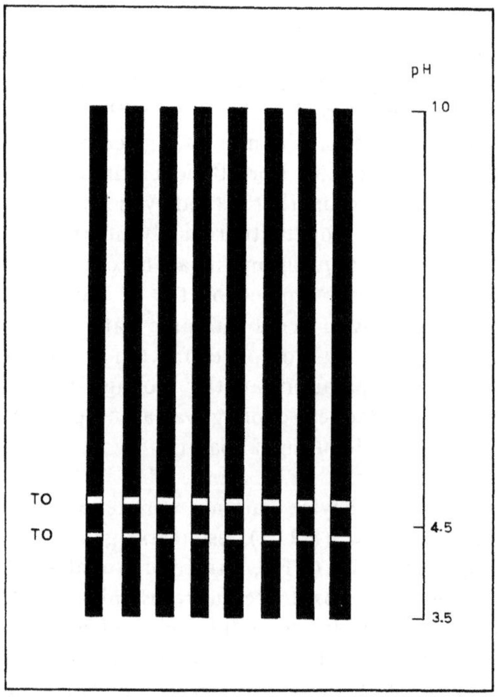

Fig. 2. Schematic drawing of the same pipe gels as in Fig. 1, but after illumination. This turns blue the nitro blue tetrazolium soaked gels, except for those regions in which tetrazolium oxidase of *Polycelis nigra* has interfered.

Table 1. Allele distribution of malate dehydrogenase isozymes in *Polycelis tenuis* and *Polycelis nigra*.

	MDH-1			MDH-2	
Bands	Upper	Middle	Lower	Upper	Lower
Polycelis tenuis					
Ambleteuse	–	–	4	–	4
Paris	–	–	20	–	20
Amsterdam	14	14	20	14	20
Polycelis nigra					
Ambleteuse	6	6	60	6	60
Amsterdam	41	41	60	41	60
Rome	–	–	25	–	25
Wales	–	–	25	–	25
Cambridge	–	–	25	–	25
Polycelis nigra-tenuis (not identified)					
Amsterdam	47	47	80	47	80

in the acidic region, albeit not exactly at the same place as in *Polycelis nigra-tenuis*.

Of particular interest is the fact that *Polycelis nigra* and *Polycelis tenuis* share the MDH bands as well as the TO bands. This is in contrast with other species, which had a completely different MDH and TO pattern. *Polycelis felina, Crenobia alpina, Planaria torva, Dendrocoelum lactum, Dugesis lugubris* and *Dugesia gonocephala* differed markedly from each other in the MDH pattern. The TO bands also differed, but being restricted to the acidic region (pH 4–6) and due to the simple structure of these bands, these differences were less distinctive.

A second interesting observation is the fact that *Polycelis tenuis* and *Polycelis nigra* showed the same allelic frequencies in the same areas. In the Hortus pond (Amsterdam) 14 of the 20 specimens of *Polycelis tenuis* showed variable bands in both MDH isozymes (70%). In the same pond *Polycelis nigra* was found: 41 out of 60 specimens showed more bands in both MDH isozymes (68%). In no other locality were *Polycelis tenuis* found that had a multiband structure. As to *Polycelis nigra,* only six out of 60 specimens found in the dune population in Ambleteuse had a multiband structure in both MDH isozymes. The rest of the populations studied had a simple banding pattern. See Table 1.

As Table 1 shows the multiband form of MDH-1 is absolutely linked with the multiband form of MDH-2; the same holds for the single band forms.

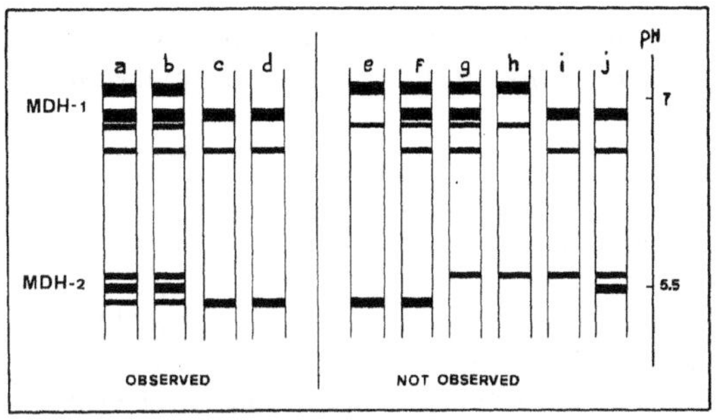

Fig. 3. The middle part of the observed and some imaginary pipe gels of *Polycelis nigra* stained for MDH. Pipes a and b are heterozygous for MDH-1 and MDH-2. MDH-1 is interpreted as a monomeric enzyme. In a diploid heterozygous animal this will give rise to two bands; the narrow bands shown always accompany the main bands and are interpreted as posttranslational products, natural or artificial. MDH-2 is a dimeric enzyme which will give rise to three bands in a diploid heterozygous animal. Pipe gels c and d show the banding pattern of animals which are homozygous for MDH-1 and MDH-2.

Never observed was the presence of the 'upper' (more alkaline) form of MDH-1 alone (e); the homozygous form of MDH-2 in combination with the heterozygous form of MDH-1 (f and g); the 'upper' (more alkaline) form of MDH-2 alone (g, h and i); or the presence of only two bands in MDH-2 (j). The absence of 3e–i is an indication that genes coding for different forms of both malate dehydrogenases do not inherit independently. The conclusion is that *Polycelis nigra* reproduces by ameiotic pseudogamy.

A second remarkable finding is the fact that the upper (more alkaline) bands in both MDH isozymes always were found in the multiband form; they never appeared alone (Fig. 3e–j).

The overall banding patterns of water soluble proteins of different flatworm species had almost no bands in common. Only the patterns of *Polycelis nigra* and *Polycelis tenuis* were very similar. Minor differences between different strains of *Polycelis nigra* were observed. It should be emphasised however that for studies on the minor differences, found in the less visible bands, the pipe gel technique is not suitable, due to non-uniform (inhomogenous) staining and destaining procedures. Some gels are more heavily stained than others, resulting in more bands. In good gels 60–80 bands are visible.

The SDS extracted proteins of *Polycelis nigra-tenuis, Planaria torva* and *Phagocata vitta* were largely similar. The SDS extracted proteins of *Dendrocoelum lactum* however, were very different. In iso-electric focusing the presence of a small amount of SDS results is a small shift of the banding pattern, due to a detoriation of the pH-gradient. This effect is restricted to the acidic region.

Discussion

The proposed interpretation of the banding pattern of malate dehydrogenase and tetrazolium oxidase is as follows. (For a review on the subject see Avise 1975.) MDH-1 is a monomeric enzyme, which is found in two forms (allozymes). A diploid organism can have two forms and so two bands. The faint bands which always accompany the strong ones are post-translational products, natural or artificial. In preliminary experiments on *Crenobia alpina, Polycelis felina* and *Planaria torva* the same phenomenon of 'shadow' bands was observed.

MDH-2 is a dimeric enzyme. A diploid organism with two different genes (alleles) for this enzyme will show three bands, the middle one twice as strong as the upper and the lower ones. An organism which is homozygous for a dimeric enzyme, however, will show only one band – the upper or the lower one.

Most populations of *Polycelis nigra-tenuis* studied consist of homozygous individuals – those with one band, but some Dutch and one French populations contained heterozygous animals as is shown in Table 1.

The two TO bands cannot be interpreted as monomeric or polymeric enzymes, because they showed no variation. In *Crenobia alpina, Polycelis felina,* and *Planaria torva* two bands were also seen, indicating that the presence of tetrazolium oxidase isozymes is usual in planarians. The fact that all these TO isozymes have a similar isoelectric point (but not exactly the same) may indicate a relationship of these planarians.

Of particular interest is the fact that *Polycelis nigra* and *Polycelis tenuis* share the MDH bands as well as the TO bands. This does not mean that both species are completely genetically identical for these four isozymes. A good deal of the amino acid replacements, caused by base pair replacements in the DNA, is not detectable with electrophoretic and iso-electrofocusing techniques (Avise 1975). But it may be concluded that *Polycelis nigra* and *Polycelis tenuis* are not only morphologically but also genetically more similar than is usual among sibling species.

In most localities *Polycelis nigra* as well as *Polycelis tenuis* were homozygous for both MDH isozymes. Moreover, both isozymes were found as the lower (more acidic) allozyme. Only in the three Amsterdam localities and in one French pond (Ambleteuse) were heterozygous animals also found. It is remarkable that the allelic frequencies of both species in these areas were almost the same. A tentative conclusion is that some sympatric *Polycelis tenuis* and *P. nigra* may still be in genetic contact. Benazzi (1963) has shown that the species were interfertile under laboratory conditions.

A peculiar feature in both species is the fact that the 'upper' (more alkaline) form of both MDH isozymes only is found in heterozygous animals and not in homozygous ones (Fig. 3e–i). This is in conflict with the Hardy-Weinberg law for populations that reproduce in a Mendelian fashion. This law states that the frequency distribution of alleles P and Q with frequencies p and q ($p + q = 1$) for PP, PQ and QQ is given by $p^2{:}2pq{:}q^2$. Evidently *Polycelis nigra* and *P. tenuis* do not form Mendelian populations. Moreover, the heterozygosity of both MDH isozymes is absolutely linked: if MDH-1 is present in three bands (two allelic forms), then MDH-2 shows two bands (also two allelic forms); similarly, a single MDH-1 band is always accompanied by a single MDH-2 band (both one allelic form) (Fig. 3a–d).

Both phenomena, the disobeyal of the Hardy-Weinberg law and the absolute linkage of two pairs of alleles, may be explained by assuming a clone-like reproduction. Clones multiply without changing the genotype. It is well known, however, that *Polycelis tenuis,* as well as *Polycelis nigra,* reproduces in nature exclusively by means of egg cocoons after mating, suggesting sexual reproduction. This paradox is overcome by the findings of Benazzi (1963; see also Benazzi & Benazzi-Lentati 1976), who has shown from karyological and cytological evidence that *Polycelis nigra* and *Polycelis tenuis* reproduce in some instance by pseudogametogenesis.

In pseudogametogenesis the sperm makes no contribution to the genotype of the progeny; it is only needed to initiate egg production. There are two types of pseudogametogenesis: one in which the maternal oocyt undergoes a meiosis, and one in which no meiosis occurs. According to Benazzi & Benazzi-Lentati (1976) the former is the case in *Polycelis* species. After meiosis the nucleus would duplicate in such a way that the egg is diploid again.

A consequence of meiotic pseudogamy is a purely homozygous progeny: in diploid animals (and all the populations in this study were diploid) the reproductive cycle goes through a haploid stage and there is no genic supplement. Evidently this is not the case here: heterozygous animals were found. According to the present findings the reproduction is ameiotic pseudogametogenesis: there is no meiosis and no haploid stage in the reproductive cycle.

A second argument for this hypothesis is the fact that the allelic forms of both MDH isozymes are absolutely linked. During the meiosis crossing-over might occur. Over many generations some individuals would show a recombination. Again, this has not been found, neither in the Amsterdam pond that was sampled regularly during two years, nor at any other locality.

There is an apparent contradiction in the indication of genic leakage between the two *Polycelis* species, which is only possible in truly sexually reproducing animals, and the hypothesis of pseudogamy. A possible explanation is that in the long term true sexual reproduction plays the important role, while over some years asyngamous reproduction dominates. It should be kept in mind that *Polycelis nigra* is found to be the same for MDH and TO over a wide area of Europe; so intraspecific divergence is not great.

The outcome of the overall protein banding patterns corroborates the taxonomic classification based on morphological characters. The water soluble proteins of *Polycelis nigra* and *P. tenuis* were almost similar, indicating a very close relationship, as might be expected. Other species differed in the water soluble proteins; even *Polycelis felina* had a very different protein banding pattern as compared to *P. nigra-tenuis*. The SDS extracted proteins, however, showed interspecific and even intergeneric similarity. The similarity of *Polycelis nigra-tenuis, Planaria torva* and *Phagocata vitta* in these experiments, combined with the dissimilarity of *Dendrocoelum lactum*, is in agreement with the taxonomic classifications based on morphological characters (Kenk 1930, 1974).

Recent findings (McConkey, Taylor & Phan 1979; Racine & Langley 1980) indicate that structural proteins, which are often found in the water insoluble fraction of the cells, are more conservative than water soluble proteins. The explanation for this phenomenon is that the evolutionary rate of free proteins is higher than that of structural proteins because of the smaller conformational limitations. A free enzyme, for instance, knows interaction with small molecules by virtue of the active site. Structural proteins are over their whole surface in close contact with neighbouring macromolecules; consequently they have less freedom to alter their amino acid sequence during evolution.

The preliminary results of the SDS extracted proteins suggest that the method might open the door to the taxonomic use of protein mapping of higher taxa.

Acknowledgements

We would like to thank Ms Elizabeth de Vries (Amsterdam), Dr. Isåbelle Martelly (Paris), and Professor T. B. Reynoldson (Bangor) for being so kind as to supply flatworms for investigation. Rob Biersma thanks Professor Ian R. Ball for his supervision of this study, which forms part of his doctoraal (M.Sc) programme in biology at the University of Amsterdam.

References

Arbuthnott, J. P. & Beeley, P. A. (Ed), 1975. Iso-electric focusing. London: Butterworths.

Avise, J. C., 1975. Systematic value of electrophoretic data. Syst. Zool. 23: 465–481.

Ayala, F. J., Tracey, M. L., Hedgecock, D. & Richmond, R. C., 1974. Genetic differentiation during the speciation process in Drosophila. Evolution, 28: 576–592.

Benazzi, M., 1963. Il problema sistematico della Polycelis del gruppo nigra-tenuis alla di recherche citologiche e genetiche. Monit. Zool. Ital. 70–71: 288–300.

Benazzi, M. & Benazzi-Lentati, G. B., 1976. Animal cytogenetics vol. 1: Platyhelmithes. Berlin: Gebrüder Borntraeger.

McConkey, E. H., Taylor, B. J. & Phan, D., 1979. Human heterozygosity: A new estimate. Proc. Natl. Acad. Sci. USA 76 (12): 6500–6504.

Johnson, G. B., 1976. Genetic polymorphism and enzyme function. In: Ed. Ayala, F. J., Molecular Evolution, Sunderland (Massachussets): Sinnauer assoc.

Kenk, R., 1930. Beitrage zum System der Probursalier (Tricladida Paludicola). Zool. Anz. 89: 289–302.

Kenk, R., 1974. Index of the genera and species of the freshwater triclads (Turbellaria) of the world. Smithsonian Contr. Zool. 183: 1–90.

LKB application note 250, 1977. Analytical electrofocusing in thin layers of polyacrylamide gels. LKB-produkter AB, Bromma (Sweden).

Racine, R. R. & Langley, C. H., 1980. Genetic heterozygosity in a natural population of Mus musculus assessed using two-dimensional electrophoresis. Nature 283: 855–857.

Thorpe, J. P., 1979. Enzyme variation and taxonomy: the estimation of sampling errors in measurement of interspecific genetic similarity. Biol. J. Linn. Soc. 11 (4): 369–386.

The ecology of the Turbellaria with special reference to the freshwater triclads

T. B. Reynoldson
Zoology Dept., Univ. College of North Wales, Bangor, Gwynedd, Wales[1]
[1] *Present address: Zoology Dept., Biological Sciences Centre, University of Alberta, Edmonton, Alberta, Canada T6G 2E9*

Keywords: Turbellaria, Tricladida, ecology, freshwater

Abstract

To attempt a complete review of turbellarian ecology in the time and space available would result in superficiality. Therefore, I have restricted this account to the four basic ecological processes which have and continue to determine flatworm distribution and abundance. These are: (1) historical or zoogeographical events which permit or prevent a species from reaching a habitat; (2) physiological limitations of the species vis à vis the habitat; (3) access to suitable energy sources and (4) the effects of competition, predation and parasitism, referred to collectively as bionomic processes.

Discussion

1 Historical events

These are seen most clearly in lake-dwelling triclads because of the semi-isolated nature of their habitat and are especially discernible in the British Isles. Both here and in Europe the impact of the last glaciation of the Pleistocene era is apparent and the classical studies of Thienemann on Continental stream species are well-known. In Britain the hypothesis which fits the facts of distribution most closely regards the four common lake species, *Polycelis nigra* (Müller), *P. tenuis* (Ijima), *Dugesia polychroa* (Schmidt) and *Dendrocoelum lacteum* (Müller), as having survived the last major glaciation in both the south of England and of Ireland (Reynoldson 1966). These species then followed the retreat of the ice slowly and have still to reach the northernmost parts of Scotland. *Polycelis nigra,* probably because of its more prevalent stream habit, has spread faster than the other three species and reached the western islands such as Anglesey, Isle of Man and Islay before they became separated from the mainland. This dispersal continues and during the last 20 years, for example, a further four lakes on Anglesey have been colonised by *P. tenuis,* mainly as a result of mans' activities. Dispersal in Ireland has been more complete, possibly due to its lesser mountain barriers and smaller size.

Within the area of their distribution ecological events of a bionomic nature are paramount in determining the distribution and abundance of these lakes species. In the British stream species the influence of historical events is no longer apparent unlike the situation on the Continent, and *Crenobia alpina* and *P. felina* live in most of the suitable habitats throughout the British Isles. They were probably less affected by the glaciation and also spread faster afterwards. They are sometimes regarded as glacial relict species.

The distribution of *Planaria torva* (Müller) in Britain is quite different and this species at present reflects historical rather than ecological events being concentrated in port areas and canals associated with the Scandinavian timber trade by which it seems to have been introduced (Reynoldson & Piearce 1979). Within these areas it is confined to highly productive habitats reflecting both historical and ecological processes, the latter by its need for

Hydrobiologia 84, 87–90 (1981). 0018-8158/81/0841-0087/$00.80.

snails. It may be expected to compete most severely with *D. polychroa* and *D. lugubris* (Schmidt).

Historical events are of lesser significance in the microturbellaria, especially the marine species, because of easier dispersal via various types of resting stage.

2 Physiological limitations

Apart from extreme conditions of temperature and oxygen concentration, lake species are not constrained by the physiological properties of natural freshwaters and differ in this respect from stream species. The ability of lake species to tolerate the natural range of chemical conditions has been demonstrated mainly by keeping them in cages in appropriate lakes. The effects of temperature on growth and reproduction are such that all lentic species are able to persist in British lakes but they show a succession of breeding peaks which may reduce to some degree the severity of interspecific competition between those species at the extremes. The nature of the substratum may also curtail distribution since triclads require stones and plants as a firm basis.

The typical stream species such as *Crenobia alpina* and *Polycelis felina* are restricted by temperature to cool, brook-type streams or high altitude/latitude lakes in Britain and Europe. Within the stream, these species are separated by different temperature optima and tolerances (Pattee 1966) in which competition for food seems to be involved (Lock & Reynoldson 1976). These two species also respond differently to stream gradient (Wright 1974; Lock 1975) with *C. alpina* being largely confined to steep gradients probably by competition from *P. felina*, particularly for *Gammarus;* the latter triclad is restricted to more shallow gradients by its lesser ability to tolerate fast flowing water. *Phagocata vitta*, a cold stenotherm, is particularly associated with ground water.

Procerodes littoralis (Maricola) is the classical species illustrating physiological limitation, being restricted to the upper seashore. The microturbellaria also demonstrate a coarse grid limitation by their separation into freshwater, marine and brackish water faunas (Karling 1974), seen also but less strikingly in triclads (Ball & Reynoldson, in press).

3 Resources

The several populations of the common species of lake triclad which have been studied in Britain (Reynoldson 1964), Canada (Boddington & Mettrick 1974) and Sweden (Herrmann 1979) are all food limited and show severe intraspecific competition. Both the total triclad populations and species spectra reflect the productivity of the lake habitat. This relation can be traced back via the three trophic levels of predator, grazer plus detritivore and primary producer to the geology of the drainage basin which in turn influences mans' agricultural activity. Much less is known of these relations in the stream triclads where increased productivity may result in physiological limits being reached. Nevertheless, Hynes (1975) has indicated in a broad way the same relation between the stream and its valley. Even less is known of the resources of the microturbellarian Orders at present.

4 Bionomic processes

These include predation, parasitism and competition. Although soft-bodied and apparently vulnerable to predation, triclads are fed upon by only a few species (Davies & Reynoldson 1971) mostly characteristic of weeded ponds such as dragonfly nymphs (Odonata) and newts, especially *Triturus helveticus*. On stony shores of lakes predation is light and chemical protection (by rhabdites, Jennings 1957) is effective. The stream species are preyed upon by stonefly nymphs and caddis larvae (Trichoptera) sufficiently to affect their distribution and abundance in some streams (Wright 1975). Parasites of lake species, mainly Sporozoa, have a negligible influence and although the ciliate *Tetrahymena* sp. causes high mortality in laboratory populations of *C. alpina* and *P. felina*, their effect on field populations is not known.

Interspecific competition is the process which exerts a major influence on distribution, and the co-existence of the common species in productive lakes is achieved by each being more efficient in feeding upon a different type of prey taxon, termed its food refuge (Reynoldson & Davies 1970). As lakes become less productive prey become both less abundant and diverse causing competition to be more severe. As a result triclad faunas become progressively reduced to a single species, usually *P. nigra* and ultimately the habitats are unable to support triclads. Triclads are not able to separate effectively on a prey size basis because of their habit of social feeding except in the case of the very large *Bdellocephala punctata* (Pallas) (Adams 1979). However, *B. punctata* has to survive a competition barrier from other species as it grows, which may account in part, for its very sporadic distribution which lacks any pattern. Spatial separation seems to play no part in reducing competition except for the possible case of young *B. punctata* but temperature-based sequential breeding of some species may contribute.

Some of the lake species are semelparous and some iteroparous. Calow & Woollhead (1977) have attributed this difference to a contrast in predator strategy. The semelparous dendrocoelid species adopt a 'sit and wait' strategy and the planariids and dugesiids a 'search out' strategy. The abundant energy saved by less locomotion is diverted to higher fecundity in *D. lacteum* and this contrast increases as food becomes scarcer. The semelparous *Planaria torva* does not fit this hypothesis, however.

Stream triclads, in contrast to lake species, reduce potential competition by spatial separation based on their different responses to temperature and stream gradient. Interspecific competition, less well established than in the lake species, also seems to be involved. Pattee's (1970) hypothesis that distribution is determined by the intrinsic rate of increase ('r') of the species at their optimal field temperature has serious drawbacks and would only seem to operate if mortality is largely density independant or if the species are competitively identical, both of which are unlikely. The 'r' values need to be determined under conditions of interspecific competition for food.

The vastly greater numbers of microturbellarian species reflects their smaller size and the consequently greater relative complexity of their habitat. They have used spatial and temporal methods of separation to reduce competition which involve responses to physical aspects of their microhabitats. For example, characteristic assemblages of species occupy different lake substrata (Kolasa 1979) and temporal contrasts in life-cycle reduce simultaneous peak demands for food (Ax 1977). However, it seems doubtful if these strategies are sufficient to account for the co-existence of so many species. Until their population dynamics and food have been studied in sufficient species the causes underlying their distribution and abundance will remain in doubt.

References

Adams, J., 1979. The population ecology of Bdellocephala punctata (Pallas), a freshwater triclad. Unpublished Ph.D. thesis, University of Liverpool.

Ax, P., 1977. Life cycles of interstitial Turbellaria from the eulittoral of the North Sea. Acta Zool. Fenn. 154: 11-20.

Ball, I. R. & Reynoldson, T. B., 1981. British Planarians. Synopses of the British Fauna (N.S.) No. 19. Cambridge University Press.

Boddington, M. T. & Mettrick, D. F., 1974. The distribution, abundance, feeding habits and population biology of the immigrant triclad, Dugesia polychroa (Platyhelminthes: Turbellaria) in Toronto Harbour, Canada. J. Anim. Ecol. 43: 681-699.

Calow, P. & Woollhead, A. S., 1977. The relationship between ration, reproductive effort and age-specific mortality in the evolution of life-history strategies - some observations on freshwater triclads. J. Anim. Ecol. 46: 765-781.

Davies, R. W. & Reynoldson, T. B., 1971. The incidence and intensity of predation on lake-dwelling triclads in the field. J. Anim. Ecol. 40: 191-214.

Herrmann, J., 1979. Population dynamics of Dendrocoelum lacteum (O. F. Müller) (Turbellaria, Tricladida) in a south Swedish lake. Arch. Hydrobiol. 85: 482-510.

Hynes, H. B. N., 1975. The stream and its valley. Verh. Int. Verein. Limnol. 19: 1-15.

Jennings, J. B., 1957. Studies on feeding, digestion and food storage in free-living flatworms (Platyhelminthes; Turbellaria). Biol. Bull. 112: 63-80.

Karling, T. G., 1974. Turbellarian fauna of the Baltic proper. Identification, ecology and biogeography. Fauna fenn. 27: 1-101.

Kolasa, J., 1979. Ecological and faunal characteristics of Turbellaria in the eutrophic lake Zbechy. Acta Hydrobiol. 21: 435-459.

Lock, M. A., 1975. An experimental study of the role of gradient and substratum in the distribution of two stream-dwelling triclads, Crenobia alpina (Dana) and Polycelis felina (Dalyell) in North Wales. Freshwat. Biol. 5: 211-226.

Lock, M. A. & Reynoldson, T. B., 1976. The role of interspecific competition in the distribution of two stream-dwelling triclads, Crenobia alpina (Dana) and Polycelis felina (Dalyell) in North Wales. J. Anim. Ecol. 45: 581-592.

Pattee, E., 1966. Coefficients thermiques et écologie de quelques planaires d'eau douce 1. Tolerance des adultes. Ann. Limnol. 2: 469-475.

Pattee, E., 1970. Coefficients thermiques et écologie de quelques planaires d'eau douce. 4. La reproduction de Dugesia gonocephala. Ann. Limnol. 6: 293-304.

Reynoldson, T. B., 1964. Evidence for intraspecific competition in field populations of triclads. J. Anim. Ecol. Jubilee Symp. Suppl. 33: 187–201.

Reynoldson, T. B., 1966. The distribution and abundance of lake-dwelling triclads - towards a hypothesis. Adv. Ecol. Res. 3: 1–71.

Reynoldson, T. B. & Davies, R. W., 1970. Food niche and co-existence in lake-dwelling triclads. J. Anim. Ecol. 39: 599–617.

Reynoldson, T. B. & Piearce, B., 1979. Predation on snails by three species of triclad and its bearing on the distribution of Planaria torva in Britain. J. Zool. Lond. 189: 459–484.

Wright, J. F., 1974. Some factors affecting the distribution of Crenobia alpina (Dana), Polycelis felina (Dalyell) and Phagocata vitta (Dugès) (Platyhelminthes) in Caernarvonshire, North Wales. Freshwat. Biol. 4: 31–59.

Wright, J. F., 1975. Observations on some predators of stream-dwelling triclads. Freshwat. Biol. 5: 41–50.

A comparative study of the food niches of lake-dwelling triclads and leeches

Johnstone O. Young
Dept. of Zoology, Univ. of Liverpool, P.O. Box 147, Liverpool L69 3BX, England

Keywords: Turbellaria, *Polycelis, Dugesia, Dendrocoelum, Erpobdella, Glossiphonia, Helobdella,* serological technique, diet, British lakes

Abstract

The food niches of four species of triclads and three species of non-parasitic leeches living in the littoral zone of British lakes of different trophic status were investigated, over one year, using a serological technique. Antisera against ten potential prey groups were employed. The basic data were adjusted to compensate for differences in predator size and seasonal changes in field temperatures. Data on two of the leech species have been presented elsewhere, but comparisons of the diet of all seven predator species are made in this paper. Much overlap in diet between the various predator species occurs, but, with one exception, each of the genera has a major food resource. The *Polycelis* spp. feed extensively on oligochaetes, *Dugesia polychroa* on molluscs, *Dendrocoelum lacteum* on *Asellus, Erpobdella octoculata* on chironomids and *Glossiphonia complanata* on molluscs. *Helobdella stagnalis* is a generalist feeder. Problems of coexistence of the leech species, and of the leech and triclad species are discussed.

Introduction

Triclads and leeches are the main predatory, invertebrate groups on the stony shores of British lakes. In Britain, the four most-common, lake-dwelling species of triclads are *Polycelis nigra* (Müller), *P. tenuis* Ijima, *Dugesia polychroa* (Schmidt, *sensu* Reynoldson & Bellamy 1970) and *Dendrocoelum lacteum* (Müller), and the three most-common, non-parasitic, lake-dwelling leech species are *Erpobdella octoculata* (L.), *Glossiphonia complanata* (L.) and *Helobdella stagnalis* (L.). All of these triclad and leech species are usually present in ion-rich, productive lakes, but with decreasing ion content *D. lacteum, D. polychroa* and *P. tenuis* disappear leaving only the one triclad and the three leech species in unproductive lakes (Reynoldson 1966; Young & Ironmonger, in press). With the exception of *P. nigra* and *E. octoculata,* the abundance of all species increases with increasing ion content and productivity; *P. nigra* is least abundant in the most productive and unproductive lakes, whilst the abundance of *E. octoculata* does not significantly change throughout the trophic range.

Triclads and glossiphoniid leeches feed on the soft-tissue and fluids of their prey, and the most satisfactory approach for determining their natural food is the use of a serological technique. Hitherto, the precipitin test has been used to investigate the general diet of the four common, lake-dwelling triclad species in a range of lakes by Reynoldson & Davies (1970), in single lakes by Pickavance (1971), Reynoldson & Sefton (1976) and Adams (1979), and to assess feeding on particular prey groups by Young *et al.* (1964), Reynoldson & Bellamy (1975) and Reynoldson & Piearce (1979a, b). Reynoldson & Davies (loc. cit.) include a discussion on the limitations of the technique. The precipitin test has also been employed by Young (1980; 1981) to provide information on the food of the glossiphoniid leeches *H. stagnalis* and *G. complanata,*

Hydrobiologia 84, 91–102 (1981). 0018-8158/81/0841-0091/$02.40.

in various lakes. In contrast to the triclads and glossiphoniid leeches, *E. octoculata* is macrophagous, and Young & Ironmonger (1979) identified its diet in British lakes by microscopic examination and identification of its gut contents. However, they suggested that soft-bodied prey such as triclads may become unrecognisable quickly as digestion proceeds, and that a serological technique should be used to check this possibility; also their technique prohibited the identification of the food of very small leeches.

The combined product of previous investigations on triclads has provided much valuable information, but in each of the studies there has been some methodological limitation. For example, in the general diet studies, a small range of prey antisera was used, and, in all studies no allowance for the influence of temperature and triclad size on the detection period of food in the gut of the triclads was made, except that of Reynoldson & Sefton (1976) who investigated the influence of two temperatures on the detection period of four prey species in the gut of *Planaria torva* (Müller), *P. tenuis*, *D. polychroa* and *D. lacteum* from Linlithgow Loch in Scotland.

In the present work, concurrent studies were carried out on the food regimes of the four triclad and three leech species in several lakes of different productivity, using a wide variety of prey antisera. The basic field data were adjusted using correction factors which allowed for differences in the titre of the prey antisera, rates of digestion of different prey organisms by each triclad or leech species, rates of digestion of the same prey by different sizes of each triclad or leech species, and rates of digestion of each prey species by each triclad or leech species at different temperatures.

Data for *H. stagnalis* and *G. complanata* have already been given in Young (1980; 1981). Here, results for the four triclad species and *E. octoculata* are presented, and compared with those obtained in previous studies. The diets of the four triclad and three leech species living in the same lakes are compared, over the same period of time.

The lakes

The lakes, with their triclad and leech faunas, are recorded in Tables 2–6. The iron-rich, productive lakes are situated in Cheshire and Shropshire, England, and the ion-poor, unproductive lakes in Gwynedd, North Wales; further details are given in Young (1973) and Young & Ironmonger (1979). Whilst the unproductive lakes support only one triclad species, *P. nigra*, the productive lakes support all four species, though *P. nigra* is absent from a few. All three leech species are found in productive and unproductive lakes, though the latter sustain only small populations of *G. complanata* and *H. stagnalis*.

In general, the productive lakes support a more diverse and more abundant standing crop of littoral organisms than the unproductive lakes, including a greater number of species of molluscs, annelids and most insect groups, with the exception of the Ephemeroptera and Plecoptera which attain greater variety and abundance in unproductive situations. Amphipoda and *Asellus* are also numerous in productive lakes whereas they are absent from unproductive localities, though in the present study *Asellus* and *Gammarus* have been recorded in small numbers from Llyn Gwynant and Llyn Nantlle respectively, but mainly from the vegetational littoral.

Methods

Field

Triclads and leeches were sampled from the stony, littoral zone at monthly (Crose Mere and Llyn-y-Gadair) or two-monthly (all other lakes) intervals over the course of one year. On collection they were subjectively categorised into small, medium or large individuals: *P. nigra* and *P. tenuis* – small, ⩽4 mm body-length; medium, 5–7 mm; large, ⩾8 mm; *D. polychroa* – small, ⩽5 mm; medium, 6–8 mm; large, ⩾9 mm; *D. lacteum* – small, ⩽7 mm; medium, 8–13 mm; large, ⩾14 mm; *E. octoculata* – productive lakes, small, ⩽59 mg body-weight; medium, 60–119 mg; large, ⩾120 mg; unproductive lakes, small, ⩽59 mg; medium + large, ⩾60 mg. Details of the three weight categories for *H. stagnalis* and *G. complanata* are given in Young (1980; 1981). After size-categorisation, specimens were placed individually into small glass vials, which were immersed immediately in frozen 'Frezella' packs or dry ice, for transport back to the laboratory where they were stored at –20 °C.

Laboratory

Full details of the preparation of the various prey antisera (see Table 2) are given in Young (1980). The titre and specificity of each of the prepared antisera were satisfactory (Tables 1 & 2 in Young, loc. cit.). Correction factors (see Introduction) were derived by separating each triclad or leech species into the size-classes outlined above and allowing numerous starved specimens to feed on an excess of a particular food item, at each of the temperatures 5, 10 and 16 °C. After feeding, the predator species was tested for positive reactions against the appropriate prey antisera at 6-hourly intervals until a negative result was obtained. The procedure was repeated for each of the prey groups considered. *D. lacteum* and *E. octoculata* would not feed readily on molluscs in the laboratory nor would *E. octoculata* on triclads, and so, on the few occasions when such prey were recorded in the gut of these predators, an average prey correction factor, appropriate for the collection temperature and predator size, was used.

The numerous data on the detection times of the various prey groups in the gut of each of the triclad species cannot be presented here because of space limitations, but the general trends obtained for *H. stagnalis* and *G. complanata* (see Young, 1980) were again observed. Thus, for each predator species, food is detected longer in the gut of large specimens and at the lower temperatures. Detection periods in the gut of any one predator species varied according to prey type. The few published data on detection periods of prey in the gut of triclads agree with the present findings.

Field temperatures were recorded during the course of the study in Crose Mere (productive) and Llyn-y-Gadair (unproductive) and are regarded as being representative of the other productive and unproductive lakes, since lakes within each of the two trophic groupings lie at a similar altitude and within the same geographical area. According to time of collection of the samples during the year, compensation for temperature was made by using the correction factor for either 5, 10 or 16 °C (see Young, 1980 for exact details).

Miniature Ouchterlony plates were used for the precipitin tests (Pickavance 1970). The gut contents of the triclads or leeches macerated in 0.9% saline were placed in the central well and a different prey antisera in each of the eight peripheral wells. Tests before and during the investigation indicated that the inclusion of a control peripheral well of 0.9% saline was unnecessary. The anti-*Asellus* and anti-Amphipoda sera were omitted from the complement of antisera used, when testing material from unproductive lakes since each of these prey groups were found in only one of the lakes sampled, and in small numbers only. When testing triclad material from all of the lakes, the anti-Tricladida serum was omitted because the predator itself would react to the antiserum. The anti-Plecoptera serum was not included when testing triclad and leech material from the productive lakes since this group is not well-represented, in terms of species and abundance, in such lakes, and well-space on the test

Table 1. Summary of the basic field data indicating the total number of animals of each triclad or leech species tested from productive (P) and unproductive (U) lakes, the numbers and percentages which gave single and multiple reactions, and the total numbers and percentages of positive reactions.

		No. tested	No. with single reaction	% with single reaction $\frac{b}{a} \times 100$	No. with double reactions	No. with triple reactions	% with multiple reactions $\frac{d+e}{a} \times 100$	Total positive reactions $b + (d \times 2) + (e \times 3)$	% positive reactions $\frac{g}{a} \times 100$
		(a)	(b)	(c)	(d)	(e)	(f)	(g)	(h)
P. nigra	P	275	169	61.5	28	0	10.2	225	81.8
	U	545	351	64.4	34	0	6.2	419	76.9
P. tenuis	P	664	434	65.4	45	3	6.8	533	80.3
D. polychroa	P	619	406	65.6	63	4	10.2	544	87.9
D. lacteum	P	512	347	67.8	46	5	10.0	454	88.7
E. octoculata	P	678	359	52.9	49	0	7.2	457	67.4
	U	690	359	52.0	42	0	6.1	434	62.9

plates was limited. For similar reasons the Anti-Ephemeroptera serum was omitted when testing *E. octoculata* from productive lakes. However, to compensate for such omissions, separate collections of all four triclad and three leech species were made at two-monthly intervals during the one-year study-period, and the triclads were tested against the anti-Plecoptera serum and the leeches against both the anti-Plecoptera serum and the anti-Ephemeroptera serum (see Table 7), since Young (1980; 1981) did not investigate feeding on stoneflies and mayflies by *H. stagnalis* and *G. complanata* in the productive lakes. In the use of each antiserum the same protocol was followed whereby the antiserum was placed in the central well and the macerated gut contents of each of eight triclads or leeches in the peripheral wells.

Results

Basic field data

The basic field results obtained for each of the triclad or leech species in the productive and unproductive lake groupings are shown in Table 1. For each of the species *P. nigra* and *E. octoculata*, analysis of the data using the χ^2 test indicated that the relative proportions of total positive reactions, single reactions and multiple reactions, where a single predator had fed on two or more types of prey, did not differ significantly ($P > 0.05$) between the productive and unproductive lake groupings. The percentage of total positive reactions was similar for each of the four species of triclads, but lower for *E. octoculata*. The same was true for the percentage of single reactions, but the percentage of multiple reactions was similar for all five species. In the unproductive lake grouping, the percentage of total positives for *E. octoculata* was similar to those recorded for *H. stagnalis* and *G. complanata* (see Young, 1980; 1981), but in the productive lakes the relative proportion of positives was higher ($P < 0.01$) for *G. complanata*, but not for *H. stagnalis*. The relative proportions of total positive reactions for both *G. complanata* and *H. stagnalis* differed ($P < 0.01$) from those for *D. polychroa* and *D. lacteum* but not for the *Polycelis* species. The relative proportions of multiple reactions did not differ significantly between the various triclad or leech species.

The value of inter-species comparisons using the basic data is doubtful, because detection times of the various prey items and at different temperatures

Table 2. The food of *Polycelis nigra* in various productive and unproductive lakes. For each prey group, the number of positive reactions is expressed as a percentage of the total number of reactions for all prey.

Lakes	No. tested	% positive reactions	% prey								
			O	M	C	As	Am	Ch	E	P	Tr
Prod.											
Crose Mere	141	83.0	65	9	0	11	4	6	0	–	5
Blake Mere	48	83.3	62	12	0	8	7	7	0	–	4
Newton Mere	86	79.1	68	7	1	10	6	4	1	–	3
Range			62–68	7–12	0–1	8–11	4–7	4–7	0–1	–	3–5
Unprod.											
Gadair	180	73.3	64	4	1	–	–	10	8	6	7
Cwellyn	77	71.4	58	8	0	–	–	6	12	10	6
Dinas	94	78.7	50	3	0	–	–	8	13	15	11
Gwynant	71	90.1	60	3	0	–	–	7	10	14	6
Mymbyr	48	68.8	56	0	1	–	–	5	16	20	2
Nantlle	75	78.7	63	8	0	–	–	8	5	2	14
Range			50–64	0–8	0–1	–	–	5–10	5–16	2–20	2–14

O = anti-Oligochaete serum; M = anti-Mollusca serum; C = anti-Cladocera serum; As = anti-*Asellus* serum; Am = anti-Amphipoda serum; Ch = anti-Chironomidae serum; E = anti-Ephemeroptera serum; P = anti-Plecoptera serum; Tr = anti-Trichoptera serum.

varied between species, and therefore Tables 2–7 and Figure 1 have been compiled using adjusted positive precipitin reactions so that results can be compared.

Composition of diet

The diets of the four triclad species and of *E. octoculata* in the various lakes investigated are indicated in Tables 2–7. Leaving aside the data for Cladocera and Tricladida for all lakes and the Ephemeroptera data for the productive lakes, because of the very low number of positive reactions obtained, χ^2 analysis of the data for each triclad or leech species indicated a significant difference in the proportions of the various prey categories eaten, in each of the lakes (Tables 2–6).

Oligochaetes were the main prey of *P. nigra* in all of the productive and unproductive lakes studied (Table 2). Substantial feeding on ephemeropteran and plecopteran nymphs was recorded in the unproductive lakes (Table 2), but not in the productive lakes (Tables 2 & 7). Within each of the trophic lake groupings, similar proportions of chironomids and trichopterans were consumed. The proportion of molluscs eaten was rather less in the unproductive than in the productive lakes, reflecting availability, though higher values were recorded from Llyn Cwellyn and Llyn Nantlle where the standing crop of gastropods is somewhat higher than in the other unproductive lakes included in the investigation (Young, unpublished data). It is pertinent to mention that the anti-Mollusca serum reacted against gastropods and bivalves, but its sensitivity to the latter group was lower (see Young, 1980). Cladocerans were poorly represented in the diet of *P. nigra* in all lakes.

The food eaten by *P. tenuis* in the productive lakes is shown in Table 3. Oligochaetes is the main food of this species, followed by *Asellus*. Molluscs were eaten to a limited extent and chironomids and trichopterans to a lesser degree. There is little evidence of feeding on plecopterans in the productive lakes (Table 7).

In all productive lakes, *D. polychroa* fed to the greatest extent on molluscs followed by oligochaetes (Table 4). Substantial feeding on *Asellus* was also recorded, whilst amphipods and chironomids were consumed in similar but lower proportions. Limited feeding also occurred on trichopterans, but feeding on ephemeropterans, cladocerans and plecopterans was minimal (Tables 4 & 7).

Asellus predominated in the diet of *D. lacteum* in all productive lakes investigated (Table 5). Oligochaetes followed by emphipods were eaten to a lesser extent in all lakes with the exception of Hatch Mere, where the reverse situation was found. The proportions of chironomids and trichopterans consumed were low. Little evidence for feeding on molluscs, cladocerans, ephemeropterans and plecopterans was found (Tables 5 & 7).

In both productive and unproductive lakes, chironomids formed the major food of *E. octoculata*, whilst oligochaetes also constituted a substantial part of the diet (Table 6). In unproductive lakes, Cladocera also featured in the food regime of this leech species, though in terms of ingested prey biomass, this small prey organism must form a relatively small part of the overall diet. In all lakes, feeding by *E. octoculata* on triclads, molluscs,

Table 3. The food of *Polycelis tenuis* in various productive lakes. For each prey group, the number of positive reactions is expressed as a percentage of the total number of reactions for all prey. Prey abbreviations as for Table 2.

Lakes	No. tested	% positive reactions	% prey							
Prod.			O	M	C	As	Am	Ch	E	Tr
Crose Mere	288	81.3	55	8	0	20	8	4	0	5
Blake Mere	60	86.7	67	6	0	13	4	6	0	4
Cole Mere	74	79.7	56	7	0	22	8	4	0	3
The Mere	61	80.3	60	10	0	14	4	8	2	2
Hatch Mere	54	79.6	60	5	0	16	7	5	2	5
Newton Mere	62	80.6	54	16	0	14	6	6	0	4
White Mere	65	70.8	63	9	2	14	4	4	0	4
Range			54–67	5–16	0–2	13–22	4–8	4–8	0–2	2–5

Table 4. The food of *Dugesia polychroa* in various productive lakes. For each prey group, the number of positive reactions is expressed as a percentage of the total number of reactions for all prey. Prey abbreviations as for Table 3.

Lakes Prod.	No. tested	% positive reactions	% prey O	M	C	As	Am	Ch	E	Tr
Crose Mere	244	89.8	20	53	0	14	6	5	0	2
Blake Mere	49	89.8	20	57	0	9	5	7	0	2
Cole Mere	73	91.2	23	56	0	11	6	3	0	1
The Mere	62	90.3	16	59	0	12	4	5	0	4
Hatch Mere	54	70.4	16	55	0	13	8	5	0	3
Newton Mere	59	86.4	21	61	0	6	4	6	0	2
White Mere	78	88.5	22	52	0	18	2	4	1	1
Range			16–23	52–61	0	6–18	2–8	3–7	0	1–4

Table 5. The food of *Dendrocoelum lacteum* in various productive lakes. For each prey group, the number of positive reactions is expressed as a percentage of the total number of reactions for all prey. Prey abbreviations as for Table 2.

Lakes Prod.	No. tested	% positive reactions	% prey O	M	C	As	Am	Ch	E	Tr
Crose Mere	136	91.9	16	0	0	68	10	4	0	2
Blake Mere	44	97.7	25	0	0	53	15	7	0	0
Cole Mere	85	85.9	20	0	0	54	18	5	0	3
The Mere	115	92.2	15	1	0	60	10	6	0	8
Hatch Mere	68	75.0	12	0	0	60	22	4	0	2
Newton Mere	64	87.5	24	1	0	58	13	3	0	1
Range			12–25	0–1	0	53–68	10–22	3–7	0	1–8

cladocerans, trichopterans, ephemeropterans and plecopterans was slight (Tables 6 & 7).

Finally, Table 7 indicates that there was little feeding on stoneflies and mayflies by *H. stagnalis* and *G. complanata* in the productive lakes. Other details on the diet of these leeches in productive and unproductive lakes are given in Young (1980; 1981), and are summarised here in Fig. 1.

Temporal changes in feeding

For each species, within each of the two trophic lake groupings, the data obtained from the various lakes were combined because seasonal information for individual lakes was sparse; six sets of data obtained in February, April, June, August, October and December were analysed. The lumping of the data is considered justifiable because within each lake grouping all of the prey taxa are represented, though abundances vary from lake to lake, and the overall diet for each species, it has been seen above, is similar for each of the lakes. χ^2 analysis of the data for each predator species indicated that in each month significant differences occurred in the utilization of prey categories. In analysis of the data from productive lakes Cladocera, Ephemeroptera and Plecoptera were omitted for all predator species, and, additionally, triclad data for *E. octoculata,* because so few positive reactions were obtained. In analysis of the data from unproductive lakes, information on Cladocera was not included for *P. nigra,* nor triclad data for *E. octoculata.*

Scrutiny of the two-monthly data suggested that temporal changes might occur in the proportions of the various prey categories eaten. However, for each species, when the association among sets of monthly data on the proportion of each prey group utilised was determined using the Kendall coefficient of concordance W, values indicated that this was not so, suggesting diet does not vary within a species with time. Young (1980; 1981) found that this was also true for *H. stagnalis* and *G. complanata.*

Overall feeding activity was lowest for all species

Table 6. The food of *Erpobdella octoculata* in various productive and unproductive lakes. For each prey group, the number of positive reactions is expressed as a percentage of the total number of reactions for all prey.

Lakes	No. tested	% positive reactions	T	O	M	C	As	Am	Ch	E	P	Tr
Prod.												
Crose Mere	205	66.3	2	21	1	1	2	1	70	–	–	1
Blake Mere	69	69.6	2	23	0	2	0	0	73	–	–	0
Cole Mere	87	67.8	4	19	0	1	2	1	72	–	–	1
The Mere	68	67.6	2	22	2	2	2	0	66	–	–	2
Hatch Mere	81	65.4	2	13	0	2	2	2	79	–	–	0
Newton Mere	65	67.7	2	18	0	2	2	0	76	–	–	0
White Mere	103	67.0	1	19	0	1	2	2	75	–	–	0
Range			1–4	13–23	0–2	1–2	0–2	0–2	66–79	–	–	0–2
Unprod.												
Gadair	215	66.0	1	22	0	14	–	–	60	1	2	0
Cwellyn	105	64.8	1	24	0	23	–	–	51	1	0	0
Dinas	89	61.8	0	16	0	13	–	–	67	2	2	0
Gwynant	92	62.0	2	21	0	16	–	–	57	0	2	2
Mymbyr	49	65.3	0	13	0	6	–	–	81	0	0	0
Nantlle	78	62.8	3	25	0	5	–	–	60	0	0	7
Ogwen	62	64.5	0	0	0	2	–	–	93	0	5	0
Range			0–3	0–25	0	2–23	–	–	51–93	0–2	0–5	0–7

T = anti-Tricladida serum. Other prey abbreviations as for Table 2.

Table 7. The number of triclads or leeches tested from various productive lakes, and the percentage which gave a positive reaction against the anti-Plecoptera and anti-Ephemeroptera sera.

Predator Prey		Lakes Crose Mere	Blake Mere	Cole Mere	The Mere	Hatch Mere	Newton Mere	White Mere	Total
P. nigra	No. tested	122	44	–	–	–	70	–	236
Plecoptera	% positive	1	0				0		0.4
P. tenuis	No. tested	201	40	60	53	41	49	47	491
Plecoptera	% positive	0	0	0	0	0	2	0	0.4
D. polychroa	No. tested	167	39	68	60	33	38	64	469
Plecoptera	% positive	0	0	0	0	0	3	0	0.2
D. lacteum	No. tested	62	23	45	79	43	38	–	290
Plecoptera	% positive	0	0	0	0	0	3	–	0.3
E. octoculata	No. tested	59	35	43	31	40	32	30	270
Plecoptera	% positive	0	0	0	0	0	0	0	0
H. stagnalis	No. tested	101	61	70	80	56	143	61	572
Plecoptera	% positive	0	0	1	1	0	0	0	0.3
G. complanata	No. tested	99	67	67	62	59	88	41	483
Plecoptera	% positive	1	0	0	0	0	1	0	0.4
E. octoculata	No. tested	64	40	46	41	34	30	20	275
Ephemeroptera	% positive	0	0	0	0	0	0	0	0
H. stagnalis	No. tested	71	33	47	79	47	63	50	390
Ephemeroptera	% positive	0	0	2	0	0	0	0	0.3
G. complanata	No. tested	66	29	31	59	64	90	38	377
Ephemeroptera	% positive	0	0	0	0	0	1	0	0.3

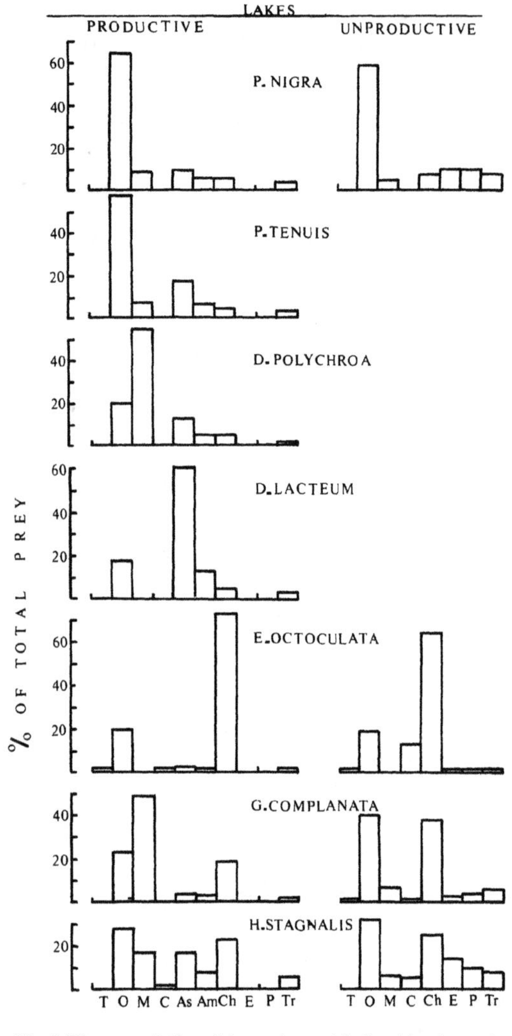

Fig. 1. The overall diet of the various triclad and leech species in productive and unproductive lakes. For each of the species the food attributable to each prey group is expressed as a percentage of total positive tests. Prey abbreviations as for Tables 2 & 6.
Data for *H. stagnalis* and *G. complanata* from Young (1980) & Young (1981) respectively.

in the winter months reflecting either reduced metabolic rates a consequence of low temperatures, or reduced prey availability. With regard to *E. octoculata,* Young & Ironmonger (1979) found a slight trough in the general level of feeding activity in late summer/autumn, and again this was evident from the present data using a serological technique. Graphical analysis of the triclad data suggested a lower feeding activity in August compared with June or October for each of the species. Young (1980; 1981) found decreased feeding activity in August and September for *H. stagnalis,* and in april for *G. complanata.*

Feeding by different size groups

Again data for the individual lakes, within each of the lake groupings, were combined for analysis. For each predator species, analysis of the year's data for all productive lakes and for all unproductive lakes, using the Kolmogorov-Smirnov one-sample test, indicated no difference in utilization of prey within any particular prey group by each size-class of predator; similar analysis of the six sets of monthly data indicated no significant differences in size-class utilization within any sampling date. For each of the species, for each of the size classes, χ^2 analysis on combined monthly samples revealed significant differences in the utilization of the various food groups in the productive and in the unproductive lakes. However, inter-weight class analysis revealed no significant difference in the proportions of the various food categories consumed by each of the size classes for both the productive lake grouping and the unproductive lake grouping. These results are similar to those observed for *H. stagnalis* and *G. complanata* by Young (1980; 1981).

The present findings for *E. octoculata* are at variance with those recorded by Young & Ironmonger (1979). They found some intraspecific resources partitioning, but that this was not consistent from lake (or group of lakes) to lake. It is conceivable that the categorisation of leeches into fewer weight classes in the present study may have masked intraspecific differences.

Discussion

Comparison with previous results

Comparison of the present results with previous findings is not straightforward. Small differences in the proportion of particular food organisms utilised may be a consequence of the omission of correction factors in earlier work and/or to the increased range of prey antisera used in the present investigation. However, evident similarities and contrasts can be highlighted.

Reynoldson & Davies (1970) and Reynoldson & Bellamy (1975) investigated the diet of triclads in a number of lakes. Though Reynoldson & Davies (loc. cit.) used only four prey antisera, viz. an anti-Oligochaeta serum, an anti-Gastropoda serum, an anti-*Asellus* serum and an anti-Amphipoda serum, their results are confirmed by the present findings on the main prey resource of each of the triclad species. Also, there is agreement with Reynoldson & Bellamy's (loc. cit.) findings from their studies on the extent of predation on Ephemeroptera and Plecoptera by the *Polycelis* species, that feeding on these prey by *P. nigra* is greater in unproductive lakes than in productive lakes, reflecting availability. However, in contrast to their results, there was no evidence that the triclad fed more on stonefly than on mayfly nymphs. They also produced evidence for some predation by *P. tenuis* on mayflies and stoneflies in three productive localities, 2.7 and 10% respectively of animals tested, but in the present study these proportions were smaller being only 0.6 and 0.4%.

Pickavance (1971), Reynoldson & Sefton (1976) and Adams (1979) included information on the diets of *P. tenuis*, *D. polychroa* and *D. lacteum* in Cole Mere, Linlithgow Loch and Bardowie Loch respectively. Because the first two studies used an anti-Tubificidae serum, the identifiable oligochaete resource was narrowed and its proportion in the overall diet was under-represented. Further, in the last two studies, the anti-Mollusca serum reacted strongly to snails but only weakly to lamellibranchs. Pickavance (loc. cit.) used an anti-*Potamopyrgus jenkensi* serum which reacted only against this snail species.

In all three investigations molluscs were the main diet of *D. polychroa*. However, amphipods and gastropods comprised the highest proportion of the overall diet of *P. tenuis* in the studies by Pickavance (loc. cit.) and Reynoldson & Sefton (loc. cit.) respectively. These results are undoubtedly a consequence of the use of an anti-Tubificidae serum (see above) which resulted in the under-representation of oligochaetes and the magnification of other prey groups. Certainly, in both studies a higher proportion of oligochaetes was recorded for *P. tenuis* than for the other triclad species investigated. Pickavance (loc. cit.) and Reynoldson & Sefton (loc. cit.) found that amphipods, followed by *Asellus*, and *Asellus* respectively were the main food resource of *D. lacteum*. Strangely, Adams (loc. cit.) found molluscs to be the main food of both *P. tenuis* and *D. lacteum* in Loch Bardowie, though only a small number of specimens of the former species were tested. He attributed these results partly to the unusual food conditions at Loch Bardowie, when samples of triclads were taken in the autumn. It is of interest that Reynoldson & Piearce (1979b), in their main study on predation of snails by *Dugesia polychroa*, *D. lugubris* (Schmidt) and *Planaria torva* (Müller), also tested *D. lacteum* for feeding on snails in Cole Mere and Linlithgow Loch and generally found a low proportion of positive reactions to snails in this triclad species; however, in one sample of *D. lacteum* substantial feeding on snails was recorded. Reynoldson & Piearce (loc. cit.) explained this anomalous result by suggesting that trampling and stone-moving activities of freshwater ecologists three hours prior to the collection of triclads for serological investigation may have damaged snails, so stimulating triclads to feed on them. Because Loch Bardowie is used for recreation and trampling of its shore must occur, it is interesting to speculate if this same phenomenon might have contributed to the observed extensive feeding on snails by these triclads. The three investigations of Pickavance (loc. cit.), Reynoldson & Sefton (loc. cit.) and Adams (loc. cit.) also extended the repertoire of prey antisera used by Reynoldson & Davies (1970), by the inclusion of an anti-Chironomidae serum (all three investigations), an anti-Cladocera serum (Pickavance), and an anti-Trichoptera serum (Adams). In these studies these prey groups formed only a minor resource. Exceptionally, Reynoldson & Sefton (1976) found a quarter of the total food intake of *P. tenuis* attributable to chironomid larvae, but again this figure, based on proportions, would be artificially exaggerated due to the detection of only tubificids in the oligochaete resource.

The present findings on the food regime of *E. octoculata* agree with the earlier observations of Young & Tronmonger (1979), based on microscopic examination of the gut contents. Because only a few reactions to the anti-Tricladida serum were obtained in their study, they suggested that these soft-bodied animals might become quickly unrecognisable as digestion proceeds. Certainly, Davies & Reynoldson (1971) investigating predation on triclads in ponds and lakes found that 10.7% of *E. octoculata* collected contained planarian remains. However, again in the present study there was little evidence of feeding on flatworms, though the percentage of positives against triclads was slightly higher than in the original study by Young & Ironmonger (loc. cit.).

Comparison of diets of triclads and leeches

It has been established that the diet of each of the triclad or leech species did not change significantly during the course of a year, and that it was consistent between size-groups. The diets of the various species are displayed in Fig. 1. Much overlap in diet is evident, and oligochaetes in particular are fed upon to a substantial degree by all the triclad and leech species. Most species, however, appear to have a 'food refuge' (*sensu* Reynoldson 1966), that is a prey group which is exploited more by a particular species than by any of the other species. From this viewpoint, again it is emphasised, that the present data on triclads agree with those of Reynoldson & Davies (1970) who used only four antisera; thus in productive lakes, the *Polycelis* spp. exploit oligochaete worms to a greater degree than the other species, *D. polychroa*, molluscs and *D. lacteum*, *Asellus*. With regard to leeches in the productive lakes, *E. octoculata* has a 'food refuge' in chironomids, *G. complanata* feeds extensively on molluscs, but *H. stagnalis* does not appear to have a 'food refuge', feeding on a variety of prey and appearing to be a generalist rather than a specialist (*sensu* MacArthur - see, for example, MacArthur & Levins 1967). In view of the wide range of antisera used which covered the common, potential prey organisms, it is unlikely that the failure to detect a 'food refuge' for this species would have been due to the omission of an important antiserum.

Further comment is required on the situations where the *Polycelis* species share their major food resource, oligochaetes, and *D. polychroa* and *G. complanata* share molluscs. Currently, Prof. T. B. Reynoldson and colleagues (private commun.) are preparing a manuscript which explains the distribution of the two species of *Polycelis* in terms of partitioning of food and different predator strategies. With regard to *Dugesia* and *Glossiphonia*, it could be that each had a specific niche within the molluscs. Each consumed oligochaetes to the same relative extent, but the relative proportion of *Asellus* and amphipods eaten was greater for *Dugesia*, whilst the reverse was true for chironomids. It is of interest that Reynoldson & Piearce (1979a) produced evidence for interspecific competition for snails in the field between both *P. nigra* and *P. tenuis*, and *D. polychroa*, which is only present in productive habitats; in the absence of *D. polychroa*, *Polycelis* returned a higher proportion of positive reactions against snails, than in its presence. It was argued that this reduction in positive reactions was not due to competition with leeches in the more productive habitats because feeding on snails by *Polycelis* in one eutrophic lake, Llyn Coron, was relatively high despite an abundant leech fauna, particularly *G. complanata*. A more detailed investigation of feeding behaviour on molluscs by *D. polychroa* and *G. complanata* is currently in progress by the present author.

In the unproductive lakes, again oligochaetes were the major food resource of *P. nigra*, just as in the productive lakes. It is of interest that the combined Ephemeroptera and Plecoptera resource in the unproductive lakes was proportionately similar to the combined *Asellus* and Amphipoda resource in the productive lakes. Chironomids again comprised the 'food refuge' of *E. octoculata* in the unproductive lakes. *G. complanata* in unproductive lakes had fed relatively more on oligochaetes and chironomids than in the productive lakes, perhaps a consequence of the reduced variety and standing crop of molluscs in the former localities. To a lesser extent the same was true for *H. stagnalis*. It is of interest that for both of these leech species, *Asellus* and amphipods together constituted a similar proportion of the overall diet as did the ephemeropterans and plecopterans in the unproductive lakes (see Fig. 1). These proportions were relatively smaller for *G. complanata* than for *H. stagnalis*.

Further comments have to be made on the observed proportions of reactions to chironomids and trichopterans. It is evident from Fig. 1 that the former were fed on to a greater extent by leeches than the triclads. Chironomids have a tough cuticle and it could be that the proboscis of glossiphoniid leeches can penetrate this more easily than the plicate pharynx of triclads; *E. octoculata* is a macrophagous leech and ingests the larvae whole.

Trichopterans were consistently fed on to a small extent by all species in both productive and unproductive lakes.

Single species studies on the population biology of each of the four species of triclads revealed periods of acute food shortage and numbers are regulated by intra-specific competition for food Reynoldson 1964; Reynoldson & Young 1965). Thus, since populations are regulated by a density-dependent food relationship, then interspecific competition for food is likely to play a decisive role in the distribution and abundance of the four triclad species. Indeed, Reynoldson & Davies (1970) and Reynoldson (1975) have produced evidence that this is the case. Co-existence of the three triclad genera is possible only if the specific 'food refuge' of each species is present in the lake. The basis of co-existence of the *Polycelis* spp. is more complicated, and will be explained in terms of food partitioning and triclad predator strategy, as stated earlier.

Unfortunately, detailed studies on the population dynamics of leech species in lakes have not been made, and the processes controlling and regulating their populations are not known. If all the species are maintained below the carrying capacity of food resources by whatever mechanism, for example, by predation on them or by density independent mortality processes, then co-existence would be possible without any partitioning of food resources. However, it would appear, whatever the situation, that *E. octoculata* and the glossiphoniid species would not compete severely by virtue of the former feeding much more on chironomids. It is possible, however, that this feeding could be genetic in its basis and not forced on *Erpobdella* by competition with the glossiphoniids. Until more is known about the population dynamics and food preferences of the two glossiphoniid species, little can be said about the possibility of competition between them.

Finally, comments must be made on the problems of co-existence of the leech and triclad species. It is clear that competition will occur between the glossiphoniids and the triclads, and that this will affect the triclads seriously. However, until knowledge on how these glossiphoniid populations are regulated is available, the degree to which competition will act on them cannot be assessed.

Acknowledgements

The author is grateful to the Natural Environment Research Council for a research grant which supported this work. He is indebted to Dr. J. Adams and Mr. J. W. Ironmonger for assistance in the production of many of the antisera, to the owners of the lakes from which samples of triclads and leeches were taken, and to Prof. T. B. Reynoldson for valuable comments on a draft of the manuscript.

References

Adams, J., 1979. The population ecology of *Bdellocephala punctata* (Pallas), a freshwater triclad. Unpublished Ph.D. thesis, Univ. of Liverpool.

Davies, R. W. & Reynoldson, T. B., 1971. The incidence and intensity of predation on lake-dwelling triclads in the field. J. Anim. Ecol. 40: 191–214.

MacArthur, R. H. & Levins, R., 1967. The limiting similarity, convergence, and divergence of coexisting species. Am. Nat. 100: 377–385.

Pickavance, J. R., 1970. A new approach to the immunological analysis of invertebrate diets. J. Anim. Ecol. 39: 715–724.

Pickavance, J. R., 1971. The diet of the immigrant planarian Dugesia tigrina Girard. II. Food in the wild and comparison with some British species. J. Anim. Ecol. 40: 637–650.

Reynoldson, T. B., 1964. Evidence for intra-specific competition in field populations of triclads. J. Anim. Ecol., Jubilee Symp. Suppl. 33: 187–201.

Reynoldson, T. B., 1966. The distribution and abundance of lake-dwelling triclads – towards a hypothesis. Adv. Ecol. Res. 3: 1–71.

Reynoldson, T. B., 1975. Food overlap of lake-dwelling triclads in the field. J. Anim. Ecol. 44: 245–250.

Reynoldson, T. B. & Bellamy, L. S., 1970. The status of Dugesia lugubris and D. polychroa (Turbellaria, Tricladida) in Britain. J. Zool., Lond 162: 157–177.

Reynoldson, T. B. & Bellamy, P., 1975. Triclads (Turbellaria: Tricladida) as predators of lake-dwelling stonefly and mayfly nymphs. Freshwater Biol. 5: 305–312.

Reynoldson, T. B. & Davies, R. W., 1970. Food niche and co-existence in lake-dwelling triclads. J. Anim. Ecol. 39: 599–617.

Reynoldson, T. B. & Piearce, B., 1979a. Feeding on gastropods by lake-dwelling Polycelis in the absence and presence of Dugesia polychroa (Turbellaria, Tricladida). Freshwater Biol. 9: 357–367.

Reynoldson, T. B. & Piearce, B., 1979b. Predation on snails by three species of triclad and its bearing on the distribution of Planaria torva in Britain. J. Zool., London 189: 459–484.

Reynoldson, T. B. & Sefton, A. D., 1976. The food of Planaria torva (Müller) (Turbellaria-Tricladida), a laboratory and field study. Freshwater Biol. 6: 383–393.

Reynoldson, T. B. & Young, J. O., 1965. Food supply as a factor regulating population size in freshwater triclads. Mitt. Int. Ver. Limnol. 13: 3–20.

Young, J. O., 1973. The occurrence of Microturbellaria in some British lakes of diverse chemical content. Arch. Hydrobiol. 72: 202–224.

Young, J. O., 1980. A serological investigation of the diet of Helobdella stagnalis (L.) (Hirudinea: Glossiphoniidae) in British lakes. J. Zool., London 192: 467–488.

Young, J. O., 1981. A serological study of the diet of British, lake-dwelling Glossiphonia complanata (L.) (Hirudinea: Glossiphoniidae). J. Nat. Hist. 15: 475–489.

Young, J. O. & Ironmonger, J. W., 1979. The natural diet of Erpobdella octoculata (L.) (Hirudinea: Erpobdellidae) in British lakes. Arch. Hydrobiol. 87: 483–503.

Young, J. O. & Ironmonger, J. W., in press. A quantitative study of the comparative distribution of non-parasitic leeches and triclads in the stony littoral of British lakes. Int. Revue ges Hydrobiol.

Young, J. O., Morris, I. G. & Reynoldson, T. B., 1964. A serological study of Asellus in the diet of lake-dwelling triclads. Arch. Hydrobiol. 60: 366–373.

Oxygen availability, redox and the distribution of some Turbellaria Schizorhynchidae and other forms

Patrick J. S. Boaden
Queen's Univ. of Belfast, Marine Biology Station, Portaferry BT22 1PF, Northern Ireland

Keywords: Turbellaria, distribution, oxygen availability, redox, phylogeny

Abstract

Records from a considerable number of meiofaunal samples are reexamined in relation to the depth within sediment distribution of various taxa especially Schizorhynchidae and the main turbellarian orders.

Frequency and density records in samples with particular oxygen flow rates and redox potentials confirm that the ranges of *Schizochilus, Proschizorhynchus* and *Neoschizorhynchus* spp. within these factors may help account for congeneric occurrence within cores. However, statistical analysis of the records is seldom able to confirm apparent differences. There is, for example, a statistically significant difference in distribution in relation to redox potential between the red schizorhynchians *Pseudoschizorhynchides ruber, Diascorhynchus rubrus* and the red acoelan *Paratomella rubra* but not in relation to oxygen availability and depth within sediment. On one beach *Gnathostomula* aff. *paradoxa* was found at Eh and O_2 availability values under which it must be assumed to be living anaerobically.

Contingency and other analyses are extended to differences between higher taxa and the results discussed in relation to habitat evolution of the Turbellaria. The suggestion that structural complexity of turbellarians has increased with increased oxygenation of the environment is tentatively supported.

Introduction

In recent years there have been a number of studies concerned with the vertical distribution of meiofauna within beach and inshore sediments. The most detailed study remains that by Fenchel (1969) relating the vertical distribution of ciliates to gradients in various abiotic and biotic factors; oxygen, oxygen rich substances (such as dissolved nitrate and phosphate and oxygen related parameters were shown to be particularly important. The vertical distribution of small metazoans in relation to sediment oxygenation has been discussed by Jansson (1967), Giere (1973), Ott & Schiemer (1973), Wieser *et al.* (1974) & McLachlan (1978). Only the first and last of these papers specifically mention Turbellaria, McLachlan demonstrating significant positive regressions of the total turbellarian population with oxygen concentration and redox potential.

More detailed information on the vertical distribution of turbellarian taxa within sand may be found in various other papers. For example, short term changes are detailed by Boaden (1968), Rieger & Ott (1971) and Boaden & Platt (1973). Seasonality of distribution, in particular life cycles in relation to temperature (Ax 1977), has been studied on the Island of Sylt by the research groups led by Professor Ax (e.g. Schmidt 1969; Ehlers 1973; Hoxhold 1974). The paper in this series by Faubel (1976) contains some statistical analysis of acoel and macrostomid distribution.

Fenchel & Riedl (1970) and Boaden (1975, 1977) have linked the proportional representation of taxa at different depths in the sediment in relation to the redox potential discontinuity layer and the under-

Hydrobiologia 84, 103–112 (1981). 0018–8158/81/0842–0103/$02.00.

lying anaerobic sand with the evolutionary history of the Metazoa. Riedl (1963) related the abundance of turbellarian orders in marine habitats with the systematic and ecological history of the class.

Various controversies have surrounded the relationships within the Turbellaria and within the Metazoa; the primitive or derived nature of the Acoela is often an integral part of such arguments (see Hyman 1951, 1959). Similar controversies may surround the primitive or derived nature of metazoan anaerobiosis (Boaden 1977, 1980; Reise & Ax 1979, 1980). It is therefore worthwhile to consider the vertical distribution of turbellarians within sand with particular attention both to oxygenation and to evolutionary relationships such as those deduced by Karling (1974).

Material and methods

The results come from analyses of over 150 sediment cores taken during May 1973 – August 1974 from six Northern Ireland beaches. The majority were from just below mid-tide level at Ballymaconell, Bangor (54°40′N 5°38′W), further details may be found in Maguire (1977). Cores were of 3.2 cm diameter and usually taken to a depth of 30 cm, at least three replicates for faunal analysis were taken on each occasion. Cores were immediately subdivided into 5 cm depth sections. Extraction of fauna in the laboratory was after 10 min narcotization in 7% $MgCl_2$ followed by shaking and decantation through 62 μm pore diameter plankton netting; this process was then repeated with a 1 min 10% ethanol wash, two seawater washes and a freshwater wash giving an extraction efficiency nearing 100% (Gray & Rieger 1971).

Field redox potential and oxygen availability were measured using a probe with six platinum wire electrodes spaced at 5 cm intervals. Readings were taken immediately before and adjacent to the core samples. Other than the multiple nature of the probe, proceedure was as in Hulings & Gray (1971).

The results of the faunal extraction of each three replicate cores are added together in some of the analyses because some of the surface fauna was so common that surface 0–5 cm material was bulked (in this case using six rather than three replicates) and subsampled. This means some analyses are made from 51 samplings and, because of some missing values, from 47 Eh and 48 O_2 availability profiles.

Results

Table 1 presents mean depth, redox and oxygen availability values for a number of selected taxa. These are means for each subcore in which the taxa occurred regardless of their abundance. These are referred to later as frequency means. The values in parentheses are means from each record for each individual and later referred to as density means. The units used are 1–6 for the depths of 0–5, 5–10 etc. cm into the sediment, corrected millivolt readings for Eh all of which means are positive and direct meter readings for oxygen availability. The latter should be multiplied by 0.7 to give μ moles of oxygen per cm^2 of electrode surface per minute. The standard differences and number of observations are given in each case. Species A could not be identified with any certainity. It is characterized by a fairly simple organisation (at about the catenulid – nemertodermatid level) and an apparently revolving statocyst; this species is included in analyses of the total Turbellaria but not of the orders.

One way analysis of variance of the density means has been carried out within each group of species containing kalyptorhynchid turbellarians shown in Table 1. Only the following are shown to be statistically significant differences: 'oxygen' means in *Schizochilus*, Eh means in the red species and all Eh means between the five orders (all at $p < 0.001$). The oxygen availability mean of Acoela is significantly different from the other orders but only at $p < 0.05$.

The mean density weighted depths of the five orders have also been calculated. These are Proseriata 2.3., Rhabdocoela 2.0, Macrostomida 1.8, Catenulida 3.1 and Acoela 3.9 units respectively. These all differ significantly at $p < 0.001$.

Contingency table analysis has been carried out for the various species listed in Table 1 to determine whether abundance in the core sections was related to redox or oxygen availability levels. This was the case for five species in relation to Eh, namely *Schizochilus marcusi* Boaden, *Neoschizorhynchus parvorostro* Ax & Heller, *Cicerina remanei* Karling, *Paratomella rubra* Rieger & Ott

Table 1. Comparative vertical distribution and habitat oxygenation of selected turbellarian taxa. Means, standard differences and number of observations from frequency records. Means in parentheses from density records. Depth in 5 cm units, Eh in mv, oxygen availability in 0.7 μ moles cm^{-2} min^{-1} units. For further explanation see text.

	Depth			Eh			O_2		
	x	s.d.	n	x	s.d.	n	x	s.d.	n
Schizochilus choriurus	2.0 (1.7)	0.8	26	252 (267)	181	22	6.9 (6.0)	6.7	23
S. marcusi	1.4 (1.4)	0.6	13	311 (311)	119	12	11.5 (11.5)	12.4	13
Proschizorhynchus bivaginatus	2.3 (2.3)	1.1	15	268 (254)	175	14	7.5 (7.3)	4.3	12
P. triductibus	1.8 (1.7)	1.1	9	321 (301)	94	8	6.3 (7.2)	3.6	8
Neoschizorhynchus brevipharynggus	1.7 (1.4)	0.7	7	341 (318)	144	7	6.8 (6.2)	3.0	7
N. longipharynggus	1.8 (1.6)	0.6	10	253 (293)	101	9	7.7 (6.8)	7.0	8
N. parvorostro	2.8 (2.7)	1.1	29	271 (293)	155	24	7.8 (7.9)	5.4	25
Carcharodorhynchus isolatus	2.3 (2.2)	0.9	18	278 (278)	222	13	5.7 (5.7)	4.0	15
Cicerina remanei	1.3 (1.1)	0.6	20	250 (266)	160	16	10.6 (11.2)	11.7	18
Pseudoschizorhynchoides ruber	2.4 (2.0)	0.9	30	229 (180)	160	26	8.8 (5.9)	9.4	27
Diascorhynchus rubrus	1.9 (1.7)	0.9	30	246 (284)	155	26	8.1 (6.4)	9.1	25
Paratomella rubra	2.8 (3.0)	1.4	30	262 (278)	165	26	6.2 (5.2)	4.6	26
Species A	3.6 (3.6)	1.4	52	228 (203)	177	47	6.2 (5.7)	5.0	44
Retronectes terpsichore	3.8 (3.3)	1.4	12	258 (243)	165	12	4.5 (8.8)	5.5	12
Schizorhynchidae	2.2 (1.9)	1.0	197	274 (244)	148	167	7.8 (7.0)	7.6	175
Schizorhynchia	2.2 (2.0)	1.1	245	272 (250)	152	207	7.7 (6.4)	7.4	215
Kalyptorhynchia	2.1 (2.0)	1.1	293	271 (254)	154	245	7.7 (6.7)	7.6	258
Proseriata	2.8 (2.3)	1.5	138	244 (203)	172	116	7.7 (9.3)	7.4	111
Rhabdocoela	2.1 (2.0)	1.2	382	262 (261)	153	321	7.7 (8.1)	7.8	331
Macrostomida	2.1 (1.8)	1.3	43	248 (274)	157	38	7.5 (8.0)	9.8	37
Catenulida	3.2 (3.1)	1.6	42	255 (245)	178	38	6.1 (7.1)	5.2	41
Acoela	3.2 (3.9)	1.5	138	228 (223)	147	118	7.7 (6.5)	7.3	116

and *Species A*, also, if zero occurrences are excluded from the tables, *Retronectes terpsichore* Sterrer & Rieger. These all showed a positive relationship with the exception of Species A which was denser at median Eh values. There is a significant relationship of four species with oxygen availability values, namely *Proschizorhynchus bivaginatus* Schilke, *N. parvorostro*, *P. rubra* and *P. terpsichore*. The latter showing decreased density at median levels, the others a positive relationship with increasing availability. The relationship of oxygen availability and *Pseudoschizorhynchoides ruber* Schilke density is also close to significance.

The frequency of occurrence of individual species at densities greater than one per core section was very low except in the 0–5 cm section and hence did not merit contingency analysis. This was however conducted for higher taxa. Results for this and for Eh and O_2 availability are shown in Table 2. Density categories were adjusted according to the taxon and ranged from 0, 1–2, 3–4, 4–5 and >5 per core section for the catenulids to 0, 1–15, 16–30 etc. for the Turbellaria *in toto*. The first figure for table size gives the number of density categories used, the second the number of factor categories. The latter for depth was always six namely 0–5, 5–10, 10–15, 15–20, 20–25 and 25–30 cm; for Eh divisions were made at <100, 101–200, 201–300, 301–400 and >400 mv; similarly O_2 availability divisions were made at 0–2.0, 2.1–4.0, 4.1–6.0 and >6.0 units.

There is a highly significant decrease in density with depth except in the Catenulida and Gnathostomulida, where depth is not significant, and in the Acoela which are denser in the middle region of the cores. Where considered statistically significant ($p(\chi^2) > 0.90$) Eh and O_2 availability is positively linked with increased density except in the Catenulida where density increases in median

Table 2. Contingency table analysis to determine whether the density of selected turbellarian taxa, Ganthostomulida and Gastrotricha varied with depth in sediment, redox potential and oxygen availability.
For further explanation see text.

	Depth		Eh		O_2	
	Size	χ^2	Size	χ^2	Size	χ^2
Schizorhynchidae	4.6	142****	4.5	13n.s.	4.6	25**
Schizorhynchia	4.6	184****	4.5	15n.s.	4.6	23*
Kalyptorhynchia	4.6	215****	4.5	20*	4.6	16n.s.
Proseriata	4.6	90****	4.5	22**	4.6	23*
Rhabdocoela	4.6	228****	4.5	23**	4.6	17n.s.
Macrostomida	4.6	91****	4.5	19*	4.6	16n.s.
Catenulida	4.6	18n.s.	4.5	28***	4.6	27**
Acoela	4.6	50****	4.5	10n.s.	4.6	24*
Turbellaria	4.6	161****	4.5	10n.s.	4.6	19n.s.
Gnathostomulida	3.6	8n.s.	3.5	25***	3.6	8n.s.
Gastrotricha	4.6	85****	4.5	20*	4.6	24*

* $p(\chi^2) > 0.9$;
** >0.95;
*** >0.99;
**** 0.999.

values of Eh, but decreased in median availability, and in the Acoela where density tends to increase with decreased oxygen availability.

Interpretation of the results through comparison of means and contingency tables tends to obscure some interesting features. For example in Table 1 the mean density derived Eh values for *Neoschizorhynchus longipharynggus* Schilke and *N. parvorostro* are both 293 mv. The former species occurred in 14 of the core sections for which Eh values were vailable, the latter in 51. There were eight cores in which both species occurred. In five of these *N. parvorostro* occurred in sections with lower Eh values than sections in which *N. longipharynggus* occurred; this situation was reversed in only one core. Thus although the mean values are identical there is good evidence that *N. parvorostro* tends to occur at lower Eh levels than its congener. Similarly in all eight cores *N. parvorostro* extended deeper into the sediment than the other species; in four of these cases the *N. longipharynggus* distribution was contained within the vertical range of *N. parvorostro* and in the other four *N. longipharynggus* was found higher in the sediment than the upper limit of *N. parvorostro*. Such results for these and other species pairs are given in Table 3. An interesting feature of such analyses is that distribution according to redox and to oxygen availability may show contrary trends.

In examining the records of depth, Eh and O_2 availability for each core the range of values in which each turbellarian order occurred can be noted. The taxa did not often extend throughout any given core. For example the Proseriata, Rhabdocoela and Macrostomida may have only been found in the upper 15 cm but both the Catenulida and Acoela be found down to 30 cm. In such a case the catenulids and acoels may be said to have a joint lowest occurrence (L). If only the Acoela had extended to the bottom of the core this may be said to be an exclusive lowest occurrence (EL) of the acoels. Lowest and exclusive lowest occurrences may also be determined with respect to redox and oxygen availability levels. These results are given in Table 4.

The % frequency of occurrence of the higher taxa at different depths, redox and oxygen availability values in the cores is listed in Table 5. Regressions of frequency against factor value have been calculated from these. Significant regressions were obtained for all the taxa except the Acoela in relation to depth but for none with either redox or O_2 values. However if Eh values in excess of 400 mv are excluded from the redox calculations, on the

Table 3. Comparative range of selected turbellarian species from cores in which they concurr. + higher, - lower occurrence in core or factor range, 0 species contained within range of other species. For further explanation see text.

	D	Eh	O_2	D	Eh	O_2	D	Eh	O_2	D	Eh	O_2
	Schizochilus choriurus vs			S. marcusi			Proschiz. bivag.			vs P. triductibus		
+	1	1	5	3	1	0	0	1	2	2	2	1
0	0	1	0	4	5	4	1	0	0	2	0	0
-	5	4	1	0	0	2	3	2	1	1	1	2
	N. brevipharynggus			vs N. longipharynggus			N. parvorostro			vs N. brevipharynggus		
+	2	2	2	1	0	1	0	1	2	2	1	1
0	0	0	0	2	2	2	1	2	1	2	2	2
-	1	1	1	0	1	0	3	1	1	0	1	1
	N. parvorostro			vs N. longipharynggus			Pseudo. ruber			vs Diasco. rubrus		
+	0	1	5	4	3	1	1	1	3	4	4	0
0	0	2	1	4	4	4	6	9	9	7	7	6
-	8	5	2	0	1	3	5	2	0	1	1	6
	Pseudo. ruber			vs Paratomella rubra			Diasco. rubrus			vs Paratomella rubra		
+	2	1	2	2	2	0	3	1	1	1	1	2
0	7	7	7	5	4	3	4	5	7	3	4	2
-	0	1	0	2	3	6	1	2	0	4	3	4

basis that some may have been due to electrode poisoning as suspected from redox profiles, significant regressions are obtained for the Macrostomida ($p < 0.001$) and the Acoela ($p < 0.01$). The significant regressions all show decrease in frequency with depth or increase with increased Eh.

The significant regressions of frequency against depth means that there is a *theoretical* depth below which there should be no occurrence of the taxon concerned. This lies at 30.8 cm for the Kalyptorhynchia compared with 69.5 for all Turbellaria. The theoretical order of disappearance with depth is

Table 4. The % of cores in which turbellarian orders occurred deepest or at the lowest redox or oxygen availability level inhabited either (L) with other orders or (EL) exclusively by that order, n, number of recorded cores inhabited.

	Depth			Eh			O_2		
	n	L	EL	n	L	EL	n	L	EL
Proseriata	45	47	16	44	41	2	42	57	5
Rhabdocoela	51	24	4	47	47	0	48	58	8
Macrostomida	23	17	0	22	36	0	22	46	0
Catenulida	28	25	4	28	36	0	28	39	4
Acoela	46	76	48	44	84	23	43	88	9

Table 5. % frequency of occurrence of turbellarian taxa in core subsections according to depth, redox potential and oxygen availability.

	Depth (5 cm units)					
	1	2	3	4	5	6
Schizorhynchidae	84	86	63	33	12	4
Schizorhychia	95	88	67	39	14	4
Kalyptorhynchia	95	90	69	37	16	8
Proseriata	90	67	61	37	20	17
Rhabdocoela	100	92	75	45	20	10
Macrostomida	58	33	20	4	6	0
Catenulida	37	33	24	27	18	13
Acoela	68	84	81	75	59	42
Turbellaria	100	100	98	90	75	58

	Redox potential (millivolts)				
	400	400–301	300–201	200–101	101
Schizorhynchidae	35	57	50	48	32
Schizorhynchia	41	64	53	48	37
Kalyptorhynchia	41	64	53	52	39
Proseriata	41	52	47	39	42
Rhabdocoela	47	71	56	59	42
Macrostomida	18	27	21	16	10
Catenulida	47	23	41	41	12
Acoela	76	82	74	73	67
Turbellaria	82	91	85	93	84

	O_2 availability (.7 μ moles cm^{-2} min^{-1})					
	10	10–8.1	8–6.1	6–4.1	4–2.1	2.1
Schizorhynchidae	50	42	62	36	55	27
Schizorhynchia	53	50	64	36	58	34
Kalyptorhynchia	59	50	64	39	60	39
Proseriata	44	50	45	36	50	36
Rhabdocoela	65	58	66	39	60	43
Macrostomida	15	21	9	14	17	25
Catenulida	29	42	13	32	40	30
Acoela	71	63	70	75	78	59
Turbellaria	94	92	91	96	82	73

Macrostomida 26.6, Rhabdocoela 31.9, Proseriata 33.3, Catenulida 45.1 and Acoela 74.4 cm, although it must be remembered the latter's regression is not significant ($p > 0.05$). It is also rather ridiculous to suggest that the Turbellaria do not occur down to the same depth as their 'subset' the Acoela.

To some extent this illustrates the pitfalls inherent in a rigid statistical approach. This is more clearly shown by consideration of other data concerning *Gnathostomula* aff. *paradoxa* Ax from an extensive survey undertaken from January–March 1977 at one of the sample localities (Doctor's Bay, Kircubbin). Here the majority of *G. paradoxa* were found at a depth of 2–6 cm in the sand at densities of between 0–48 per cm^2 of beach surface. Forty-eight measurements of redox and of oxygen availability were taken adjacent to cores but no significant regression of either factor with *Gnathostomula* density could be demonstrated. However all values fell in the ranges of -10 to -210 mv and of 0.1 to 3.0 μ moles of O_2 cm^{-2} min^{-}. It is only possible to suppose that the species exists anaerobically under such conditions and that the lack of regression of density with the factors is because their range falls within the optimum for the species. Significant correlations can only be expected where the range of factor being studied extends towards limiting levels.

Discussion

The analyses confirm the vertical partitioning of the interstitial habitat with regards to closely related species as for example noted by Hoxhold (1974) and Boaden (1977). The latter lists *Neoschizorhynchus brevipharynggus* Schilke, *Proschizorhynchus triductibus* Schilke and *Schizochilus marcusi* as typical aerobic sand forms, *N. longipharynggus, Paratomella rubra, Proschizorhynchus bivaginatus* and *Pseudoschizorhynchoides ruber* as redox potential discontinuity layer forms, *N. parvorostro* as a black sand form and *Diascorhynchus rubrus* Boaden and *S. choriurus* Boaden as more euryoxic.

The *Schizochilus* partitioning is confirmed by depth, Eh and particularly oxygen availability records and that of *Proschizorhynchus triductibus* by depth and Eh. *Neoschizorhynchus parvorostro* is found deeper in the sediment than the other listed kalyptorhynchids although *Nematorhynchus parvoacumine* Schilke was found deeper (frequency means depth 3.25, Eh 166, O_2 availability 7.58 units). The relative depth distribution of the *Neoschizorhynchus* triad is confirmed but the redox and availability readings suggest that *N. parvorostro* is euryoxic rather than a black sand form.

Pseudoschizorhynchoides ruber is apparently more 'at home' in deeper sediment and lower redox than *D. rubrus* although concurrence records suggest the latter is more adaptable to low oxygen availability. *Paratomella rubra* concurrence records suggest that it inhabits a wider range of conditions than the two compared species, although *Diascorhynchus rubrus* is more surface swelling. Thus categorization of these red species is confirmed.

Experimental values for oxygen consumption by aerobic sand turbellarians obtained at Portaferry (ElHag, unpublished) can be used to calculate the mean flow rate through the body surface required to sustain their respiration rate. This is in the order of 0.02 μ moles $cm^{-2}\,min^{-1}$. Assuming a reasonably efficient oxygen transport system through the body wall, none of the availability values recorded in the general field work would appear to be limiting. This is with the exception of one value from the 25–30 cm part of the cores taken in January 1974, which part contained several nematodes and gastrotrichs (*Dolichodasys* sp.), 1 acoel and 1 seriate. The 25–30 cm sections were within the black sand layer on all sampling occasions but their mean availability was 3.0 μ moles $O_2\,cm^{-2}\,min^{-1}$. This includes some high values which may have been due to electrode poisoning but even excluding these a mean value of 2.5 μ moles is obtained. McLachlan (1978) gives a figure of about 0.6 μ moles as the lowest value even in reducing conditions on the Port Erin beach.

Redox potential values have been regarded as a better indication of the oxygenation state than have oxygen availability values (Giere 1973; McLachlan 1978). It appears however that Eh readings are also likely to exaggerate actual oxygenation since Revsbech *et al.* (1980) have shown that in a sediment in which dissolved oxygen was absent below 1 cm the redox state was positive to depths of 3 cm or more. In addition Jørgensen (1977) has demonstrated reducing sites within generally oxic sediment, although it is equally true that oxidized sites may be found within reduced layers (e.g. Reise & Ax 1979).

It is likely that the boundary between oxic and anoxic layers of sediment should be drawn somewhere between the +200 and +110 mv levels, that is between the expected change from the ferric to ferrous state and the level of Desulphovibrio anaerobiosis. Even bearing this and the possible raising of mean values due to electrode poisoning in mind, the values presented in this paper indicate general anaerobic occurrence in only two of the Kalyptorhynchids, *Pseudoschizorhynchoides ruber* and *Nematorhynchus parvoacumine*. However in other taxa, for example Species A from 21 core sections with Eh less than 200 mv and 13 of these with Eh $<$100 mv, Catenulida 15 $<$ 200, 6 $<$ 100, Acoela 50 $<$ 200, 32 $<$ 100 (one attenuated black species of which had a frequency mean of 77 mv), *Coelogynopora* sp. 20 $<$ 200, 14 $<$ 100 and to a lesser extent *Schizochilus choriurus* 10 $<$ 200, 9 $<$ 100, there appears to be a substantial anaerobic capacity.

Table 6. Ranking of turbellarian orders with regard to depth in sediment, redox potential and oxygen availability. Values derived from frequency and density records and from the number of joint (L) or exclusive (EL) lowest occurrences – see text. Low values imply occurrence in deeper layers, lower redox or lower oxygen availability than high values.

	Frequency			Density			L			EL		
	D	Eh	O_2	D	Eh	O_2	D	Eh	O_2	D	Eh	O_2
Proseriata	3	2	4	4	5	4	2	3	3	2	2	3
Rhabdocoela	4	5	5	3	4	5	4	2	1	3	–	2
Macrostomida	5	3	2	5	3	2	5	4	4	4	–	–
Catenulida	1	4	1	2	2	1	3	5	5	–	–	4
Acoela	2	1	3	1	1	3	1	1	2	1	1	1

It is not necessarily correct to link the comparative depth occurrence range of redox value and range of oxygen availability inhabited directly to the respiratory requirements of the taxa concerned since other factors such as predation, food resources or temperature preferenda may be governing distribution. However the balance between oxidation state and organic input does appear to be a dominant factor in the microbial and meiofaunal ecology of sediments (Fenchel 1969). Since the input of oxygen, oxidised substances and organic matter must be primarily from the sediment surface it is not suprising to find strong vertical distribution patterns within the substratum.

Table 6 ranks the frequency, density, lowest and exclusively lowest occurrences of the five turbellaria orders considered in relation to the three factors examined. Low ranking corresponds to the furthest depth into the sediment, to the lowest redox and the lowest oxygen availability values or mean values. These rankings vary considerably although the Proseriata and Macrostomida never appear first and the Acoela never appear last. The taxa are listed in order of Karling's (1974) Alternative 1 phylogeny of the Turbellaria. Is it merely coincidence that the density/Eh rankings match this?

If means are taken of the rankings for each order the overall ranking becomes.

Acoela < Catenulida < Proseriata < Rhabdocoela < Macrostomida. This ranking also applies if the exclusively lowest category is omitted and also if only depth rankings are considered. In these and most other combinations of ranking, the Macrostomida come higher on the list than in Karling's phylogenetic series; they however come joint second in O_2 and Eh rankings omitting the EL category.

General examination of the records presented in this paper show that the occurrence of the Acoela is relatively independant of depth in sediment, Eh and oxygen availability compared with that of the Macrostomida and, to a slightly lesser extent, the Catenulida. This may be seen from the frequency, regression and contingency tables. The catenulid distribution is less oxic than that of the macrostomids. If Species A is included with the Catenulida (giving frequency means of 3.4 depth, 240 Eh and 6.2 oxygen availability units) this is even more marked. The catenulid distribution corresponds to their general occurrence in sand with a high detritus content noted by Sterrer & Rieger (1974) although the mean Eh values are somewhat higher than expected for a group said (l.c.) to occur mainly in the vicinity of the 'black' or redox discontinuity layer. However, as stated, all the samples of the 25–30 cm layer taken during the general survey were within the black sand layer and nearly half of these had redox values above 200 mv, the mean being 182 mv.

The Proseriata and Rhabdocoela both emerge as groups which are very common in the upper sand layers but decline with depth, decreasing Eh and decreasing oxygen availability. This decline is less marked in the Proseriata which have a fairly strong deep component especially of young and/or unidentified *Coelogynopora* spp. In the habitats investigated the Rhabdocoela were dominated by the Schizorhynchia (63.8% of the order) as indeed were the Turbellaria by the Rhabdocoela (35% of the class) and there is little doubt that the majority of these taxa occurred in the fairly well oxygenated environment of the upper sand layers.

If we assume that Karling's (1974) sequence of turbellarian orders is ranked in order of increasing departure from the primitive condition and we assume that at least some members remain represented in their original habitat (primary representation – see Riedl 1963), what do the records presented here imply about the evolution of Turbellaria in or into new habitats? Other assumptions are required before an answer to this and the previous question can be attempted. The chief of these is that sand was not colonized separately by the separate orders, this implying that the turbellarian stem form occurred interstitially. This is not unreasonable in view of the extraordinary diversity of the Turbellaria in marine sands. A valid alternative assumption is that the orders may have colonized sand separately but they did so at sediment depths and oxygenation characteristics corresponding to their original habitat(s).

If these things are so, then it is possible to argue the present distribution from at least two different view points. These differences stem from whether (a) it is adjudged that older less derived groups are more likely to have remained in habitats closer to the original and are less likely to occur in new habitats than the derived groups or whether (b) it is adjudged more likely that the younger more derived groups will have formed in the original habitat and displaced the older groups to new habitats.

According to the 'conservative' alternative (a) the Catenulida and Acoela envolved in the least oxic conditions but the Macrostomida must have originated in conditions that were at least oxidizing if not fully oxic.

According to the 'radical' alternative (b) the Catenulida, Acoela and Macrostomida all evolved in oxic conditions but the advent of the Rhabdocoela and Proseriata, and perhaps the Macrostomida themselves, 'forced' catenulids and acoels into anoxic or micro-oxic habitats.

Competition theory predicts that with intraspecific competition niche width, that is the range of a particular resource used by the species, should increase but with inter-specific competition niche width should narrow (see Pianka 1974). It seems to me that the former circumstances should lead to production of the more derived (apomorphic) species away from the original resource-mean but the latter lead to plesiomorphic (i.e. less derived) species being restricted to nearer the resource-mean. In other words the conservative alternative (a) is the more likely direction of habitat adoption.

Conclusion

The apparent relationship of phylogeny with habitat oxygenation shown by sand turbellarians is possibly a consequence of evolutionary movement from plesiomorphic anoxic or micro-oxic conditions to apomorphic oxic conditions.

The evidence presented is by no means conclusive. However the author hopes it will encourage further work on the relationship between turbellarian ecology and systematics and thus be a fitting tribute to Tor Karling's 70th birthday.

Acknowledgements

The work was conducted with assistance from a Natural Environmental Research Council (U.K.) grant. The author also wishes to thank Cathy Maguire, ElHag and Nick Johnston for access to unpublished results.

References

Ax, P., 1977. Life cycles of interstitial Turbellaria from the eulittoral of the North Sea. Acta. Zool. fenn. 154: 11–20.

Boaden, P. J. S., 1968. Water movement – a dominant factor in interstitial ecology. Sarsia 34: 125–136.

– 1975. Anaerobiosis, meiofauna and early metazoan evolution. Zool. Scripta 4: 21–24.

– 1977. Thiobiotic facts and fancies (Aspects of the distribution and evolution of anaerobic meiofauna). Mikrofauna Meeresbod. 61: 45–63.

– 1980. Meiofaunal thiobios and 'the Arenicola negation': case not proven. Mar. Biol. 58: 25–29.

– & Platt, H. M., 1973. Daily migration patterns in an intertidal meiobenthic community. Thalassia jugosl. 7: 1–12.

Ehlers, U., 1973. Zur Populationsstruktur interstitieller Typhloplanoida und Dalyellioida (Turbellaria, Neorhabdocoela). Mikrofauna Meeresbod. 19: 1–105.

Faubel, A., 1976. Populationsdynamik und Lebenszyklen interstitieller Acoela und Macrostomida (Turbellaria). Mikrofauna Meeresbod. 56: 1–107.

Fenchel, T., 1969. The ecology of marine microbenthos IV. Ophelia 6: 1–182.

– & Riedl, R. J., 1970. The sulfide system: a new biotic community underneath the oxidized layer of marine sand bottoms. Mar. Biol. 7: 225–268.

Giere, O., 1973. Oxygen in the marine hygropsammal and the vertical distribution of oligochaetes. Mar. Biol. 21: 180–189.

Gray, J. S. & Rieger, R. M., 1971. A quantitative study of the meiofauna of an exposed sandy beach, at Robin Hood's Bay, Yorkshire. J. Mar. Biol. Assoc. U.K. 51: 1–19.

Hulings, N. C. & Gray, J. S., 1971. A manual for the study of meiofauna. Smithson. Contr. Zool. 78: 1–83.

Hyman, L. H., 1951. The Invertebrates. Volume II. Platyhelminthes and Rhynchocoela. New York: McGraw-Hill, 550 pp.

– 1959. The Invertebrates. Volume V. Smaller Coelomate Groups. New York: McGraw-Hill, 783 pp.

Jansson, B. O., 1967. The availability of oxygen for the interstitial fauna of sandy beaches. J. Exp. Mar. Biol. Ecol. 1: 123–143.

Jørgensen, B. B., 1977. Bacterial sulfate reduction within reduced microniches of oxidized marine sediments. Mar. Biol. 41: 7–17.

Karling, T. G., 1974. On the anatomy and affinities of the turbellarian orders. In: Biology of the Turbellaria. Ed. Riser, N. W. & Morse, M. P., New York: McGraw-Hill, pp. 1–16.

Maguire, C., 1977. Meiofaunal community structure and vertical distribution: a comparison of some Co. Down beaches. In: Biology of Benthic Organisms, Ed. Keegan, B. F., O'Ceidigh, P. & Boaden, P. J. S., Oxford: Pergamon Press pp. 425–431.

McLachlan, A., 1978. A quantitative analysis of the meiofauna and chemistry of the redox potential discontinuity zone in a sheltered sandy beach. Est. Coast. Mar. Sci. 7: 275–290.

Ott, J. & Schiemer, F., 1973. Respiration and anaerobiosis of free living nematodes from marine and limnic sediments. Neth. J. Sea. Res. 7: 233–243.

Pianka, E. R., 1974. Evolutionary Ecology. New York: Harper & Row, 356 pp.

Reise, K. & Ax, P., 1979. A meiofaunal 'thiobios' limited to the anaerobic sulphide system of marine sand does not exist. Mar. Biol. 54: 225–237.

– 1980. Statement on the thiobios-hypothesis. Mar. Biol. 58: 31–32.

Revsbech, N. P., Sørensen, J., Blackburn, T. H. & Lomholt, J. P., 1980. Distribution of oxygen in marine sediments measured with microelectrodes. Limnol. Oceanogr. 25: 403–411.

Riedl, R., 1963. Probleme und Methoden der Erforschung des litoralen Benthos. Zool. Anz. Suppl. 26: 505–567.

Rieger, R. & Ott, J., 1971. Gezeitenbedingte Wanderungen von Turbellarien und Nematoden eines Nordadriatischen Sandstrandes. Vie Milieu 22: 425–427.

Schmidt, P., 1969. Die quantitative Verteilung und Populationsdynamik des Mesopsammons an Gezeitensandstrand der Nordseeinsel Sylt. II. Quantitative Verteilung und Populationsdynamik einzelner Arten. Int. Rev. ges. Hydrobiol. 54: 95–174.

Sterrer, W. & Rieger, R., 1974. Retronectidae – a new cosmopolitan marine family of Catenulida (Turbellaria). In: Ed. Riser, N. W. & Morse, M. P., Biology of the Turbellaria, New York: McGraw-Hill, pp. 63–92.

Wieser, W., Ott, J., Schiemer, F. & Gnaiger, E., 1974. An ecophysiological study of some meiofauna species inhabiting a sandy beach at Bermuda. Mar. Biol. 26: 235–248.

Sand dwelling Turbellaria from the Netherlands Delta area

Paul M. Martens & Ernest R. Schockaert
Dept. SBM, Limburgs Universitair Centrum, B-3610 Diepenbeek, Belgium

Keywords: Turbellaria, meiobenthos, fauna composition, taxonomy, Netherlands Delta area

Abstract

Sand dwelling Turbellaria from the Delta of the Rivers Rhine, Meuse and Scheldt have been investigated. Thirty-eight samples taken from littoral and sublittoral stations in the Grevelingen, Eastern and Western Scheldt have been analysed.

Thirty-three species were recorded (Acoela were not considered); twenty-four of them are new for the area and seven new species are described.

Density and diversity of Turbellaria were higher in the Eastern Scheldt than in the Western Scheldt or in the Lake Grevelingen. A maximum density of 82 ind./100 cm^3 was noted. A tentative calculation on relative abundance of the representatives of the different Turbellaria orders is established. Proseriata seem to be dominant in the localities studied.

Introduction

The benthic communities of the Netherlands Delta area are intensively studied at the Yerseke Delta Institute for Hydrobiological Research and at the Zoological Institute of the University of Ghent. Turbellaria have been disregarded sofar and the present study is an attempt to fill this gap in our knowledge of the meiofauna of the area. Only Den Hartog (in a number of publications issued from 1963 to 1968) and Boaden (1976) have investigated the turbellarian fauna in the Delta area. Den Hartog sampled in intertidal habitats such as salt march creeks and ponds and sandy mud flats, and focussed his attention mainly on the Trigonostomida. Boaden made an inventory of the sanddwelling species in littoral and sublittoral localities. The total number of species observed by both authors was 65, but no data on density and relative abundance are available.

Material and methods

Sublittoral samples were taken in 20 different localities: 14 in the Lake Grevelingen, two in the Eastern Scheldt and four in the Western Scheldt; seven eulittoral stations in the Western Scheldt were also sampled. Localities in the Eastern Scheldt and Grevelingen are indicated by the code letters of the nearest buoy, and in the Western Scheldt by the coordinates.

Some of the new species had also been found in North Sea localities by the senior author (E.R.S.) and some by D. Degadt and provisorily described by him (1973, unpublished). These data are incorporated in the descriptions below.

The total number of samples was 38, most of them taken in the second half of 1979, while some samples from the Western Scheldt were taken in December 1978. The eulittoral samples where taken with a hand-held plastic corer of 10.2 cm^2. For sublittoral sampling, different methods were used involving a meiosticker (Govaere & Thielemans

Hydrobiologia 84, 113–127 (1981). 0018-8158/81/0842-0113/$03.00.

1979), Reinecker box-corer (Farris & Crezee 1976), hand-cores held by SCUBA divers, and the Van Veen grab. Sediment analysis was performed on subsamples (Heip *et al.* 1979). Classification of the sediment was according to the Wentworth scale (Buchanan & Kain 1971) and the degree of sorting according to Wolff (1973). The characterisation of the sediment is indicated by letters symbols in Table 1.

Turbellaria were extracted from the sediment using the $MgCl_2$ method (Boaden 1963b, Sterrer 1968) for the qualitative approach, and the Uhlig icewater method (Uhlig 1964, 1966) for quantitative data. Since samples taken with different methods were to be compared an approximatively equal amount of sediment was processed, and Turbellaria were counted alive under the dissecting microscope and identified under the microscope in squeezed preparations. Whole mounts were prepared with lactophenol from the squeezed individuals; other representatives of species available in sufficient numbers were fixed with Bouin's fluid, sectioned and stained with Masson's trichrome, Erhlich's hematoxylin or Heidenhain's ironhematoxylin, if necessary using eosin or erythrosin as counterstain. Typematerial of the new species is deposited in the zoological collection of the department S.B.M. of the Limburgs Universitair Centrum, Diepenbeek (Belgium). Figures without scale are freehand drawings.

Compositions of the Turbellaria fauna

Observations

Thirty three species were identified and they are listed in Table 1. Acoela, Nemertodermatida and Catenulida were not identified to the species level but their total number was recorded in order to establish the relative abundance of the different turbellarian orders as given in Table 2.

In Lake Grevelingen only seven samples of 19 (from 14 stations) contained Turbellaria. Only five different Neoophora species and relatively many Acoela were observed. The total number of individuals in all samples was 82. In the Western Scheldt 11 samples of the 13 (from 11 stations) contained Turbellaria. In the 11 positive samples a total of 58 individuals was observed. Apart from the Acoela, they belonged to seven different species. In the Eastern Scheldt four of the six samples from the two stations contained 217 individuals of Acoela, Nemertodermatida and 27 different identified species altogether. The other two samples were very rich in Turbellaria but not quantitatively processed. The maximum density we noted was 82 ind./100 cm^3 in a sublittoral station in the Eastern Scheldt (018: 26.10.79).

Discussion

It appears from Tables 1 & 2 that both the density of the Turbellaria and species diversity is much higher in the Eastern Scheldt than in the other two parts of the Delta area. Similar results were obtained in studies on other meiofauna taxa (Heip *et al.* 1979; Van Damme *et al.* 1981). It seems that the high degree of pollution in the Western Scheldt has had a deleterious effect on the composition of its bottom fauna and the silting of the bottom and oxygen deficiency that occured in lake Grevelingen after its closure in 1971 may be held responsible for the low species diversity and population densities in this habitat.

The composition of the Turbellaria fauna of sandy beaches has been studied by Sopott (1973), Ehlers (1973) and Hoxhold (1974) on the North Sea island Sylt and by Gray & Rieger (1971) at Robin Hood's Bay, Yorkshire. On the beaches of Sylt Proseriata are the dominant group (31–53%), followed by Acoela (21–33%) and Neorhabdocoela (15–28%). Within the Neorhabdocoela, the Kalyptorhynchia are most abundant with 46–56%, closely followed by the Typhloplanoida with 39–48%. At Robin Hood's Bay Acoela and Neorhabdocoela are the dominant taxa, Proseriata representing only 7–27% of the turbellarian fauna, while within the Neorhabdocoela 67–88% of the individuals belong to the Dalyellioida. In the Delta as a whole, Proseriata and Acoela are about equally well represented and approximatively one fifth of the turbellarian fauna consists of Neorhabdocoela and 90% of these are Kalyptorhynchia. It can be dirived from the data in Table 2 that the relative abundance of these three main taxa is different in the three parts of the Delta studied and that seasonal fluctuations may occur.

No conclusion can be made from this variety of statements at this moment, except perhaps that the

Table 1. Occurrence of Turbellaria species in Western Scheldt, Eastern Scheldt and lake Grevelingen.

	Grevelingen	Western Scheldt	Eastern Scheldt
Retronectes sp.			c
Bradynectes sterreri Rieger, 1971			b
Paromalostomum fusculum Ax, 1952			c
Haplopharynx rostratus Meixner, 1938			c
Carenscoilia bidentata Sopott, 1972			bc
Cirrifera dumosa Sopott, 1972	d		
Cirrifera aculeata (Ax, 1951)		Bc	bc
Archilopsis unipunctata (Fabricius, 1826)	d		
Promonotus marci Ax, 1954			bc
Archimonocelis oostendensis sp.n.			bc
Parotoplana capitata Meixner, 1938		d	
Parotoplana papii Ax, 1956			b
Kataplana germanica Meixner in Ax, 1951			bc
Philosyrtis coomansi sp.n.		A	bc
Ciliopharyngiella constricta sp.n.			b
Subulagera rubra sp.n.			c
Rogneda hibernica (Southern, 1936)	d		
Scanorhynchus forcipatus Karling, 1955			c
Toia ycia Marcus, 1952	p		
Cicerina remanei Meixner, 1928			b
Paracicerina deltoides sp.n.			b
Gnathorhynchus conocaudatus Meixner, 1929			bc
Schizorhynchoides karlingi sp.n.			c
Schizorhynchoides symmetricus sp.n.			b
Schizochilus choriurus Boaden, 1963	x		c
Neoschizorhynchus parvarostro Ax & Heller, 1970		b	c
Thylacorhynchus caudatus Meixner, 1928		e	
Carcharodorhynchus sp.			b
Limirhynchus danicus Schilke, 1970		B	b
Cheliplanilla caudata Meixner, 1938			b
Celiplanilla rubra Schilke, 1970			c
Rhinipera remanei Meixner, 1928			b
Diascorhynchus rubrus Boaden, 1963		A	bc

a : fine sand, less well sorted
b : fine sand, well sorted
c : fine sand, very well sorted
d : medium sand, well sorted
e : medium sand, very well sorted
p : periphyton
x : no data on the sediment
Capital letter: eulittoral station

present sampling and processing procedures are insufficient to provide reliable data for populations studies, whilst it cannot be denied that environmental factors strongly affect the fauna composition.

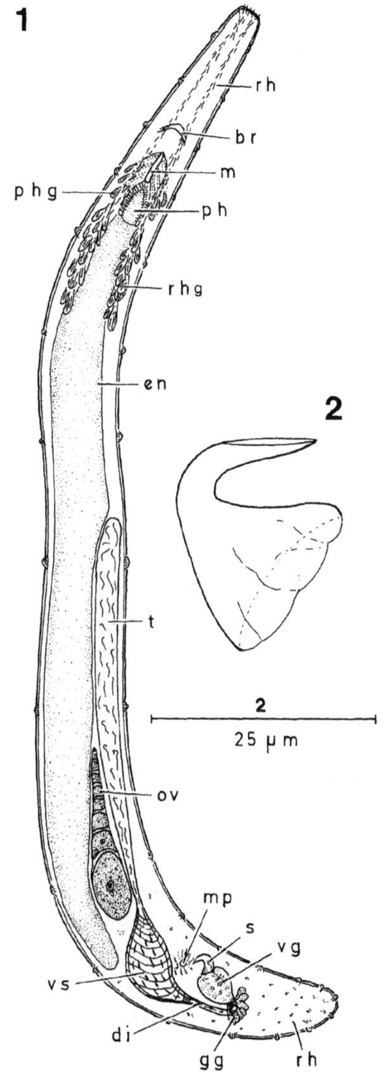

Fig. 1–2. Bradynectes sterreri. 1. General organisation (from living animal). 2. Stylet (from whole mount).

Descriptions and discussions of the new species and some known species

Bradynectes sterreri Rieger, 1971 (Figs. 1–2 & 24)

New locality: Eastern Scheldt, near buoy 013, fine sand at 3–4 m depth, 24.08.79.

Material: several animals studied alive two of them in whole mount and one serially sectioned.

Table 2. Percentage of numbers of different turbellarian orders in the Grevelingen, Western Scheldt, Eastern Scheldt and the individual Eastern Scheldt stations.

	Greve-lingen	Western Scheldt	Eastern Scheldt	Total	Eastern Scheldt			
					Station 018		Station 013	
					26.10.79	21.12.79	26.10.79	21.12.79
Turbellaria n	82	58	217	357	89	33	75	20
Acoela	62%	14%	29%	34%	20%	27%	38%	30%
Nemertodermatida	0%	0%	1%	1%	0%	3%	1%	0%
Catenulida	0%	0%	10%	6%	24%	6%	0%	0%
Macrostomida	0%	2%	7%	4%	4%	21%	4%	5%
Proseriata	27%	57%	35%	37%	35%	27%	37%	35%
Neorhabdocoela	11%	27%	18%	18%	17%	15%	19%	30%

Description and discussion

In his original description of the species, Rieger (1971) discerned three different phena: the Kristineberg-form, the Carolina-form and the Robin Hood's Bay-form. A fourth form, the Sylt-form was added to the list by Faubel (1974). The distinction of these forms is based on stylet shape and size, location of the gonads, and presence or absence of a sphincter around the vas deferens. The present specimens differ from all the above mentioned forms in that the ovaria are between testis and gut, there is no sphincter at the vas deferens and the stylet measures 24 μm at its convex side and 26 μm at its concave side.

Archimonocelis oostendensis sp. n. (Figs. 3–5 & 25)

Localities:

1. Oostende (Belgium), eulittoral zone of the sandy beach at the Memorial Monument, fine sand with detritus (not analysed), 25.07.72 (Typelocality).
2. North Sea, at N lat. 51°12′32″ and E. long. 2°48′52″, fine sand at 15 m depth, 28.09.72 and at N lat. 52°02′12″ and E long 3°47′00″, medium sand at 18 m depth, 13.09.72, (De Gadt).
3. Eastern Scheldt, near buoy 013, fine sand at 4 m depth, 02.08.79.

Material: Several animals studied alive from all localities, five of them in whole mount (one designated as holotype).

Description

The living animals are about 2.5 mm long, filiform and colourless, without eyes. The anterior tip is provided with sensory bristles and the caudal end with adhesive papillae. The cnidosacs are placed at regular distances in a medio-dorsal row. The topography of the organs is shown in Fig. 3.

Except for the gonads, all the genital organs are situated in the hindmost part of the animal. A vagina surrounded by glands and a prepenial bursa occur in front of the male organs. The female duct ends in the posteriorly situated female pore with the usual cement glands.

A single row of 16–25 testis follicles occur. The seminal vesicles however are paired. The copulatory organ is directed forwards with caudal seminal vesicles, an interposed prostate vesicle and a stylet, 49–53 μm long, 3.5–5 μm wide at its centre and 7–9 μm at the proximal opening, with a subterminal opening, and ending in a blunt massive extension 6–8 μm in length (measurements on five specimens). The stylet is surrounded by bristles, 35–40 μm long, all but two roughly of the same type (0.7 μm thick). The two bristles nearest to the stylet are some 1.1 μm thick, with a subterminal large and flattened lateral hook. Adjacent to the copulatory organ this far described, an accessory organ can be observed. It consists of a bundle of accessory glands of which the exit is encircled by eight needles: five of them are very fine, 18 μm long (average in the holoptype) and of the same shape as the majority of needles around the stylet. The other three needles are 26–31 μm long and of three different construction-types as can be seen in Fig. 5.

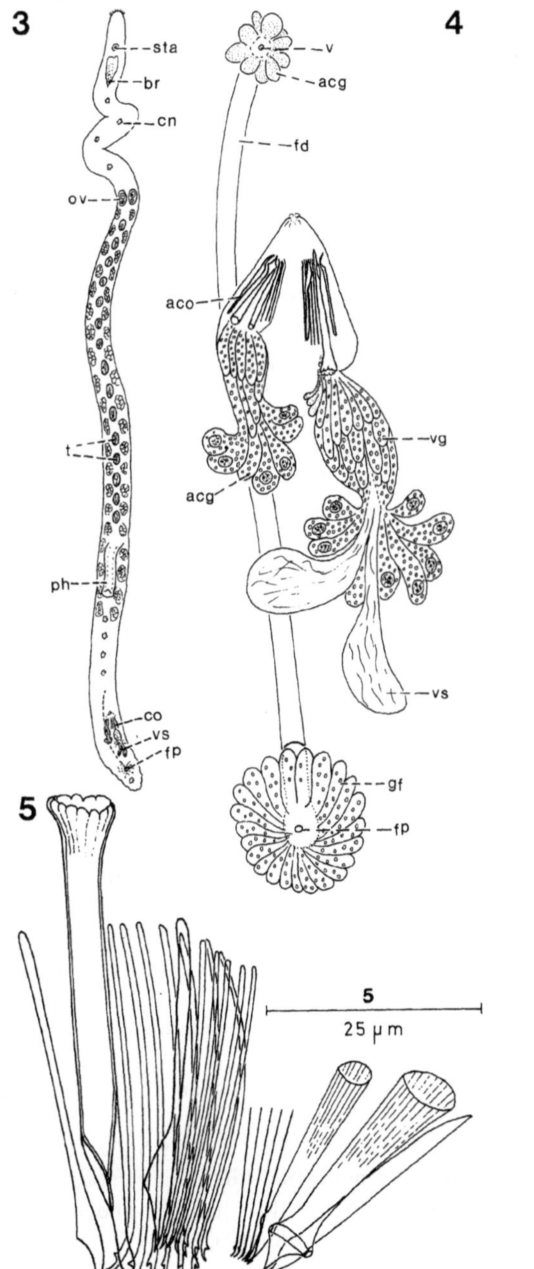

Fig. 3–5. Archimonocelis oostendensis. 3. General organisation. 4. Copulatory organ. (Both from living animal.) 5. Cuticular apparatus (from whole mount).

Discussion

Within the genus *Archimonocelis* (Family Monocelididae, Proseriata), *A. semicircularis, A. coronata* and *A. oostendensis* are evidently closely related, the presence of an accessory glandular organ in the copulatory organ being considered as a synapomorphy. An obvious evolutionary trend within the genus is the specialisation and diversification of the bristles, features that may have arisen more than once as in *A. bathycola, A. helfrichi, A. semicircularis* and *A. oostendensis.* A specialised character apparently unique to *A. oostendensis* is the distinct separation of the accessory organ from the copulatory organ.

Identification key to species of the genus *Archimonocelis*

1. Unpaired seminal vesicle — *A. kionocystis* Karling, 1966
   Paired seminal vesicle — 2.
2. Without accessory organ — 3.
   With accessory organ — 5.
3. With hooks — 4.
   With bristles — *A. bathycola* (Westblad, 1952)
4. Stylet surrounded by 4 hooks — *A. helfrichi* Karling *et al.* 1972
   Stylet surrounded by a crown of hooks — *A. mediterranea* Meixner, 1938
5. All bristles of the same type — *A. coronata* Karling, 1966
   Bristles of different types — 6.
6. Bristles around the accessory organ not separated from the others; 2 different types — *A. semicircularis* Karling, 1966
   Bristles around the accessory organ separated from these around the stylet; 6 different types — *A. oostendensis* sp. n.

Philosyrtis coomansi sp. n.* (Figs. 6–8)

Localities:

1. Eastern Scheldt, near buoy 013, fine sand at 3–4 m depth, 24.08.79, 26.10.79, 21.12.79 (Typelocality); near buoy 018, fine sand at 3–4 m depth, 24.08.79.

* Dedicated to Prof. Coomans, director of the Laboratorium voor Morf. & Syst. der Dieren of the University of Ghent.

2. Western Scheldt, at N Lat. 51°22′42″ and E long. 03°52′00″, fine sand at 3–4 m depth, 12.12.79.

Material: Several animals studied alive, two of them conserved as whole mounts (one of them designated as holotype) and two specimens sectioned.

Description

The species moves rapidly and is extremely haptic, colourless and transparent, with a length less than 1 mm (0.6 mm in whole mount). The cephalic lobe is clearly set off and provided with long sensory bristles. A short row of cilia occurs at both sides behind the cephalic lobe. Adhesive papillae and bristles occur on the lateral side of the animal at more of less regular distances and at the tailtip. Three types of rhabdites are present: large isolated rhabdites all over the body, numerous grouped needle-like rhabdites between brain and pharynx and six parallel rows at the anteriour end. The following generic characters (Ax 1956) could be checked in sectioned material: dorsal unciliated epidermis with nuclei, ventral ciliated epidermis of the creeping sole with insunk nuclei, pharynx lumen ciliated.

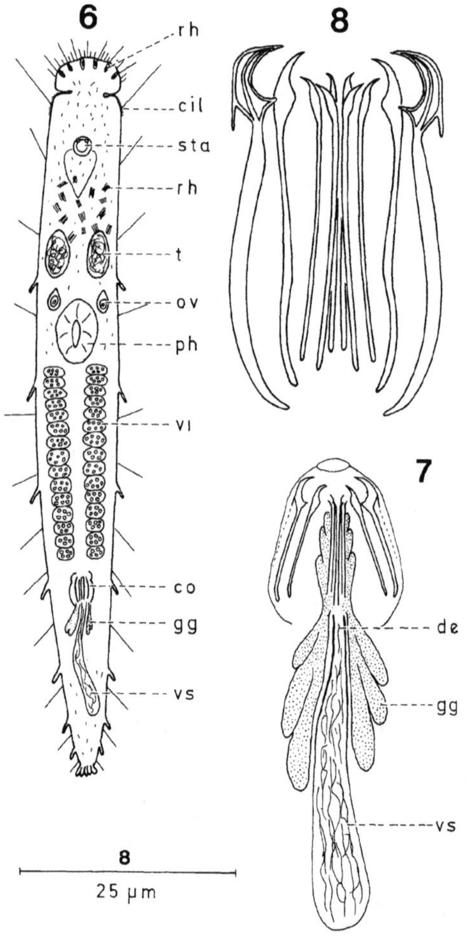

Fig. 6–8. Philosyrtis coomansi. 6. General organisation. 7. Copulatory organ. (Both from living animal.) 8. Cuticular apparatus (from whole mount).

The location of the pharynx (50 µm diameter), gonads and copulatory organ and common genital pore can be seen in Fig. 6. The caudally situated and elongated seminal vesicle is surrounded in its distal part (with muscular wall) by well developed prostate glands which do not form a distinct prostate vesicle (Fig. 7). The cuticular apparatus consists of three different types of needles: a central group of 6 straight needles, 32–35 µm long, two needles of 40 µm long at both side, which are stouter than the former ones and two large lateral spines of 42 µm. The latter are somewhat bent and have a sickle-like terminal end with a small backward projection for the attachment of the muscles surrounding the whole apparatus.

Discussion

Ph. coomansi differs from *Ph. sanjuanensis* Ax & Ax, 1967; *Ph. santacruzensis* Ax & Ax, 1974 and from *Ph. rutilata* Sopott, 1976 in that the cuticular elements in these species are separated in two groups. *Ph. fennica* Ax, 1954 and *Ph. eumeca* Marcus, 1950 are different from *Ph. coomansi* for they have a large number of cuticular needles and from *Ph. rotundicephala* Sopott, 1972 in that this species has only one kind of cuticular elements.

Ciliopharyngiella constricta sp. n. (Figs. 12 & 13)

Locality: Eastern Scheldt, near buoy 013, fine sand at 3–4 m depth, 24.08.79 (Typolocality).

Material: One animal studied alive and prepared in whole mount (holotype).

Description and discussion

The individual studied is a slender, colourless

and very translucent animal. Its organisation agrees in most respects with that of *Ciliopharyngiella intermedia* Ax, 1952: a pharynx rosulatus slightly behind the body middle, a single testis some distance in front of it, an avoid copulatory bulb (functioning as prostate vesicle with two kinds of prostate secretions), a pair of extra-capsular elongated seminal vesicles, a common genital pore and a bursa posteriorly to it. The seminal duct, running axially through the prostate vesicle is somewhat expanded to form an intracapsular seminal vesicle. The ovovitellaria, with posterior egg-producing compartment, are situated at both sides of the pharynx extending in equal parts in front and behind the latter. The stylet is 80 μm long and depicted in fig. 13. Its most conspicuous feature is an abrupt narrowing at about 1/3 from its distal end. The main characters in which the animal studied differs from *C. intermedia* are tabulated below:

C. intermedia Ax, 1952	*C. constricta* n. sp.
– dark grey, very opaque	– colourless, very translucent
– stylet 100–120 μm long without constriction	– stylet 80 μm long, constricted
– vitellaries completely in front of the pharynx	– vitellaries in part behind the pharynx

Though the above description is based on only one individual, these differences justify in our opinion the erection of a new species.

Subulagera rubra sp. n. (Figs. 9–11 & 26)

Localities:

1. North Sea, at N lat. 51°50′50″ and E long. 03°37′15″, medium sand at 18 m depth, 28.09.72 (Typelocality) and at N lat. 52°36′30″ and E long. 03°18′15″, medium sand at 27 m depth, 12.09.72 (Degadt).
2. Eastern Scheldt, near buoy 018, fine sand at 4 m depth, 24.08.79.
3. Germany, Helgoland, eulittoral, fine sand, Aug. 1974 and Sylt, many localities in eu- and sublittoral. (leg. Dr. U. Ehlers, Göttingen).
4. France, Atlantic coast, sandy beach of Arcachon, Sept. 1974 (leg. Dr. U. Ehlers, Göttingen).
5. Boaden, 1977, (p. 52, 53, Fig. 5: *Subulagera* sp. locality not mentioned).

Material: Several animals studied alive from the North See stations and from the Eastern Scheldt. Three whole mounts (one of them designated as holotype) and four specimens serially sectioned (all from the North Sea). Four sectioned specimens from Sylt (Dr. Ehlers).

Description

The living animals are 1.5–2.2 mm long and about 0.1 mm broad, slowly moving and without eyes. The colour is red (pigment is dissolved in the mesenchyme) with yellow-brown spots in the rear-end. Two kinds of dermal rhabdites (4 μm and 2μm) and two kinds of adenal rhabdites occur, the latter produced in numerous and well developed glands in the front end of the animal. The pharynx (110 μm diameter) is situated slightly before the middle of the body. The epidermis is cellular and nucleated. Habitus and topography of the species is depicted in Fig. 9. A reconstruction of the genital organs is shown in Fig. 11. The paired testes are small, lying just behind the brain and in front of the gut. The large paired seminal vesicles, posterior to the pharynx, are surrounded by weak spirally arranged muscles. The lining epithelium is slightly thickened where the seminal vesicles join each other at their entrance into the copulatory bulb. This copulatory bulb is pear-shaped (about 140 μm long and 65 μm diameter) and surrounded by three muscle layers: a thin inner layer of which the fibres are mainly longitudinal, and two outer thicker layers which form spirals at right angles to each other. The bulb (actually the prostate vesicle) is filled with strands from only one type of glands which large cellbodies are lying extracapsular. Proximally and distally in the bulb some spaces between the prostate secretion and the bulb wall is left and filled with a loose tissue. The axial seminal duct is lined by a very thin epithelium (without nuclei) and surrounded by fine longitudinal muscles.

The cuticular apparatus (Fig. 10, 26) is extremely complicated as in *S. mucronata* Ehlers, 1974. As far as its construction could be analysed and understood it consists of a funnellike and slightly cuved stylet (12 μm at its proximal end, tapering to a diameter of 2–2.5 μm and 29–32 μm long). The proximal rim of the funnel seems to be reflected to form a kind of collar or sheath. The dorsolateral

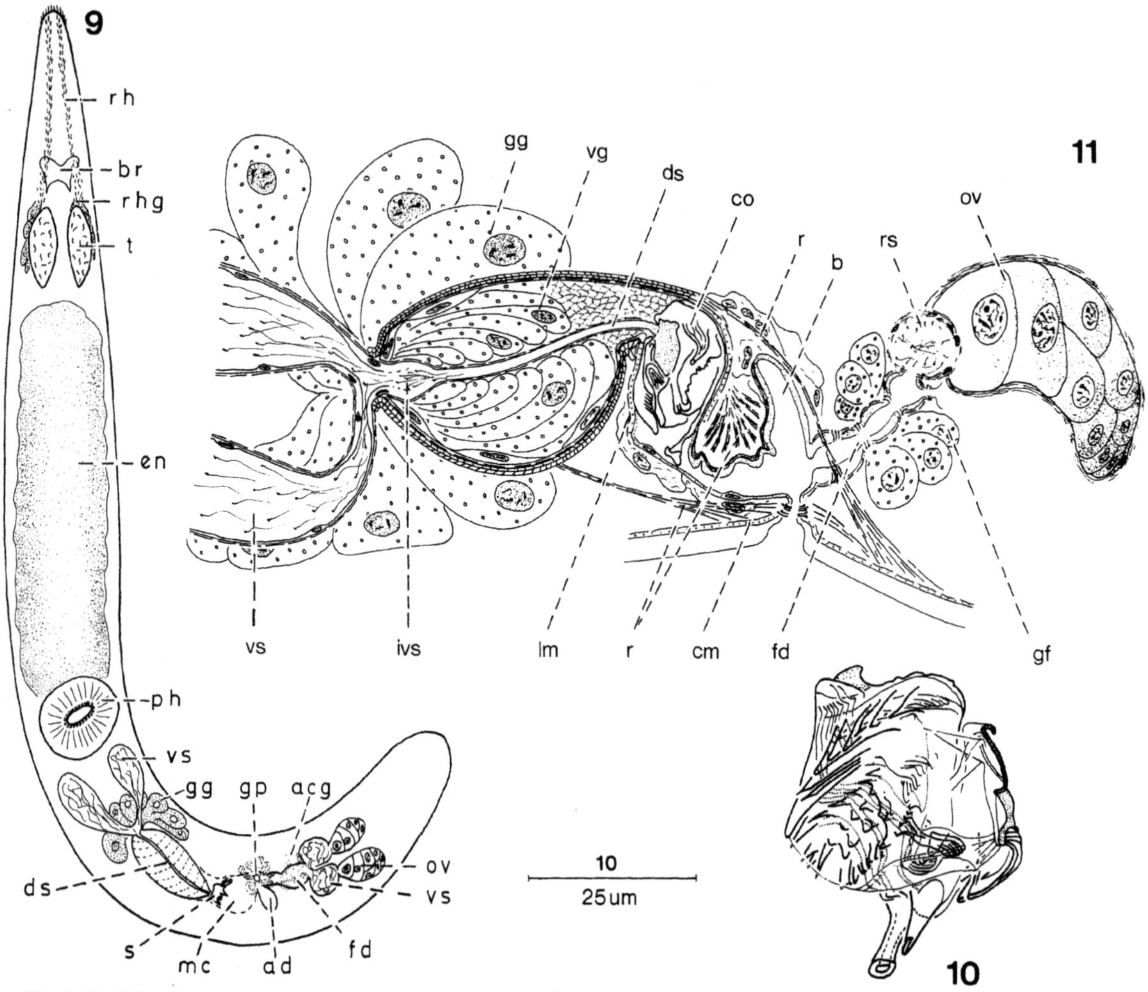

Fig. 9–11. Subulagera rubra. 9. General organisation (from living animal). 10. Cuticular apparatus (from whole mount). 11. Sagittal reconstruction of the genital apparatus, viewed from the left (from serial sections).

side (left) of this collar is ornamented with 4–5 rows of some 10 small spines and one row of about 12–15 larger flat and triangular spines; at the left side the collar bears one strong and solid spine accompanying the stylet's distal end. The right side of the collar shows numerous parallel and sinuous folds or ridges. The maximum diameter is 33 μm and the maximum hight (without stylet) is about 31 μm.

The individuals from the North Sea and from the Eastern Scheldt, all collected in the late summer, had not reached full female maturity and had very small ovaries. Vitellaria were not observed clearly, neither in the living nor in the sectioned material. According to Degadt's preliminary description, vitellaria could be found mainly posteriorly to the pharynx. The individuals from Sylt had reached full sexual maturity and in Dr. Ehlers' drawings vitellaries were depicted also in front of the pharynx. The female duct, which enters the common atrium through its posterior wall, is short and surrounded by weak longitudinal and some circular muscles forming consecutive weak sphincters. The epithelium of the female duct is high with insunk nuclei and surrounded by large basophilous glands. At its proximal end the female duct bifurcates into two seminal receptacles which can be considered as part of the oviducts: the ovaries are indeed directly connected to their seminal receptacle and envelopped in the same tunica.

The genital atrium contains the copulatory stylet

in its anterior part. Furthermore it forms a large mediodorsally directed diverticle, an atrial bursa, in its posterior part. The ventral wall is lined by a flattened and nucleate epithelium. The epithelium of the dorsal wall is also thin, without nuclei and with a very thick basement membrane which stains heavily in iron hematoxylin. Numerous muscles are attached to the anterior wall of the bursa. Several protractor muscles of the bulb, and some dorso-ventral muscles, run around the atrium. The genital pore is guarded by a sphincter of some 4–5 muscle fibres.

Discussion

The genus *Subulagera* was introduced by Ehlers (1974) for *S. mucronata*. In this species the seminal receptacles have an excentric position and enter the oviducts from the side. It also has a copulatory bursa that is connected to the distal part of the female duct. In spite of these differences the new species is provisionally considered as a representative of the genus *Subulagera*. In our opinion, the construction of the stylet in both species is basically the same, while the differences mentioned evidently are a shift in relative position of homologous structures. From the study of sectioned paratypes of *S. mucronata* the pharynx construction may also show some structural differences in both species. The taxonomic importance of all these differences is difficult to evaluate at this moment: many monospecific genera have been described within the poorly diagnosed subfamilies Brinkmaniellinae (Den Hartog 1964b) and Adenorhynchinae (Ax & Heller 1970). We agree with Karling *et al.* (1972), Rieger (1974), Ehlers (1974) and even Luther (1962), that a thorough revision of the family Promesostomidae is necessary.

Paracicerina deltoides sp. n. (Figs. 14, 15, 27 & 28)

Locality: Eastern Scheldt, near buoy 013, fine sand at 3–4 m depth, 21.12.79, (Typolocality).
Material: One animal studied alive and prepared as a whole mount (holotype).

Description

The habitus and the general organisation of the individual studied (Fig. 14) agrees with that of *Paracicerina globulosa* (Brunet 1973). But it must be mentioned that our specimen has a pair of eyes and a terminal girdle of ten adhesive papillae. The testes are paired and lie anteriorly to the pharynx just before the vitellaria. The cirrus is 30–35 μm long and 18–24 μm broad (Fig. 15). Its spines are not uniform: the proximal spines are longer than the distal ones while distally the cirrus is surrounded by a crown of triangular hook shaped spines (6–8 μm long). The bursa, with its two cuticular spermducts, is situated in front of the copulatory organ.

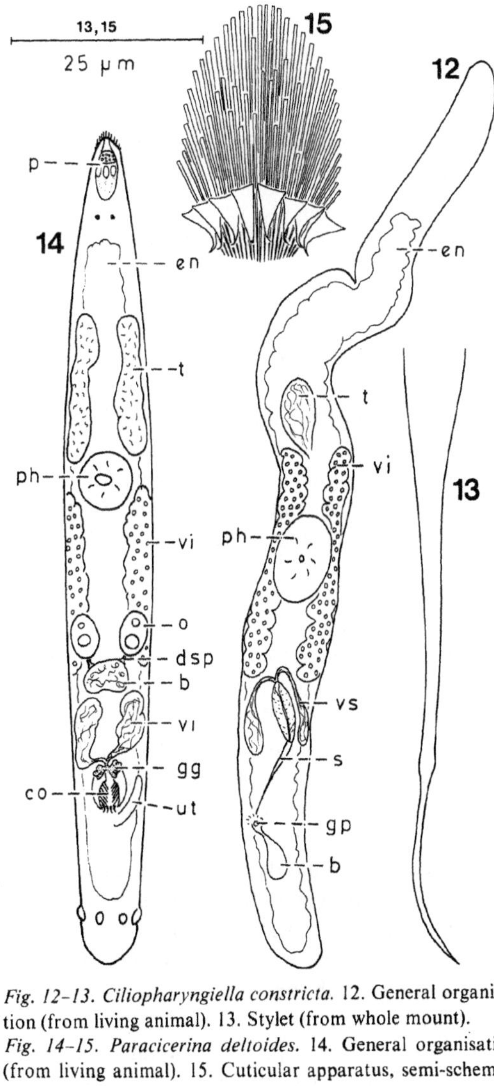

Fig. 12–13. Ciliopharyngiella constricta. 12. General organisation (from living animal). 13. Stylet (from whole mount).
Fig. 14–15. Paracicerina deltoides. 14. General organisation (from living animal). 15. Cuticular apparatus, semi-schematically (from whole mount).

Discussion

Since the above description is based on only one individual as is Brunet's (1973) description for *P. globulosa* it is with great hesitation that a new taxon is introduced. The main differences between both specimens are:

P. globulosa Brunet, 1973	*P. deltoides* sp.n.
- without eyes	- with eyes
- no adhesive papillae	- a girdle of ten adhesive papillae
- bursa behind copulatory organ (?)	- bursa in front of the copulatory organ
- cirrus with nearly uniform spines (Brunet 1973 p. 23 fig. 23b)	- cirrus with more spines and a distal crown of large curved spines.

More material of the mediterranean as well as of the Delta population is undoubtedly necessary.

Schizorhynchoides symmetricus sp. n. (Fig. 16–18)

Localities:

1. Eastern Scheldt, near buoy 013, fine sand at 3–4 m depth, 24.08.79 (Typelocality).
2. Bredene (Belgium), eulittoral zone of the sandy beach, fine sand (not analysed), 28.08.68.

Material: Several animals studied alive, three of them in whole mount (all from the Eastern Scheldt, one of them designated as holotype).

Description

The slender and colourless animals are about 1 mm (in whole mounts) and provided with a tail. A terminal girdle of six adhesive papillae occur. Eyes and sensory bristles are absent. The proboscis is 90–120 μm long or about 1/10th of the total body length. As in the other representatives of the genus *Schizorhynchoides* the pharynx is situated slightly behind the body middle and is preceded by two pairs of testes of which the left ones lie somewhat more anteriorly than the right ones.

The paired elongated seminal vesicles fuse to a seminal duct which runs axially through the copulatory bulb. The prostate vesicle fills the proximal part of this bulb. The cuticular organ consists of a slightly curved stylet (29–31 μm long) lying in a V-shaped cuticular sheath (34–40 μm long).

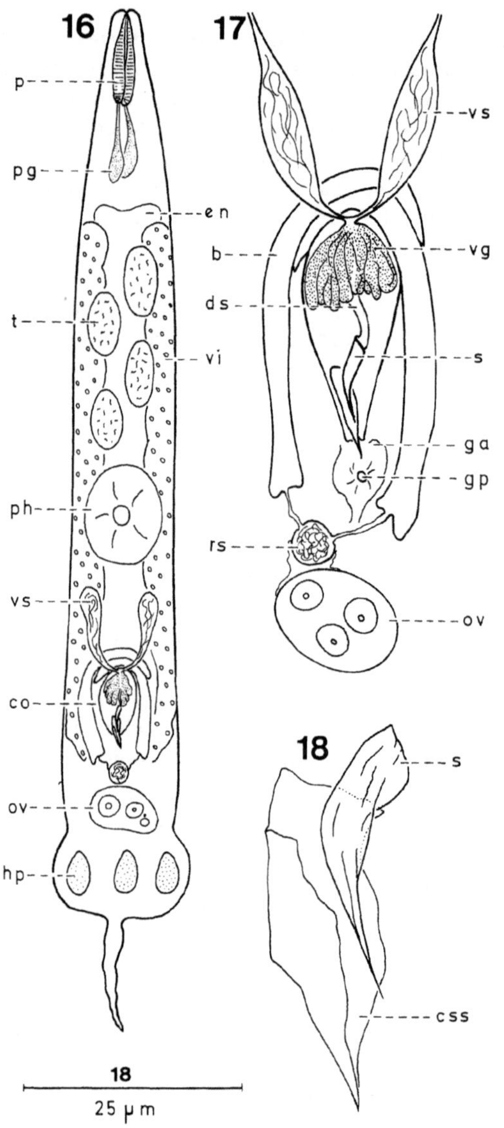

Fig. 16–18. Schizorhynchoides symmetricus. 16. General organisation. 17. Copulatory organ. (Both from living animal.) 18. Cuticular apparatus (from whole mount).

A little spherical seminal receptacle in front of the unpaired ovary is connected with two large elongated bursae by two non cuticular ducts. The slender bursae lie on either side of the copulatory bulb, with their anterior part reflected over the latter. No external vagina was observed.

Discussion

See the discussion below on *S. karlingi* n. sp..

*Schizorhynchoides karlingi** n. sp. (Figs. 19–23 & 29)

Locality: Eastern Scheldt, near buoy 018, fine sand at 3–4 m depth, 24.08.79, (Typelocality).
Material: Three specimens studied alive, two conserved as whole mounts (one of them designated as holotype).

Description

The slender, red, and blind animals are about 2 mm long, have some sensory bristles at the anterior tip and a rounded tail end. The proboscis is about 1/12 th of the total body length. No rhabdites were observed. The topography of the pharynx and genital organs can be derived from Fig. 16. The genital pore, surrounded by glands, is found in the last quarter of the body, the male organ anteriorly, the female organs posteriorly of it. The seminal duct, arising from the confluence of the paired seminal vesicles, run axially through a large copulatory bulb (ca 100 μm long). This duct is expanded to form a relatively small prostate vesicle, the glandbodies lying inside and outside the bulb. The proximal half of the cuticular organ is in part enclosed in the bulb and directly connected to the prostate vesicle. The cuticular apparatus consists of a thin walled cylindrical sheath (33 μm long and 21 μm wide with a proximal expansion), that is ornamented with little bumps and longitudinal folds, and provided on its inner side with five spines. One of the spines arises from the proximal expansion. It is 33 μm long and roughly circular in cross section. This spine probably functions as a sperm draining stylet. The other spines are arranged in pairs on the distal part of the sheath: a lateral proximal pair (39 μm) and a mediolateral distal pair (33 μm). These four spines are clearly triangular in cross section.

The unpaired, caudally situated ovary is connected to the bursa by a weakly cuticularised spermduct. It is funnel shaped, ending in a spherical widening at the side of the bursa were accessory glands occur.

Discussion

Ten species are now known in the genus *Schizorhynchoides.* In most of them, the copulatory stylet is surrounded by a kind of sheath that may be: (1) gutterlike (V-shaped in section) as in *S. meixneri* Boaden, 1963a, *S. rarus* Ax, 1951, *S. symmetricus* sp. n., *S. canaliculatus* L'Hardy, 1963; and most probably also in *S. aculeatus* L'Hardy, 1963, or (2) cylindrical and spiny as in *S. coronostylus* Boaden, 1963a and *S. karlingi* sp. n.; or (3) cylindrical and slightly or not cuticularised as in *S. diplorchis* Meixner, 1928, *S. globulosus* Brunet, 1970 and probably also in *S. macrostylus* L'Hardy, 1963 (with thickened (muscular?) sheath: 'pénis . . . armé d'une canule cuticulaire'). Only in *S. spirostylus* Boaden, 1963a, with a very long stylet, has such a sheath not been described.

The organisation of the female apparatus is poorly known in most species. From the information available it seems much like that in other Schizorhynchidae and even other Schizorhynchia. There are always present: a single caudally situated ovary (with or without adjacent seminal receptacle) and a non-muscular syncytial bursa in from of it. The bursa is connected to the ovary (or seminal receptacle) by (1) a single spermduct in *S. coronostylus, S. karlingi, S. globulosus, S. macrostylus* and *S. meixneri* (in the latter two species only observed *in sectioned material); (2) paired sperm ducts in S. duplorchis, S. spirostylus* and *S. symmetricus;* (3) unknown in *S. aculeatus, S. canaliculatus* and *S. rarus.* On the other hand the bursa opens (1) to the exterior by a vagina in *S. diplorchis* (vagina anteriorly to the male pore), *S. coronostylus* and *S. rarus* (vagina posteriorly to the male pore); or (2) in the atrium as in *S. meixneri* and *S. globulosus*; or (3) has no opening of its own in *S. macrostylus*; while (4) the exact situation is unknown in the other species but most probably without external vagina. The taxonomic weight of this character is of minor importance: an external vagina can arise more than once independently, as can be seen in other genera (e.g. in *Thylacorhynchus, Proschizorhynchus*). The bursal complex in *S. symmetricus* is rather peculiar and necessitates closer consideration: the presence of two sperm-

* Dedicated to Prof. Karling, honoured in this Symposium and volume.

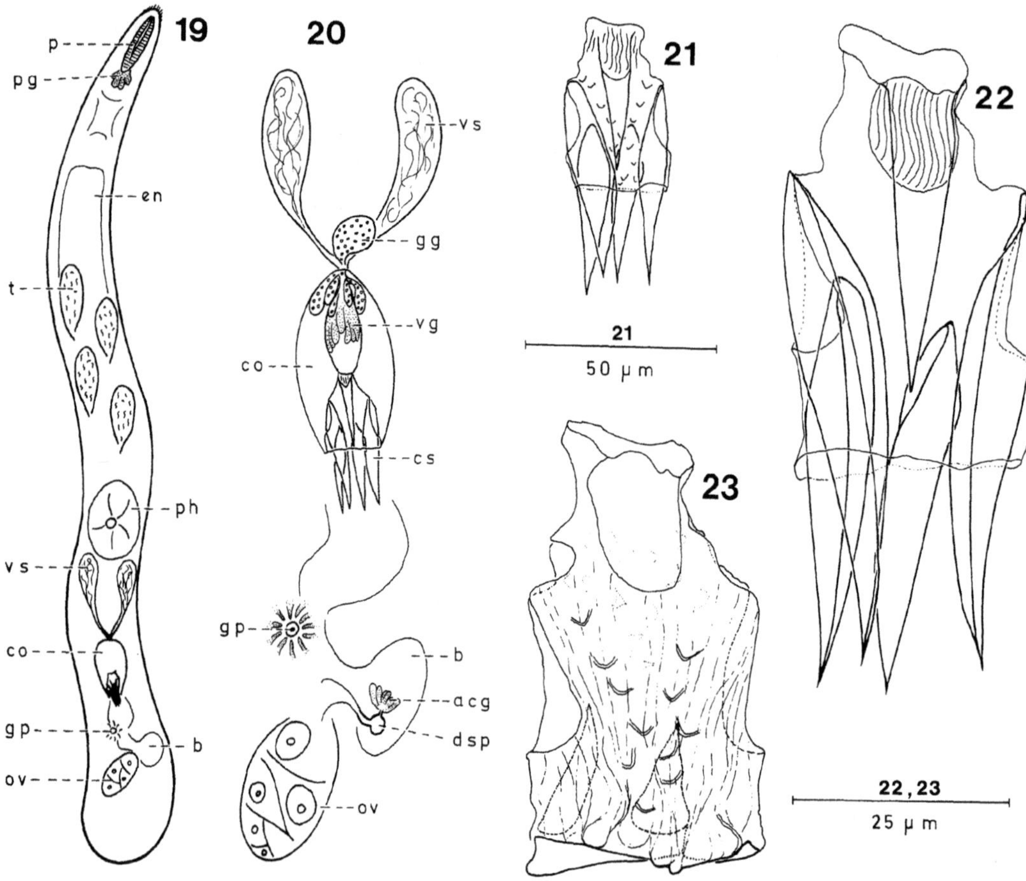

Fig. 19–23. Schizorhyncoides karlingi. 19. General organisation. 20. Genital organs (both from living animals). 21. Cuticular apparatus at low magnification. 22. Spines of the cuticular apparatus. 23. Sheath of the cuticular apparatus. (21–22 from whole mount.)

ducts in *S. duplorchis* and in *S. spirostylus* might be considered of a remnant from an ancestor with paired ovaries (for discussion on related problem: see Karling 1955) (so a plesiomorphy) but the duplication of the spermducts in *S. symmetricus* is evidently the result of the duplication of the bursa, (both apomorph characters). A bilobed bursa has been mentioned for *S. canaliculatus* and *S. meixneri* but in neither of these species does a duplication of the bursa occur. From the above discussion we can conclude: (1) that *S. symmetricus* is closely related to *S. meixneri* and *S. canaliculatus* (stylet construction, bilobed or duplicated bursa) while *S. aculeatus* and *S. rarus* may be included in this species-group (stylet construction), (2) that *S. karlingi* is related to *S. coronostylus* and *S. globulosus* and perhaps *S. macrostylus* (stylet construction and unpaired spermduct) (3) that *S. duplorchis* and *S. spirostylus* form a third group of related species (paired spermducts, stylet construction?).

Identification key to the species of the genus *Schizorhynchoides*

1. Cuticular parts of the copulatory organ consisting of two simple stylet-like pieces (stylet + gutterlike sheath) 2.
   Cuticular organ otherwise 6.
2. External vagina present (behind male pore) *S. rarus* Ax, 1963
   No external vagina 3.
3. Two large bursae anteriorly directed and next to the male copulatory bulb *S. symmetricus* n. sp.
   A single, sometimes bilobed bursa 4.
4. Animals with elongated caudal end but without a little tail *S. aculeatus* L'Hardy, 1963
   With distinctly off set tail 5.

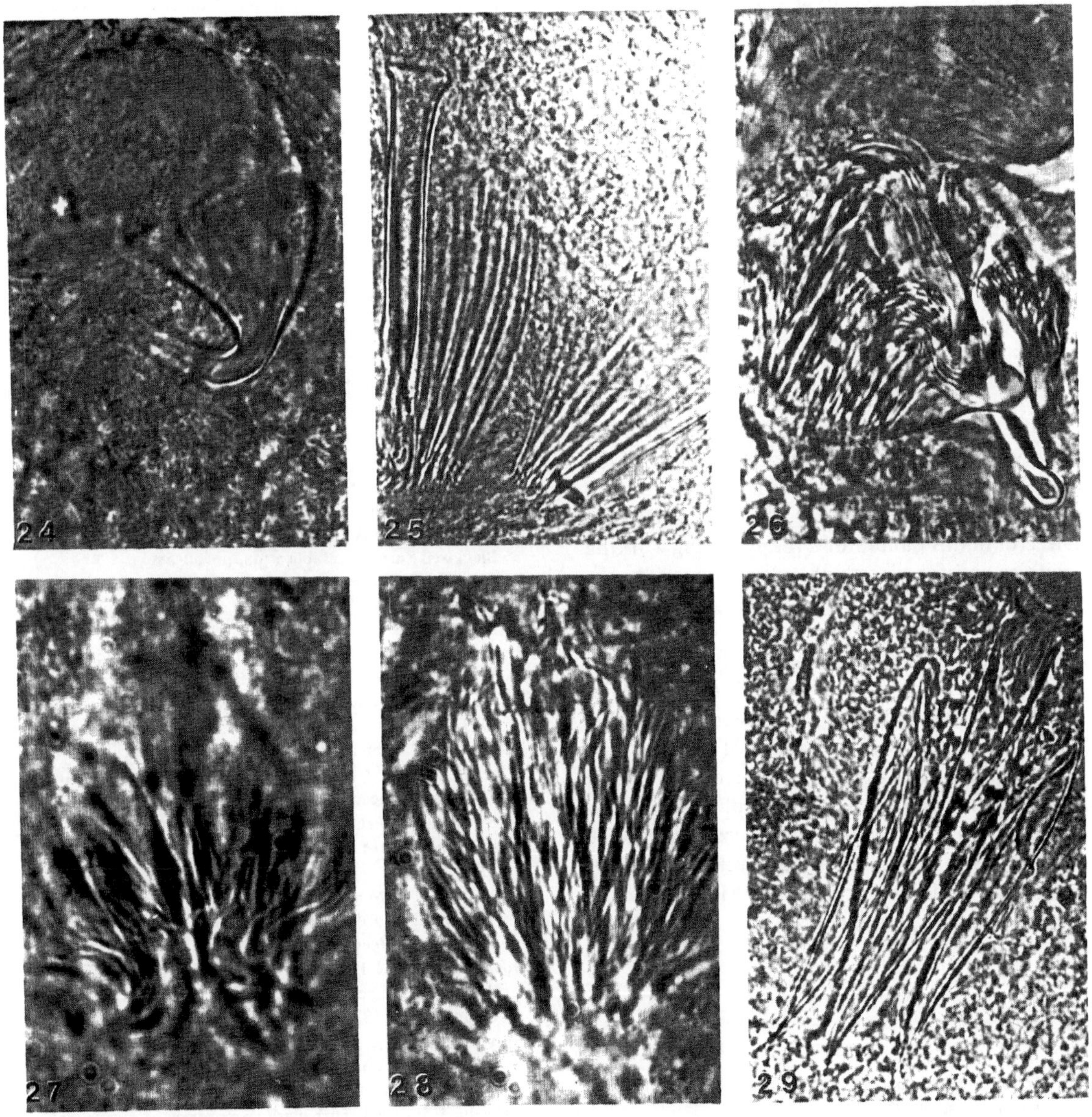

Fig. 24–29. 24. *Bradynectes sterreri;* stylet. 25. *Archimonocelis oostendensis;* cuticular apparatus. 26. *Subulagera rubra;* cuticular apparatus. 27–28. *Paracicerina deltoides;* cuticular apparatus, distal view (27) and proximal view (28). 29. *Schizorhinchoides karlingi;* cuticular apparatus. (All fotographs are from whole mounts.)

5. Stylet and gutterlike sheath straight, 25–27 μm and 35–40 μm respectively
   S. canaliculatus L'Hardy, 1963
   Stylet slightly curved gutterlike sheath very thin, 28–30 μm and 25 μm respectively
   S. meixneri Boaden, 1963
6. Stylet large, spiralised or with spiral ornamentation 7.
   Stylet small with cuticular sheath (weak or folded or spiny) 8.
7. Large stylet with spirally arranged ridge. A single, weakly cuticularised spermatic duct
   S. macrostylus L'Hardy, 1963
   Stylet forms spiral. Paired cuticular spermatic ducts *S. spirostylus* Boaden, 1963

8. Stylet surrounded by a weakly cuticularised sheath. Two spermatic ducts
S. diplorchis Meixner, 1928
Stylet surrounded by a spiny or folded sheath. An unpaired spermatic duct 9.
9. Stylet straight, sheath with spines 10.
Stylet curved, sheath with longitudinal folds
S. globulosus Brunet, 1970
10. Sheath with four spines *S. karlingi* n. sp.
Sheath with numerous small spines
S. coronostylus Boaden, 1963

Acknowledgements

Part of the work was done at the Laboratorium voor Morfologie en Systematiek der Dieren of the State University of Ghent. The junior author (P.M.M.) acknowledges Prof. Dr. Coomans, Director of this Laboratory, for providing working facilities and for his valuable discussions. Thanks are also due to staffmembers of the Laboratory of Ghent for technical assistance more specially Drs. K. Willems, Drs. R. Herman and Mr. W. De Craemer. We are also endebded to the Delta Institute of Hydrobiological Research (Yerseke, Netherlands) and Mr. A. Sandee of this Institute for sampling as well as to Dr. U. Ehlers (Göttingen) for providing sectioned material of *Ciliopharyngiella intermedia* and the *Subulagera* species. The help of Mr. and Mrs. Withofs-Ieven of the Limburgs Universitair Centrum and of Mr. L. Snoeckx was very much appreciated too.

References

Ax, P., 1951. Die Turbellarien des Eulittorals der Kieler Bucht. Zool. Jb. Syst. 80: 277–378.

Ax, P., 1952. Ciliopharyngiella intermedia nov. gen. nov. spec., Repräsentant einer neuen Turbellarien-Familie des marinen Mesopsammon. Zool. Jb. Syst. 81: 286–312.

Ax, P., 1954. Die Turbellarien des Küstengrundwasser am Finnische Meerbusen. Acta zool. fenn. 81: 1–54.

Ax, P., 1956. Monographie der Otoplanidae (Turbellaria). Morphologie und Systematik. Akad. Wiss. Lit. Mainz. Abh. Math.-naturw. Kl. 13: 499–796.

Ax, P. & Ax, R., 1967. Turbellaria Proseriata von der Pazifikküste USA (Washington). I. Otoplanidae. Z. Morph. Tiere 61: 215–254.

Ax, P. & Ax, R., 1974. Interstitielle Fauna von Galapagos. V. Otoplanidae (Turbellaria, Proseriata). Akad. Wiss. Lit. Mainz. Mikrofauna Meeresboden 27: 1–28.

Ax, P. & Heller, R., 1970. Neue Neorhabdocoela (Turbellaria) vom Sandstrand der Nordsee-Insel Sylt. Akad. Wiss. Lit. Mainz. Mikrofauna Meeresboden 2: 55–98.

Boaden, P. J. S., 1963a. The interstitial Turbellaria Kayptorhynchia from some North Wales Beaches. Proc. Zool. Soc. Lond. 141: 173–205.

Boaden, P. J. S., 1963b. The interstitial fauna of some North Wales beaches. J. Mar. Biol. Assoc. U.K. 43: 79–96.

Boaden, P. J. S., 1976. Soft meiofauna of sand from the Delta region of the Rhine, Meuse and Scheldt. Neth. J. Sea Res. 10: 461–471.

Boaden, P. J. S., 1977. Thiobiotic Facts and Fancies. In: Eds., Sterrer, W. & Ax, P., The Meiofauna Species in Times and Space. Akad. Wiss. Lit. Mainz. Mikrofauna Meeresboden 61: 45–63.

Brunet, M., 1970. Turbellariés Schizorhynchidae de sables infralittoraux de la côté marseillaise. Cah. Biol. Mar. 2: 279–306.

Brunet, M., 1973. La famille des Cicerinidae (Turbellaria, Kalyptorhynchia). Zoologica Scripta 2: 17–31.

Buchanan, J. B. & Kain, J. M., 1971. Measurement of the Physical and Chemical Environment. In: Eds., Holme, N. A. & Mc Intyre, A. D., IBP Handbook N° 16. Oxford: Blackwell Scientific Publication, pp. 30–58.

Degadt, D., 1973. Bijdrage tot de faunistiek van de Turbellaria in de Noordzee. Licentiaatsverhandeling Rijksuniversiteit Ghent, 73 pp.

Ehlers, U., 1973. Zur Populationsstruktur interstitieller Typhloplanoida und Dalyellioida (Turbellaria, Neorhabdocoela). Akad. Wiss. Lit. Mainz. Mikrofauna Meeresboden 19: 1–105.

Ehlers, U., 1974. Interstitielle Typhloplanoida (Turbellaria) aus dem Litoral der Nordseeinsel Sylt. Akad. Wiss. LitK Mainz. Mikrofauna Meeresboden 49: 1–102.

Farris, R. A. & Crezee, M., 1976. An improved Reineck box for sampling coarse sand. Int. Revue ges. Hydrobiol. 61: 703–705.

Faubel, A., 1974. Macrostomida (Turbellaria) von einem Sandstrand der Nordseeinsel Sylt. Akad. Wiss. Lit. mainz. Mikrofauna Meeresboden 45: 1–32.

Govaere, J. C. R. & Thielemans, L. K., 1979. A simple device for taking undisturbed sediment cores in shallow water. Biol. Jb. Dodonaea 47: 137–139.

Gray, J. S. & Rieger, R. M., 1971. A Quantitative Study of the Meiofauna of an Exposed Sandy Beach, at Robin Hood's Bay, Yorkshire. J. Mar. Biol. Assoc. U.K. 51: 1–19.

L'Hardy, J. P., 1963. Turbellariés Schizorhynchidae des sables de Roscoff. I. Les genres Carchorodorhynchus, Schizohilus et Schizorhynchoides. Cah. Biol. Mar. 4: 459–472.

Hartog, C. den, 1963. The distribution of the Marine Triclad Uteriporus vulgaris in the Netherlands. Proc. Kon. Ned. Akad. Wetensch. C 66: 196–204.

Hartog, C. den, 1964a. Proseriate flatworms from the Deltaic area of the rivers Rhine, Meuse and Scheldt. I and II. Proc. Kon. Ned. Akad. Wetensch. C 67: 10–34.

Hartog, C. den, 1964b. A preliminary revision of the Proxenetes group (Trigonostomidae, Turbellaria). I-III. Proc. Kon. Ned. Akad. Wetensch. C 67: 371–407.

Hartog, C. den, 1965. A preliminary revision of the Proxenetes groups (Trigonostomidae, Turbellaria). IV–V. Proc. Kon. Ned. Akad. Wetensch. C 68: 98–120.

Hartog, C. den, 1966a. A preliminary revision of the Proxenetes group (Trigonostomidae, Turbellaria). VI–X. Proc. Kon. Ned. Akad. Wetensch. C 69: 97–163.
Hartog, C. den, 1966b. A preliminary revision of the Proxenetes group (Trigonostomidae, Turbellaria). Supplement. Proc. Kon. Ned. Akad. Wetensch. C 69: 557–570.
Hartog, C. den, 1968. An analysis of the Gnathorhynchidae (Neorhabdocoela, Turbellaria) and the position of Psittacorhynchus verweyi nov. gen. nov. sp. in this family. Proc. Kon. Ned. Akad. Wetensch. C 71: 335–345.
Heip, C., Herman, R., Bisschop, G., Govaere, J. C. R., Holvoet, M., Van Damme, D., Vanosmael, C., Willems, K. R., and De Coninck, L. A. P., 1979. Benthic studies of the Southern Bight of the North Sea and its adjacent continental estuaries. Progress Report I. In: Geconcerteerde Onderzoeksacties Oceanologie. Programmatie van het Wetenschapsbeleid, Brussel 133–163.
Hoxhold, S., 1974. Zur Populationsstruktur und Abundanzdynamik interstieller Kalyptorhynchia (Turbellaria, Neorhabdocoela). Akad. Wiss. Lit. Mainz. Mikrofauna Meeresboden 41: 1–134.
Karling, T. G., 1955. Studien über Kalyptorhynchien (Turbellaria) V. Der verwandtschaftkreis von Gyratrix Ehrenberg. Acta zool. fenn. 88: 1–39.
Karling, T. G., 1966. Marine Turbellaria from the Pacific Coast of North America. IV. Coelogynoporidae and Monocelididae. Ack. f. Zool. 18: 493–528.
Karling, T. G., Mack-Fira, V. & Dörjes, J., 1972. First report on marine microturbellarians from Hawaii. Zoologica Scripta 1: 251–269.
Luther, A., 1962. Die Turbellarien Ostfennoskandiens. III. Neorhabdocoela I. Dalyellioida, Typhoplanoida: Byrsophlebidae und Trigonostomidae. Fauna Fennica 12: 1–71.
Marcus, E., 1950. Turbellaria Brasileiros (8). Bolm Fac. Filos. Ciênc. Univ. S. Paulo, Zool. 15: 5–192.
Meixner, J., 1928. Aberrante Kalyptorhynchia (Turbellaria, Rhabdocoela) aus dem Sande der Kieler Bucht (1). Zool. Anz. 77: 229–253.
Rieger, R. M., 1971. Bradynectes sterreri gen. nov., spec. nov., eine neue psammobionte Macrostomide (Turbellaria). Zool. Jb. Syst. 98: 205–235.
Rieger, R. M., 1974. A new group of Turbellaria-Typhloplanoida with a probiscis and its relationship to Kalyptorhynchia. In: Eds. Riser, N. W. & Morse, M. P., Libbie H. Hyman memorical volume; Biology of the Turbellaria, New York: Mc Graw-Hill Company, pp. 23–62.
Sopott, B., 1972. Systematik und Ökologie von Proseriaten (Turbellaria) der deutschen Nordseeküste. Akad. Wiss. Lit. Mainz. Mikrofauna Meeresboden 13: 1–72.
Sopott, B., 1973. Jahrezeitliche Verteilung und Lebenszyklen der Proseriata (Turbellaria) eines Sandstrandes der Nordseeinsel Sylt. Akad. Wiss. Lit. Mainz. Mikrofauna Meeresboden 15: 1–106.
Sopott, B., 1976. Interstitielle Macrostomida und Proseriata (Turbellaria) von der französischen Atlantikküste und den Kanarichen Inseln. Akad. Wiss. Lit. Mainz. Mikrofauna Meeresboden 60: 1–35.
Sterrer, W., 1968. Beiträge zur Kenntnis der Gnathostomulida. I. Anatomie und Morphologie des Genus Pterognathia Sterrer. Ark. Zool. 22: 1–125.
Uhlig, G., 1964. Eine einfache Methode zur Extraktion der vagilen, mesopsammalen Mikrofauna. Helgol. Wiss. Meeresunters. 11: 178–185.
Uhlig, G., 1966. Untersuchungen zur Extraktion der vagilen Mikrofauna aus marinen Sedimenten. Zool. Anz., Suppl. 29: 151–157.
Van Damme, D., Herman, R., Sharma, Y., Holvoet, M. & Martens, P., 1981. Benthic studies of the Southern Bight of the North Sea and its adjacent continental estuaries. Progress Report II. Fluctuations of the meiobenthic communities in the Westerschelde estuary. In: Geconcerteerde Onderzoeksacties Oceanologie. Programmatie van het Wetenschapsbeleid, Brussel: 131–170.
Wolff, W. J., 1973. The estuary as habitat. An analysis of data on the soft-bottom macrofauna of the estuarine area of the rivers Rhine, Meuse and Scheldt. Zool. Verh., Leiden 126: 1–242.

Abbreviations

acg : accessory glands
aco : accessory organ
ad : atrial diverticle
b : bursa
br : brain
cil : cilia
cm : circular muscle
cn : cnidosac
co : copulatory organ
cs : cuticular spines
css : cuticular stylet sheat
de : ejaculatory duct
di : ductus intervesicularis
ds : seminal duct
dsp : spermatic duct
en : enteron
fd : female duct
fp : femal pore
ga : genital atrium
gf : glands in female duct
gg : glands
gp : genital pore
hp : adhesive papillae
ivs : intra capsular seminal vesicle
lm : longitudinal muscle
m : mouth
mp : male pore
ov : ovary
p : proboscis
pg : proboscisglands
ph : pharynx
phg : pharyngial glands
r : retractor muscle
rh : rhabdites
rhg : rhabdite glands
rs : seminal receptacle
s : stylet
sta : statocyst
ut : uterus
t : testis
v : vagina
vg : prostate vesicle
vi : vitellary
vs : seminal vesicle

Notes on the biology of some psammophile Turbellaria of the Black Sea

Galena-Vanzetti Murina
Institute of Biology of the Southern Seas AS Uk.SSR, Nakhimov 2, Sevastopol 335000, U.S.S.R.

Keywords: Turbellaria, *Cercyra*, *Pseudomonocelis*, ecology, reproduction, feeding

Abstract

The biology of two marine turbellarians, *Cercyra hastata* (Tricladida) and *Pseudomonocelis ophiocephala* (Proseriata) has been studied over a period of five years. They are the main components of the biocoenosis of the saccocirrus sand of Sevastopol Bay. These species have a significant role in the processes of secondary production and transformation of the organic matter in the coastal zone.

Introduction

Faunistically, the turbellarian fauna of the Black Sea is well known but knowledge of biological aspects is lacking. We have studied, over a five-year period, the abundance, life-cycle and feeding of two turbellarians which form the main components of the biocoenosis of the *Saccocirrus* sand of Sevastopol Bay (Murina 1974, 1975, 1976). The species are a marine triclad, *Cercyra hastata* O. Schmidt, 1861, and a proseriate, *Pseudomonocelis ophiocephala* O. Schmidt, 1861 and the results of these studies are here summarised.

Occurrence and abundance

Cercyra hastata occurs in the littoral zone throughout the year. It achieves its maximum density, 55 000 ind./m^2, in spring, the largest mature specimen, 5 × 1 mm, being collected in March. The mean annual density was 15 750 ind./m^2 and the mean biomass 5.78 g/m^2.

Pseudomonocelis ophiocephala also occurs throughout the year reaching a maximum density of 115 800 ind./m^2 in August. The mean annual density is 14 640 ind./m^2 with an average biomass of 2.49 g/m^2. The largest specimen, 7 × 0.5 mm, was found in May.

Reproduction

The breeding regimes of the two species have been studied under laboratory conditions and data on seasonal variation in cocoon production and number of young per cocoon obtained. Both are highly fecund species.

Observations on *Cercyra hastata* were carried on throughout the year on laboratory populations simulating natural conditions with a temperature range of 9–24 °C. The mean number of individuals hatching per cocoon was 1.4 in spring, 1.5 in summer, 1.6 in autumn, and 1.7 in winter. Daily cocoon production per worm was 0.4 in spring and summer and 0.5 in autumn and winter. The size of the hatchlings varied in length from 0.5 to 0.85 mm and in width from 0.15 to 0.4 mm, the largest worms occurring among the spring broods. One worm, during a life-span of six months, may produce 128 offspring. The largest animals attained a relative fecundity of 5–6%, representing a wet weight of eggs of 0.038 mg for a bodyweight of 0.71 mg.

Hydrobiologia 84, 129–130 (1981). 0018-8158/81/0842-0129/$00.40.

Incubation time is affected by the temperature and varied from 13 days in summer to 35 days in winter. The mean breeding rate, in the sense of Edmonson (1960) was 1.4 in spring, 4.2 in summer, 3.8 in autumn, and 1.4 in winter. Individual triclads became mature at a length of 2.5 to 3 mm one month after hatching. The triclads breed throughout the year and can produce seven generations during their life-span. In aquarium cultures the maximum longevity was seven months. Newly hatched worms weight 0.03 mg, the mean size for adults was 2.5 to 3.0 mm with a bodyweight of 0.56 mg, and the largest triclads, 3.95 to 4.35 mm, weighed 0.71 mg.

The data for *Pseudomonocelis ophiocephala* were obtained from aquarium cultures maintained at 13.5 to 25 °C. Mean daily cocoon production was 0.52, varying from 0.3 in winter to 0.8 in summer. The size of the hatchlings varied from 0.80 to 0.95 mm in autumn and winter to 0.95 to 1.3 mm in spring and summer. Occasionally two worms hatched from a single cocoon. In the course of its life-span one worm may produce 96 offspring and relative fecundity may attain 3%.

The incubation period varied from 10 days in summer to 22 days in winter. The mean breeding rate was 1.9 in spring, 4.7 in summer, 4.8 in autumn, and 0.8 in winter. In summer cocoon production began in worms at the age of 21 days with a bodylength of 2.5 mm, and in winter at 52 days for a bodylength of 3.7 mm. Longevity was 10 to 11 months during which 4–5 generations were produced.

Feeding

Both species are typical predators and they attack amphipods, oligochaetes, harpacticoids and isopods. Daily food intake, or daily ration, was estimated from feeding experiments in which the difference between the weight of food offered and that remaining at the end of the day was measured. The daily intake varied greatly depending on such factors as the season, the age of the worms and the type of food offered.

The mean daily ration of juvenile *C. hastata* weighing 0.08 mg was 0.09 mg at 12–13 °C. For *P. ophiocephala* 3.5–5.5 mm long it was 0.05 mg at 10–15 °C. The maximal daily ration was 0.25–0.28 mg following 2–3 days of starvation of the worms.

After satiation the turbellarians may survive in the laboratory without further food for one month. *Cercyra hastata* can consume up to 7.6 times its own bodyweight. Such a high ability for food ingestion by these psammophile turbellarians may be considered an adaptation to the large seasonal fluctuations in the food resources of this habitat. Studies using P^{32} as a tracer have established that faecal production begins 1–2 h after feeding and continues for at least two days.

According to Greze (1967) one individual amphipod of the species *Gammarus insensibilis*, a preferred food item, yields 2.86 mg of available food. In one year one mature *C. hastata* may consume 98.5 mg, or about 34 specimens, and one young about 36.5 mg or 13 amphipods. The mean annual density of *C. hastata* has been estimated as 15 750 ind./m^2 and the ratio of mature to young worms is approximately 3:7. Thus, it may be calculated that about 304 000 ind./m^2 of amphipods are consumed each year by the triclads. Similar estimates for *P. ophiocephala* indicate a loss of amphipods to this predator of about 589 000 ind./m^2/year. The high values of these parameters is a clear indication of the importance of these dominant species of turbellarians in the utilisation of the secondary production and the transformation of organic matter in the coastal zone of the Black Sea.

References

Edmonson W. T., 1960. Reproductive rates of rotifers in natural populations. Mem. Ist. Ital. Idrobiol. 12: 21–77.

Greze I. I., 1967. O. kolichestve chitina i kaltsita in pantsiryakh bokoplavov (Amphipoda, Gammaroidea). Zool. Zh. 46: 11.

Murina V. V., 1974. K. biologii chernomorskoy turbellarii Cercyra hastata. V kn. Ekologicheskie issledovania donnykh organismov. Naukova dumka, Kiev, 32: 3–32.

Murina V. V., 1975. Pazmnozhenie i rost psammofil'noy turbellarii Pseudomonocelis ophiocephala v Chernom more. Biologia morya, Naukova dumka, Kiev, 33: 105–116.

Murina, V. V., 1976. Izochenie biologii turbellarii v usloviyakh izmenyaushchejsya solenosti, V kn. Ekspeditsionnye issledovaniya v Sredizemnom more v sprele-iyune i avguste-sentebre 1974 g. Naukova dumka, Kiev: 26–34.

On some commensal Turbellaria of the Canadian East Coast

Lesly C. Fleming[a], Michael D. B. Burt[a] and G. B. Bacon[b]
[a] *Department of Biology, University of New Brunswick, Fredericton N.B., E3B6El Canada* [b] *Research and Productivity Council, Fredericton, N.B., Canada*

Keywords: Turbellaria, Crustacea, Mollusca, commensalism

Abstract

The brachyurans *Chionoecetes opilio* and *Hyas araneus* collected from the east coast of Canada harbour two species of commensal turbellarians. *Ectocotyla hirudo* (Levinsen) and *E. multitesticulata* Fleming & Burt are found on the gills and branchial chambers of male and female crabs. The molluscs *Crassostrea virginica* and *Mytilus edulis* collected from several locations where they are cultured along the Northumberland strait carry two commensal eulecithophoran turbellarians, both on the gills, viz., *Urastoma cyprinae* (Graff) and *Paravortex gemellipara* (Linton), the latter being a new host record. Aspects of the biology and life-history of these turbellarians are discussed, especially in relationship to the biology of their hosts.

Introduction

Two species of *Ectocotyla* (Seriata, Monocelididae) and one species of *Peraclistus* (Seriata, Monocelididae) have been described as ectocommensals of the queen crab, *Chionoecetes opilio* (O. Fabricius) and of the spider crab, *Hyas araneus* (Linnaeus), collected off the east coast of Newfoundland and off the Bay of Chaleur, New Brunswick (Fleming & Burt 1978a,b). Moreover, one species each of *Urastoma* (Eulecithophora, Hypotrichinidae) and *Paravortex* (Eulecithophora, Grafillidae) are known as commensals found on the gills of both oysters, *Crassostrea virginica* (Gmelin), and mussels, *Mytilus edulis* Linnaeus, cultured in many areas along the Northumberland Strait coast of New Brunswick. These turbellarians found on molluscs appear not to be year round commensals and they have been found free-living in nature (Marcus 1951, Westblad 1955). The turbellarians associated with the crabs, however, are year round commensals and appear to have a much closer relationship with their hosts.

In the present paper we consider a number of aspects of the general biology of these species, a topic often sparsely covered in the literature on ectocommensal Turbellaria.

Materials and methods

Specimens of *Ectocotyla* were obtained from the gills and surrounding carapace of *C. opilio* and *H. araneus*, the crabs being caught off the east coast of Newfoundland, and from the Bay of Chaleur, New Brunswick. Specimens of *Peraclistus* were recovered from the egg cases and the abdomen of the same host species. Most of the worms examined were removed from the live crabs immediately after collection, but some crabs were maintained for 2–3 months at 0–4 °C in aerated tanks of artificial sea water (Aquarium Systems Inc.) during which time examinations for worms were performed.

To collect specimens of *Paravortex* and of *Urastoma* monthly samples of oysters were taken from Baie Ste. Anne, Richibucto, Buctouche,

Hydrobiologia 84, 131–137 (1981). 0018-8158/81/0842-0131/$01.40.

1

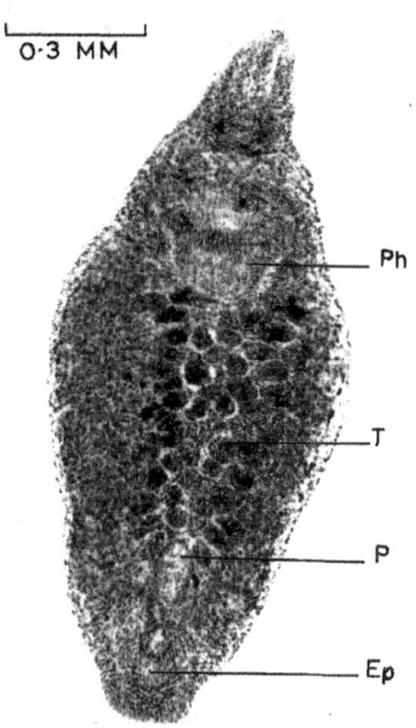

2

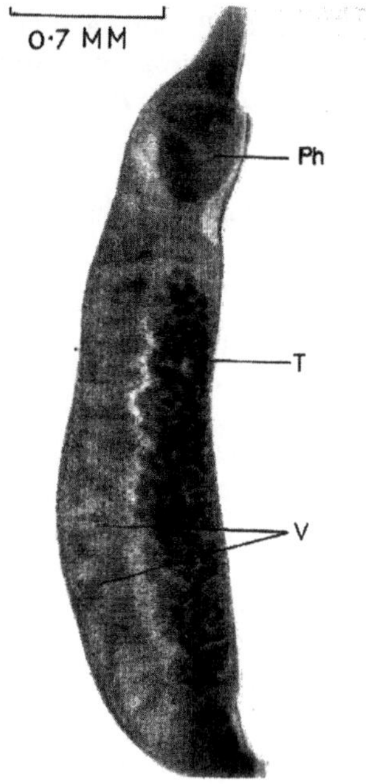

3

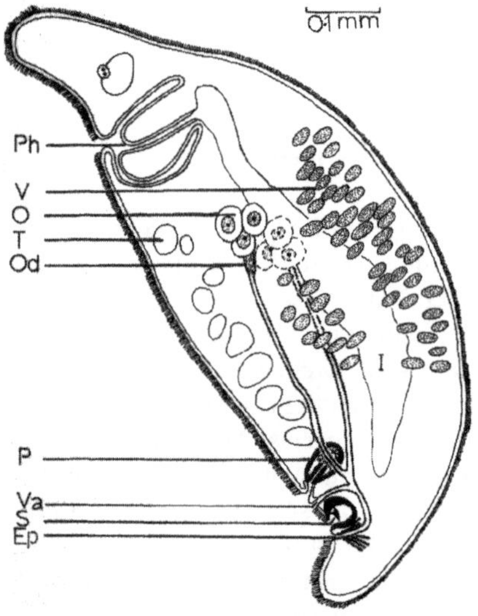

4

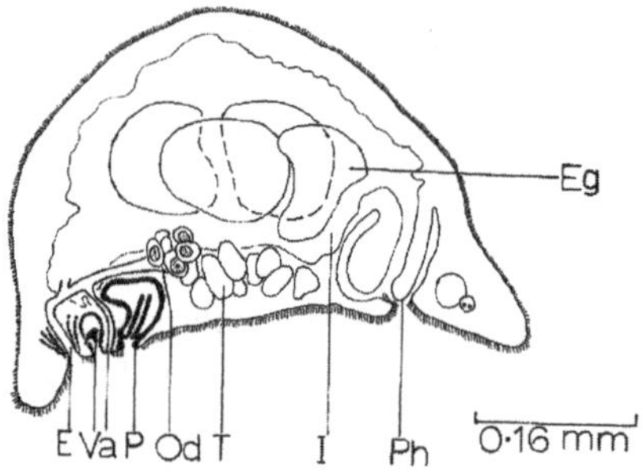

Fig. 1. Ectocotyla hirudo. Whole mount.
Fig. 2. Ectocotyla multitesticulata. Whole mount.
Fig. 3. E. hirudo. Composite diagram.
Fig. 4. E. hirudo. Composite diagram to show eggs in the intestine.

Cocagne and Shediac, N.B. during the period June to October 1978, and in May 1979. A further sample of oysters was collected from Shediac in December 1978.

Samples of 20 mussels were collected from the same sites as the oysters in the months August to October 1978 and in May 1979. Additional samples were collected from Cocagne in February 1979, and from Shediac in December 1978 and February 1979.

Both the oysters and the mussels were examined as soon as possible, usually within 48 h, after they were brought into the laboratory, and they were kept refrigerated, with their shells closed, until examination. Any turbellarians removed from the oysters and mussels that were required for histological study were fixed in hot 5% formalin sea water. Further histological techniques were as described by Fleming & Burt (1978a).

Life history

The Ectocotyla species

In both *Ectocotyla hirudo* and *E. multitesticulata* (Figs. 1 & 2) the young worms hatch from the encapsulated eggs between February and June. At this time there are undifferentiated cells evident in all regions of the body of the worm. The spicule of the accessory organ is absent on hatching but it is one of the first things to develop after the worm leaves the egg. The testes also begin to develop immediately and the penis must begin differentiation soon after hatching as very few young worms are seen without it.

The male reproductive organs continue their development until the autumn (Oct.-Nov.). The testes then have sperm within the follicles and some of it is stored in the muscular seminal vesicle (Fig. 3). During these months the worms copulate after which sperm is stored in the oviducts of the partner (Fig. 3) which must have developed prior to this time although they are not visible until the entry and storage of the sperm.

The number of testes in each worm is largest in the autumn but after copulation the number decreases as they dedifferentiate and those that do remain have but little sperm.

Most worms do not contain eggs during the period immediately after hatching (from spring to autumn). If eggs are present there are usually only 2–4 although on rare occasions up to eight have been seen. After copulation egg production increases to the maximum found for each species. The largest number found in *E. hirudo* was 21, and in *E. multitesticulata* was 33 eggs. During this period the vitellaria continue to develop. They are visible, initially, as thin dorso-lateral bands down either side of the body, but they eventually occupy much of the surrounding region of the gut.

The eggs, fertilized by the stored sperm, pass down the oviduct into the gut where the yolk is added from the vitellaria (Fig. 4). The fertilized developing eggs are released in capsules from February to June via the common oviduct which extends from the gut to the egg-laying pore. The worm then presumably dies later in the summer.

The released eggs remain attached to the host until the following spring. Examination of the egg cases throughout the year reveals little development between the time of egg release and October and it seems that there may be a period of dormancy during that time. Close examination of eggs after October shows the worm to be developing. Eggs stained with toluidine blue and cleared with lactophenol showed the worm to be at an advanced stage of development in April. Evidence of hatching has been found from February until June under natural conditions, and worms have been observed emerging from eggs, under laboratory conditions, in March, May and June. Eggs released by *Ectocotyla*, under laboratory conditions, appeared one every four days for a period of three weeks.

The number of turbellarians associated with recently moulted crabs is much lower than that found on crabs which had moulted several months previously. All the worms found on recently moulted crabs were at least mature with respect to the male reproductive system and turbellarian eggs were never present on such crabs.

Peraclistus oofagus

The life cycle of *Peraclistus oofagus* appears, from the data available, to be similar to that of *Ectocotyla*. Regular monthly samples, however, were not available as female *C. opilio*, on which *P. oofagus* are found, are not caught by commercial trawlers from whom most crab samples were

5

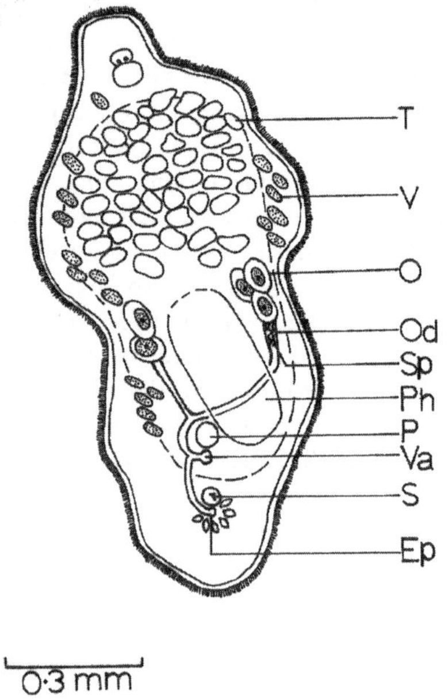

6

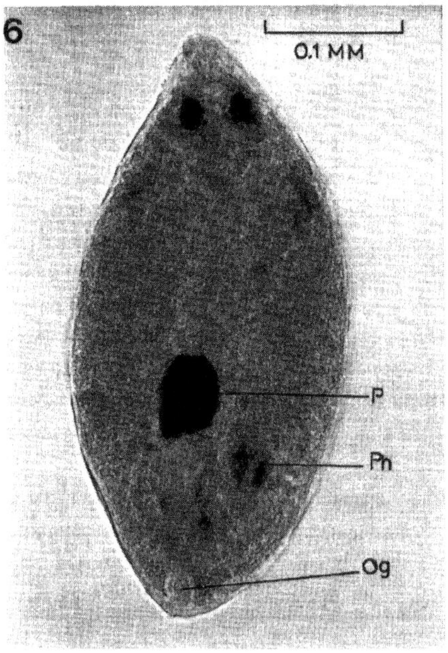

7

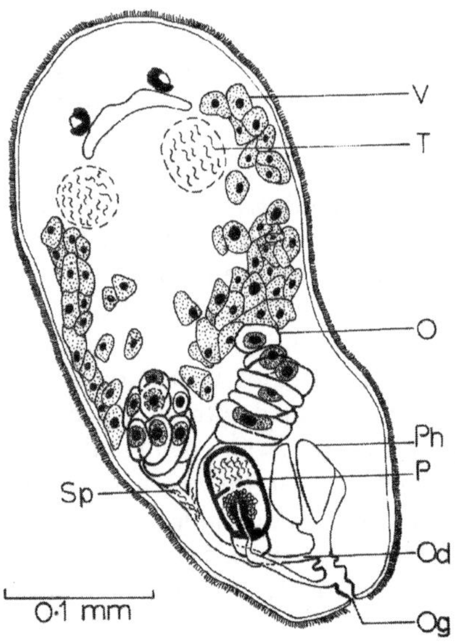

8

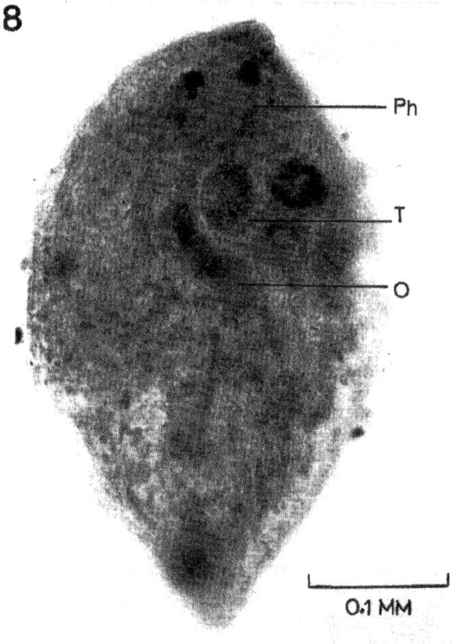

Fig. 5. Peraclistus oofagus. Composite diagram.
Fig. 6. Urastoma cyprinae. Whole mount.
Fig. 7. Urastoma cyprinae. Composite diagram to show sperm fertilizing eggs.
Fig. 8. Parovortex gemellipara. Whole mount.

obtained. *Hyas araneus*, caught in Canadian waters, also provided specimens of *P. oofagus* (Fig. 5) for this study. Examination of those specimens that were available, collected in March, May and November, indicated that the differences in gut structure during the period of egg release are similar to those found in *Ectocotyla* species (Fleming & Burt 1978a).

Peraclistus oofagus and its eggs are also found on the gills and branchial chamber of both male and female *Hyas araneus* collected in Scotland, although they have not been found in these locations on the same host from Canada. There are not sufficient data to show if the worms move freely from the eggs to the gills of the crabs, but the worms have been seen exteriorly on the carapace and legs of the crabs by Steinbock (1932) Hyman (1964) and by us. It is probable that *P. oofagus*, like the species of *Ectocotyla*, can also survive for some time without its host.

Urastoma cyprinae

This species (Fig. 6) was found on the gill surfaces and amongst the gill filaments of *Crassostrea virginica* and *Mytilus edulis* and in the surrounding water during the examination. *Urastoma cyprinae* is a protandrous hermaphrodite. Juvenile specimens have neither male nor female copulatory organs but specimens found in December (two in one oyster) have developed testes but no traces of developing eggs. Worms found from May to June have well developed eggs and vitellaria and much sperm in the penis. Sperm may also be observed fertilizing the eggs in the oviducts (Fig. 7). In July the majority of the worms recovered are juveniles. In this study the number of worms decreased sharply after July and did not increase again until the following May. From May to July, however, *U. cyprinae* was found in large numbers on the oysters collected (only one specimen was found on a mussel).

Paravortex gemellipara

This species (Fig. 8), like *U. cyprinae*, is found on the gill surfaces and among the gill filaments of oysters and mussels and in the surrounding water during the examination. It was recovered in June, August, September, October and December. This turbellarian is also a protandrous hermaphrodite and the single specimens recovered in October and in December both had well developed testes and ovaries which were just beginning to develop. Female worms, with encapsulated young, were found in June, August and September. In July and August both juveniles and worms with developing testes were found.

Both *U. cyprinae* and *P. gemellipara* were found in higher numbers in the summer months, when the oysters and mussels were spawning, than at other times. Almost without exception they were found in females, or in oysters and mussels whih could not at that time be sexed. One mussel with developing male gonads, and collected from Baie du Village (Richibucto) in September 1978, carried two fully mature specimens of *P. gemellipara*.

Table 1 shows the distribution, intensity of

Table 1. Prevalence, intensity, and distribution of *Urastoma cyprinae* and *Paravortex gemellipara*.

		June 1978	July 1978	August 1978	Sept. 1978	Oct. 1978	Dec. 1978	Feb. 1979	May 1979
Baie Ste Anne	Mussels	N/C	N/C	–	–	–	N/C	N/C	–
	Oysters	2P/1 Oy	–				N/C	N/C	–
Richibucto	Mussels	N/C	N/C	–	–	–	N/C	N/C	–
	Oysters	1U/1 Oy	–				N/C	N/C	–
Bouctouche	Mussels	N/C	–	–	–	–	N/C	N/C	–
	Oysters	–	–	–	–	3U/3 Oy	N/C	N/C	12U/4 Oy
Cocagne	Mussels	N/C	–	N/C	–	1P/1 Mu	1P/1 Mu	–	–
	Oysters	–	1P/1 Oy	–	–	–	2U/1 Oy	N/C	37U/7 Oy
Shediac	Mussels	N/C	–	1P/1 Mu	–	–	N/C	–	1U/1 Mu
	Oysters	22U/Oy 1P	48U/12 Oy	2U/2 Oy	–	–	N/C	–	3U/2 Oy

N/C = no collection; P = *Paravortex;* U = *Urastoma;* Oy = Oysters infected; Mu = Mussels infected.

infection and the prevalence of *U. cyprinae* and *P. gemellipara* in both mussels and oysters.

Discussion

Investigations during the past few years have shown that many proseriates are capable of reproducing only during a particular period of their life-cycle. Hartog (1974), Giesa (1966), and Soppott (1973) have described life-cycles studied under natural conditions and the two last named have also studied colonies in culture.

Sopott separated the proseriates she studied into four types: (1) those that reproduce in the spring but have a dormant period within the egg during the summer, (2) those that reproduce in the spring but in which there is no dormant period, (3) those that reproduce in the winter, and (4) those that can produce eggs the year round or which can live for several years. These data concern free-living forms inhabiting eulittoral beaches. Here the temperature fluctuates between 0 and 22 °C and other factors, such as salinity or availability of food, may affect the reproductive cycles of the population.

Ectocotyla, on the other hand, is found relatively deep in the sea where the temperature varies only from 0 to 5 °C and it would appear that its life-cycle is modified more by that of its host. Even so, in the *Ectocotyla* species it is possible for there to be mature females, newly released eggs and developing males at the same time because while one generation is hatching from the eggs the other is preparing and releasing new eggs.

It is clear that both species of *Ectocotyla* can survive as free-living organisms, separated from their crab host, for up to three months at least. This is clearly important as all the hard parts of the crab are lost at each moult. Consequently, any turbellarians associated with the branchial chamber or with the abdominal flap will also be 'shed'. Similarly, if any unhatched turbellarian eggs are attached to the parts shed in the moult, then presumably it will be more difficult for the newly emerged worms to find a new crab host.

When the life-cycles of the crab host and its associated turbellarians are compared there appears to be a close correlation between them. The turbellarian eggs are released by adult worms from February to June and, once laid, they remain up to twelve months before the young emerge. Moulting of crabs takes place between June and September. Consequently, when the crab sheds its carapace, the turbellarians are sufficiently well developed to be able to find their way to the same or another crab host, although judging from the numbers found on recently moulted crabs, not all the worms previously associated with the crabs find their way back. As those worms which are found on recently moulted crabs are at least male mature, it would appear that some advantage accrues if the worms are relatively well developed when the crab moults.

C. opilio has been recorded (Watson 1971) partially buried in the mud on the sea bottom presumably as a protection against predators. Under these circumstances, the turbellarians may be capable of finding a new host, as *Ectocotyla* has been recorded by both Levinsen (1879) and Hyman (1944), and seen during this study, on the surface of the crab.

Crabs kept in holding facilities for longer than a year lost their turbellarians entirely. It is not known, however, whether the turbellarians died due to an increase in water temperature above the optimum, to 10–16 °C during that time, or whether they left the crabs to find a non-moulting host. As the crabs were mature, and had not moulted since they were captured, moulting was presumably imminent. Movement from one crab to another would not be difficult if the crabs are stationary, as Watson (1971) suggests, and there is some reason to believe that the turbellarians respond to conditions present in a crab about to moult by leaving the crab and perhaps seeking another non-moulting or post-moulting host. Passano (1960) has studied the moulting of brachyurans and he states that the proecdysis stage, preceeding the actual moult, occupies nearly 25% of the time from one moult to the next. During this period, which would be about six months of a two year cycle, there are many chemical changes and towards the end of proecdysis the animal ceases feeding and retreats to a safe place. These, or other, factors could signal the forthcoming moult to the turbellarians on the gills and branchial chamber of the crab.

Once the eggs are released onto the gills and branchial chamber they remain there for nine to twelve months, possibly with a period of dormancy between the time of release and the autumn.

Urastoma cyprinae was first found in the mantle

cavity of *Cyprina islandica* and it has since been recorded from *Mytilus edulis, Crassostrea virginica, Solen vagina* and *Modiolus modiolus* (see Burt & Drinnan 1968). The presence of this worm in great numbers on oysters during the summer, and its absence in the winter months, suggests that its life cycle is in some way linked to that of the mollusc host as described by Yonge (1966). The fact that between May and August only mature female worms and juveniles were to be found provides further evidence that their reproductive processes could be linked with the spawning process of the oyster. It is possible that spawning permits the effective release of juvenile worms into the surrounding water along with the expulsion of oyster eggs since female oysters expel their eggs with considerable force.

There was no evidence of encapsulated eggs of *Urastoma* (such as are found in *Ectocotyla, Peraclistus* and other turbellarians). It can only be supposed that development of the juveniles occurs somewhere in or on the mollusc host as very young worms were found on the gills.

The *Peraclistus* and *Paravortex* species were not collected regularly enough for them to be discussed in detail here. The latter we have identified as *Paravortex gemellipara* (Linton 1910) (cf. Ball 1916). This species originally was described from the mantle cavity of the ribbed mussel, *Modiolus demissus* and our findings constitute a new host record. Currently the genus *Paravortex* comprises four species (see Pike & Burt 1981), only *P. gemellipara* being known from North America (Freeman 1957).

Abbreviations in the figures

A	Atrium
An	Antrum
C	Capsule
CO	Common Oviduct
E	Embryo
Ep	Egg-laying pore
GP	Genital Pore
I	Intestine
O	Ovary
Od	Oviduct
Og	Ora-genital Pore
Ph	Pharynx
Sp	Sperm
SV	Seminal Vesicle
T	Testes
Va	Vagina
Y	Yolk

References

Ball, S. C., 1916. The development of Paravortex gemellipara. J. Morphol. 27: 453-558.

Burt, M. D. B. & Drinnan, R. E., 1968. A microturbellarian found in oysters off the coast of Prince Edward Island. J. Fish. Res. Bd. Can. 25: 2521-2522.

Fleming, L. C. & Burt, M. D. B., 1978a. A revision of the genus Ectocotyla Hyman, 1944 (Seriata, Monocelididae) associated with the crabs Chionoecetes opilio (O. Fabricius) and Hyas araneus (Linnaeus). J. Fish. Res. Bd. Can. 35: 1223-1233.

Fleming, L. C. & Burt, M. D. B., 1978b. On the genus Peraclistus (Turbellaria, Proseriata); with redescription of P. oofagus (Friedman). Zool. Scr. 7: 81-84.

Freeman, R. F. H., 1957. Paravortex scrobiculariae (Graff) in Great Britain. Nature (London) 180: 1213-1214.

Giesa, S., 1966. Die Embryonalentwicklung von Monocelis fusca Oersted (Turbellaria, Proseriata). Z. Morph. Okol. Tiere. 57: 137-230.

Hartog, C. den, 1974. Proseriate flatworms from the deltaic areas of the rivers Rhine, Meuse and Sheldt. I and II. Proc. Kon. Ned. Akad. Wetensch. 67: 10-34.

Hyman, L. H., 1944. Marine Turbellaria from the Atlantic coast of North America. Am. Mus. Novit. 1266: 1-15.

Hyman, L. H., 1964. North American Rhabdocoela and Alloeocoela 7. A new seriate alloeocoel, with corrective remarks on alloeocoels. Trans. Am. Microsc. Soc. 83: 248-251.

Levinsen, G. M. R., 1879. Bidrag til Kundskab om Grönlands Turbellariefauna. Vid. Medd. Nat. For. Kbvn. 1879-80: 165-204.

Linton, E., 1910. On a new rhabdocoele commensal with Modiolus plicatulus. J. Exp. Zool. 9: 371-386.

Marcus, E., 1951. Turbellaria Brasileiros (9). Bolm. Fac. Filos. Cienc. Univ. Sao Paulo. (Zool.) 16: 5-215.

Passano, L. M., 1960. Moulting and its control. Physiology of the Crustacea. Vol. 1. Metabolism and Growth. Ed. T. H. Waterman. Academic Press, New York. pp. 474-536.

Pike, A. W. & Burt, M. D. B., 1981. Paravortex karlingi: sp. nov. from Cerastoderma edule L., in Britain. (This volume).

Sopott, B., 1972. Systematik und Okologie von Proseriata (Turbellaria) der deutschen Nordseekuste. Mikrofauna des Meersbodens. 13. 72pp.

Sopott, B., 1973. Jahreszeitliche Verteilung und Lebenszyklen Proseriata (Turbellaria) eines Sandstrandes der Nordsee-insel Sylt. Mikrofauna des Meersbodens. 15. 106pp.

Steinbock, O., 1932. Die Turbellarien des arktischen Gebietes. Fauna Arctica, Jena. 6: 295-342.

Watson, J., 1971. The Atlantic snow (queen) crab. Fisheries Fact Sheet, Department of the Environment, Ottawa.

Westblad, E., 1955. Marine 'Alloeocoels' (Turbellaria) from North Atlantic and Mediterranean coasts. I. Ark. Zool. 7(24): 491-526.

Yonge, C. M., 1966. Oysters. Collins, London. 209 pp.

New England Coelogynoporidae

Nathan W. Riser
Marine Science Institute, East Pt., Nahant, MA 01908, U.S.A.

Keywords: Turbellaria, Proseriata, Coelogynoporidae, taxonomy, New England

Abstract

Eleven species of the proseriate turbellarian family Coelogynoporidae have been encountered between Cape Cod, Mass. and the southern shore-line of New Brunswick, Canada. The distributions of *Coelogynopora schultzii, C. biarmata* and *Cirrifera cirrifera* are reported. Four new species belonging to the genus *Coelogynopora,* one new species of *Cirrifera* and a species belonging to a new genus are described. Two species remain undescribed. Biological observations on a laboratory-reared colony of *C. biarmata* maintained since 1978 are reported.

Introduction

Our group began a survey of intertidal interstitial animals from Cape Cod to the southern shoreline of Canada in the late 1960's. In addition, our divers have established a few sampling stations in sandy bottom areas near Nahant, in Massachusetts Bay. The majority of coelogynoporid proseriate turbellarians have been encountered in coarse sand above the mid-tide level at intertidal stations.

Material and methods

Specimens for sectioning were anaesthetized with Xylocaine and fixed in Hollande's cupri-picri-formal-acetic. The individuals were dehydrated in a graded alcohol series, pre-stained in acid fuchsin in 95% alcohol and embedded in polyester wax. Sections were cut at 4–10 μm and stained either with Heidenhain's Azan technique or Heidenhain's iron haematoxylin counterstained with 0.2% azophloxin. Type specimens were deposited in the American Museum of Natural History.

Observations and descriptions

Genus Coelogynopora

C. schultzii Meixner, 1938 (Figs. 1A, B)

This species has been encountered near ground water in the high intertidal from Nahant, Ma. to Blacks Harbour, New Brunswick. It has occurred most frequently between the summer and winter berms, but at St. Andrews, N. B., where the high tide line extends up the cliff faces, the species is common at the base of the cliffs. Under these conditions, it may be found in the upper 5 cm of the substrate. It also is found throughout the region close to the surface above the high tide line where the sand/gravel substrate shallowly covers beds of clay. The species rarely moves in the twisted 'J' characteristic of the genus. Their sluggish behaviour and flaccid appearance is similar to that of certain species of the nemertinean genus *Ototyphlonemertes.* Epidermal glands are primarily of the granular vermiform type.

C. biarmata Steinböck, 1924 (Figs. 1D, E)

The species is commonly found below mid-tide

Hydrobiologia 84, 139–145 (1981). 0018-8158/81/0842-0139/$01.40.

line in very coarse poorly sorted beaches of the north shore of Massachusetts Bay. Sexual specimens range between 8 and 10 mm in length and 0.6–0.7 mm in maximum breadth while gliding fully extended. The paired accessory spines are bent at almost right angles (Fig. 1D) until late in the reproductive season at which time they become straighter with the apex sharply pointed and the basal portion rugose. One, three, and four accessory spines are occasionally encountered. Those that are unpaired are straight or only slightly flexed.

On 17 July 1978, specimens were obtained from Canoe Beach, Nahant, Ma. by baiting with crushed *Mytilus edulis.* They were maintained in small covered Petri dishes on a sea-water table. Two weeks after isolation, the animals were fed small pieces of gonad and digestive gland of *Littorina littorea.* When the dish was cleaned the following day, it was found that a massive egg laying had occurred. All worms were removed from the dish and a laboratory colony has been established from this single group of eggs. The eggs are stalked and operculated (features which appear to be characteristic of the family). The capsules are usually attached to one another *in vitro.* Very few are encountered individually attached to the bottom of the dish. Capsule diameter ranges from 0.19–0.21 mm and the length from 0.21–0.25 mm. The stalk is short, 0.15–0.2 mm in length. The appearance of the capsules is similar to that of *C. gynocotyla* except that the operculum is smaller than figured by Sopott (1973). Rotation of the ciliated embryos becomes apparent at about ten days and flexed vermiform larvae are distinctive at 17 days. Most eggs hatch in 24–30 days, but some from a single 24-h egg laying period still contain motile vermiform larvae at 35 days. Newly hatched worms are exceedingly thin and about 0.8 mm long. The gut epithelium is filled with digestive vacuoles. Weekly feeding with *Mytilus* gonad or digestive gland/gonad of *L. littorea* has been instituted beginning one week after hatching. Food is initially passed forward in the gut. Feeding ceases and the pharynx is retracted when the post-pharyngeal intestine is only partially packed. Defecation begins at about 90 h and the gut is normally cleared by all members of the colony at 108 h. Egg laying in the laboratory at ambient sea-water temperature +1 °C has begun in March and extended into October. Each year since the culture was established, adults begin to die off in July with less than 40% remaining into September at which time the death rate increases so that almost all have died by November. It appears that the life-span is 14–18 months.

Coelogynopora hymanae n. sp. (Fig. 1I)*

Length to 6 mm. Large oval epidermal flask glands concentrated laterally; scattered dorsally; in two rows ventrally. Sphaerulous glands on dorsal and ventral surfaces, interspersed with flask glands in the ventral rows. Pharynx at anterior end of posterior one-fourth of body length. Precerebral chordoid gut present. Anterior one-third of accessory spines thinner than basal portion and moderately flexed; apex slightly indented and drawn out into two lateral points; total length 70–80 μm. Eight penis spines; two 75–83 μm long with a triangular guard 25 μm below tip; six pointed spines slightly curved toward apex, 95–105 μm long.

The chordoid gut diverticulum begins at the posterior end of the ganglion and extends just beyond the statocyst. The plan of the atrial complex is typical for the genus. The ovovitelline ducts enter the female genital duct ventrally. The female genital duct within the bursal complex is shorter than the bursal stalk in all sectioned specimens. Shell glands are weakly developed. However, there is an extensive development of cement glands opening through the ventral epidermis immediately behind the atrial pore.

C. hymanae has been collected in small numbers in the high intertidal in coarse sand/fine gravel lenses in poorly sorted beaches between the north shore of Massachusetts Bay and Bocabec, New Brunswick. A large uniform population was present in March, 1980, approximately 8 m from the cliff face at Brandy Cove.

Type locality: Brandy Cove, St. Andrews, N.B. Canada.

Type: AMNH No. 930.

C. juxtaforcipis n.sp. (Figs. 1G, J)

Length of sexual animals 3.5–4.8 mm when gliding. Contorted vermiform glands in epidermis. Precerebral region long. Mouth and pharynx at posterior end of middle one-third of body length. Precerebral chordoid gut present. Genital atrium in

*This species is named for Dr. Libbie H. Hyman who applied her unique form of pressure to induce the author to investigate marine turbellaria.

middle of posterior one-third of body length. Accessory spines thin, flat, 0.09 mm long with pointed apex, a strong guard 0.013 mm behind tip and a crest extending from tip of apex to region of guard along opposite surface. Ten penial spines fused at bases; four slender spines 0.064 mm long, with strongly hooked tips; 4 more robust spines 0.44 mm long with serrated curved tips and a small guard; and two short spines 0.025 mm long associated like forceps or a clamp (to which the trivial epithet refers), with guards about in the middle of the length.

C. juxtaforcipis occurs in well sorted coarse sand. It was first encountered in a small lense on Forty Steps Beach at Nahant, and in 1972 was collected by Tor Karling on Preston Beach, Swampscott, Ma. It is common in the proper substrate on the beaches of Plum Island, Ma., but in the pre-impact study following the Argo Merchant oil spill off Nantucket, dense populations were encountered fully ripe in the upper intertidal of Dionis Beach, Nantucket.

Type locality: Dionis Beach, Nantucket Island, Ma.

Type: AMNH No. 932.

C. poaceaglandis n.sp. (Figs. 1C, F; 2A, E)

To 7.0 mm in length when gliding, thin threadlike. Pharynx near beginning of posterior one-fourth of body length. Precerebral chordoid gut long. Paracnids in two orderly rows down each side of body, a variable pattern on dorsal and ventral surfaces from anterior end of pharynx to caudal end of body. Small clear glands less than 6 μm in diameter abundant in epidermis. Adhesive pads widely spaced on sides of body between statocyst and pharynx, numerous posterior to genital atrium. Accessory spines 60–70 μm long pointed with flared beak-like hood. Six pairs of penial spines, two spines 114–120 μm long slightly bent at tip, tip rounded; six spines 85–90 μm long with guards 16 μm below tip, four spines 102–106 μm long with guards 16 μm below tips.

The paracnids (Figs. 1C; 2A, E) 24–32 μm long, 12–15 μm in diameter, are characteristic of this species in the western North Atlantic. They are similar to those described for *C. axi* Sopott, 1972 and *C. gallica* Sopott-Ehlers, 1976. The small clear glands of living animals occur on all surfaces of the body and contain block-like masses of acidophilic material. They lie in the parenchyma with necks reaching through the epidermis, and are densely packed at the posterior end of the body. All penial spines are paired and there are two size classes of spines with guards. The species fits into the complex of *C. hangoensis* Karling, 1955 – *C. gallica* and appears to be extremely close to the latter species. The accessory spines are an intergrade between those of *C. gallica* and *C. hangoensis.*

Type locality: Plum Island, Mass.

Type: AMNH No. 933.

C. erotica n.sp. (Figs. 1H, K)

Gliding sexual animals 7.5–9. mm long. Vermiform glands in epidermis. A few stereocilia on anterior end, 2–3 on each side between apex of head and statocyst. Accessory spines, 12 per bundle, hooked, 60–90 μm long with a hatchet-like bicuspid guard. Penial spines of three types; large with guards near tip, small with guard near middle, and long, slender with guard extending from posterior third almost to tip.

The species has only been encountered in the lower intertidal of poorly sorted beaches. Isolated individuals have been found between Anisquam, Mass. and Pagan Pt., St. Andrews, N.B.

Type locality: Liberty Point, Robbinston, Maine.

Type: AMNH No. 935.

Cirrifera cirrifera Sopott, 1972 (Fig. 2H)

This is the most frequently encountered species of the family in western North Atlantic beaches. It occurs in poorly sorted as well as well sorted substrates. Major concentrations are seaward of the beach surface where the ground water leaches out as recorded by Sopott (1973) at Sylt. The reproductive cycle also follows the pattern reported by Sopott (1973). At Pagan Point, St. Andrews, N.B. in February 1980, the top 4 cm of substrate of the *Cirrifera* zone at one station was 13% by volume *C. cirrifera* eggs. The species attains a very large size in the Bay of Fundy where reproducing specimens range from 16–20 mm in length and 0.7–0.9 mm in maximum width when fully extended.

Cirrifera xanthoderma n.sp. (Fig. 2I)

Gliding sexual individuals 14–18 mm long and 0.8–1. mm wide. Chartreuse colour from densely packed flask cells 20 μm in diameter. Precerebral

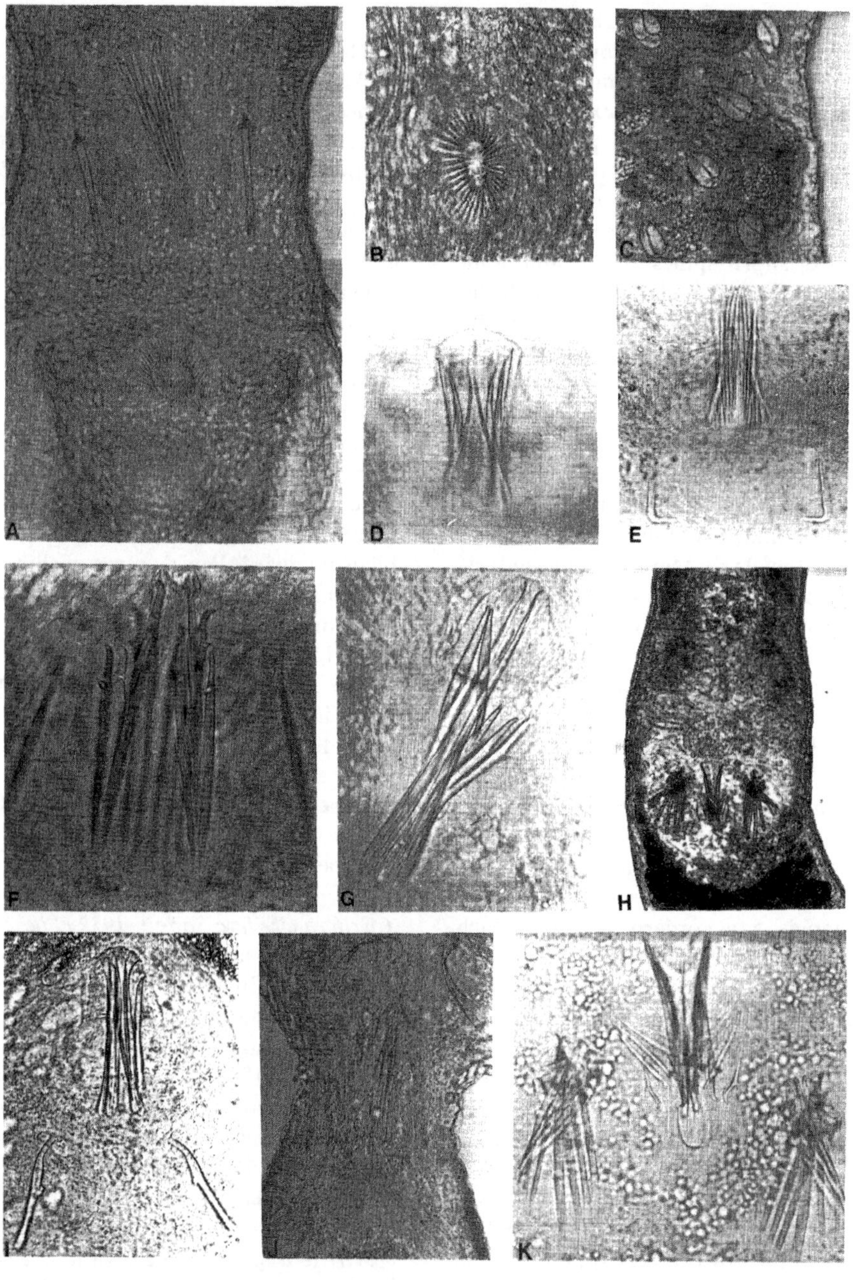

Fig. 1 A. *Coelogynopora schultzii*, penial and accessory spines, solar organ, and seminal vesicles.
B. *C. schultzii*, solar organ.
C. *C. poaceaglandis*, paracnids, surface view.
D. *C. biarmata*, penial spines, ventral view.
E. *C. biarmata*, penial and accessory spines, dorsal view.
F. *C. poaceaglandis*, penial and accessory spines.
G. *C. juxtaforcipis*, penial spines.
H. *C. erotica*, penial and accessory spines.
I. *C. hymanae*, penial and accessory spines.
J. *C. juxtaforcipis*, penial and accessory spines.
K. *C. erotica*, penial and accessory spines.

gut diverticulum absent. Pharynx in posterior one-third of body. Ovaries at end of testes row some distance anterior to pharynx. Atrial complex posterior to middle of post pharynx region. Terminal part of cirrus bulbous and uniformly armed with large tripod-like spines 12–16 μm long with curved bifid tips; stalk of cirrus densely armed with small scimitar-shaped spines. Seminal vesicles paired, ejaculatory duct enters cirrus from behind.

The species is readily identifiable by its chartreuse colouration. The flask glands touch one another and form a continuous layer beneath the epidermis of the dorsal surface. In sectioned material, some of these gland cells are empty and others contain a finely granular acidophilic secretory product. The flask cells of the other species reported here are all empty, with the same histological procedures. Thus, two different gland cells appear to be present. The flask cells fill most of the space between the dorsal epidermis and the internal organs. The dorsal musculature is reduced, and the ventral musculature is strongly developed. In addition, the species differs from *C. cirrifera* in the armature and orientation of the cirrus. The ejaculatory duct does not arise anterior to the cirrus, but joins it posteriorly as in *C. dumosa* Sopott 1972. The pharyngeal glands extend from the ovaries to the tip of the pharynx where there necks disrupt the musculature and ciliated epidermis. The basophilic granules in the necks of the glands occupy most of the space between the two walls of the pharynx. A rod of basophilic cells extends from the cerebral region emptying at the anterior tip of the body. The posterior end of the rod lies between the gut and cerebrum when the head is contracted. Contracted living animals have the body configuration of placobdellid leeches. The species has the same breeding season as *C. cirrifera* and the egg capsule is similar in size, 325–385 μm in diameter, and appearance. Freshly collected specimens frequently contain setae of the polychaet, *Microphthalmus cf. listensis*.

C. xanthoderma has been collected on Plum Island, Ma. in addition to the type locality. On both beaches, it has a spotty distribution but occurs in abundance where the dominant sand particle size is between 0.5 and 1.0 mm. (Manomet – 0.4–0.8% < .5 mm, 22–28% > 1. mm; Plum Island – 0.8–1.% < .5 mm, 15% > 1. mm).

Type locality: Manomet Beach, Manomet, Mass.
Type: AMNH No. 936.

Macroatrium n.g.

Coelogynoporidae with a bursa and a cirrus. Seminal vesicles posterior to gonopore.

Macroatrium setosum n.g. n.sp. (Figs. 2B, C, D, F, G)

Gliding sexual individuals 6.–8. mm long. Epidermal bacillary glands packed with elongated rhabdites, single precerebral row along both sides of midline, others scattered. Cephalic stereocilia long and numerous. Precerebral chordoid diverticulum almost reaching anterior end of body. Precerebral region short. Statocyst close in front of cerebrum. Pharynx large and globular. Ovaries in front of pharynx base at distance equal to precerebral region of animal. Testes and vitellaria extend ventrally approximately 2/3rds distance between ovaries and anterior end of body. Mouth at anterior end of posterior 1/4th of body length. Genital pore about 1/2 distance between mouth and posterior end of body. Atrial complex large, occupying middle 1/3rd of post-oral length. Cirrus bipartite, approximately 20 taenoid hooks 22–25 μm long with guards, in ventral wall between regions; spinous portion muff-like with curved spines 9–12 μm long; post-hook region with glandular epithelium.

The statocyst is located close to the cerebral ganglion (Figs. 2C, F). The latter structure has a tri-lobed fan shape under pressure with the two lateral lobes almost as long as the middle lobe. Sections demonstrate that the middle lobe is encapsulated except at its lateral sides. The precerebral chordoid diverticulum (Fig. 2F) extends almost to the anterior end of the body. Bacillary glands are abundant throughout the parenchyma of the body. Anterior to the cerebrum they are scattered on the ventral and lateral surfaces but form two discrete rows, one to either side of the chordoid diverticulum in the dorsum. Each of the cells in these rows lies parallel to the epidermis. Posteriorly they extend between the gut diverticulae almost half-way through the body and are concentrated in the region behind the gut (Figs. 2B, G). The secretion from these glands forms an amorphous mass when extruded from the cells. Adhesive pads

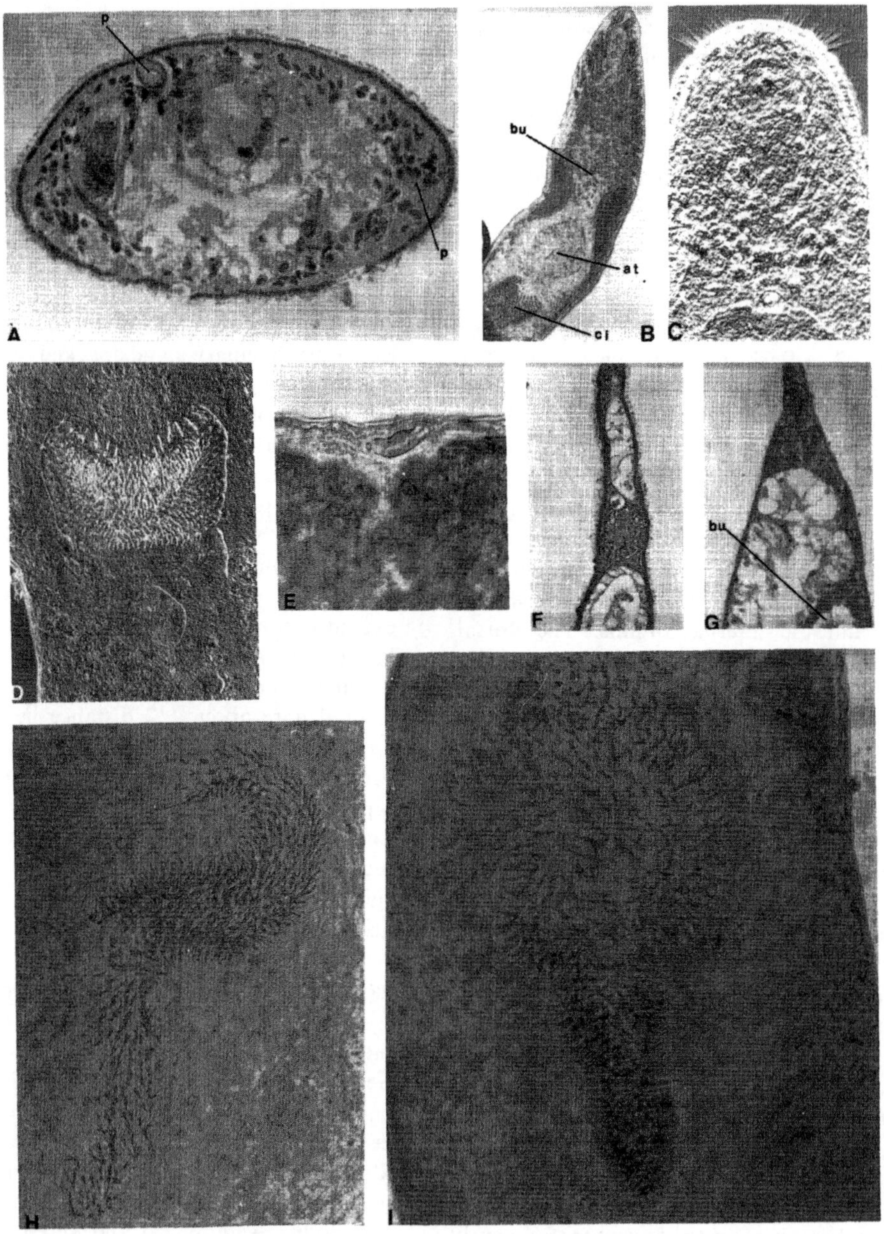

Fig. 2 A. Cross-section of *C. poaceaglandis*, p = paracnids.
B. *Macroatrium setosum*, atrial complex, at = atrium, bu = bursa, ci = cirrus.
C. *M. setosum*, anterior end with apical and lateral bristles, and statocyst immediately in front of cerebrum.
D. *M. setosum*, cirrus.
E. *C. poaceaglandis*, paracnid in lateral view.
F. L-section of anterior end of *M. setosum*.
G. L-section of posterior end of *M. setosum*, bu = bursa.
H. Cirrus of *Cirrifera cirrifera* with anterior ejaculatory duct.
I. Cirrus of *C. xanthoderma* with posterior ejaculatory duct.

occur from the cerebral region to the posterior end of the body on the ventral surface, being most abundant postpharyngeally. The opening into the genital atrium is very large, and the bursa is about the same size as the atrium (Fig. 2B). The epithelium of the bursal wall (Fig. 2G) consists of basophilic secretory cells. Acidophilic glands discharge into the female genital canal near its opening into the atrium. A genito-intestinal canal is absent. The seminal vesicles extend about half-way back along the sides of the bursa.

M. setosum has been collected in the low intertidal from a number of beaches in Massachusetts. It has also been obtained from sand samples at 6 meters depth along the north shore of Massachusetts Bay.

Type locality: Dionis Beach, Nantucket, Ma.

Type: AMNH No. 937.

Discussion

Coelogynopora schulzii, C. biarmata, and *Cirrifera cirrifera* are now recorded from both sides of the North Atlantic. At the present time, no North Atlantic species of the family have been encountered in the Pacific (Ax & Ax 1974, Ax & Sopott-Ehlers 1979, Karling 1966, Tajika 1978). The long period of development of the eggs may account for the distribution of some of the species in the North Atlantic which could have been transported in sand ballast. However, this leads to some confusion because both sides of the Atlantic also seem to contain endemic species. Two species belonging to the genus *Coelogynopora* have been encountered in addition to those reported in this paper, but in insufficient numbers to allow for an adequate description.

References

Ax, P. & Ax, R., 1974. Interstitielle fauna von Galapagos VII. Nematoplanidae, Polystyliphoridae, Coelogynoporidae (Turbellaria, Proseriata). Mikrofauna Meeresbodens 29: 613–638.

Ax, P. & Sopott-Ehlers, B., 1979. Turbellaria Proseriata von der Pazifikküste der USA (Washington) II. Coelogynoporidae. Zool. Scripta 8: 25–35.

Karling, T. G., 1966. Marine turbellaria from the Pacific coast of North America IV. Coelogynoporidae and Monocelididae. Ark. Zool. (2) 18: 493–528.

Sopott, B., 1972. Systematik und Ökologie von Proseriaten (Turbellaria) der deutschen Nordseeküste. Mikrofauna Meeresbodens 13: 169–236.

Sopott, B., 1973. Jahreszeitliche Verteilung und Lebenszyklen der Proseriata (Turbellaria) eines Sandstrandes der Nordseeinsel Sylt. Mikrofauna Meeresbodens 15: 255–358.

Tajika, K. I., 1978. Zwei neue arten der gattung Coelogynopora Steinböck, 1924 (Turbellaria, Proseriata) aus Hokkaido, Japan. J. Fac. Sci. Hokkaido Univ. (6) 21: 295–316.

Physiological adaptations to entosymbiosis in three species of graffillid rhabdocoels

J. B. Jennings
Dept. of Pure and Applied Zoology, Baines Wing, Univ. of Leeds, Leeds LS2 9JT, England

Keywords: Turbellaria, Graffillidae, entosymbionts, physiology, nutrition, respiration

Abstract

Three entosymbiotic graffillid rhabdocoels (*Paravortex scrobiculariae, P. cardii* and *Graffilla buccinicola*) from marine bivalve and gastropod molluscs show several physiological adaptations to their life-style which are intimately related to the nutritional physiology and ecology of their respective hosts.

All three species feed on their hosts' partially digested food plus the cellular debris released at the end of the hosts' own digestive cycle. *G. buccinicola* supplements this diet by actively removing intact cells from the host's digestive epithelium. Host enzymes, ingested with the food, are utilized for digestion within the flatworms' gut; there is concomitant reduction in the types and amounts of endogenous enzymes and the gastrodermal gland cells characteristically found in free-living species are absent.

Food reserves in the three species consist mainly of glycogen, following the pattern seen in other entosymbiotic flatworms (Turbellaria, Digenea, Cestoda); in *P. scrobiculariae* this primary adaptation, believed to be linked in all entosymbiotes to the ready availability of food and to high fecundity, probably has a secondary function, in relation to anaerobic respiration, of the type found in cestodes.

Other adaptive features, closely correlated with host ecology, are the occurrence of a physiologically active haemoglobin in the brain and pharynx of *P. scrobiculariae* and, in this species and *P. cardii*, the differential occurrence and distribution of dehydrogenase systems concerned with aerobic respiration (Krebs cycle), the pentose phosphate shunt and anaerobic respiration (glycolysis).

P. scrobiculariae and *P. cardii* are viviparous and the normal provision of yolk for embryonic nutrition is supplemented by direct passage of materials to the later larvae from the parental gut, thus facilitating extended development of the larva until its birth as an immature miniature adult.

Introduction

Structural and life cycle modifications occurring in symbiotic Turbellaria as adaptations to life upon or within other organisms are now relatively well known and have been reviewed in detail elsewhere (Jennings 1971, 1974a). Comparable physiological adaptations, however, are less well known apart from information on the nutrition of ectosymbiotic temnocephalids and triclads, and entosymbiotic umagillids and graffillids (Jennings 1977, 1980). Of these four groups the graffillids appear to be the most modified in that the types and amounts of endogenous enzymes involved in digestion and absorption are much reduced, when compared with those produced by free-living turbellarians, and the flatworms depend largely on enzymes of host origin ingested with the food (Jennings & Phillips, 1978). This occurs in all entosymbiotic graffillids so far studied and is related to the characteristic pattern of digestive physiology common to their molluscan hosts. In contrast, adaptive features in respiratory physiology vary from species to species in degree of

Hydrobiologia 84, 147–153 (1981). 0018-8158/81/0842-0147/$01.40.

elaboration and apparent importance in the life of the entosymbiotes. These variations are intimately related to the ecology of the particular host species, and especially to the effect this has on the host's feeding and digestive cycles, its behaviour during the twelve hour tidal cycle and the consequent variations in the amounts of oxygen available in the entosymbiotes' microhabitat.

In reviewing these various adaptive features it is convenient to consider first nutritional physiology, since this has received most attention and, in any case, nutritional relationships are the basis of most, if not all, symbioses. This will include an account of certain features of nutrition in the developing larvae of the two viviparous species. Finally, respiratory modifications for entosymbiosis will be discussed.

Species reviewed

1. *Paravortex scrobiculariae* (Graff). Viviparous. From the intestine of the bivalve mollusc *Scrobicularia plana*, occurring in British waters in estuarine mudbanks at or near highwater mark.

2. *Paravortex cardii* Hallez. Viviparous. From the main ducts of the digestive gland of the bivalve *Cerastoderma edule*, occurring in Britain on sandy bottoms around low water mark.

3. *Graffilla buccinicola* Jameson. From the digestive glands of the sub-littoral gastropods *Buccinum undatum* (from Scottish waters) and *Neptunea decemcostata* (from the coast of Maine, U.S.A.).

Adaptations in nutritional physiology

1. The adult

Feeding, gut structure, digestion and food storage in *P. scrobiculariae*, *P. cardii* and *G. buccinicola* have been studied in detail by Jennings & Phillips (1978). All three species feed on their hosts' partially digested food which they abstract from the lumen of the digestive gland ducts, and in *G. buccinicola* this diet is supplemented by intact digestive cells which are sucked off from the digestive epithelium. The gut in the three species is of the usual dalyellioid type, with an anterior doliiform pharynx and saccate intestine. The gastrodermis, however, is unusual in that the club-shaped gland cells which are a characteristic feature of most free-living Turbellaria (Jennings 1974b) are absent; the intestine is lined by a single layer of monotypic columnar cells whose cytoplasm generally contains a variety of inclusions shown by histochemical methods to be heterolysosomes ('food vacuoles') in which phagocytosed food particles are undergoing digestion and absorption. In *P. scrobiculariae* and *P. cardii* the columnar cells are often so distended by these intracellular vacuoles that opposing faces of the gastrodermis may touch at irregular intervals. This occlusion of the intestinal lumen subdivides it into large, temporary extracellular vacuoles. Food particles trapped in these undergo digestion prior to being engulfed by the columnar cells for completion of digestion intracellularly. In *G. buccinicola* there are no extracellular 'vacuoles' and the gastrodermis is at all times a regular organised layer of cells.

Absence of histologically distinct gland cells has been confirmed in histochemical and ultrastructural studies. In contrast, the free-living graffillid *Pseudograffilla arenicola* has a gastrodermis composed of well differentiated columnar phagocytes and club-shaped secretory cells (Jennings, unpublished observations on preserved material kindly supplied by Drs. Beate and Ulrich Ehlers). The clear implication that this difference in gastrodermal structure is related to the differences in diet and life style has been confirmed by histochemical studies which show that the bulk of the enzymes responsible for both extra- and intracellular digestion in the entosymbiotic species are, in fact, of host origin and are ingested along with the food. Ready availability of such exogenous enzymes, manifestly of the types best suited for the digestion of the particular food materials ingested by the host (bacteria, diatoms, unicellular algae and organic detritus in the microphagous hosts of *P. scrobiculariae* and *P. cardii*, and animal tissues in the carnivorous hosts of *G. buccinicola*), has apparently led to the loss from the graffillids' gut of many similar endogenous enzymes and certainly of specific gland cells adapted for their extracellular secretion.

Enzymes produced in the digestive glands of the two microphagous hosts (*S. plana* and *C. edule*) include non-specific esterases, A- and C-esterases, lipase, arylamidases, acid phosphatase, β-glucuronidase and β-galactosidase. All these enzymes have been demonstrated in the gut contents of graffillids living in these hosts and so, too, have heterolysosomes which contain the enzymes together with granules rich in iron and calcium. Such heterolysosomes are formed within the hosts' digestive cells and are eventually released, when the cells break down at the end of a digestive cycle, as 'excretory spheres.' In the hosts the contained enzymes initiate extracellular digestion in the first phase of the next digestive cycle. The inorganic contents of the spheres are easily demonstrable, histologically, and thus form convenient markers proving the origin of the graffillids' gut contents.

Graffillids from these two hosts when starved by being isolated *in vitro* in filtered seawater show greatly reduced histochemical reactions for the enzymes listed above, indicating only low levels of endogenous enzymes. Two enzymes, β-galactosidase and arylamidase, were never detected in such individuals although they were readily demonstrable in others freshly removed from their hosts. It would appear, then, that the graffillids are entirely dependent on the hosts for these particular enzymes. Graffillids fed *in vitro* on enzymically inert foods (starch grains and heat-inactivated bacteria, algae or yeast cells) similarly showed only low levels of enzymic activities in the gut, confirming the dependence on host enzymes.

A similar overall pattern occurs in *G. buccinicola,* living in the carnivorous and scavenging gastropods *Buccinum undatum* and *Neptunea decemcostata.* Here, again, histochemical techniques have shown that the graffillid is dependent on its hosts for both food and digestive enzymes. An interesting further adaptation is seen in this particular symbiosis, however. The hosts, taking largely proteinaceous food of animal origin, show great emphasis on proteases in their spectrum of digestive enzymes whereas in the microphagous bivalves, and their symbiotes, carbohydrases predominate. *G. buccinicola* is obviously adapted to this type of diet in that, unlike the two *Paravortex* species, it possess some endogenous arylamidases. It still depends on its hosts, though, for the bulk of these enzymes required for digestion of the protein-rich diet.

In all three species the dependence on the hosts for provision of digestive enzymes, as well as of food, is correlated with the type of digestive physiology characteristic of bivalve and gastropod molluscs (ably reviewed by Owen 1966, 1972 & 1974, & Purchon 1977). A salient feature of this is the occurrence of extracellular digestion by enzymes arising intracellularly and released by cell disintegration, or cytoplasmic abstriction, at the end of intracellular digestion of an earlier meal. Without this feature, it is doubtful whether this particular type of entosymbiosis could have arisen.

In inter-tidal bivalves feeding and extra- and intracellular digestion show a pronounced cyclical pattern; phases in the cycle are cued by tidal factors and this is seen most clearly, of the host species considered here, in *Scrobicularia plana.* Living in mud near highwater mark, feeding can occur only during the relatively short period of submergence; during this period *P. scrobiculariae* is to be found only in the intestine but it later migrates anteriorly into the digestive gland to feed during the digestive phase of the host's cycle, presumably because by this time the host's food is partially digested and host enzymes are available for ingestion. *P. cardii* lives permanently in the main ducts of the digestive gland and it is not clear why *P. scrobiculariae* should not do likewise. The host of *P. cardii,* however, living around low water mark, does not show the clearly delineated phases in its feeding and digestive cycles found in *S. plana* and this would appear, somehow, to be the decisive factor.

The much-reduced capacity to secrete their own digestive enzymes means that these three entosymbiotes have a high degree of metabolic dependence on their hosts and they are unable to survive away from them for more than a few days. This is in marked contrast to ectosymbiotes such as the temnocephalids and bdellourid triclads, which produce the normal complement of digestive enzymes and can survive for up to several months, and even breed, away from their hosts (Jennings 1977). There is an interesting contrast, too, with the entosymbiotic nemertean *Malacobdella grossa;* this species lives within the mantle cavity of its bivalve host and merely abstracts suspended food from the water brought into the cavity by the host's feeding current (Gibson & Jennings 1969). It does not utilize any other component of its microphagous host's feeding mechanism, nor any part of its digestive physiology. Consequently, it produces the full spectrum of enzymes needed for digestion and assimilation of its diet, unlike the graffillids. Thus the type of entosymbiosis shown by *M. grossa* can be regarded, specifically, as entocommensalism since nutritional aspects of the relationship are restricted to a sharing of the same food. In the graffillids, though, their entosymbioses consist more of entoparasitism than pure entocommensalism, and in *G. buccinicola* active abstraction of living host cells classifies this species firmly as an entoparasite.

2. Nutrition in the embryo and larva

Both *Paravortex scrobiculariae* and *P. cardii* are viviparous; the embryos are usually twinned and develop within capsules in the mesenchyme until they have functional eyes, nervous and alimentary systems. At this point they break from their capsules, pass through the parental gastrodermis into the intestine and eventually are voided from the mouth in what is believed to be the archaic turbellarian manner of giving birth (see Hyman 1951).

Development follows the pattern described by Ball (1916) for *P. gemellipara*, with yolk globules provided with the fertilised egg nourishing the growing embryo until the eyes, pharynx and intestine are clearly visible and the epidermal cilia functional. At this stage, when the young can be regarded as larvae and are moving freely, by ciliary action, within their capsules it is possible to demonstrate both soluble and particulate nutriments, identical with those in the parental gut, in the capsular fluid and larval gut (Jennings & Phillips 1978). At these sites reactions were obtained for inorganic iron and calcium present either in solution or confined in particles identical with those in the heterolysosomes ('excretory spheres') taken in by the adult from the host digestive gland. Further, yeast cells and starch grains fed to the parent in experimental feeding procedures were demonstrated at both sites. The mechanism of the implied transfer across the parental gastrodermis and larval capsule wall remains unknown, but the capacity to achieve this at the time when the larval gut is becoming functional is presumably of some advantage. It may well allow the young *Paravortex* to develop beyond the capacity conferred by the normal provisioning of the egg by the yolk glands, thus enhancing its chances of survival during transfer to a new host or establishment in that of its parent. Such larval feeding may, of course, continue within the parental gut before birth but this has not been demonstrated.

No such modifications of embryonic or larval nutrition have been reported for *G. buccinicola*, but it is perhaps noteworthy that the yolk glands and vitellaria are extremely well-developed so that the eggs on laying are well supplied with nutriments.

3. Food reserves

The principal food reserve in all three species is glycogen, which is laid down in the mesenchyme, gastrodermis and vitellaria (Jennings & Phillips 1978). Quantitative data are not available but histological and histochemical techniques suggest that the gyclogen is present in amounts comparable to those found in other entosymbiotic flatworms (see Mettrick & Jennings 1969; Calow & Jennings 1974). Very little lipid is present, in contrast to the situation in free-living turbellarians (Jennings 1974b). It would appear from this that the entosymbiotic graffillids follow the pattern observed in all other entosymbiotic flatworms, and that the emphasis on glycogen storage is another fundamental adaptation to their life style. The significance of glycogen storage, and the relative lack of emphasis on lipid, in terms of reproductive strategies, 'r' and 'K' selection and the evolution of entosymbiotic life styles have been discussed at length elsewhere (Calow & Jennings 1974; Jennings & Calow 1975; Jennings 1977) and will not be explored further here. Suffice it to say that it has been argued that high glycogen levels reflect a reliable and adequate food supply, permitting channelling of all available resources into reproduction ('r' strategy) and involving high energy turnover, rather than preservation of the adult by lipid storage to tide over periods of food shortage and resulting in production of fewer but 'fitter' progeny ('K' strategy). In the entosymbiotic graffillids glycogen storage is thus a primary adaptation to entosymbiosis in terms of reproduction; in one species (*P. scrobiculariae*) it will be argued later that it is also a secondary adaptation in relation to anaerobic respiration.

Adaptations in respiratory physiology

1. The occurrence of haemoglobin in Paravortex scrobiculariae

P. scrobiculariae differs from the other entosymbiotic graffillids in possessing a red pigment associated particularly with the brain and pharynx but present also in smaller amounts elsewhere in the body. The pigment has been shown by spectrographic, cytochemical and ultrastructural techniques to be a physiologically active haemoglobin capable of reversible reactions with oxygen which are dependent on environmental oxygen tension (Phillips 1978). Individuals treated *in vitro* with carbon monoxide survived for as long as untreated specimens. Such treatment induces a permanent cherry red colouration, shown spectrographically to be due to irreversible formation of a carboxyhaemoglobin, and this type of non-lethal effect is generally regarded as proof that the haemoglobin is not essential for survival under normal aerobic conditions (Prosser 1973). It is believed, therefore, that the adaptive significance of this haemoglobin, in *P. scrobiculariae,* lies in its utilization during the anterior migration of the graffillid from the host's intestine into the digestive gland to feed. It will be remembered that this migration occurs after the period of total submergence by the host at high water, during the period when the host's ctenidial filtration system is closed down, extra- and intracellular digestion in the digestive gland are in full operation and when paradoxically, oxygen will be least freely available. It is at precisely this time that the graffillid is most active and, therefore, in most need of oxygen, especially in the brain and pharynx. The relatively small amount of haemoglobin present in *P. scrobiculariae* would not seem to be sufficient for it to act as an oxygen store, but it may be adequate for a process of 'facilitated diffusion' of the type postulated by Wittenberg (1970). This would allow the graffillid to compete successfully, against the host's tissues, for such oxygen as is available within the host's body, blood or in the residual water of the mantle cavity, the host lacking any comparable haemoglobin or haemocyanin (Morton 1967).

Thus the haemoglobin of *P. scrobiculariae* is probably a further adaptation to entosymbiosis, arising in this species because of the particular nature of its host's ecology and nutritional physiology. *P. cardii* and *G. buccinicola,* living in hosts whose habitats are permanently well-aerated, are not subjected to changes in oxygen tension and, understandably, lack haemoglobin.

The occurrence of haemoglobin has been reported in a few other, albeit free-living, rhabdocoels. It is present, for example, in *Phaenocora unipunctata* and *P. typhlops* (Crompton & Smith 1963; Young & Harris 1973) and both these species, significantly, are mud-dwellers living under conditions of low or variable oxygen tension. Another mud-dweller, the nematode *Enoplus brevis* has a pharyngeal haemoglobin comparable to that of *P. scrobiculariae* which presumably facilitates the action of the pharyngeal pump; the closely related species *E. communis,* however, lacks any such pigment but lives in more aerated situations (Atkinson 1975).

2. The differential occurrence and distribution of dehydrogenase systems in Paravortex scrobiculariae and P. cardii

A considerable number of dehydrogenases can now be demonstrated histochemically (Pearse 1972) and these include some from each of the three major energy-producing pathways (the Krebs tricarboxylic acid cycle, the pentose phosphate shunt and glycolysis). Positive or negative histochemical reactions for such dehydrogenases, application of semi-quantitative methods of interpretation and the use of appropriate controls, therefore, permit deductions to be made regarding the probable presence or absence of these pathways and their relative importance in the life of the animal.

Using this approach Jennings & LeFlore (1979) have shown that in both *P. scrobiculariae* and *P. cardii* dehydrogenases associated with the aerobic Krebs cycle occur in most parts of the body and, as would be expected, are most evident in areas of high metabolic activity such as the musculature, alimentary system, brain and reproductive system. At these sites positive histochemical reactions were obtained, in adults, embryos and larvae, for succinic dehydrogenase (SDH), malate dehydrogenase (MDH) and isocitrate dehydrogenase (ICDH).

Similarly, both species showed strong positive reactions for dehydrogenases indicative of the presence of the pentose phosphate shunt. 6-phosphogluconate dehydrogenase (6-PGDH) and glucose-6-phosphate dehydrogenase (G-6-PDH) were particularly evident in embryos and larvae of all ages from early gastrulae to late larvae complete with eyes and active epidermal cilia. The pentose phosphate shunt is known to be associated with synthesis of nucleotides and nucleoproteins; under normal conditions *Paravortex* embryos grow rapidly within the parent and stages in mitotic division are commonly found in histological preparations. Thus the occurrence and ease of demonstration of 6-PGDH and G-6-PDH in the embryos most probably reflects the rapid synthesis of nuclear material during growth and tissue differentiation. In the adult, moderate reactions for these two enzymes were obtained in the brain and pharynx and weak to very slight reactions occurred in the gastrodermis.

Other dehydrogenases sharing a common occurrence and distribution in the two species are glutamate dehydrogenase (GDH), β-hydroxybutarate dehydrogenase (β-HBDH), nicotinamide-adenine dinucleotide diaphorase (NADHD) and nicotinamide-adenine dinucleotide phosphate diaphorase (NADPHD). GDH occurs in the vitellaria and gastrodermis; this enzyme is concerned with amino acid synthesis from keto acids and thus its presence at these sites is understandable. Similarly β-HBDH, associated with fatty acid metabolism, occurs in the vitellaria in both species and NADHD and NADPHD, involved in many metabolic processes, occur in most parts of the body and especially in the epidermis.

In marked contrast to this common distribution of Krebs cycle, pentose phosphate shunt and other dehydrogenases concerned with general metabolism there are significant differences in the occurrence and distribution of those involved in anaerobic glycolysis. In *P. scrobiculariae* the brain, pharynx and vitellaria of adults, and the brain and pharynx of advanced larvae, contain large amounts of easily demonstrable α-glycerophosphate dehydrogenase (α-GPDH), lactate dehydrogenase (LDH) and alcohol dehydrogenase (ADH). In *P. cardii,* though, these enzymes occur, in very small amounts only, in the adult epidermis and vitellaria.

It would appear from this difference in the distribution of α-GPDH, LDH and ADH, between the two species, that glycolysis is of little importance in *P. cardii,* whereas in *P. scrobiculariae* it seems that this method of energy release approaches in importance that of the Krebs cycle. The occurrence of these enzymes in the brain and pharynx is of especial significance, paralleling as it does the distribution of haemoglobin at these sites and taking into consideration the pattern of feeding and digestion found in this species. The migration of *P. scrobiculariae* from its host's intestine into the digestive gland, the ingestion of food and the early stages of digestion all occur after the tide has receded from the host's habitat and during the period, therefore, when oxygen supplies will be minimal. The adaptive value in this situation of some capacity for anaerobic release of energy in the brain and pharynx, supplementing or possibly for a time replacing aerobic processes facilated by the presence of haemoglobin, is obvious.

Glycolysis is common in many Digenea and most, if not all, Cestoda (Read 1968; von Brand 1946 & 1973) and its adaptive significance in these flatworms, living under variable or deficient oxygen levels, is equally apparent. The significance of glycogen storage in these entosymbiotes and in the three graffillid species, as a primary adaptation to entosymbiosis and the assured food supply conferred by this life style, has been mentioned earlier (p. 150). It has also been argued that the presence of large quantities of glycogen, in the Digenea and especially the Cestoda, constituted a pre-adaptation for glycolysis in that the substrate was already available and allowed colonization of further entosymbiotic habitats in which deficiency or lack of oxygen was a potentially limiting factor (Jennings 1977). A broadly comparable situation is seen here in *P. scrobiculariae*; glycogen is readily available for glycolysis provided the necessary enzyme systems can be evolved and thus the graffillid can be most active at critical periods when, due to peculiarities in its hosts' ecology, activity could be expected to be suppressed due to oxygen deficiency. In this context, then, storage of glycogen in *P. scrobiculariae* constitutes a secondary but most important physiological adaptation to entosymbiosis.

References

Atkinson, H. J., 1975. The functional significance of the haemoglobin in a marine nematode Enoplus brevis (Bastian). J. Exp. Biol. 62: 1-9.

Ball, S. J., 1916. Development of Paravortex gemellipara. J. Morphol. 27: 453–557.
Brand, T. von, 1946. Anaerobiosis in Invertebrates. Biodynamica, Normandy, Missouri. 328 pp.
Brand, T. von, 1973. Biochemistry of Parasites. 2nd. edn. New York: Academic Press, 499 pp.
Calow, P. & Jennings, J. B., 1974. Calorific values in the phylum Platyhelminthes: the relationship between potential energy, mode of life and the evolution of entoparasitism. Biol. Bull. 147: 81–94.
Crompton, D. W. T. & Smith, M. H., 1963. Haemoglobin in a flatworm. Nature, London 197: 1118–1119.
Gibson, R. & Jennings, J. B., 1969. Observations on the diet, feeding mechanism, digestion and food reserves of the entocommensal rhynchocoelan Malacobdella grossa O. F. Müller. J. Mar. Biol. Assoc. U.K. 49: 17–32.
Hyman, L. H., 1951. The Invertebrates: Platyhelminthes and Rhynchocoela. The acoelomate Bilateria II. New York: McGraw-Hill, 550 pp.
Jennings, J. B., 1971. Parasitism and commensalism in the Turbellaria. In: Ed. Dawes, B., Advances in Parasitology, 9, New York: Academic Press, pp. 1–32.
Jennings, J. B., 1974a. Symbioses in the Turbellaria and their implications in studies on the evolution of entoparasitism. In: Ed. Vernberg, W. B., Symbiosis in the Sea, Univ. South Carolina Press, Columbia, South Carolina, pp. 127–160.
Jennings, J. B., 1974b. Digestive physiology of the Turbellaria. In: Riser, N. W. & Morse, M. P. (eds.), Biology of the Turbellaria, pages 173–197. McGraw-Hill Book Co., New York.
Jennings, J. B., 1977. Patterns of nutritional physiology in free-living and symbiotic Turbellaria and their implications for the evolution of entoparasitism in the phylum Platyhelminthes. In: Karling, T. G. & Meinander, M. (eds.), The Alex Luther Centennial Symposium on Turbellaria. Acta Zool. fenn. 154: 63–79.
Jennings, J. B., 1980. Nutrition in symbiotic Turbellaria. In: Eds., Smith, D. C. & Tiffon, Y., Nutrition in the Lower Metazoa, Oxford: Pergamon Press, pp. 45–56.
Jennings, J. B. & Calow, P., 1975. The relationship between high fecundity and the evolution of entoparasitism. Oecologia 21: 109–115.
Jennings, J. B. & LeFlore, W. B., 1979. Occurrence and possible adaptive significance of some histochemically demonstrable dehydrogenases in two entosymbiotic rhabdocoels (Platyhelminthes: Turbellaria). Comp. Biochem. Physiol. 62B: 301–304.
Jennings, J. B. & Phillips, J. I., 1978. Feeding and digestion in three entosymbiotic graffillid rhabdocoels from bivalve and gastropod molluscs. Biol. Bull. 155: 542–562.
Mettrick, D. F. & Jennings, J. B., 1969. Nutrition and chemical composition of the rhabdocoel turbellarian Syndesmis franciscana, with notes on the taxonomy of S. antillarum. Can. J. Zool. 26: 2669–2679.
Morton, J. E., 1967. Molluscs. 4th edn. London: Hutchinson and Co., 244 pp.
Owen, G., 1966. Digestion. In: Eds., Yonge, C. M. & Wilbur, K. M., Physiology of Mollusca, vol. 2, New York: Academic Press, pp. 53–96.
Owen, G., 1972. Lysosomes, peroxisomes and bivalves. Sci. Prog. 60: 299–318.
Owen, G., 1974. Feeding and digestion in the Bivalvia. In: Ed., Lowenstein, O., Advances in Comparative Physiology and Biochemistry, 5, New York: Academic Press, pp. 1–35.
Pearse, A. G. E., 1972. Histochemistry: theoretical and applied. 3rd edn. Edinburgh: Churchill Livingstone.
Phillips, J. I., 1978. The occurrence and distribution of haemoglobin in the entosymbiotic rhabdocoel Paravortex scrobiculariae (Graff) (Platyhelminthes: Turbellaria). Comp. Biochem. Physiol. 61A: 679–683.
Prosser, C. L., 1973. Respiratory functions of blood. In: Ed., Prosser, C. L., Comparative Animal Physiology, pages 317–361. New York: Academic Press.
Purchon, R. D., 1977. The Biology of the Mollusca, 2nd edn. Oxford: Pergamon Press, 560 pp.
Read, C. P., 1968. Intermediary metabolism of flatworms. In: Eds. Florkin, M. & Scheer, B. T., Chemical Zoology 2, New York: Academic Press, pp. 327–357.
Wittenberg, J. B., 1970. Myoglobin-facilitated oxygen diffusion: role of myoglobin in oxygen entry into muscle. Physiol. Rev. 50: 559–636.
Young, J. O. & Harris, J. H., 1973. Another instance of haemoglobin in a flatworm Phaenocora typhlops (Vejdovsky 1880). Freshwater Biol. 3: 85–88.

The diet of three species of umagillid neorhabdocoel turbellarians inhabiting the intestine of echinoids

George L. Shinn
Dept. of Zoology, Univ. of Washington, Seattle, WA 98195, U.S.A.
and Friday Harbor Labs, Friday Harbor, WA 98250, U.S.A.

Keywords: Turbellaria, Umagillidae, symbiosis, nutrition, *Syndisyrinx, Syndesmis*

Abstract

Three species of umagillid turbellarians were found to have different nutritional relationships with their echinoid hosts: *Syndisyrinx franciscanus* ingests host intestinal tissue and ciliates that are symbiotic in the intestine of the host; *Syndesmis dendrastrorum* consumes intestinal tissue and materials that have been ingested by the host; an umagillid that closely resembles *Syndesmis echinorum* subsists entirely on host intestinal tissue.

Introduction

Neorhabdocoel turbellarians of the family Umagillidae live symbiotically in various echinoderms and sipunculans. This family has more symbiotic species than any other family of Turbellaria (Jennings 1971), but our understanding of associations between umagillids and their hosts is fragmentary.

Syndisyrinx franciscanus is found in the northeastern Pacific in several species of *Strongylocentrotus* and in *Lytechinus anamesus* (Lehman 1946; Orihel 1952; Barnes 1969). This worm and *'Syndesmis' antillarum* (probably also a member of the genus *Syndisyrinx*) are the only umagillids whose diets have been reported in the literature. Mettrick & Jennings (1969) recovered *Syndisyrinx franciscanus* from the coelom and intestine of *Strongylocentrotus purpuratus*, and found that ciliates, which also live as symbionts in *Str. purpuratus*, constitute its main or only food. Holt & Mettrick (1975) found only ciliates in the intestine of *Syndisyrinx franciscanus* that had been taken from the digestive tract of *Strongylocentrotus purpuratus*. In both studies it was concluded that *Syndisyrinx franciscanus* is a harmless commensal.

Syndesmis antillarum has been reported from the coelom of *Diadema antillarum* (Powers 1936; Mettrick & Jennings 1969) and from the coelom and intestine of *Lytechinus variegatus* (Westblad 1953; Jennings & Mettrick 1968); both of these echinoids are found in the Caribbean. According to Jennings & Mettrick (1968) *S. antillarum* subsists largely on ciliates that live in the coelom, but it consumes some host coelomocytes. They concluded that *S. antillarum* is a relatively harmless commensal with slight tendencies toward parasitism.

There are several reasons for suspecting that nutritional relationships between umagillids and their hosts are more involved than revealed in these studies. Some umagillids, such as *Syndesmis dendrastrorum* from the irregular echinoid *Dendraster excentricus*, inhabit hosts that do not harbor symbiotic ciliates (Orihel 1952; Berger 1964). These umagillids must consume either host tissue or material ingested by their hosts. Studies by protozoologists have revealed that echinoid-inhabiting ciliates live in the intestine but not in the coelom of their hosts; ciliates reported from the coelom have probably escaped from the gut during dissection (Beers 1948; Berger 1964; Powers 1936). If umagillids do indeed inhabit the coelom of echinoids, they would have to consume host tissue or take up nutrients from the coelomic fluid. Finally, some variety may be expected in the nutritional biology of umagillids considering that nearly 40 species in 16 genera have been described (Jennings 1971).

The diets of *Syndisyrinx franciscanus, Syndesmis dendrastrorum,* and an umagillid that resembles *Syndesmis echinorum** were studied by examining the contents of food vacuoles in gastrodermal cells of live worms. The locations of the umagillids in their hosts, and of symbiotic ciliates in the hosts, were determined through careful dissections. Specimens of *Syndisyrinx franciscanus* were artificially introduced into the perivisceral coelom to determine if they can survive there.

Material and methods

Specimens of *Syndisyrinx franciscanus* and the umagillid that resembles *Syndesmis echinorum* were obtained from the regular echinoids *Strongylocentrotus pallidus* and *Str. droebachiensis;* these were collected subtidally near San Juan Island, Washington, U.S.A. Freshly-collected urchins were dissected following the general procedure of Beers (1948). The peri-oral membrane and muscles attached to the Aristotle's lantern were cut and the lantern was pulled part way out of the body. Several drops of perivisceral coelomic fluid were pipetted from the urchin and examined, using a compound microscope, for the presence of ciliates. The rest of the coelomic fluid was poured into a Petri dish and scanned with a dissecting microscope. In order to examine the lining of the perivisceral coelom for the presence of symbionts, parts of the test were carefully chipped away. Next, the gut was removed intact, and then opened. As umagillids were found they were transfered to chilled filtered sea water so that they would not ingest foreign material before being examined. With the objective of making an adequate survey of the distribution of worms within hosts, urchins were dissected until 15 of each species had been examined that contained at least ten *Syndisyrinx franciscanus.*

* Studies on the systematic status of this species are in progress.

Syndesmis dendrastrorum were obtained from specimens of the irregular echinoid *Dendraster excentricus* collected intertidally at False Bay, San Juan Island, Washington, U.S.A. The hosts were dissected following the procedure of Smith (1973).

Live umagillids were compressed, dorsal side up, on a slide and examined with the aid of a compound microscope. Food vacuoles in gastrodermal cells were usually visible through the nearly transparant body wall. The gut contents of worms from freshly-collected hosts were compared to squashed pieces of intestinal tissue of the host and also to the contents of the intestine of the host. *Syndisyrinx franciscanus* from echinoids that had been starved in the laboratory for $2\frac{1}{2}$ months were similarly examined.

Umagillids and pieces of tissue from the intestines of their hosts were fixed following the procedure of Reed & Cloney (1977), embedded in Epon, and sectioned at 1 μm for light microscopy.

To determine if *Syndisyrinx franciscanus* can survive outside of the digestive tract, specimens were artificially introduced into the perivisceral coelom of ten *Strongylocentrotus pallidus* and ten *Str. droebachiensis.* The worms were collected from urchins of the species into which they would be introduced. They were rinsed in Millipore-filtered sea water, then transferred to either new Millipore-filtered sea water or to coelomic fluid that had been removed with a syringe from an experimental urchin two days before, Millipore-filtered, and kept frozen until the experiment. A small hole was drilled through the experimental urchin's test near the periproct, between the two rows of tube feet of one of the ambulacra (in this area the body wall borders the perivisceral coelom). Five worms were pipetted into each urchin along with very small amounts of one or the other of the last fluids in which they were rinsed. The holes in the tests were plugged with corks, and the urchins were fed pieces of the kelp *Nereocystis luetkeana.* Two urchins of each species were dissected 1, 7, 14, 60, and 90 days later. Worms recovered from the coelom of these urchins were examined immediately with dissecting and compound microscopes. Several specimens were preserved in Bouin's fixative, embedded in paraffin, and sectioned. Controls were treated the same as experimental urchins except that no worms were involved, and 0.5 ml aliquots of either Millipore-filtered sea water, or Millipore-filtered coelomic fluid that had been removed from the same urchin and kept frozen for two days, were introduced into them.

Results

Syndisyrinx franciscanus was found only in the second half of the intestine of *Strongylocentrotus pallidus* and *Str. droebachiensis.* Forty-one urchins were examined and found to contain between 0 and 186 (avg. 29) worms. The worms were usually crawling on the intestinal epithelium, although some were among the contents of the intestine. Symbiotic ciliates were present throughout the intestine but were not found in the coelom of any of the urchins.

All *Syndisyrinx franciscanus* that were experimentally introduced into the perivisceral coelom of urchins became enclosed by a dense layer of aggregated coelomocytes (Fig. 1). The worms became necrotic within seven days when pipetted in with sea water, but they survived unharmed for at least as long as two months when introduced along with coelomic fluid. Worms introduced with coelomic fluid contained large numbers of cells that had been ingested from the inside of the layer of coelomocytes. Urchins that were dissected three months after worms were introduced into their coelom contained neither worms nor aggregates of coelomocytes. Some whispy aggregates of coelomocytes were present for about two weeks in the coelomic fluid of controls into which sea water had been pipetted. Controls into which their own coelomic fluid had been introduced did not develop coelomocyte aggregates.

Ciliates were the most easily-recognized objects in the intestine of *Syndisyrinx franciscanus* (Figs. 2 & 3). *Cyclidium stercoralis, Madsenia indomitus, Euplotes tuffraui,* and a species of *Cryptochilidium* were sometimes found alive within worms that had eaten them, and could be identified. These ciliates are common inhabitants of *Strongylocentrotus pallidus* and *Str. droebachiensis.* Even in relatively late stages of digestion, ciliates could be recognized because the pellicle is one of the last parts to be digested. Ciliates in food vacuoles were often smaller than those of the same species in the intestine of the host; they were probably the products of recent divisions. Of 100 *Syndisyrinx fransciscanus* from freshly-collected hosts whose intestinal contents were quantitatively recorded, 37 contained no ciliates, 31 contained up to 20 ciliates, and 32 contained from 20 to several hundred ciliates. Worms that had ingested large numbers of ciliates were found in only a few of the urchins that harbored *S. franciscanus.*

The intestinal lumen and food vacuoles of *Syndisyrinx franciscanus* from freshly-collected hosts commonly contained amorphous, orange-yellow material. This filled the intestinal lumen of some flatworms (Fig. 4), but in most it was distributed as small masses in the food vacuoles (Fig. 5). The material consisted of cellular debris with clumps of red granules – as many as several hundred clumps in a single worm – scattered through it (Fig. 4). This was the most abundant food in 53 of the 100 *S. franciscanus* whose intestinal contents were quantitatively recorded; it was not present, however, in 5 of the worms. All specimens of *S. franciscanus* contained some food that had undergone so much digestion that it could not be identified.

The intestine of starved *Strongylocentrotus pallidus* and *Str. droebachiensis* contained no food after three weeks. Zero to 32 (avg. 21) adult *Syndisyrinx franciscanus* were taken from each of four urchins that were starved for 2½ months. The worms appeared normal and their gut contained yellow material with red granules, as described above, but contained no ciliates. The starved hosts harbored few ciliates.

When viewed with transmitted light, the intestinal epithelium of *Strongylocentrotus pallidus* and *Str. droebachiensis* is orange-yellow to light-yellow, depending on the urchin and the region of the intestine. The apices of epithelial cells in the first half of the intestine are packed with round, light-yellow, refractile granules (Figs. 7 & 8). The epithelium is pervaded by red coelomocytes that contain red granules identical to those seen in the yellow amorphous material in the intestine of *Syndisyrinx franciscanus* (Fig. 6).

The umagillid that resembles *Syndesmis echinorum* was present in the first half of the intestine of seven of the freshly-collected urchins that were examined. Nineteen worms of this species were recovered and were found to contain a single type of food. This was an orange-yellow floculent material consisting of lipoid droplets, numerous particles, and large numbers of round, light-yellow, refractile

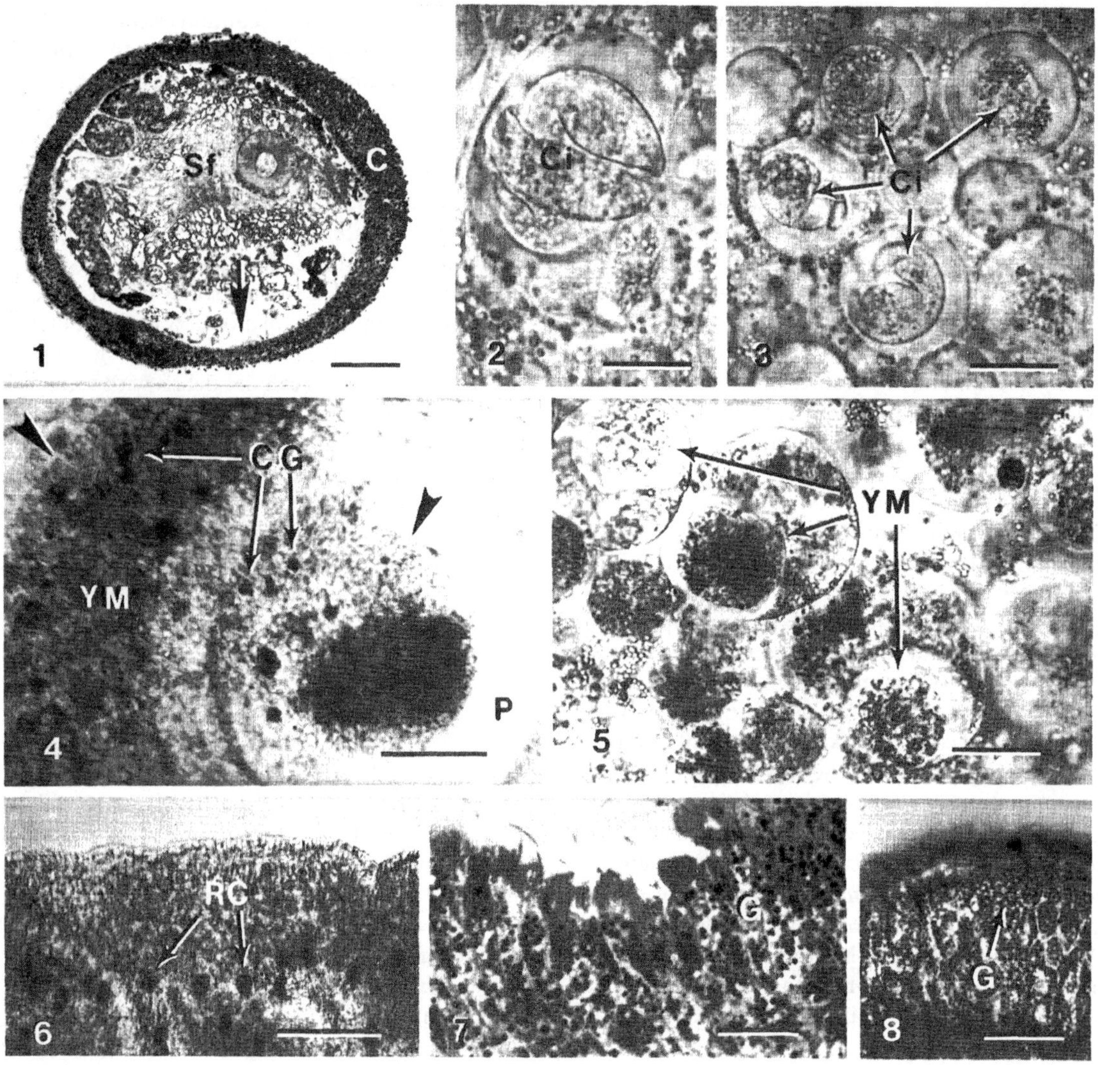

Fig. 1. Frontal section of *Syndisyrinx franciscanus* (Sf) encapsulated by a layer of coelomocytes (C). The worm's epidermis has been destroyed (arrow) and some other parts of the specimen are necrotic. Paraffin section stained with iron hematoxylin. Bar = 250 μm.

Figs. 2 & 3. Ciliates (Ci) in food vacuoles of live *Syndisyrinx franciscanus*. Bars = 32 μm.

Fig. 4. *Syndisyrinx franciscanus* containing recently-ingested amorphous, orange-yellow material (YM) and clumps of red granules. (CG). The limit of the gut lumen is indicated by arrowheads. P, pharynx. Bar = 60 μm.

Fig. 5. Gastrodermal cells of *Syndisyrinx franciscanus* dissociated by compressing the worm; the food vacuoles contain amorphous masses of relatively undigested orange-yellow material (YM). Bar = 32 μm.

Fig. 6. Portion of the second half of the intestine of *Strongylocentrotus pallidus*, showing numerous red coelomocytes (RC) in the epithelium. Bar = 60 μm.

Figs. 7 & 8. Apices of epidermal cells of the first half of the intestine of *Strongylocentrotus pallidus*, showing numerous granules (G) that are refractile in life. Fig. 7 shows a 1 μm Epon section stained with methylene blue-azure II; Fig. 8 shows a piece of fresh tissue. Bars = 20 μm.

granules (Figs. 9 & 10). These granules were identical to those in epithelial cells from the first half of the intestine of host urchins (Figs. 7 & 8). A few red granules identical to those in red coelomocytes in the intestinal epithelium of *Strongylocentrotus pallidus* and *Str. droebachiensis* were present among this material in some worms.

Syndesmis dendrastrorum was found only in the digestive tract of *Dendraster excentricus*. Zero to 47 (avg. 14) worms were taken from 20 freshly-collected hosts. Yellow material, similar to that described from the umagillid that resembles *S. echinorum* but containing smaller refractile granules and occaisional clumps of red-orange granules, was the most abundant food, or the only food, in 73% of the 46 *S. dendrastrorum* examined. Other contents of the gut of *S. dendrastrorum* were fairly discrete objects (Figs. 12, 14, 15 & 16). They were often amber or green in color. One common type of amber object (Fig. 15) was revealed in Epon sections to consist of concentric layers of an apparently non-cellular material (Fig. 16). Some other materials consisted of intact animal cells (Fig. 13). Green material (Fig. 11) did not fluoresce when observed with UV light following the procedure of Williamson (1977) and its pigment did not diffuse away. The various objects were seldom abundant in any one worm, although worms from a single host consistently contained similar-appearing objects. Symbiotic ciliates were not present in either the coelom or the intestine of *Dendraster excentricus*.

Intestinal cells of *Dendraster excentricus* are yellow in color and their apical portions are packed with small refractile granules. Coelomocytes containing red or, more commonly, red-orange granules are abundant among them.

Discussion

The three species of umagillids studied were found only in the intestine of their hosts. *Syndisyrinx franciscanus* and the species that resembles *Syndesmis echinorum* are also reported to occur in the perivisceral coelom (Orihel 1952; Mettrick & Jennings 1969). Because *Syndisyrinx franciscanus* has not been found naturally within coelomocyte capsules, observations that it becomes encapsulated when artificially introduced into the perivisceral coelom suggest that it does not normally inhabit the coelom. *Syndisyrinx franciscanus* was found restricted to the intestine by Lehman (1946) and Barnes (1969). *Syndisyrinx franciscanus* is destroyed when introduced to the coelom along with small amounts of sea water. If specimens were occasionally to escape into the coelom accompanied with fluids from the intestine, it is likely that they would also be destroyed.

Orihel (1952) reported that the umagillid that resembles *Syndesmis echinorum* occurs in the coelom of *Strongylocentrotus droebachiensis* when it is abundant in a particular host. Barnes (1969) found it in the intestine, but not in the coelom of *Str. purpuratus* and *Str. franciscanus*, even when it was present in large numbers. His observations on its life history further suggest that it lives only in the intestine. In my experience, this species is restricted to the intestine, but I did not find it in large numbers.

In this study, *Syndisyrinx franciscanus* was found to subsist on ciliates and on masses of cellular debris that were probably derived from the host intestinal epithelium. These amorphous masses are the same colour as intestinal cells and contain red granules that are unique to coelomocytes in the intestinal epithelium. They were, moreover, the only food of *S. franciscanus* living in hosts that were starved for 2½ months. Besides host intestinal tissue, ciliates were the only food objects available to the worms in starved hosts, and ciliates would have been recognizable in food vacuoles because of their distinctive shapes. The persistence of *S. franciscanus* in starved hosts suggests that this species can survive on a diet of host tissue alone.

It appears that populations of *Syndisyrinx franciscanus* inhabiting different host species have different nutritional relationships with their hosts. According to Mettrick & Jennings (1969) and Holt & Mettrick (1975) worms taken from *Strongylocentrotus purpuratus* subsisted only on ciliates. Relatively few of the *Syndisyrinx franciscanus* that I examined had eaten large numbers of ciliates. Some species of echinoid-inhabiting ciliates, including *Cyclidium stercoralis* and *Madsenia indomitus*, exhibit periodic population blooms (Beers 1948; Berger 1964). The small size of many of the ciliates ingested by the worms that I examined, and my observations that in only certain urchins does *S. franciscanus* consume ciliates in large

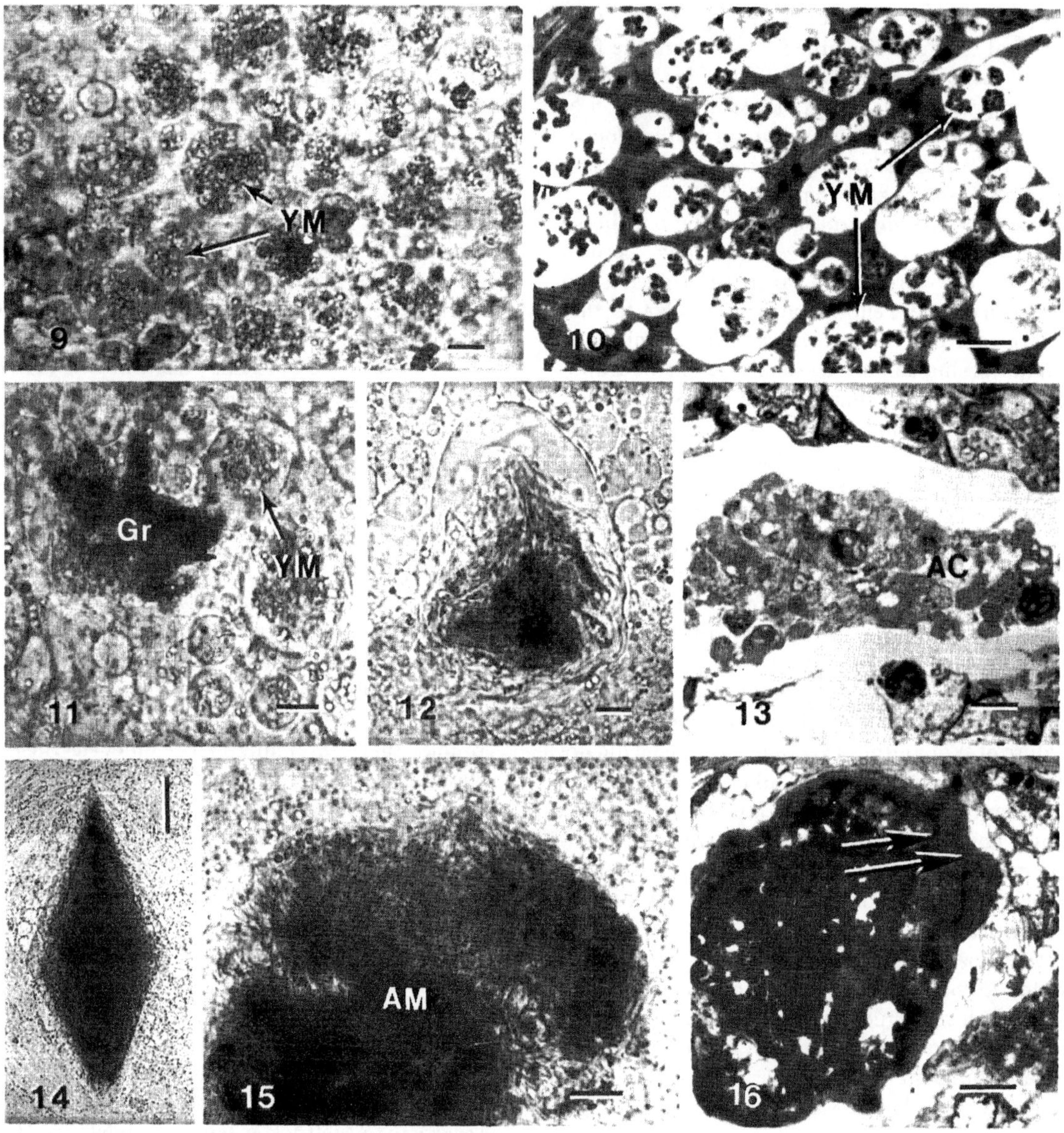

Figs. 9 & 10. Amorphous orange-yellow material containing many refractile granules (YM), in food vacuoles of the umagillid that resembles *Syndesmis echinorum*. Fig. 9 shows food vacuoles of a live specimen; Fig. 10 shows a 1 μm Epon section stained with methylene blue-azure II. Bars = 20 μm.

Figs. 11 to 16. Contents of the digestive tract of *Syndesmis dendrastrorum*.

Fig. 11. Green material (Gr), and amorphous yellow material (YM) in food vacuoles of a live specimen. Bar = 10 μm.

Fig. 12. Unidentified amber object in food vacuole of live specimen. Bar = 10 μm.

Fig. 13. Longitudinal section through intestine, showing a mass of intact animal cells (AC) in the lumen. 1 μm Epon section stained with methylene blue-azure II. Bar = 5 μm.

Fig. 14. Unidentified object found only in worms taken from one host. Bar = 250 μm.

Fig. 15. Unidentified amber material (AM) in live specimen. Bar = 10 μm.

Fig. 16. 1 μm Epon section through material like that shown in Fig. 15, showing layers of non-cellular material (arrows). Methylene blue-azure II stain. Bar = 10 μm.

numbers, suggest that the flatworms may take advantage of such blooms. Between blooms the worms may feed predominately on host tissue.

The diet of the umagillid that resembles *Syndesmis echinorum* consisted of cellular debris that probably originated from the host intestinal epithelium. The debris in the gut of the worms contained large numbers of refractile granules that were identical to refractile granules present in and characteristic of cells of the first half of the host's intestine. Pieces of brown algae were the only objects in the host intestine that had a somewhat similar orange colour. The algal cells did not contain the same refractile granules, however, and the cellular debris in the worms' gut contained nothing resembling algal cell walls.

Syndesmis dendrastrorum consumes a variety of foods. Yellow-coloured masses of cellular debris were almost routinely present in food vacuoles and probably consisted of host tissue; they contained refractile granules characteristic of host intestinal cells. Other objects could not be precisely identified. Some, such as intact animal cells that were distinct from host intestinal cells, were probably selected by the worms from among the food in the host's intestine. The host, *Dendraster excentricus*, is a detritivore and ingests a variety of small organisms and organic debris. Green objects in the food vacuoles of *S. dendrastrorum* were probably not composed of plant material. Unlike materials that contain chlorophyll, they did not show red fluorescence when illuminated with UV light. The green colour in them did not diffuse out during digestion as has been described for diatoms consumed by acoel turbellarians (Kozloff 1972). Tissues of *Dendraster excentricus* commonly turn green when damaged (unpublished observations). The reason for this is not known, but the green material in *Syndesmis dendrastrorum* may have been host tissue. Some objects ingested by *S. dendrastrorum* had very distinctive shapes or appearances and should have been recognizable among contents of the hosts' intestine. The nature and origin of these objects remain enigmatic.

It is not known whether umagillids remove material from the intact intestinal epithelium or whether the host tissue they ingest represents the remains of cells sloughed from the epithelium during normal cell-replacement activities, such as occur in the intestine of vertebrates (Leblond & Walker 1956). Proliferation and sloughing of cells are not mentioned in morphological studies of echinoid digestive tracts (Tokin & Filimonova 1977; Fugi 1961; Reisman 1965; Lewis 1964; Boolootian 1964), but these processes may occur. While lesions in the epithelium were not seen when the intestine of hosts that contained large numbers of umagillids were examined with a dissecting microscope, large masses of host tissue were present in some worms. Red coelomocytes are not strictly part of the intestinal epithelium and would presumably not be involved in turnover of epithelial cells. Thus, the presence of granules from red coelomocytes among intestinal tissue ingested by *Syndisyrinx franciscanus* suggests that this species erodes cells from the intact epithelium. It is possible that *Syndesmis dendrastrorum* and the umagillid that resembles *Syndesmis echinorum* obtain host tissues in a somewhat different way than *Syndisyrinx franciscanus;* red coelomocytes that were abundant throughout the intestinal epithelium of the hosts were not common in the tissue ingested by these two species. Even if the host tissue eaten by these umagillids merely consists of cells sloughed from the epithelium, it is possible that much of the contents of these cells would be digested and absorbed further on in the digestive tract. If this is the case, the umagillids are essentially taking nutrients from their hosts if they consume sloughed off cells.

Epidermal uptake of nutrients evidentally occurs in at least one umagillid, *Fallacohospes inchoatus*, which has no digestive tract (Kozloff 1965). In Epon sections studied with the light microscope, epidermal cells of this species closely resemble those of the umagillids studied here (unpublished observations). It is possible that all three of the species with which I have worked also take up nutrients through the epidermis, but this has not been established.

Umagillids are often considered to be harmless commensals; none of the species of umagillids studied have been shown to harm their hosts. Physical damage is not a qualification for being a parasite, however. Results of this study strongly suggest that umagillids compete with their hosts for nutrients, if they don't cause physical damage, and thus have adverse metabolic affects. Umagillids are apparently parasitic on their hosts, but to varying degrees.

Acknowledgements

I thank the people at Friday Harbor Labs for their help, including especially, Eugene Kozloff, Steve Stricker, Chris Reed, Robert Fernald, Craig Young, and Ed Ruppert.

References

Barnes, A. T., 1969. Two endoparasitic turbellarians from California echinoids. Ph.D. Thesis. Univ. of California, Santa Barbara.

Beers, C. D., 1948. The ciliates of Strongylocentrotus drobachiensis: incidence, distribution in the host, and division. Biol. Bull. 94: 99–112.

Berger, J., 1964. The morphology, systematics, and biology of the entocommensal ciliates of echinoids. Ph.D. Thesis. Univ. of Illinois, Urbana.

Boolootian, R. A., 1964. A histological study of the food canal of Strongylocentrotus franciscanus. Helgoland. wiss. Meeresunters. 11: 118–127.

Fugi, A., 1961. Studies on the biology of the sea urchin IV. Histological observation of the food canal of Strongylocentrotus intermedius. Bull. Fac. Fish. Hokkaido Univ. 11: 195–202.

Holt, P. A. & Mettrick, D. F., 1975. Ultrastructural studies of the epidermis and gastrodermis of Syndesmis franciscanus (Turbellaria: Rhabdocoela). Can. J. Zool. 53: 536–549.

Jennings, J. B. & Mettrick, D. F., 1968. Observations on the ecology, morphology and nutrition of the rhabdocoel turbellarian Syndesmis franciscanus (Lehman 1946) in Jamaica. Caribb. J. Sci. 8: 57–69.

Jenning, J. B., 1971. Parasitism and commensalism in the Turbellaria. In: Dawes, B. Ed., Advances in Parasitology 9: 1–32. New York: Academic Press.

Kozloff, E. N., 1965. Desmote inops sp.n. and Fallacohospes inchoatus gen. and sp.n., umagillid rhabdocoels from the intestine of the crinoid Florometra serratissima (A. H. Clark). J. Parasitol. 51: 305–312.

Kozloff, E. N., 1972. Selection of food, feeding, and physical aspects of digestion in the acoel turbellarian Otocelis luteola. Trans. Am. Microsc. Soc. 91: 556–565.

Leblond, C. P. & Walker, B. E., 1956. Renewal of cell populations. Physiol. Rev. 36: 255–276.

Lehman, H. E., 1946. A histological study of Syndisyrinx franciscanus gen. et sp.nov., an endoparasitic rhabdocoel of the sea urchin, Strongylocentrotus franciscanus. Biol. Bull. 91: 295–311.

Lewis, J. B., 1964. Feeding and digestion in the tropical sea urchin Diadema antillarum Philippi. Can. J. Zool. 42: 549–557.

Mettrick, D. F. & Jennings, J. B., 1969. Nutrition and chemical composition of the rhabdocoel turbellarian Syndesmis franciscanus, with notes on the taxonomy of S. antillarum. J. Fish. Res. Bd. Can. 26: 2669–2679.

Orihel, T. C., 1952. Entocommensal rhabdocoels from echinoids of Puget Sound. Master's Thesis, Univ. of Washington, Seattle.

Powers, P. B. A., 1936. Studies on the ciliates of sea urchins. A general survey of the infestations occuring in Tortugas echinoids. Papers Tortugas Lab. Carnegie Inst. Wash. 29: 295–326.

Reed, C. G. & Cloney, R. A., 1977. Brachiopod tentacles: ultrastructure and functional significance of the connective tissue and myoepithelial cells in Terebratalia. Cell Tiss. Res. 185: 17–42.

Reisman, A. W., 1965. The histology and anatomy of the intestinal tract of Dendraster excentricus, a clypeasteroid echinoid. Master's Thesis, Univ. of California, Los Angeles.

Smith, N. S., 1973. A new description of Syndesmis dendrastrorum (Platyhelminthes, Turbellaria), an intestinal rhabdocoel inhabiting the sand dollar Dendraster excentricus. Biol. Bull. 145: 598–606.

Stott, F. C., 1955. The food canal of the sea urchin Echinus esculentus and its functions. Proc. Zool. Soc. London 125: 63–86.

Tokin, I. B. & Filimonova, G. F., 1977. Electron microscope study of the digestive system of Strongylocentrotus droebachiensis (Echinodermata: Echinoidea). Mar. Biol. 44: 143–155.

Westblad, E., 1953. New Turbellaria parasites in echinoderms. Ark. Zool. Ser. 2. 5: 269–288.

Williamson, C. E., 1977. Fluorescence identification of zoochlorellae: a rapid method for investigating alga-invertebrate symbioses. J. Exp. Zool. 202: 187–194.

Reproductive biology of Dugesia sanchezi, a fresh-water planarian from Chile

Mario Benazzi
Institute of Zoology, Univ. of Pisa Via A. Volta 4, 56100 PISA (Italy)

Keywords: Turbellaria, *Dugesia*, karyology, reproduction

Abstract

Dugesia sanchezi reproduces in nature both sexually and by fission. Laboratory cultures have shown that fissioning is controlled by genetic factors. From sexual specimens randomly mated, 105 sexual and five fissiparous offspring were obtained. The former produced almost exclusively sexual descendents while the latter gave rise to fissiparous individuals, although successively many of them became sexual. These ex-fissiparous specimens show a high fertility and their offspring were almost all fissiparous, with the same characteristics displayed by the preceeding fissiparous generation. Some hypotheses are advanced in order to attempt a genetical interpretation of these results.

Introduction

Dugesia sanchezi Hyman, 1959 was first collected by Dr. P. Sanchez in coastal hillside streams and brooks of Central Chile. He observed that this planarian reproduces sexually and the stalked capsules are readily found in nature and also readily laid in laboratory cultures. However, the worm may also undergo the usual transverse fission behind the pharynx (cf. Hyman 1959).

I was able to confirm these statements (Benazzi 1978) on specimens collected in the vicinity of Concepcion (Chile) and to attempt an analysis of some genetical aspects of the reproductive mechanisms.

It is opportune to recall that this planarian belongs to the subgenus *Girardia* Ball, 1974 which comprises numerous species distributed in North, Central and South America. In many aspects these species of the New World are similar to the species of the Old World belonging to the subgenus *Dugesia* (see Ball 1974). In both subgenera, sexuality represents the normal condition but in several species fissioning frequently occurs and in some strains it is the sole method of reproduction. Analogies between the species of the two subgenera can be recognized also in the chromosome complements. Most species possess a basic number of eight chromosomes, and so far as relative length is concerned they can be ranked in a rather similar way. Polyploid and heteroploid forms related to asexual reproduction are known. (For a summary of the karyology of the genus *Dugesia* cf. Benazzi & Benazzi Lentati 1976).

D. sanchezi has 2n = 16, n = 8; in sexual specimens gametogenesis is normal and in the fissiparous individuals the diploid complement of 16 chromosomes appears to be constant; no heteroploid variations seem to arise in connection with asexual reproduction.

The main purpose of my research was to try to establish the determinism of the fissioning, taking into consideration the results of my previous investigations on forms of the '*Dugesia gonocephala* group' which showed a strong genetic control of this phenomenon. In fact, from crosses between sexual individuals of certain populations I obtained both sexual and fissiparous specimens in a ratio

Hydrobiologia 84, 163–165 (1981). 0018–8158/81/0842–0163/$00.60.

close to 1:1. This result suggests the existence of fission-controlling factors, presumably corresponding to Mendelian genes, which inhibit the sexual evolution and determine the transverse division. (Benazzi, 1974). In the fissiparous strains, some individuals may reach sexuality; the frequency and the fertility of these ex-fissiparous specimens vary greatly in the various populations, probably owing either to different frequencies in the fission-controlling factors or to differences in their penetrance or expression (Benazzi 1974).

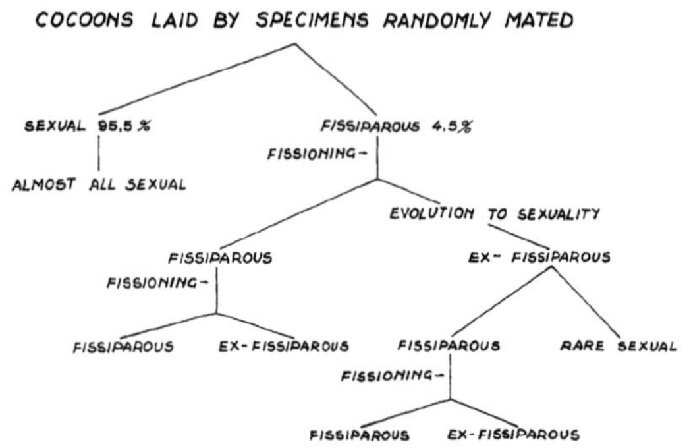

Fig. 1. Schematical representation of the reproductive patterns of *Dugesia sanchezi.*

Results

The first data on *D. sanchezi* were obtained in 1977 starting from cocoons laid in the laboratory by specimens randomly mated; 105 offspring reached sexuality while five underwent fission. In the following year the former produced almost exclusively sexual offspring, while the latter gave rise to fissiparous descendents which multiplied quickly. However, many of these descendents successively became sexual and in 1978 laid a great number of fertile cocoons. The characteristics of the generation born from these cocoons have shown that the genetic constitution of the ex-fissiparous specimens is very peculiar. Their fertility is higher than that of the individuals which never fissioned. In fact, 56 cocoons laid by the ex-fissiparous specimens produced 300 offspring, most of which were normal although a certain number appeared anomalous and not viable. Instead, from 66 cocoons laid by the individuals which never divide only 200 young were obtained. Therefore, the average number of newborn per cocoon was 5.33 and three respectively. Unpublished observations by Dr. Sanchez (*in litt.* February 7, 1979) showed for this species 'a mean of about four youngs per capsule.'

The high fertility of the ex-fissiparous specimens deserves to be emphasized because in many other planarian species the ex-fissiparous specimens frequently show a low fertility or total sterility (cf. Benazzi 1974).

A second interesting point regarding the ex-fissiparous specimens concerns the reproductive patterns of their offspring which were almost all fissiparous. However, after a period of fissioning many of them reached sexuality and again produced fissiparous offspring. Therefore, these offspring manifested the same characteristics of their parents (Fig. 1).

Discussion

The data now summarized suggest some conclusions. Fissioning in *D. sanchezi,* as in other planarian species, appears to be genetically determined. The first results obtained in the laboratory from sexual specimens randomly mated showed that a segregation between sexual *vs.* asexual descendents occurs, although the ratio was very low (95.5:4.5). We are, therefore, inclined to think that fission-controlling factors are also present in sexual individuals but at a very low frequency. The phenotypic manifestations of these factors, namely the inhibition of the development of the sexual apparatus and the induction of fission, occur only in individuals with a determined genotype, which may perhaps correspond to a high frequency of these factors. In the fissiparous individuals, or in some of them, the inhibiting action, for unknown reasons, is overcome at a certain moment; however, the genetic constitution remains unchanged in the ex-fissiparous specimens and determines the reappearance of fission in almost all their offspring.

References

Benazzi, M., 1974. Fissioning in planarians from a genetic standpoint. In: Eds., Riser, N. W. & Morse, M. P., Biology of Turbellaria. New York: McGraw-Hill, pp. 476–492.

Benazzi, M., 1978. Karyological and genetic data on the planarian Dugesia sanchezi from Chile. Rd. Acc. Naz. Lincei, Ser. 8, 64: 299–301.
Benazzi, M. & Benazzi Lentati, G., 1976. Platyhelminthes. In: Ed. John, B., Animal Cytogenetics. Borntraeger, Berlin-Stuttgart, p. 1–182.
Hyman, L. H., 1959. On two fresh-water planarians from Chile. Am. Mus. Novit. No. 1932: 1–11.

Contrasting power of the factors for fission and sexuality in a polyploid planarian

Giuseppina Benazzi Lentati and Mario Benazzi
Institute of Zoology, Univ. of Pisa Via Volta 4, 56100 Pisa (Italy)

Keywords: Turbellaria, *Dugesia*, reproduction, karyology

Abstract

One of the biotypes of the planarian *Dugesia benazzii* is triploid in the somatic line, hexaploid in the female line owing to a chromosome set doubling, and diploid in the male line due to a haploid set elimination. In a population of this biotype only 50% of the oocytes are hexaploid, the others being triploid as a results of the lack of set doubling; the male line is always diploid. After a long period of laboratory culture most of the individuals became asexual and fissiparous. Almost all the oocytes of the few specimens which have remained sexual showed triploid complement; B-chromosomes also appeared. These events represent the manifestation of a new genetic background which act upon the two germ lines in different ways and moments. These topics are discussed.

The fresh-water triclad *Dugesia benazzii* from Sardinia, Corsica and Capraia, belonging to the '*Dugesia gonocephala* group', presents three biotypes differentiated by the ploidy level and by the type of development. One of these biotypes is triploid in the somatic line, hexaploid in the female line (owing to a chromosome set doubling in the cells differentiating into oogonia) and diploid in the male line (as a result of a haploid set elimination in the cells destined to be transformed into spermatogonia). The development of the hexaploid oocytes is pseudogamic, and this fact explains the reason for chromosome set doubling in the female line; in fact, the oocytes are synaptic, and after the maturation divisions give rise to triploid eggs, from which triploid embryos develop (Fig. 1).

Some populations of this biotype from Corsica show a peculiar characteristic, namely, simultaneous presence in each individual of triploid and hexaploid oocytes; in these specimens, therefore, the chromosome set doubling does not occur in all oocytes (Benazzi 1968). Other studies on polyploid planarians have demonstrated that the chromosome set doubling is genetically determined (Benazzi Lentati 1966, 1970; Benazzi Lentati & Mezzani 1974). Therefore, the difference between the individuals of the normal triplo-hexaploid biotype showing only hexaploid oocytes and the individuals of the populations in which the chromosome set doubling occurs only in a percentage of the oocytes may be attributed to variations in the frequencies of the genes responsible for the chromosome doubling (Benazzi & Giannini 1970).

The chromosome cycle of the populations with two types of oocytes is as follows (Fig. 2): (1) diploid zygotes develop from the triploid oocytes

Fig. 1. Schematic representation of the triplo-hexaploid biotype: chromosome cycle in the female line. The triploid neoblasts (24 chromosomes) produce hexaploid oocytes (24 bivalents) from which through pseudogamy triploid pseudozygotes derive.

Hydrobiologia 84, 167–169 (1981). 0018-8158/81/0842-0167/$00.60.

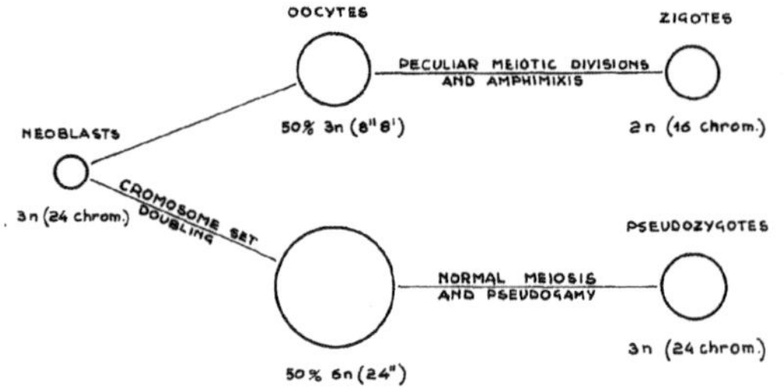

Fig. 2. Schematic representation of the population from Roccapina, showing both triploid (8 bivalents and 8 univalents) and hexaploid (24 bivalents) oocytes. The former produce by amphimixis diploid (16 chromosomes) zygotes, the latter give rise through pseudogamy to triploid pseudozygotes.

after a peculiar maturation process followed by amphimixis; the great majority of them, however, die very early for unknown reasons. (2) triploid pseudozygotes develop from the hexaploid pseudogamous oocytes; they reproduce the triploid somatic line and allow the maintenance of the original chromosome characteristics (Giannini & Puccinelli 1969).

One of these populations, coming from the locality of Roccapina, was characterized by a ratio between triploid and hexaploid oocytes near to 1:1 (Fig. 3). However, after about ten years of laboratory culture, we found that most of the specimens had lost the sexuality and had become fissiparous. Only 11 specimens were still sexual and two others returned to sexuality after a period of fissioning. Evidently in this population fission-controlling genes were present although repressed, and they manifested their effect only after a long period of laboratory conditions.

Karyological investigations accomplished on these sexual specimens have shown that while the male line is always diploid, the oocytes are almost all triploid. As a matter of fact, five individuals during three reproductive periods laid many cocoons with complessively 60 oocytes, 57 of which were triploid and three hexaploid; five individuals produced 23 oocytes all triploid; the other three individuals never laid cocoons. Therefore, the frequency of the triploid oocytes is about 94%, while in the first years of laboratory culture it was, as already stated, about 50%.

It can now be pointed out that the fission-controlling genes are present also in the sexual animals. In fact, from successive inbreeding, or from crosses with individuals of the diploid biotype, they produced about 80% sexual specimens and 20% fissiparous specimens.*

Another interesting fact evidenced by karyological examination is the appearance in the somatic and germ lines of B-chromosomes (Benazzi Lentati

* This great number of individuals was an unexpected event because previous research, as already mentioned, had shown that the offspring of the triploid oocytes die in the first stages of development. We are not able, at present, to suggest an interpretation of this result.

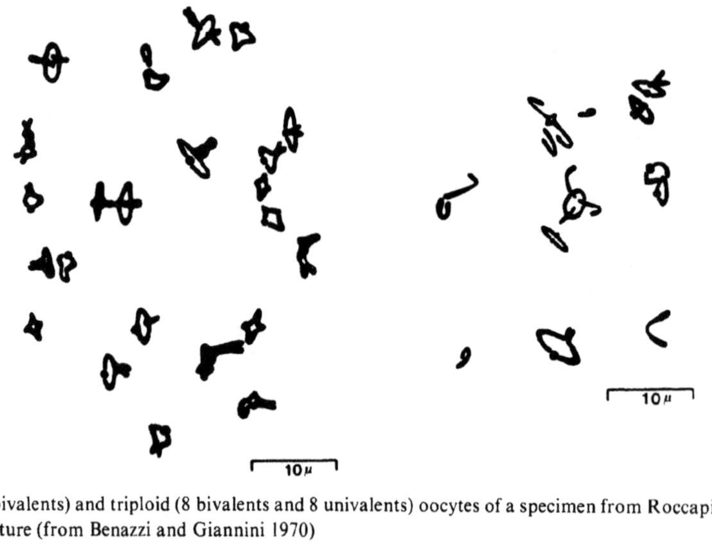

Fig. 3. Hexaploid (24 bivalents) and triploid (8 bivalents and 8 univalents) oocytes of a specimen from Roccapina examined in the first years of laboratory culture (from Benazzi and Giannini 1970)

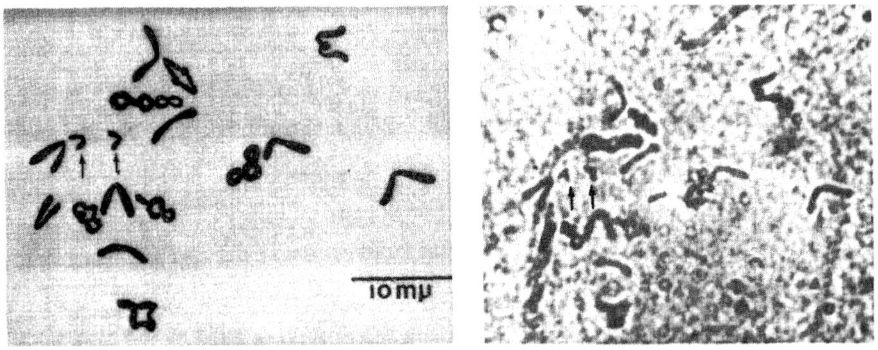

Fig. 4. Triploid oocyte of a specimen from Roccapina examined after a long period of laboratory culture: two B-chromosomes (arrows) are present (at right microphotograph, at left the corresponding drawing).

& Deri 1977) (Fig. 4).

These two events, i.e. the almost complete absence of the chromosome set doubling in the female line and the rise of B-chromosomes (even if the origin and the possible action of the latter are completely unknown) could be attributed to a new genetic constitution. This seems to influence the drastic reduction of the hexaploid oocytes more than the loss of sexuality, because the formation of triploid oocytes lasts for a long time. This statement may be surprising, but it is in agreement with the results of previous research, according to which the chromosome set doubling normally occurs in the female territory, but it is not necessarily correlated with the differentiation of the female cells (Benazzi Lentati 1976). Also the presence in some populations of two types of oocytes differentiated by the chromosome complement (triploid and hexaploid, respectively) prove that the chromosome set doubling may be independent of the differentiation of the female germ cells.

With regard to the male line it is known that in the triplo-hexaploid biotype the cells destined to be transformed into spermatogonia become diploid owing to a haploid set elimination. In the population analyzed in this study, elimination and spermatogenesis with formation of haploid sperms occur normally, as shown by the chromosome complement (always diploid) of the offspring born from the crosses between these individuals, acting as male, and individuals of the diploid biotype. Therefore, while in the female line the inhibition of the chromosome set doubling occurs prior to the total disappearance of the sexuality, in the male line the elimination of a haploid set takes place only when the sexuality is completely lost.

Summing up, the new genetic constitution which has arisen in correlation with activation of the fission-controlling genes and the appearance of B-chromosomes, acts differently upon the two germ lines, with regard to time and modality. Thus, the two events strictly connected with the synaptic polyploid cycle are not mutually correlated.

References

Benazzi, M., 1968. Popolazioni di Dugesia benazzii della Sardegna e della Corsica di probabile origine ibrida. Atti Assoc. Genet. Ital. 13: 117–124.

Benazzi, M. & Giannini, E., 1970. Ricerche su popolazioni della planaria Dugesia benazzii con due tipi di ovociti:analisi statistica delle frequenze e considerazioni di ordine genetico. Riv. Biol. 63: 145–157.

Benazzi Lentati, G., 1966. Amphimixis and pseudogamy in fresh-water triclads:experimental reconstitution of polyploid pseudogamic biotypes. Chromosoma 20: 1–14.

Benazzi Lentati, G., 1970. Gametogenesis and egg fertilization in planarians. Int. Rev Cytol. 27: 101–179.

Benazzi Lentati, G., 1976. On the relationship between the chromosome set and the sexual differentiation of the neoblasts in planarians. Acta Embryol. Exp., issue 3: 299–317.

Benazzi Lentati, G. & Deri, P., 1977. Insorgenza di B-chromosomi in individui poliploidi di Dugesia benazzii (triclade paludicolo) allevati in laboratorio. Rendic. Acc. Naz. Lincei, Ser. 8, 62: 848–851.

Benazzi Lentati, G. & Mezzani, V., 1974. Research on the hereditary mechanism of polyploidy in planarians. Rend. Acc. Naz. Lincei Ser. 8, 57: 294–300.

Giannini, E. & Puccinelli, I., 1969. Ciclo cromosomico ed ovogenesi in popolazioni di Dugesia benazzii con due tipi di ovociti. Atti Soc. Tosc. Sci. Nat. Pisa, Mem., Ser. B, 76: 150–166.

The embryological development of Turbellaria Kalyptorhynchia (Rhabdocoela)

J. P. L'Hardy
Laboratoire de Biologie Animale, Université du Maine, 72017 Le Mans-Cedex, France

Keywords: Turbellaria, Rhabdocoela, Kalyptorhynchia, embryology

Abstract

The ectolecithal egg mass of most Turbellaria Kalyptorhynchia is formed by two eggs and about hundred vitelline cells. The fertilization of the primary oocyte takes place in the genital atrium and maturation occurs in the uterus during the formation of the egg-shell. After its sclerotinization, the egg-capsule is fixed on a substratum by the secretion of the uterine glands.

The cleavage of the oligolecithal egg is total and unequal. The two blastomeres AB and C appear successively at the anterior pole and on the side of the macromere. The cell multiplication, which almost exclusively occurs in the D-quadrant, gives the morula where some micromeres and two endodermic mother-cells can be distinguished.

The fragmentation of the yolk mass round each of the two embryos is followed by the formation of a perivitelline embryonic epithelium constituted from micromeres migrating through the yolk by means of amoeboïd movements. Gastrulation takes place by centrifugal migration of the blastomeres following the yolk absorption by vitellophages of endodermic origin.

Subsequently, the undifferentiated endodermic cells make a temporary embryonic intestine round the yolk-filled vitellophages that are gathered in the body-cavity of the embryo.

During gastrulation a small pharyngeal blastema appears in the ventral part of the embryo. It will form later the whole digestive tract. The other blastomeres gather in a cephalic blastema which will give the nervous system, the subepidermal muscle cells and the proboscis. The definitive epidermis differentiates from the perivitelline embryonic epithelium.

Hydrobiologia 84, 170 (1981). 0018-8158/81/0842-0170/$00.20.

The problem of cell totipotency, dedifferentiation and transdifferentiation in Turbellaria

Vittorio Gremigni
Istituto di Zoologia e Anatomia Comparata, Via Volta 4, I-56100 Pisa, Italy

Keywords: Turbellaria, Tricladida, cell differentiation, regeneration

The term *cell totipotency* in the strict sense defines the capability of a single cell to give rise to a whole organism, but this term can also be defined more generally as the capability of a cell to differentiate into any cell type of a multicellular organism. *Cell dedifferentiation* refers to the capability of a specialized cell to return to an undifferentiated state and possibly become totipotent again. *Cell transdifferentiation* is the ability of a cell to first dedifferentiate and then redifferentiate along another line.

The term totipotency in the strict sense is applicable to a very small number of cells including the egg or zygote in all organisms, as well as several somatic cells in some organisms, as for example Angiosperms, where a single somatic cell is capable of regenerating a complete plant (review by D'Amato 1977). Transplantation experiments which placed somatic cell nuclei into the cytoplasm of enucleated eggs have demonstrated that animal nuclei are also totipotent (review by Mc Kinnel 1978).

Examples of cell totipotency, even when considered in a wider sense are still limited to a small population of cells in most animals. In some sexually reproducing animals, apart from the egg or zygote, practically no other cells should be considered totipotent if their blastomeres undergo a precocious determination and only give rise to one or a few cell types. This situation was first described in the classical studies on the nematode *Ascaris* and the insects *Miastor, Sciara* and *Mayetiola* where there is a precocious segregation of the germ and somatic cell lines during the early cleavage divisions (review by Beams & Kessel 1974). In contrast, blastomeres and embryonic cells are thought to be totipotent up to the onset of their differentiation in those organisms lacking a precocious segregation of cell lines. In asexually reproducing animals a complete organism can develop from part of the original somatic tissues, hence in this case several cells are thought to be able to maintain or reacquire totipotency.

One widely held belief is that cell totipotency is strictly correlated with a morphologically undifferentiated state. This notion is not always valid, indeed it has been demonstrated that undifferentiated cells are often not totipotent and vice-versa. For instance, stem cells of each tissue appear undifferentiated but they only differentiate into one cell type, while differentiated cells can acquire totipotency if they first dedifferentiate and then gain the ability to transform into all cell types. To demonstrate this fact I again cite the case of Angiosperms in which a single specialized cell can give rise to a complete plant.

It is generally held that a precocious segregation of germ and somatic lines does not occur in Turbellaria, and particularly in freshwater triclads which reproduce by gametes. A different situation, that of precocious segregation, was hypothesized by Melander (1963) who, on the basis of a cytological study carried out mainly in the species *Polycelis nigra*, suggested that future female and male germ cells can be recognized and they are therefore already predetermined in the early stages of embryonic development. Support for this hypothesis is mainly based on the following observations: meiotic prophasic oocytes have two bivalents containing one positively heteropycnotic chromocentre with

Hydrobiologia 84, 171–179 (1981). 0018-8158/81/0842-0171/$01.80.

an attached nucleolus. According to Melander the oocytes would represent the descendants of certain blastomeres having two chromosomes with a negatively heteropycnotic, secondary constriction. Similarly, spermatocytes which lack conspicuous chromocentres, would represent the descendants of certain blastomeres which lack secondary constrictions. Likewise somatic cells would probably be descendants of blastomeres which have a variable number of secondary constrictions. However, some observations call this hypothesis in question: Melander himself writes that the secondary constrictions of blastomeres (which would correspond to oocyte chromocentres with attached nucleoli) do not form nucleoli, since these organelles are formed from satellites located on different chromosomes. In addition, different blastomeres which have the same euploid number of chromosomes would have a variable number of secondary constrictions. In any case, to the best of my knowledge, Melander's data and hypothesis have not been confirmed to date.

The differences in the size and morphology of blastomeres which arise in early embryogenesis have not been correlated with precocious differentiation of somatic and germ lines (Mattiesen 1904; Fulinsky 1914; Carlè 1935; Seilern-Aspang 1958; Le Moigne 1966; Koscielsky 1966). These differences are the consequence of unequal mitotic divisions and seem to occur in a small percentage of blastomeres which become closely associated during the formation of the provisional embryonic pharynx. Also many free embryonic cells apparently retain their original undifferentiated state in later stages of development and most researchers believe that a number of these cells remain undifferentiated and possibly totipotent for long periods of time, perhaps throughout the life of the worm.

In Turbellaria the problem of cell totipotency, dedifferentiation and transdifferentiation has been mainly investigated during reparative regeneration of triclads. In the laboratory the use of regeneration is favored over the controlled study of embryogenesis because the later is not yet technically feasible. The central question in planarian regeneration concerns the origin of the undifferentiated cells which form the blastema from which any injured or removed part of the body is rebuilt. Two main theories have been proposed to answer this question. The first, the neoblast theory, claims that embryonic reserve cells, the so-called neoblasts, always remain scattered throughout the parenchyma of adult worms and are the only cells capable of dividing and migrating to form the blastema. This theory was definitively formulated by the French school of Wolff, Dubois, Lender and their colleagues and it is presently supported by most scientists studying planarian regeneration, although at times it has appeared in slightly modified forms (review by Brøndsted 1969; Gremigni 1974; Baguña 1975). The second, the dedifferentiation theory, claims that embryonic, reserve cells do not exist or in any way take part in reparative regeneration, and that the blastema is formed from dedifferentiated cells. This theory which was formulated at the beginning of this century by Lang (1912) and Steinmann (1925) has been recently reproposed by some American researchers, in particular by Hay (1968) and Coward (1969).

Much indirect evidence has been gathered by many authors in support of the first or the second theory. In support of the neoblast theory I cite the excellent experiments which demonstrated the existance of undifferentiated cells within the parenchyma of adult worms and their ability to divide and migrate to the wound where the blastema will form. Much of the research was detailed, well documented and beleivable, but in my opinion, none of this studies can completely counter the criticisms offered by the dedifferentiation school of thought, specifically these studies cannot exclude the possibility that undifferentiated cells present in the parenchyma are at least in part derived from the dedifferentiation of specialized cells.

On the other hand, the well documented findings supporting the existance of dedifferentiation of several cell types cannot prove (a) that free undifferentiated cells or 'beta cells' according to the nomenclature introduced by Hay & Coward (1975) serve solely for cell replacement and not for reparative regeneration as well; (b) that at least some beta cells are not actually embryonic, reserve cells.

I believe that these theories are then not mutually exclusive and that the best way to try to resolve this controversial question is to attempt to distinguish embryonic from differentiated somatic or germ cells by means of specific markers which remain evident throughout the life cycle or at least remain present long enough to conduct sufficient experimental procedures.

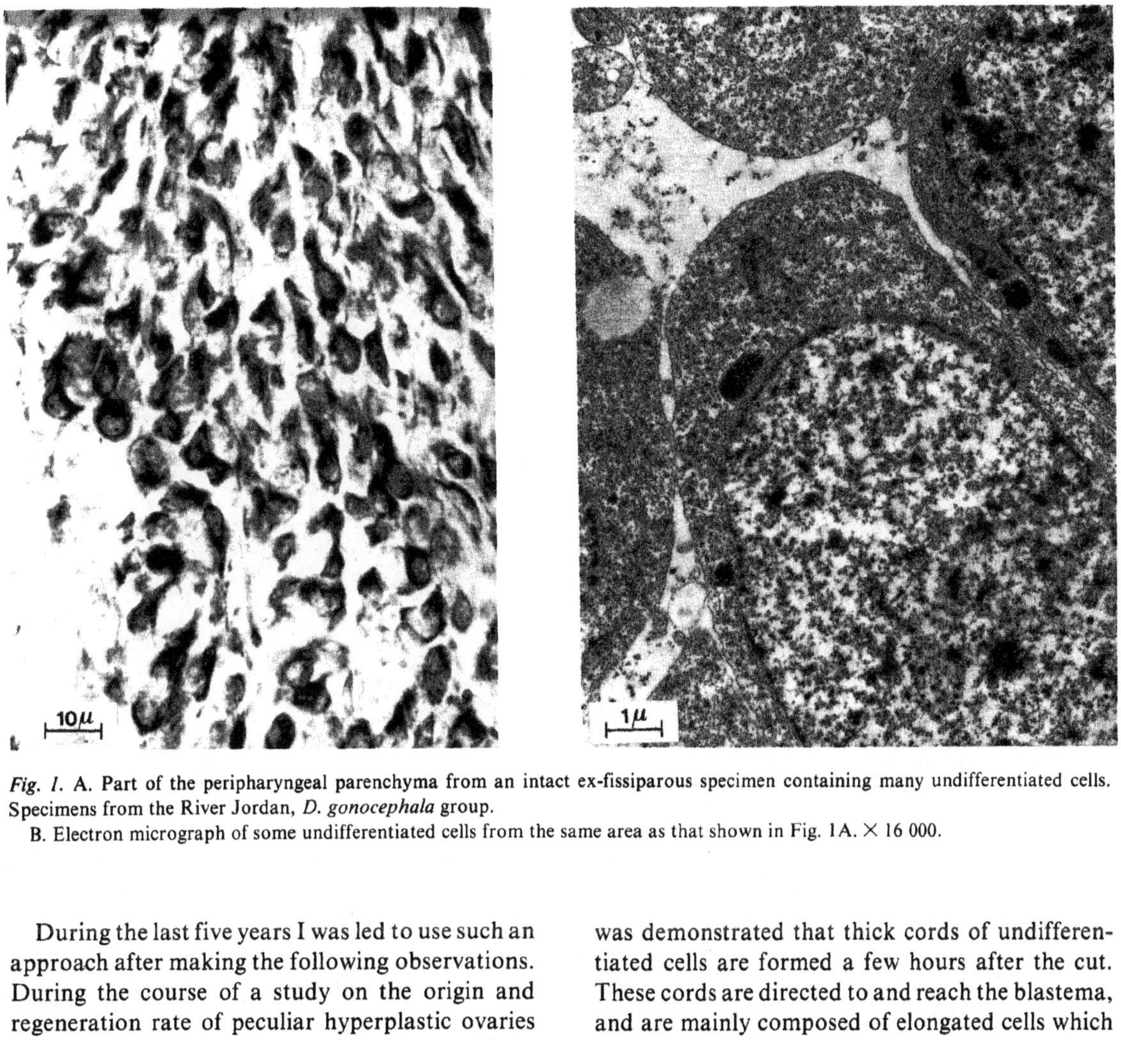

Fig. 1. A. Part of the peripharyngeal parenchyma from an intact ex-fissiparous specimen containing many undifferentiated cells. Specimens from the River Jordan, *D. gonocephala* group.

B. Electron micrograph of some undifferentiated cells from the same area as that shown in Fig. 1A. × 16 000.

During the last five years I was led to use such an approach after making the following observations. During the course of a study on the origin and regeneration rate of peculiar hyperplastic ovaries present in some ex-fissiparous specimens of the *Dugesia gonocephala* group it was noted that a great number of apparently undifferentiated cells were scattered throughout the body, particularly in the peripharyngeal area (Fig. 1) (Banchetti & Gremigni 1973). Transection experiments demonstrated that the regeneration rate of the head and ovary in the ex-fissiparous specimens is unusually high when compared with that of normal specimens of the same species. This fast rate of regeneration was attributed to the extraordinarily high number of undifferentiated cells present in the parenchyma, which were also seen dividing before and after the cut (Gremigni & Banchetti 1972). In another series of experiments (Banchetti & Gremigni 1973) performed with the same ex-fissiparous specimens, it was demonstrated that thick cords of undifferentiated cells are formed a few hours after the cut. These cords are directed to and reach the blastema, and are mainly composed of elongated cells which have their long axis directed toward the wound and which contain many microtubules arranged parallel to the long axis of the cell. When four to five cuts were repeated at the same level at two day intervals the regeneration rate remained fairly constant and the cell cords were still thick. However, when the interval between two consecutive cuts was seven days the regeneration rate progressively decreased and the cell cords were no longer present after the second or third cut. The interpretation of these results was that undifferentiated cells lying free in the parenchyma, actually play a major role in blastema formation by first migrating to the cords and then into the blastema. Moreover it was proposed that the two day interval between each cut was too short for all the cord cells to reach the blastema, so that the cords remained thick after successive cuts and the remaining cord cells were able to take part in subsequent blastema formation. For this reason the regeneration rate of the head or tail remained fairly constant. With a seven day interval between cuts, most of the cord cells had time enough to reach the blastema, but this interval was insufficient to allow for complete replacement of free parenchymal cells in the stump. In this manner the cords progressively disappeared and consequently a lower regeneration rate was obtained after the second or third cut. Thus these results and their proposed interpretation were in support of the neoblast theory. But histological observations of regenerants showed that young male or female germ cells also apparently leave the gonads after transection and take part in cell cord and blastema formation (Fig. 2). Thus I developed the idea that both undifferentiated, reserve cells and differentiated cells which are not irreversibly specialized, i.e. young germ cells in our experiments, do participate in reparative regeneration.

But I was not satisfied with these results as the proof was somewhat indirect. Thus I planned subsequent studies which took advantage of a triplo-hexaploid biotype of the *Dugesia lugubris-polychroa* group. The planarians used were collected at lake Iseo in northern Italy and posses a natural and stable karyological marker which allows investigators to distinguish embryonic and somatic cells from male and female germ cells (Fig. 3). In sexually mature specimens blastema formation was induced from stumps which either contained or lacked gonads. Karyological analysis of the squashed blastemas formed from stumps devoid of gonads used as controls showed that all dividing blastema cells had 12 chromosomes. In the blastemas formed from stumps containing testes about 95% of the dividing cells had 12 chromosomes, the other 5% had either 8 chromosomes or a variable number of chromosomes between 8 and 12 associated with clear evidence of chromosome elimination. Thus 5% of the cells were interpreted as being differentiated or differentiating male germ cells respectively (Gremigni & Puccinelli, 1977). In the blastemas formed from stumps containing ovaries about 98% of the dividing cells had 12 chromosomes, the other 2% had 24 chromosomes, the latter being interpreted as differentiated female germ cells (Puccinelli & Gremigni 1978). No evidence was obtained concerning the source of the triploid cells and they may either represent embryonic, reserves cells or dedifferentiating somatic cells or both. An ultrastructural analysis showed that the blastema contained both undifferentiated and dedifferentiat-

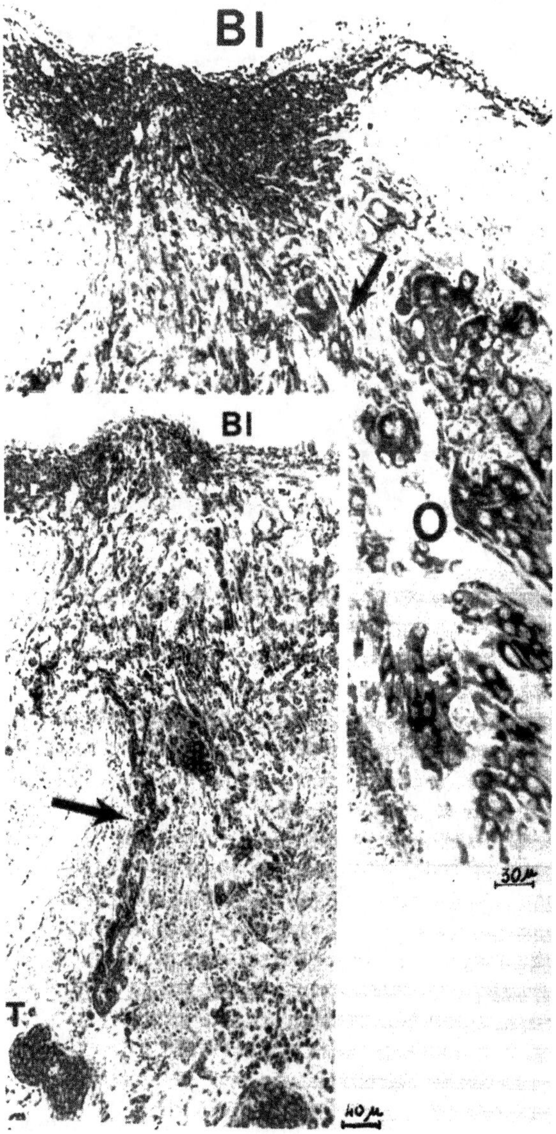

Fig. 2. Sections of the blastema regions from regenerating fragments. The arrows indicate cell cords which are directed towards the blastema. Young germ cells from the ovary (O) or testes (T) seem to take part in cord and blastema formation. Specimens from the Jordan river, *D. gonocephala* group.

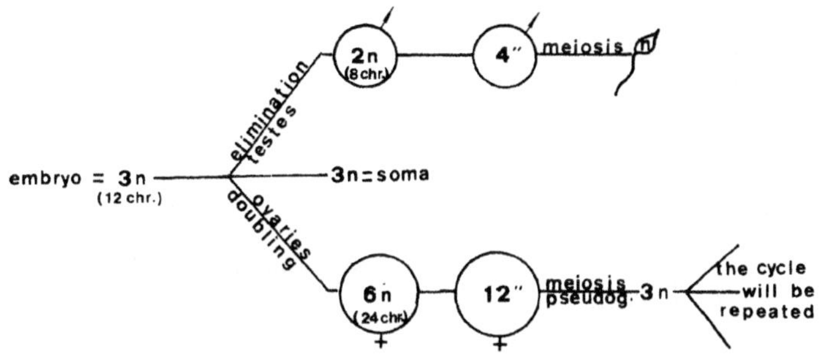

Fig. 3. Scheme of the chromosome cycle of a triplo-hexaploid biotype. Iseo population, *D. lugubris-polychroa* group.

ing cells thus suggesting that both types of cells take part in blastema formation (Gremigni & Puccinelli 1977).

From the above experiments it was demonstrated that young male and female germ cells can readily take part in blastema formation by leaving the gonads which subsequently either decrease in size or completely disappear when a transection is performed in their vicinity. Young germ cells are probably gametogonia and primary gametocytes, since it was noted that secondary spermatocytes, spermatids and mature sperms degenerated after transection at the level of the testes.

At this point in my work another question arose: what is the role of germ cells which migrate into the blastema during regeneration? In other words, are blastema cells that come from gonads actually totipotent, being able to differentiate into any cell type, or are they unipotent, being destined to redifferentiate again into the original cell type? To resolve this question I used two different experimental techniques (karyological and cytophotometric) on the same triplo-hexaploid specimens. By the karyological method we first sought to demonstrate whether young male or female germ cells take part in regenerating somatic tissues. Thus a series of experiments was done to induce these cells to take part in blastema formation and the subsequent rebuilding of the cephalic or caudal areas, zones which never contain gonads. The karyological analysis showed that either some diploid ($2n = 8$) or hexaploid ($6n = 24$) cells coming from the testes or ovaries respectively are present in the fully regenerated cephalic or caudal areas which are devoid of gonads (Gremigni, Miceli & Puccinelli 1980). These results strongly support the hypothesis that germ cells take part in rebuilding somatic tissues. Similar results were also obtained using prolonged starvation instead of transection to induce dedifferentiation and migration of germ cells (Gremigni & Puccinelli 1979). By this method we reproduced a type of regeneration similar to a physiological or repetitive regeneration which occurs in some planarians having seasonal alternations between the sexual and asexual condition. Starvation, in fact, produces progressive decrease and disappearance of gonads which are then rapidly rebuilt after the specimens are renourished.

However, these results were not conclusive in demonstrating the occurrence of cell transdifferentiation, since mitotic cells are widely held as not being differentiated cells. Thus a different karyological approach was attempted, using the planarians. Sexually mature specimens were transected just behind ovaries and the posterior stump containing testes was allowed to rebuild an anterior fragment which included new ovaries (Fig. 4). The regenerated specimens were allowed to reach sexual maturity and their deposited eggs were karyologically analyzed. It was found that, in addition to a very high number of oocytes (386) with 12 bivalents as normally occurs, 14 oocytes containing eight bivalents were also produced by regenerated specimens. On the contrary, all oocytes (400) from untransected, control specimens had 12 bivalents (Fig. 5). These data suggest that after transection some young male germ cells with 8 chromosomes migrate to the blastema and can differentiate into female germ cells, thus doubling their chromosome number and giving rise to tetraploid oocytes with eight bivalents, if they encounter the new ovarian field. This fact only exceptionally occurs in natural

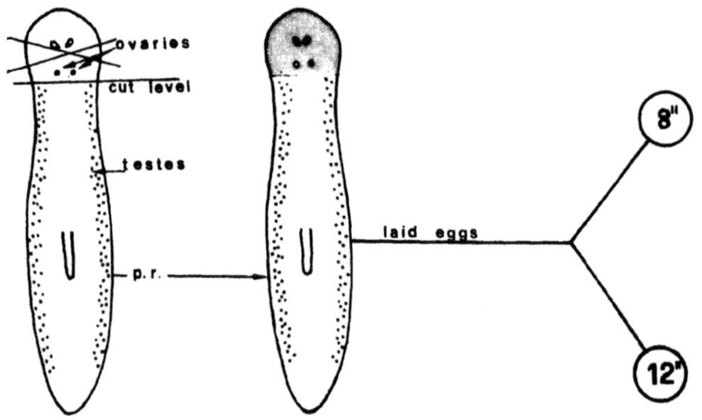

Fig. 4. Scheme of the experiments performed to induce ovary regeneration from a posterior fragment containing testes. Iseo population, *D. lugubris-polychroa* group.

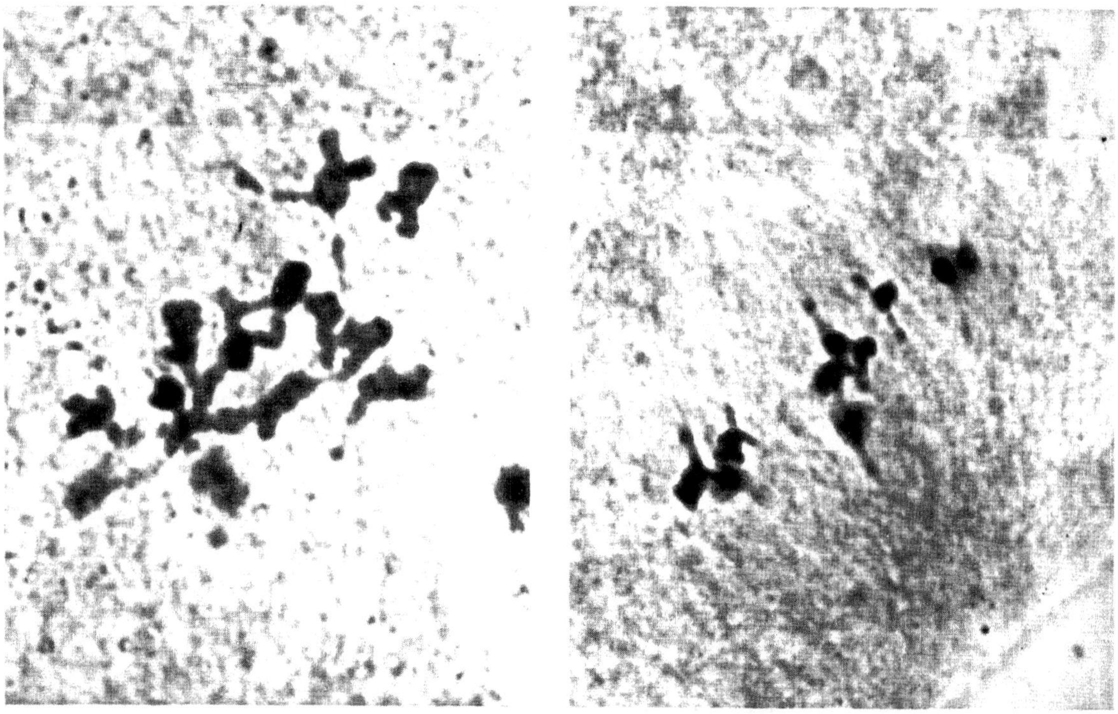

Fig. 5. A meiotic oocyte with 12 bivalents (A) produced by an intact specimen and a similar meiotic oocyte with eight bivalents (B) produced by a regenerated specimen. Iseo population, *D. lugubris-polychroa* group. × 2600.

biotypes of the *D. lugubris polychroa* group (Benazzi-Lentati 1976).

Utilizing a cytophotometric analysis it was established that the Feulgen-DNA content of diploid male germ cells is fairly constant and clearly less than that of embryonic and somatic triploid cells (Fig. 6A) (Gremigni, Miceli & Picano 1980). Then the Feulgen-DNA content of unmistakably specialized somatic cells was analyzed prior to and after regeneration. Muscle cells of the pharynx were chosen for these experiments because their nuclei are easily distinguished from the other pharyngeal

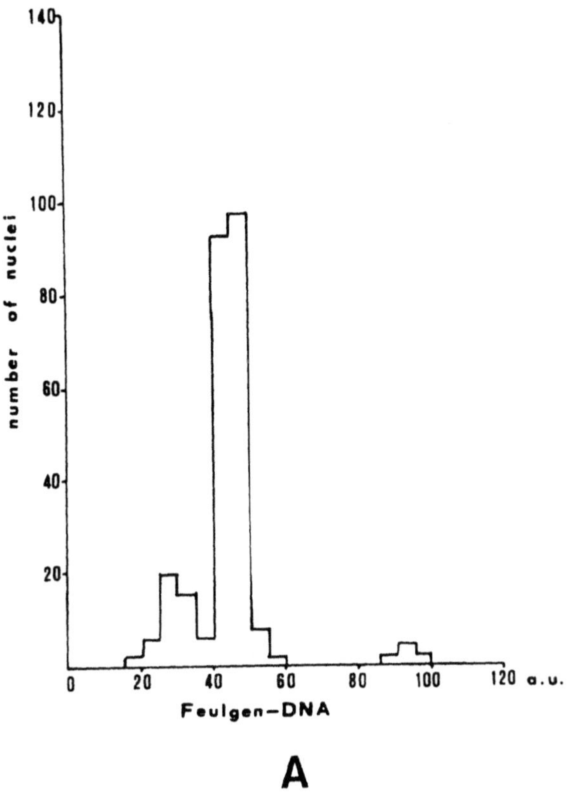

Fig. 6. A. Diagram representing the Feulgen-DNA content of cell nuclei from a parenchymal area containing testes. The small peak on the left represents diploid male germ cells, the peak on the right represents triploid somatic or reserve cells.

cell nuclei. Muscle cell nuclei are characteristically elongated while the others are roundish. The cytophotometric analysis showed that muscle cell nuclei from the pharynx of intact, control animals constantly had a triploid Feulgen-DNA content equivalent to that of the embryonic and somatic cell nuclei previously analyzed (Fig. 6B). Muscle cell nuclei of the pharynx regenerated from stumps containing testes had either a triploid Feulgen-DNA content (95% of the cells) or a diploid Feulgen-DNA content (5% of the cells) equivalent to that previously observed in diploid male germ cells (Fig. 6C) (Gremigni & Miceli 1980).

Thus both karyological and cytophotometric methods directly demonstrated that young male germ cells which took part in blastema formation can redifferentiate into different cell types, specifically female germ cells or pharyngeal muscle cells.

A question remains to be answered. Can a young germ cell be considered a differentiated cell or not? The following observations show that it is at least reasonable to consider gametogonia and primary gametocytes partially differentiated because the male germ cells are formed only when a haploid chromosome complement is eliminated from undifferentiated, embryonic cells, and the female germ cells are only formed when undifferentiated, embryonic cells double their chromosome number. Therefore I suggest that the transformation of young germ cells, as demonstrated in the above experiments, should be considered a peculiar form of transdifferentiation. I also suspect that a similar phenomenon may occur in some somatic cells although technical difficulties have so far prevented direct experimental demonstration of this.

The problem of cell totipotency, dedifferentiation and transdifferentiation has been inadequately investigated during embryogenesis or regeneration in other Turbellaria. However, recent investigations carried out by Moraczewski (1977) suggest that the cellular basis of regeneration may not be dissimilar to that found in triclads. Indeed, asexual propagation in catenulids seems to involve undifferentiated cells which according to the author are mainly derived from a dedifferentiation process. Further investigations are needed to clarify these points in other turbellarians.

To conclude, recent evidence suggest that for triclads and probably most turbellarians as well, one should consider blastomeres and embryonic cells totipotent until their differentiation, since no precocious determination of the germ and somatic lines seem to occur. Cell dedifferentiation seems to be a common phenomenon which most likely involves several types of partially specialized cells and is used both for physiological and reparative regeneration. Blastema cells should be considered totipotent as they most likely differentiate into any cell type depending on the field influence they encounter (Wolff, Lender & Ziller-Sengel 1964). Thus blastema cells which derive from dedifferentiation of somatic or germ cells can undergo transdifferentiation and this event probably occurs more often than previously supposed in triclads and other turbellarians.

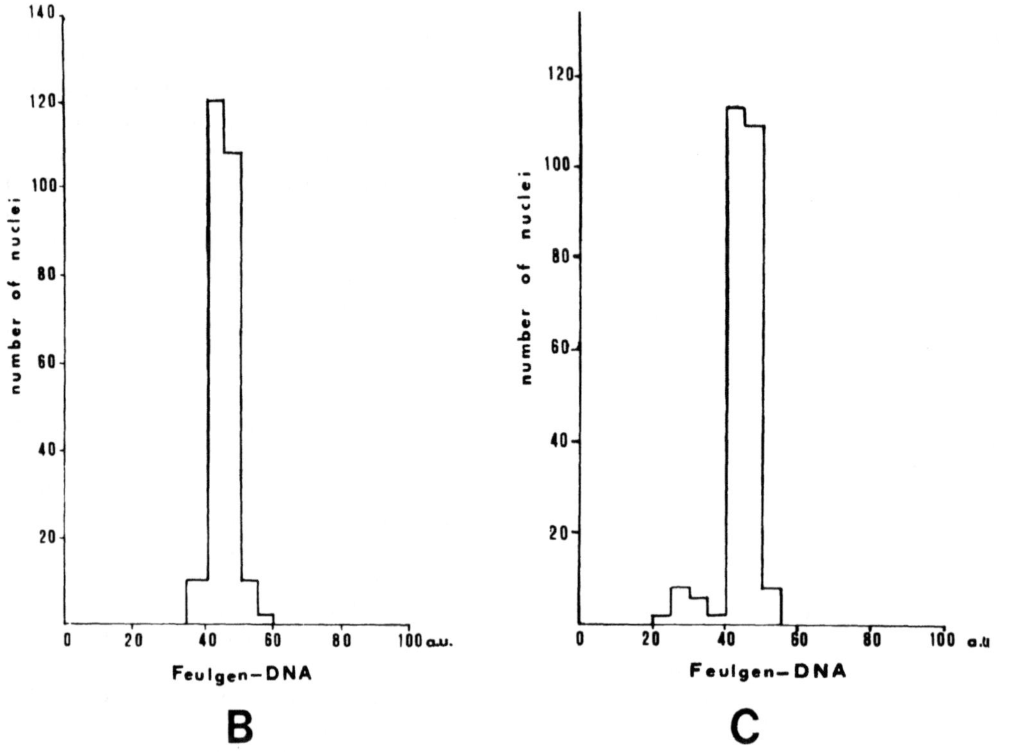

Fig. 6. B–C. Diagrams representing the Feulgen-DNA content of muscle cell nuclei from an intact (B) or a regenerated (C) pharynx. In diagram B the solitary peak represents triploid cell nuclei. In diagram C the small peak on the left represents diploid cell nuclei, while the high peak on the right represents triploid cell nuclei. Iseo population, *D. lugubris-polychroa* group.

References

Baguña, J., 1975. Mitosis in the intact and regenerating planarian Dugesia mediterranea n. sp. II Mitotic studies during regeneration and a possible mechanism of blastema formation. J. Exp. Zool. 195: 65–79.

Banchetti, R. & Gremigni, V., 1973. Indirect evidence for neoblast migration and for gametogonia dedifferentiation in ex-fissiparous specimens of Dugesia gonocephala s.l.. Accad. Naz. Lincei 55: 107–115.

Beams, H. W. & Kessel, R. G., 1974. The problem of germ cell determinants. Int. Rev. Cytol. 39: 413–479.

Benazzi-Lentati, G., 1976. On the relationship between the chromosome set and the sexual differentiation of the neoblasts in planarians. Acta Embryol. Exp. 3: 299–317.

Bröndsted, H. V., 1969. Planarian Regeneration. Oxford: Pergamon Press.

Carlé, R., 1935. Beiträge zur Embryologie der Landplanarien. I Frühentwicklung, Bau und Funktion des Embryonalpharynx. Z. Morph. Ökol. Tiere 29: 527–558.

Coward, S. J., 1969. Regeneration in planarians: some unresolved problems and questions. J. Biol. Psych. 11: 15–19.

D'Amato, F., 1977. Nuclear cytology in relation to development. Cambridge: University Press.

Fulinsky, B., 1914. Die Entwicklungs geschichte von Dendrocoelum lacteum. I Teil: Die erste Entwicklungsphase vom Ei bis zur Embryonalpharynxbildung. Bull. Acad. Sci. Cracovie, ser. B: 148–187.

Gremigni, V., 1974. The origin and cytodifferentiation of germ cells in the planarians. Boll. Zool. 41: 359–377.

Gremigni, V. & Banchetti, R., 1972. The origin of hyperplasia in the ovaries of ex-fissiparous individuals in Dugesia gonocephala s.l. Accad. Naz. Lincei 53: 477–485.

Gremigni, V. & Miceli, C., 1980. Cytophotometric evidence for cell 'Transdifferentiation' in planarian regeneration. Wilhelm Roux's Arch. 188: 107–113.

Gremigni, V., Miceli, C. & Picano, E., 1980. On the role of germ cells in planarian regeneration: II Cytophotometric analysis of the nuclear Feulgen-DNA content in cells of regenerated somatic tissues. J. Embryol. Exp. Morphol. 55: 65–76.

Gremigni, V., Miceli, C. & Puccinelli, I., 1980. On the role of germ cells in planarian regeneration: I A karyological investigation. J. Embryol. Exp. Morphol. 55: 53–63.

Gremigni, V. & Puccinelli, I., 1977. A contribution to the problem of the origin of blastema cells in planarians: a karyological and ultrastructural investigation. J. Exp. Zool. 199: 57–72.

Gremigni, V. & Puccinelli, I., 1979. Sdifferenziamento e migrazione di cellule germinali in planarie tenute a digiuno: indagine cariologica. Ric. Sci. Educ. Perm. suppl. 6: 115–116.

Hay, E. D., 1968. Dedifferentiation and metaplasia in vertebrate and invertebrate regeneration. In: Ed. Ursprung, H., The stability of differentiate state. New York: Springer.

Hay, E. D. & Coward, S. J., 1975. Fine structure studies on the planarian Dugesia. I Nature of the 'neoblast' and other cell types in noninjured worms. J. Ultrastruct. Res. 50: 1–21.

Koscielski, B., 1966. Cytological and cytochemical investigation on the embryonic development of Dendrocoelum lacteum O. F. Müller. Zool. Pol. 16: 83–102.

Lang, A., 1912. Über Regeneration bei Planarien. Arch. Mikr. Anat. 79: 361–426.

Le Moigne, A., 1966. Étude au microscope électronique des cellules d'embryons de Polycelis (Turbellarié, Triclade) au début de leur dévelopment. C.R. Acad. Sc. 263: 550-553.

Mattiesen, E., 1904. Ein Beitrag zue Embryologie der Süsswasserdendrocoelen. Z. wiss. Zool. 77: 274–361.

McKinnel, R. G., 1978. Cloning. Nuclear transplantation in Amphibia. Univ. Minnesota Press.

Melander, Y., 1963. Cytogenetic aspects of embryogenesis in Paludicola, Tricladida. Hereditas (Lund) 49: 119–166.

Moraczewski, J., 1977. Asexual reproduction and regeneration of Catenula (Turbellaria, Archoophora). Zoomorphologie 88: 65–80.

Puccinelli, I. & Gremigni, V., 1978. Sulla presenza nei blastemi di planarie di cellule provenienti dal territorio gonadico femminile. Accad. Naz. Lincei 63: 588–592.

Seilern-Aspang, F., 1958. Entwicklungsgeschichtliche Studien an Paludicolen Tricladen. Arch. Entwm. 150: 425–480.

Steinmann, P., 1925. Das verhalten der Zellen und Gewebe im regenerienden TricladenKörper. Verh. Naturforsch. Ges. 36: 133–162.

Wolff, E., Lender, T. & Ziller-Sengel, C., 1964. Le rôle de facteurs auto-inhibiteurs dans la régénération des Planaires. Rev. Suisse Zool. 71: 75–98.

Quantitative analysis of cell types during growth, degrowth and regeneration in the planarians Dugesia mediterranea and Dugesia tigrina

J. Baguñá and R. Romero
Dept. de Genètica, Facultat de Biologia, Univ. de Barcelona, Gran Via 585, B-7, Spain

Keywords: Turbellaria, *Dugesia,* cell types, growth, degrowth, regneration, maceration-technique

Abstract

A method of tissue maceration (dissociation) of planarian tissues into single cells was used to characterize the basic cell types in the planarians *Dugesia mediterranea* and *Dugesia tigrina,* and to determine the total cell number and distribution of cell types during growth, degrowth and regeneration.

Using this method, 13 basic cell types have been determined for both species. The total number of cells increases with body length and volume whereas the distribution of cell types is only slightly affected. Growth and degrowth occur mainly through changes in total cell number leaving cell distribution only moderately affected. During regeneration, an increase in neoblast density in the blastema followed later on by increases in nerve cells are the more significant changes detected.

These results are discussed in relation to mechanisms of cell renewal, blastema formation and maintenance of tissue polarity.

Introduction

The understanding of growth, degrowth and regeneration in planarians requires data on abundance and distribution of the basic cell types in the animal. Obtaining such data in histological sections is rather difficult because planarian parenchyma is a loosely organized tissue composed of several cell types with numerous and interdigitating processes that make it impossible to establish with certainty the cell limits. Moreover, neoblasts and nerve cells are difficult to distinguish using conventional histological techniques, gastrodermal and globet cells are difficult to count, and some cell types such as flame cells, muscle cells and pigment cells are almost impossible to recognize.

To overcome these difficulties we have used the method of maceration of planarian tissues into single cells. In such macerations each cell type is easily distinguishable and its abundance can be determined by counting.

In this paper we describe 13 basic cell types and the criteria for distinguishing them in the planarians *Dugesia mediterranea* and *Dugesia tigrina.* The total number of cells and the percentages of the different cell types for standard animals of different sizes are also described. Further, the distribution of cell types in different regions along the cephalo-caudal axis, and the changes that occur in total cell number and cell distribution in growing, starving and regenerating animals are described.

Material and methods

1. Material

The planarians used in this work belong to the asexual races of the species *Dugesia mediterranea* (Benazzi *et al* 1975) and *Dugesia tigrina* (Girard). The animals were reared in petri dishes and kept in

Hydrobiologia 84, 181–194 (1981). 0018–8158/81/0842–0181/$02.80.

filtered pond water at 17 °C in the dark. They were fed weekly on beef liver. All the experiments were done at 17 °C.

2. Methods

2.1. Total body volume determination (V)

The body volume of animals of different sizes for each species was determined according to Baguñà (1976a).

2.2. Maceration technique

The technique used is a modification of David's technique for *Hydra* (David 1973) and has been described elsewhere (Baguñà 1973, 1976a). Briefly, it consists in macerating whole animals in a solution containing methanol: glacial acetic acid: glycerol: distilled water (3:1:2:14) at 8–10 °C. After 24–48 h, the test tube is shaken very gently until a suspension of single cells is obtained. The cell suspension is fixed by addition of 0.1 volume of 20% formaldehyde and/or 0.1 volume of 1% osmium tetroxide and spread on microscope slides to dry on a leveled surface. To examine cells a drop of water and a coverslip were placed on the slide and examined by phase contrast optics (×40 and ×63 objectives).

2.3. Counting of total cell number (C)

The total number of cells for whole animals of different sizes was obtained by macerating a known number of identical animals (usually five) in a known volume of maceration solution and determining the cell concentration with an improved Neubauer cell counter (0.1 mm depth).

To estimate the individual variability in cell number, ten planarians of 5 mm and ten planarians of 7 mm of the species *Dugesia tigrina* were macerated individually and the total cell number determined. The mean cell number per planaria was found to be 314 100 ± 49 800 (standard deviation) for 5 mm long individuals, and 554 250 ± 69 600 for 7 mm long animals. Samples of five to ten animals of identical length were macerated together. The mean cell number per planaria was found to be 342 100 ± 8 240 and 523 350 ± 14 380 for 5 and 7 mm long animals respectively. All the experiments reported in this paper were done according to the last method.

2.4. Total number and distribution of cell types

Total number and distribution for each cell type were calculated from the total number of cells per animal (or region) and the percentage of a given cell type in a macerate of that animal (or region). The data presented are the average of ten experiments using five organisms per experiment.

Samples of planarians of the same length macerated individually and samples of 5–10 planarians macerated together were examined to determine the variability of different cell types among individual animals. The differences between both kind of samples were small, and were not reduced by pooling the organisms.

Though classification of most cell types in planarians is not difficult, there are cells difficult to assign to a specific cell type. Unfortunately such cells are not uncommon being mainly cells in the process of differentiation, a process enhanced by the high turnover of cells in planarians and by the fact that neoblasts are the stem-cells of most differentiated cell types. The percentage of these cells, labelled as 'unclassified' attains 10% in *Dugesia tigrina* and around 5% in *Dugesia mediterranea*. The higher value in the former is due to the difficulties in classifying flame, rhabdite and striped cells in this species as compared to their easy recognition in *Dugesia mediterranea.*

3. Experimental procedures

3.1. Degrowth experiments

One week-starved animals of 7 and 11 mm length of both species were kept fasting for three months at 17 °C. At different intervals in time five animals of each group were macerated together and the total cell number and total number and distribution of cell types measured.

3.2. Growth experiments

One week-starved *Dugesia tigrina* of 5 and 7 mm length were used. The animals were fed on an artificial mixture of sonicated beef liver and starch from the start of the experiment and every seven days during a period of three months. The artificial food was used because previous results (Romero, unpublished observations) showed a very slow digestion of raw beef liver by planarians that left many intact liver cells even three days after feeding.

At 1, 3, and 6 days after each feeding, five animals

of each group were removed, macerated together and their total cell number and total number and distribution of cell types measured.

3.3. Regeneration experiments

Animals of 11 mm length of both species, starved for a week, were used. The animals were cut at postcephalic level and the posterior part was kept regenerating in filtered pond water at 17 °C. At different intervals in time (1, 2, 3, 5, 9 and 15 days), five animals were killed and the blastema (B) and a narrow region behind it (Post-blastema, PB) were cut with a fine scalpel and macerated separately. The total number and distribution of cell types were then measured.

Results

1. Maceration and identification of basic cell types

Using the maceration technique, planarian cells have morphological characteristics which help to identify most of them. It should be pointed out that 5 to 10% of the cells cannot be properly classified because they share characteristics of more than one cell type or because they are in the process of differentiation being between a neoblast and some terminal differentiated cell type.

We give now a description of the cell types as well as the criteria to identify them in phase contrast optics.

Neoblasts (Figure 1a). Small cells (8–15 μm), rounded or pear-shaped, with a large nucleus, scanty cytoplasm and usually a short process of 5–20 μm that may contain a lipid droplet. All these characteristics agree with the cell type described by light and electron microscopy.

Nerve cells (Figure 1b). Though several nerve cell types have been described with light and electron microscopy in planarians, they are morphologically so similar in phase contrast that we include them in a single group. Nerve cells have a small cell body (8–20 μm) and long (up to 500 μm), thin and branched processes, some of them beaded. The nucleus is large and a prominent nucleolus is usually present. Some nerve cells present vacuoles and/or lipid droplets in the cytoplasm near the cell body. The nerve cells are unipolar (50–60%), bipolar (20–30%) and rarely multipolar (less than 10%).

Epidermal cells (figure 1c). Regular-sized cuboidal or slightly columnar cells (20–40 μm long, 20–30 μm wide). Epidermal cells from the ventral size bear cilia at their distal end whereas cells from the dorsal side do not have them being taller and narrower. There is a small nucleus at the basal end with highly condensed chromatin. The cytoplasm is usually filled with cigar-shaped bodies or rhabdites alligned parallel to the proximo-distal axis. The basal end is usually highly indented.

Fixed parenchyma cells (Figure 1d). Highly polymorphic cells of large size (up to 100 μm in diameter) with numerous processes generally lost during the maceration procedure. Nucleus and cytoplasm pale in phase contrast optics and filled with vacuoles, intracellular spaces and lipid droplets of different sizes, numerous mitochondria and occasionally pigment granules.

Gastrodermal cells (Figure 1e). Tall, columnar cells (60–80 μm long, 15–20 μm wide) with the proximal end highly indented. Nucleus of medium size, placed in the proximal half of the cell and usually with a nucleolus. Cytoplasm filled with a heterogeneous group of food vacuoles and lipid droplets, highly refractile. During starvation the droplets and vacuoles disappear progressively.

Goblet or club cells (Figure 1f). Elongated cells (40–60 μm long, 10–15 μm wide) with a narrow and indented proximal end and a wider and rounder distal end filled with vacuoles separated by thin strands of cytoplasm. The nucleus, rather small, is usually obscured by the vacuoles.

Rhabdite cells (Figure 1g). Middle-sized, round or rectangular cells (30–40 μm long, 20–30 μm wide) with a variable number of rhabdites in different stages of formation. Nucleus with highly condensed chromatin. It is found in the parenchyma and it is presumed to be the precursor of the epidermal cells. Only described for *Dugesia mediterranea*.

Muscle cells (Figure 1h). Big, elongated cell (up to 150–200 μm long, 5–10 μm wide) with a central

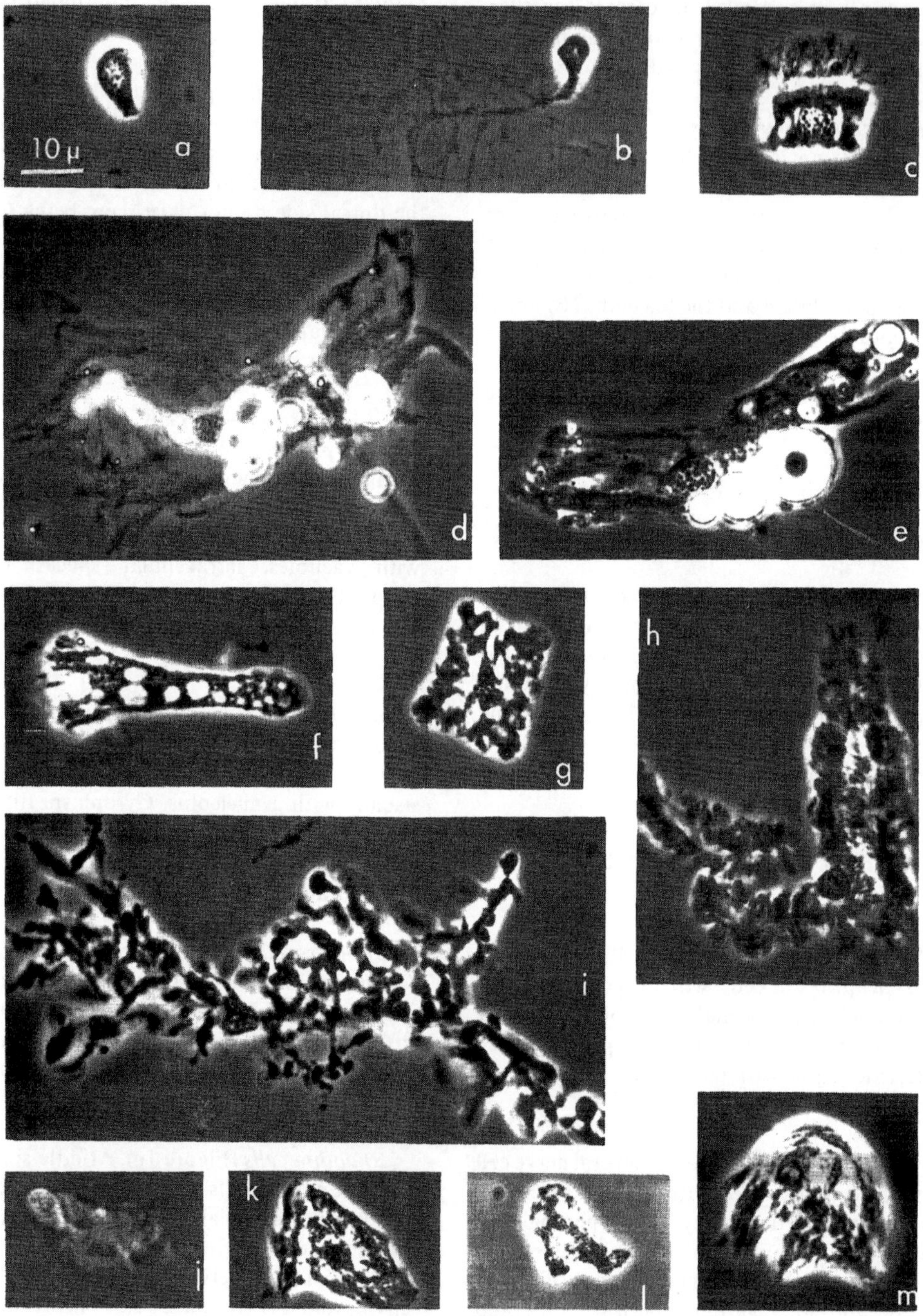

Fig. 1a–m. Basic cell types in the planarians *Dugesia mediterranea* and *Dugesia tigrina.* (a) neoblast; (b) nerve cell; (c) epidermal cell; (d) fixed parenchyma cell; (e) gastrodermal cell; (f) goblet cell; (g) rhabdite cell; (h) muscle cell; (i) pigment cell; (j) flame cell; (k) basophilic cell; (l) striped cell; (m) acidophilic cell. All photographs ×640.

fibrilar stem flanked with a double fringe of cytoplasmic extension filled with mitochondria. Large nucleus with highly condensed chromatin situated in a large cytoplasmic extension linked to the stem by a thin neck.

Pigment cells (Figure li). Large and highly branched cells (up to 100–130 μm in diameter) with the cytoplasm filled with pigment granules. The cytoplasmic processes are highly branched ending with bulbous terminals. Large nucleus, usually centrally located, with highly condensed chromatin.

Flame cells (Figure lj). Elongated cells (20–30 μm long, 8–15 μm wide) with a nucleus placed in the basal end and a centrally located tuft of long cilia. Cytoplasm scanty and homogeneous without inclusions or vacuoles. Only described for *Dugesia mediterranea.*

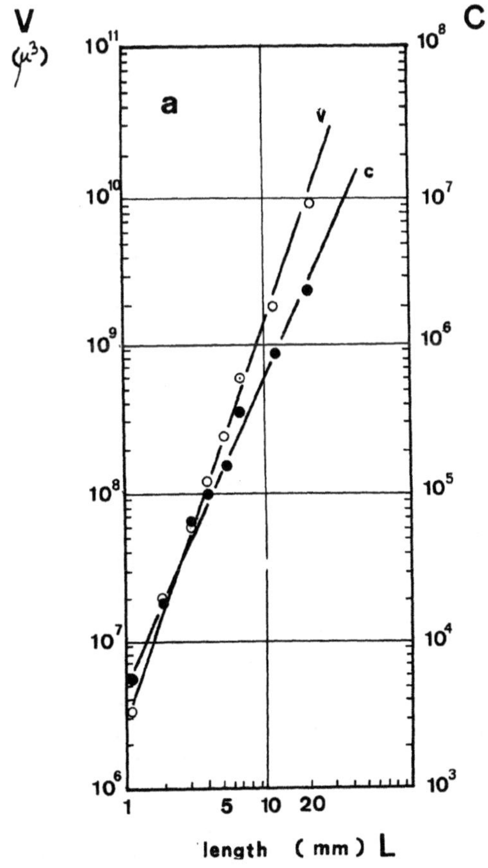

Fig. 2. Total body volume (V) (fixed state) and total cell number (C) as a function of body length (L) in *Dugesia mediterranea.*

Basophilic cells (Figure lk). Middle-sized to large cells (up to 50–70 μm in diameter) of variable form bearing some processes of different length which are usually lost during the maceration procedure. Cytoplasm filled with large and polymorphic granules heterogeneous in density giving a foamy appearance to the cell. Nucleus of middle size, often barely visible.

Striped cells (Figure 11). Cells of regular size (20–30 μm in diameter), ovoid to rectangular, with a cytoplasm very rich in endoplasmic reticulum arranged in concentric rings around the nucleus giving a striped appearance under phase contrast optics. The nucleus is small (6–8 μm) with finely dispersed chromatin. Most probably these cells are not terminal differentiated cells but they are in the process of differentiation from neoblasts to several types of parenchyma gland cells (rhabdite, basophilic, acidophilic, . . .). Only described for *Dugesia mediterranea.*

Acidophilic cells (Figure 1m). Middle sized cells (30–60 μm in diameter), usually oval or pear-shaped. Cytoplasm homogeneous, highly refractile in phase optics, and filled with numerous granules. Nucleus generally not visible.

2. Total body volume (V) and total cell number (C) as a function of the animal body length (L)

2.1 The 'standard' planarian

Since feeding conditions and temperature may affect cell number and cell distribution in planarians, we define a 'standard' planarian as an animal of a given length starved for a week at 17 °C. This is the period of starvation most commonly employed in regeneration and growth experiments.

2.2. The relationship between length (L), volume (V) and total cell number (C)

The relationship between the length (L) of a planarian with the total body volume (V) (fixed state) and the total cell number (C) for both species is shown in Figures 2 and 3. For similar-sized animals the total number of cells is higher in

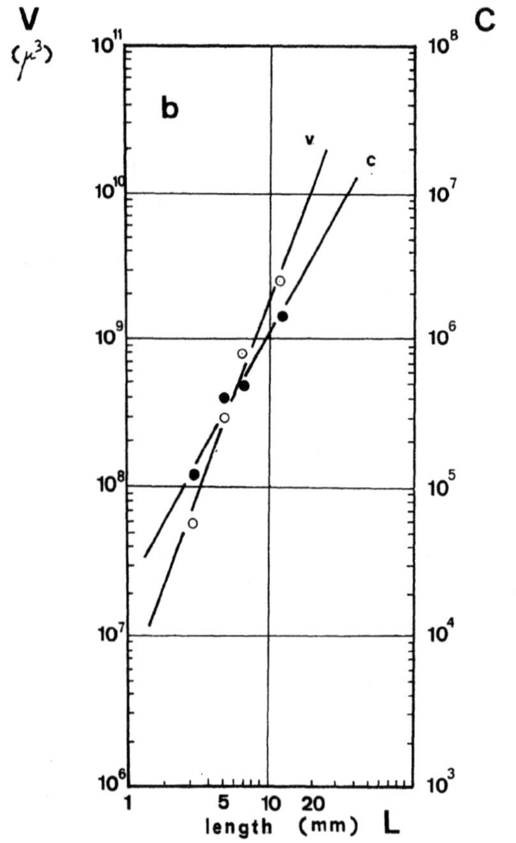

Fig. 3. Total body volume (V) (fixed state) and total cell number (C) as a function of body length (L) in *Dugesia tigrina*.

Dugesia tigrina than in *Dugesia mediterranea*. Since the body volumes are rather similar, the mean cell size (V/C) in *Dugesia tigrina* is lower than in *Dugesia mediterranea*. In both species the total number of cells rises with length at a slower rate than it does with total body volume, this trend being slower for *Dugesia mediterranea*. This means that a longer animal has in general a lower density since the mean size of its cells (V/C) is greater than that of a smaller animal. Correcting for the degree in tissue shrinkage (due to fixation) along the three body axis (Lange 1967) the live volume could be as much as eight times those given for fixed tissue. This means that the mean cell size of a 16 mm long *Dugesia mediterranea* is about 29×10^3 μm^3 (equivalent to a cuboidal cell of 31 μm each side) as compared to a volume of 12×10^3 μm^3 (equivalent) to a 23 μm sided cuboidal cell) for a 4 mm long animal.

However, this increase in body volume due to hypertrofy of cells is small compared to the hyperplasia or increase in cell number. Hence, most of the increase in body volume must have been due to cellular proliferation.

3. Total number and distribution of cell types in standard planarians

The total number and distribution of the 13 basic cell types in standard *Dugesia mediterranea* of, 4, 7, 11 and 16 mm in length, and ten basic cell types in standard *Dugesia tigrina* of 5, 7 and 11 mm in length, are shown in Tables 1 and 2 respectively.

It is interesting to note the decrease in neoblast

Table 1. Distribution of cell types in standard individuals of *Dugesia mediterranea* of different size.

Cell type		Body length (mm) 4	7	11	16
Neoblast	%	32	28	23	18
	#	35 200	95 200	190 900	324 000
Nerve	%	20	16	11.5	8
	#	22 000	54 400	95 400	144 000
Epidermal	%	14	13	13	11
	#	15 400	44 200	107 900	198 000
Fix. Parq	%	11.5	15	20.5	25
	#	12 650	51 000	170 000	450 000
Acidoph	%	4.5	5.5	6.5	9
	#	4 900	18 700	53 900	162 000
Basoph	%	2.5	4	5.5	8
	#	2 700	13 600	45 600	144 000
Pigment	%	1	1.5	1.5	1.5
	#	1 100	5 100	12 400	27 000
Muscular	%	2	2.5	2.5	2
	#	2 200	8 500	20 750	36 000
Gastrod	%	2.5	4	5.5	6
	#	2 700	13 600	45 600	108 000
Globet	%	0.5	1	1	1.5
	#	550	3 400	8 300	27 000
Flame	%	1	1	1	1.5
	#	1 100	3 400	8 300	27 000
Rhabdite	%	2.5	2	2	2
	#	2 700	6 800	16 600	36 600
Striped	%	1	1.5	1.5	1.5
	#	1 100	5 100	12 400	27 000
'Unclass.'	%	5	5	5	5
	#	5 500	17 000	41 500	90 000
Total Cell Number (C)		110 000	340 000	830 000	1 800 000

% percentage of total cells
# total number

Table 2. Distribution of cell types in standard individuals of Dugesia tigrina of different size.

		Body size (mm)		
Cell type		5	7	11
Neoblast	%	34	32	28
	#	115 600	166 400	372 400
Nerve	%	23	22	21
	#	78 200	114 400	279 300
Epidermal	%	13	11.5	10
	#	44 200	59 800	133 000
Fix. Parq	%	8	10.5	15
	#	27 200	54 600	199 500
Acidoph	%	3	3	4
	#	10 200	15 600	53 200
Basoph	%	2.5	3	3.5
	#	8 500	15 600	46 550
Pigment	%	1.5	1.5	1.5
	#	5 100	7 800	19 950
Muscular	%	1.5	1.5	1
	#	5 100	7 800	13 300
Gastrod	%	3	4	5
	#	10 200	20 800	66 500
Globet	%	0.5	1	1
	#	1 700	5 200	13 300
'Unclass.'	%	10	10	10
	#	34 000	52 000	133 000
Total Cell Number (C)		340 000	520 000	1 330 000

% percentage of total cells
# total number

density with the animal's growth; this trend is more evident in *Dugesia mediterranea* than in *Dugesia tigrina* and could be of importance in explaining the differences in the regenerative abilities of animals of different length. On the other hand there is a significant decrease in nerve cell density during growth in *Dugesia mediterranea* though this trend is not statistically significant for *Dugesia tigrina.* Both results agree with measures made at histological level (Lange, 1967, for *Dugesia polychroa;* Bagunà, 1973, 1976a, and unpublished results, for *Dugesia mediterranea).* There is too, for both species, a decrease in epidermal cell density during growth; this result is in agreement with the decrease in the ratio surface/volume as long as the animal increases in length.

Conversely, there are increases in cell percentages for the fixed parenchyma cells, both gland cells, and gastrodermal and goblet cells, the rest of the cells being mainly unchanged. Of special interest are the increases in fixed parenchyma and gastrodermal cells because as body volume increases an even greater increase in the relative volume or percentages of cells and tissues involved in food digestion and transport is necessary in organisms like planarians that lack an efficient transport system (Baguñà & Ballester 1978). This trend is evident for both species though more marked for *Dugesia mediterranea.*

Although actually present, flame, rhabdite and striped cells of *Dugesia tigrina* were difficult to visualize and count; therefore we placed them as 'unclassified' cell types. In *Dugesia mediterranea* (as in *Dugesia polychroa,* Baguñà unpublished results) those cells are easily seen and counted, though no statistically significant variations have been found between organisms of different length.

4. Antero-posterior distribution of cell types in 'standard' planarians

Table 3 gives the distribution of cell types in each region (1 to 6 along the body. 1: head; 2: post-cephalic region; 3: prepharyngeal region; 4: pharyngeal region minus pharynx; 5: post-pharyngeal region; 6: tail; Phx: pharynx) for a standard 11 mm long *Dugesia mediterranea.* Results for different sized *Dugesia mediterranea* and for *Dugesia tigrina* are rather similar, not being included for the sake of clarity. Because of difficulties to classify, pharyngeal cells are also not included though its total cell number was counted.

From the data in Table 3 we can draw the following conclusions: (a) the head (region 1) of the organism, is, as expected, high in nerve cells and also in neoblasts. Parenchyma and gastrodermal cells, especially in longer animals, are barely represented. (b) the tail (region 6) show similar characteristics to the head though the density in neoblasts and nerve cells is lower. Compared to the middle regions (2–4) parenchymal and gastrodermal cells have lower densities. We have not detected morphological differences between head and tail nerve cells.
(c) the middle region of the body (regions 2 to 4) show a uniform distribution of cell types. Neoblasts and nerve cells are at their lowest levels, and

Table 3. Distribution of cell types in different regions of a standard 11 mm long *Dugesia mediterranea.*

		Body region							
Cell type		1	2	3	4	5	6	phx	Total
Neoblast	%	28.5	23	20	19	21.5	24	-	23
	#	31 350	33 350	30 000	22 800	26 875	21 600		179 400
Nerve	%	20	11.5	6.5	6	8	10	-	11
	#	22 000	16 675	9 750	7 200	10 000	9 000		85 800
Epidermal	%	20.5	11.5	10.5	10.5	10.5	21	-	14
	#	22 550	16.675	15 750	12 600	13 125	18 990		109 200
Fix. Pareq	%	10.5	23.5	26.5	22	23	19	-	20.5
	#	11 550	34 075	39 750	26 400	28 750	17 100		159 900
Acidoph	%	4.5	5	8	13.5	9	6.5	-	7.5
	#	4 950	7 250	12 000	16 200	11 250	5 850		58 500
Basoph	%	2	5.5	6	7.5	5	2	-	5
	#	2 200	7 975	9 000	9 000	6 250	1 800		39 000
Pigment	%	1,5	1	1	1	1	1.5	-	1
	#	1 650	1 450	1 500	1 200	1 250	1 350		7 800
Muscular	%	2.5	2	2.5	2.5	2.5	1.5	-	2
	#	2 750	2 900	3 750	3 000	3 125	1 350		15 600
Gastrod	%	1.5	6.5	7.5	7	7.5	5	-	5.5
	#	1 650	9 425	11 250	8 400	9 375	4 500		42 900
Globet	%	0.5	1	1.5	1.5	1.5	1	-	1
	#	550	1 450	2 250	1 800	1 875	900		7 800
Flame	%	1	1.5	1	1	1	1	-	1
	#	1 100	2 175	1 500	1 200	1 250	900		7 800
Rhabdite	%	2.5	1.5	1.5	2	2	1.5	-	2
	#	2 750	2 175	2 250	2 400	2 500	1 350		15 600
Striped	%	1.5	1.5	1.5	1.5	1.5	1	-	1.5
	#	1 650	2 175	2 250	1 800	1 875	900		11 700
'Unclass'	%	3	5	6	5	6	5	-	5
	#	3 300	7 250	9 000	6 000	7 400	4 500		39 000
Total Cell Number ($\times 10^3$)		110	145	150	120	125	90	35	780

\- Data not measured
phx pharynx
% percentage of total cells
\# total number

parenchyma and intestinal cells attain their highest levels. The other cell types do not precent variation that are statistically significant.

5. The effects of degrowth on cell number and cell distribution

The effects of degrowth by starvation on cell number and cell distribution are shown in Figure 4. For the sake of clarity we only show the results for 7 mm long *Dugesia tigrina,* other body lengths giving similar results.

There is an increase in cell number during the first week (up to the third day) due to cell proliferation after feeding (Baguñà 1974; Romero & Baguñà, unpublished data). A week after, the cell number levels off and from then on decreases steadily reaching a number of 60 000 cells three months later. The final length and cell number are those typical of a 2 mm long organism.

The changes in the cell distribution brought about by starvation are much less extreme. Comparing the percentages at the start and at the end of the experiment, there are slight, but statistically

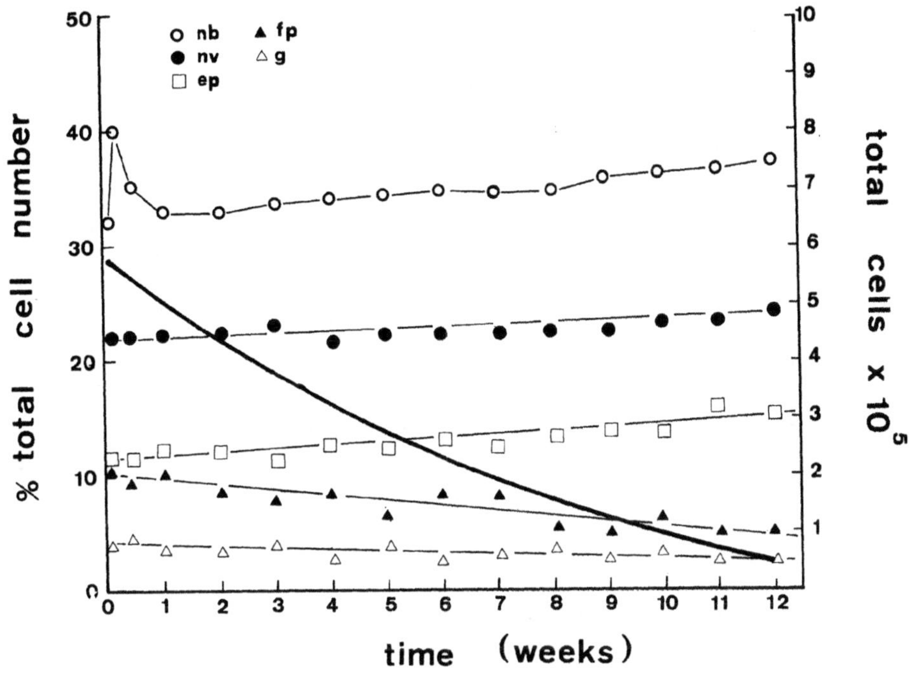

Fig. 4. Effect of degrowth by starvation on total cell number (thick line) and distribution of the main five cell types in *Dugesia tigrina*.

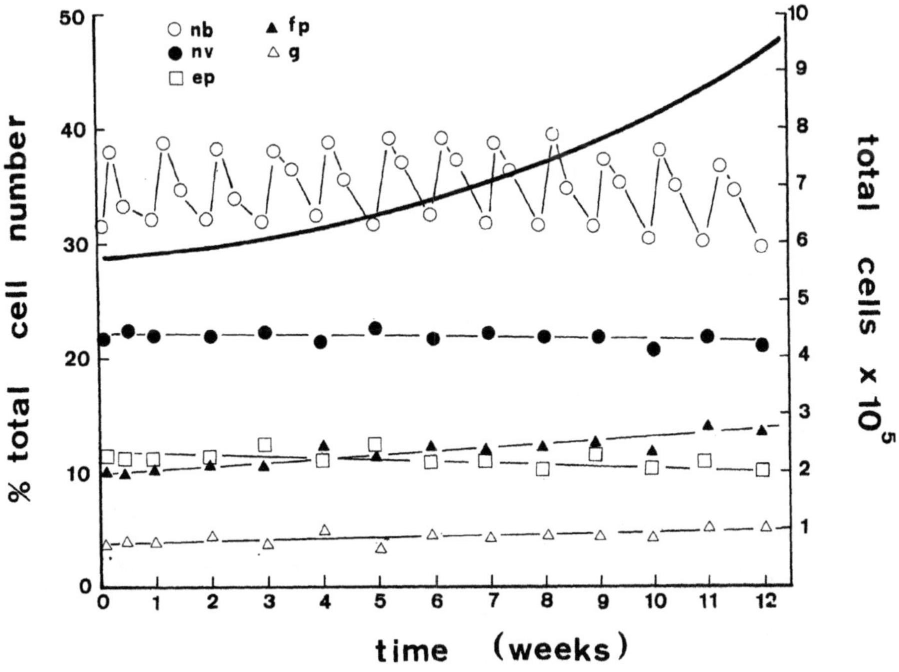

Fig. 5. Effect of growth by feeding on total cell number (thick line) and distribution of the main five cell types in *Dugesia tigrina*.

significant, increases in neoblasts and epidermal cells and a significant decrease in fixed parenchyma cells. Other changes in cell types are not statistical significant.

Overall, these results suggest that the effect of starvation is very marked in the number of cells per animal. We estimate that a 7 mm long starving *Dugesia tigrina* loses about 7 000 cells/animal/day when kept at 17 °C (Romero & Baguñà, unpublished data). The changes in cell composition are less severe. Apparently, during degrowth, there are adjustements in cell composition that make a 5 mm long starving organism very similar in cell composition to a 5 mm long growing animal. This could be relevant in explaining the 'rejuvenatory' effects brought about by starvation as has been suggested by several authors (see Brønsted 1969 and Reynoldson 1966 for references).

6. The effects of growth in cell number and cell distribution.

The effects of growth by feeding on cell number and cell distribution are shown in Figure 5. Only results for a 7 mm long starving *Dugesia tigrina* individuals are presented.

From the beginning of the experiment there is a continuous and steady increase in cell number. This increase is markedly steeper up to three days after feeding due to a sudden but transient burst of cell proliferation. This pattern repeats after each feeding. After three months an animal fed weekly has 950 000 cells, which is characteristic for a 9 mm long organism.

Growth by feeding results in only slight changes

Table 4. Cell distribution during head regeneration in *Dugesia mediterranea* (11 mm in length).

				Time (days)				
Cell type		0		1	2	5	9	15
Neoblast	%	23	B	50.5	59	47	39	33.5
			PB	25	34	37	34	29
Nerve	%	11.5	B	13.5	15.5	22.5	24	24.5
			PB	12	10.5	15.5	20	20
Epiderm	%	11.5	B	16.5	16.5	14	14	17.5
			PB	13.5	10.5	10.5	12	11
Fix. Parq	%	23.5	B	9	3.5	5	7	8
			PB	21.5	14	10	12	15
Acidoph	%	5	B	2.5	1.5	2.5	5	5
			PB	4.5	7	6	5.5	6
Basoph	%	5	B	3	1	1.5	2	2
			PB	5.5	6	4.5	3	4
Pigment	%	1	B	0.5	0	0	0	0.5
			PB	1.5	1	1.5	1	1
Muscular	%	2	B	1	1	1.5	2	2
			PB	2	2	2	2	2
Gastrod	%	6.5	B	0.5	0	0	0	0
			PB	3	4	2	2.5	2.5
Globet	%	1.5	B	0	0	0	0	0
			PB	1	1	0.5	0.5	0.5
Flame	%	1.5	B	0	0	0	0	0.5
			PB	1.5	1.5	1.5	1	1
Rhabdite	%	1.5	B	0	0	0	0.5	1
			PB	1.5	2	2	1.5	1.5
Striped	%	1.5	B	0	0	0	0.5	0.5
			PB	1.5	1.5	1	1	1.5
'Unclass.'	%	5	B	3	2	6	6	5
			PB	6	5	6	4	5

in cell composition that reverse the trend shown by starvation. There is a significant decrease in neoblast and a significant increase in fixed parenchyma cells. The cell composition at the end of the experiment is similar to the one of a standard 9 mm long animal. It is interesting to point out the weekly but transient increase in neoblast percentage due to cell proliferation. This increase levels off 3–4 days after feeding, being sustained in organisms fed twice weekly (Romero, unpublished results). These results reinforce again the suggestion that neoblasts are the only planarian cells endowed with mitotic power.

7. Changes in cell distribution during anterior regeneration

Table 4 gives the distribution of cell types for blastema (B) and post-blastema (PB) during anterior regeneration in a 11 mm long *Dugesia mediterranea*. Only data for this length is shown for the sake of clarity and because other lengths gave similar results. The data at 0 days are the values for region 2 of an 11 mm long intact organism (see Table 3) and serve as an internal control.

The results obtained show a great increase in neoblast density in blastema due to cell proliferation (Baguñà 1976b) reaching up to 60% of total cell number at two days of regeneration. Later, neoblasts decrease in density as long as different cell types begin appearing or increasing in number. There is too a slight but significant increase in nerve cells at two days and especially from five days on, and an increase in epidermal cells due probably to epidermal contraction during wound closure. Conversely, fixed parenchyma cells, acidophilic and basophilic cells and gastrodermal and goblet cells decrease in percentage. This could be due either to cell lysis, cell dedifferentiation, or cell dilution by neoblast proliferation. Though the phenomenon of cell dedifferentiation could be accepted as a likely but unproven possibility, the unclear evidence backing it and the high rates of mitosis seen after cutting (Baguñà 1976b) suggest that the decreasing percentages of these differentiated cells are mainly due to cell dilution, and to a minor extent to cell lysis.

The results obtained for control regions (post-blastema, PB) show that besides slight increases in neoblast and nerve cell densities from the second day on and transient decreases in some differentiated cells (e.g. fixed and secretory parenchyma cells), most cell types remain mainly unchanged. However, it is interesting to point out that gastrodermal and goblet cells dissapear from the blastema at 1–2 days of regeneration while diminishing in density at post-blastema.

Discussion

1. The maceration technique and quantitative cell analysis in planaria.

The maceration technique employed here is a modification of David's technique to macerate *Hydra* tissues (1973). Planarians present some difficulties to macerate since they are not simple bilayered organisms and because their parenchyma is a loose arrangement of several cell types with long and intermingled processes. The modifications introduced (the use of methanol, lower temperatures, and longer periods of maceration) has enabled the preservation, to a reasonable extent, of most of the nerve and parenchyma cell processes.

Despite these difficulties, the maceration technique offers several advantages over standard histological and electron microscopy sections: (1) it provides single cell suspensions good for quantitative analysis of the different cell types; (2) the relative speed and ease of preparation make cell kinetic studies of growth and degrowth possible; (3) when ^{3}H-thymidine autoradiography in planarians becomes possible, maceration would be the method of choice to measure cell turnover of particular cell types in intact and regenerating organisms. Also, this method, when combined with proper histochemical staining methods, would allow a better classification of specific cell types.

2. Cell numbers and distribution

The number of cells per animal and, to a lesser extent, the cellular composition depend on the body length and on feeding regime. It is of interest to point out the decrease in neoblasts and nerve cells and the increase in parenchymal and gastrodermal cells as organisms increase in length. The early phenomena, as stressed by Lange (1968), could be relevant when trying to explain the decreasing rates of regeneration as organisms increase in size.

The antero-posterior distribution of cell types show decreasing gradients for neoblasts and nerve cells, with a slight increase in tail regions. Data on neoblasts agree with published results from histological sections on *Dugesia polychroa* (Lange 1967). The decreasing gradient in nerve cell density could be related to the mechanisms that define polarity in intact and regenerating worms. Indeed, nerve cells have been identified as the source of morphogenetic factors that control polarity and regeneration in *Hydra* (Schaller & Gierer 1973). Although evidence for a similar mechanism operating in planarians is still lacking it still remains as a very likely possibility.

The high percentage of neoblasts found in intact worms of every length studied (20–35% of total cells) is rather surprising even if we consider them as the stem-cell of all differentiated cell types. In fact, most renewing cell systems (epidermis, gut epithelia, haemopoietic system, . . .) are functionally (and often structurally) divided in three main compartments: a small one composed of slowly-proliferating stem-cells, a bigger one composed of amplyfying and rapidly-proliferating predifferentiated cells, and a terminal compartment of functional differentiated cells. It is possible that, similarly to those systems, planarian neoblasts comprise two different populations: a small one representing uncommited multipotential cells (true neoblasts) and a bigger one representing cells in different stages of determination but retaining the capacity of division.

3. Quantitative cell analysis of growth and degrowth in planarians

Growth and degrowth in planarians are continuous processes whose balance depends on feeding and temperature. From a quantitative point of view crude estimates on the magnitude of increases and reductions in body volume and length have been considered by Abeloos (1930) and Reynoldson (1966) respectively. A better estimate have been given by Calow (1977) using plan area as an index of body size. The maceration technique used here has the advantages of looking at growth and degrowth as changes in cell number and of measuring the changes in cell type distribution during these processes.

In planarians, growth and degrowth occur by increase or decrease in cell number, though changes in the mean cell volume also occur, bigger animals having bigger cells. The decrease in cell number brought about by starvation (approx. 7000 cells/animal/day for a 7 mm long *Dugesia tigrina* kept at 17 °C, Romero & Baguñà, unpublished data) results from more cells being lost by cell death (approx. 11 000 cells/animal/day) than produced by cell proliferation (approx. 4 000 cells animal day). Contrary to expectations, cell proliferation is maintained during starvation (Baguñà 1976a) probably because most tissues must be renewed for proper functioning and survival. This means that, besides epidermal and gastrodermal cells, most dead cells from the parenchyma must be re-cycled by living cells. This phenomenon, together with the use of stored energy (mainly fat and glycogen), must sustain the metabolic rate needed to maintain the proliferative rate for proper cell renewal.

The increase in cell number brought about by feeding is the result of more cells being born by cell proliferation than lost by cell death. Surprisingly, growth by feeding does not change significantly the death toll rate (Romero & Baguñà, unpublished data), and the increasing cell number is mainly the result of an increased cell proliferation rate caused by feeding (Baguñà 1974a). In summary, planarians are organisms in a continuous state of cell turnover. Since growth and degrowth affect cell type distribution only slightly, since neoblasts are the only cell type endowed with mitotic power, and because the cell death rate is maintained despite changing feeding regimes, we suggest that a fixed programme of cell determination and differentiation through cell lineage is operating throughout the organism's lifetime.

4. Quantitative cell analysis of planarian regeneration

One of the main problems of planarian regeneration that remains is the never ending controversy over neoblast versus metaplasia in blastema formation. Supporters of the first theory ('neoblast theory') invoke the existence of a permanent population of reserve cells (neoblasts) that give rise to blastema formation through migration and proliferation (Wolff & Dubois

1947). The second theory ('dedifferentiation theory') claims that neoblasts do not exist at all and that the source cells for blastema are specialized cells that dedifferentiate to an undifferentiated state (Hay 1968; Coward 1969).

Recently, some attempts to bridge both theories have been advanced suggesting that both types of cells (neoblasts and dedifferentiated cells) can take part in blastema formation (Gremigni & Miceli 1980; Gremigni 1981). Though the technique used (chromosomal markers) offer new avenues for re-evaluating the problem of blastema formation in planarians, the results obtained are open to discussion and the conclusions formulated by the authors are questionable.

The results obtained using the maceration technique offer little help in solving this problem. From the results it is clear that neoblasts accumulate rapidly from the very beginning of regeneration. Since during this period (0–24 hours of regeneration) a burst of cell proliferation has been detected in both species (Baguñà 1976b for *Dugesia mediterranea;* Saló, unpublished data for *Dugesia tigrina)* it is sound to suggest that the increase in neoblast percentage is due to cell proliferation and not to cell dedifferentiation. Moreover, the decrease in percentage of some differentiated cells (fixed parenchyma cells, acidophilic and basophilic, gastrodermal and goblet cells), held as a proof of massive cell dedifferentiation, could be interpreted as well as a process of cell dilution by proliferating neoblasts. Indeed, post-blastema regions show very small changes in cell distribution, contrary to what is expected of a massive cell dedifferentiation near the wound.

It is also evident that from 2–3 days of regeneration neoblasts decrease in the blastema while differentiated cells begin to appear or increase in percentage. These results are in agreement with the first published results of a quantitative analysis of blastema cells with electron microscopy (Hori 1978). This work emphasizes the role of neoblasts as a replacement cells both in intact and regenerating animals, and stresses once more the lack of evidence for cell dedifferentiation in the blastema and regions behind it.

To summarize, from the results obtained using the maceration technique and from previous results of mitotic analysis during regeneration (Baguñà 1976b) we suggest that blastema formation in planarians is mainly due to local proliferation of neoblasts near the wound. Cell dedifferentiation, if actually present, would give only a limited percentage of blastema cells. The main problems of planarian regeneration will have a clear answer in the future if we can answer first these two key questions: (1) what is the role of the neoblasts in the intact worm?, and (2) from where do they come?

Abbreviations

nb: neoblasts
nv: nerve cells
ep: epidermal cells
fp: fixed parenchyma cells
g: gastrodermal cells

References

Abeloos, M., 1930. Recherches expérimentales sur la croissance et la régénération chez les planaires. Bull. Biol. Fr. Belg. 64: 1–130.

Baguñà, J., 1973. Ph. D. Thesis. University of Barcelona, Spain.

Baguñà, J., 1974. Dramatic mitotic response in planarians after feeding and a hypothesis for the control mechanism. J. Exp. Zool. 190: 117–122.

Baguñà, J., 1976a. Mitosis in the intact and regenerating planarian Dugesia mediterranea n.sp. I. Mitotic studies during growth, feeding and starvation. J. Exp. Zool. 195: 53–65.

Baguñà, J., 1976b. Mitosis in the intact and regenerating planarian Dugesia mediterranea n.sp. II. Mitotic studies during regeneration and a possible mechanism of blastema formation. J. Exp. Zool. 195: 65–80.

Baguñà, J. & Ballester, R., 1978. The nervous system in planarians: peripheral and gastrodermal plexuses, pharynx innervation and the relationship between central nervous system structure and the acoelomate organization. J. Morphol. 155: 237–252.

Benazzi, M., Baguñà, J., Ballester, R., Puccinelli, I & Del Papa, R., 1975. Further contributions to the taxonomy of the Dugesia lugubrispolychroa group, with description of Dugesia mediterranea n.sp. Boll. Zool. 42: 81–89.

Brønsted, H. V., 1969. Planarian Regeneration. Pergamon Press Ltd, London.

Calow, P., 1977. The joint effect of temperature and starvation on the metabolism of triclads. Oikos 29: 87–92.

Coward, S. J., 1969. Regeneration in planarians: some unresolved questions and problems. J. Biol. Psychol. 11: 15–19.

David, C. N., 1973. Quantitative method for maceration of Hydra tissue Wilhelm Roux' Archiv. 171: 259–268.

Gremigni, V., 1981. The problem of cell totipotency, dedifferentiation and transdifferentiation in Turbellaria. In: Schockaert, E. R. & Ball, I. R. Turbellaria. Proc. Third Int. Symp. Hydrobiologia. This volume.

Gremigni, V. & Miceli, C., 1980. Cytophotometric evidence for cell 'transdifferentiation' in planarian regeneration. Wilhelm Roux' Archiv 188: 107–113.

Hay, E. D., 1968. Dedifferentiation and metaplasia in vertebrate and invertebrate regeneration. In: Ed. Ursprung, H., The Stability of the Differentiate State. Berlin, Heidelberg, New York: Springer, pp: 85–108.

Hori, I., 1978. Possible role of rhabdite-forming cells in cellular succession of the planarian epidermis. J. Electr. Microsc. 27: 89–102.

Lange, C. S., 1967. A quantitative study of the number and distribution of neoblasts in Dugesia lugubris (Planaria) with reference to size and ploidy. J. Embryol. Exp. Morphol. 18: 199–213.

Lange, C. S., 1968. A possible explanation in cellular terms of the physiological ageing of the planarian. Exp. Geront 3: 219–230.

Reynoldson, T. B., 1966. Preliminary laboratory experiments on recruitment and mortality in triclad populations. Verh. int. Ver. Limnol. 49: 1621–1631.

Schaller, H. & Gierer, A., 1973. Distribution of the head-activating substance in Hydra and its localization in membranous particles in nerve cells. J. Embryol. Exp. Morphol. 29: 39–52.

Wolff, E. & Dubois, F., 1947. La migration des cellules de régénération et les facteur qui la provoquen chez les triclades. C. R. Acad. Sci. 224: 1387–1388.

Relationship between variations of cAMP, neuromediators and the stimulation of nucleic acid synthesis during planarian (Polycelis tenuis) regeneration

Isabelle Martelly, Raphael Franquinet & Albert Le Moigne
Laboratoire de Biologie Animale, Université de Paris Val de Marne, Avenue du Général de Gaulle, Créteil 94010 Cedex, France

Keywords: Turbellaria, regeneration, cAMP, neuromediators

Abstract

Regenerating planarians of the species *Polycelis tenuis* have been studied with respect to the two distinct phases which occur during regeneration, viz., the first 24 h involving cellular activation, and the following days when differentiation of the blastema occurs after a period of cellular proliferation. We have studied particularly the biochemical events that control regeneration with respect to the nature of the signals which induce cellular activation, the membrane receptors of these signals, and the consequences of these signals for the cellular metabolism of DNA, RNA and proteins. The roles of neurohormones such as serotonin, dopamine, and noradrenaline are assessed, and a provisional model of the process of cellular activation is proposed which takes account of all the information that is now available concerning planarian regeneration.

Introduction

Several observations lead us to assume that two distinct phases occur during the regeneration process of planarians. The first period, between 0 and 24 h after sectioning, corresponds to the cellular activation that is necessary for the subsequent growth of the blastema and its differentiation. Indeed, actinomycin-D, which completely inhibits RNA synthesis in adult planarians (Martelly & Le Moigne 1979), also completely inhibits regeneration when applied during these first 24 h. The inhibitor is less effective when applied during the following day, and after 48 h it loses its effect (Le Moigne & Martelly 1974). The same result was obtained with X-ray irradiation (Gabriel 1968). After 48 h of regeneration differentiation of the blastema commences (Sauzin 1967a, b; Sauzin-Monnot 1973; Spiegelman & Dudley 1973). The most important period occurs within the first 24 h following section and it may be even the first 6 h (Moraczewski, private communication).

In the present study we have endeavoured to determine the biochemical events that induce regeneration with special reference to the signals that induce cellular activation, the membrane receptors of these signals, and the consequences of these signals on the cellular metabolism of DNA, RNA, and proteins. These problems have been studied on regenerating planarians of the species *Polycelis tenuis* Iijima. They have the advantages of reproducing sexually while also possessing full regenerative powers. Some aspects of the problems have also been studied using planarian cells in culture.

Materials and methods

Preparation of the cellular suspension

The planarians were starved for six days before being dissociated into cells, by means of a tight fitting pestle (Dounce), in a culture medium to which was added benzalkonium (1% DigestEUR, Eurobio). The dissociated cells were centrifuged for 5 min at 500 g and washed five more times in the

Hydrobiologia 84, 195–201 (1981). 0018–8158/81/0842–0195/$01.40.

same medium. Finally the cells were suspended in the culture medium without added benzalkonium.

Composition of the medium and culture conditions

The medium consisted of 1/3 199 medium with Hanks salts (Gibco) and 2/3 MEM (Eurobio), and for each litre of medium there was added 0.367 g KCl, 0.337 g $CaCl_2$, 0.094 g $MgSO_4$, 0.05 g K_2HPO_4, 10 000 Units of penicillin, 10 mg streptomycine and 0.5 g polyvinylpyrrolidine (Sigma). The pH was adjusted to 7.25 with $NaHCO_3$. The cultures were maintained at 20 °C for 3 h, 18 h, and 38 h. Hormones and their antagonists were added from the beginning and were renewed every 24 h. The concentrations used were 0.1 mM for serotonin (Calbiochem), noradrenaline (Sigma), and dopamine (Sigma). Other products such as methiothepine (Rhone-Poulenc), fluphenazine (Squibb), propranolol (I.C.I.) and dibutyryl cyclic AMP (dbcAMP) (Sigma) were adjusted to 10 μM.

Other techniques used have been described in previous papers from our laboratory and these are referred to below.

Results

A. Signals activating the cells

In the regeneration of many animals the role of the nervous system has been established. During planarian regeneration the number of neurosecretory cells increases (Lender & Klein 1961) and neurosecretions have an activatory effect on RNA synthesis (Sauzin-Monnot 1976). The composition of the neurosecretion, however, is still in question and we search for the existence of some well known neuromediators which possibly could have some function in the course of regeneration. Some of them can easily be ascertained and their presence in planarians has been established by Welsh (1972), Franquinet (1979) and Franquinet & Le Moigne (1979). These neuromediators are serotonin, dopamine and noradrenaline.

Quantitative variations of the neurohormones. The quantitative variations of serotonin and the catecholamines, dopamine and noradrenaline, during planarian regeneration are shown in Fig. 1. These hormones can be localised in the tissues by microfluorometric techniques, and they are seen mainly in the cerebral ganglia and in the nerve trunks (Franquinet & Catania 1979).

Effect of the neurohormones on regeneration. The effects of the hormones on regeneration are shown in Table 1. Serotonin and dopamine have an activating effect whereas noradrenaline probably does not. All their antagonists, however, have an inhibitory effect, especially propranolol, the beta-antagonist of noradrenaline. It may be concluded that biogenic amines do exist in planarians and can influence the rate of regeneration. Their concentrations change during the regeneration process.

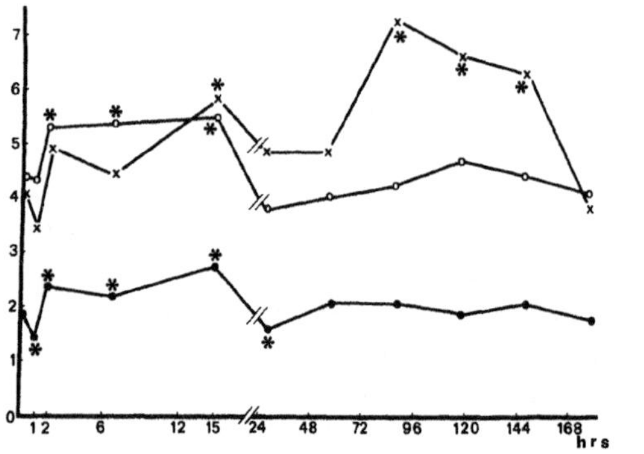

Fig. 1. Variations of the concentrations of neurohormones during regeneration. Abscissa: duration of regeneration (h) ordinate: Serotonin (●), Noradrenaline (O) and Dopamine (x) in μg/g fresh weight.

Table 1. Effect of neurohormones and their antagonists on adenylate cyclase activity and on regeneration.

Hormones and antagonists	Adenylate cyclase	Regeneration
Serotonin	++ Ca^{++} indep.	++
methiotepine, nigergoline	—	—
Dopamine	++ Ca^{++} depend.	++
fluphenazine, haloperidol	—	—
Noradrenaline	+ Ca^{++} depend.	?
propranolol (β antagonist)	0	—
phenoxybenzamine, phentolamine (α antagonist)	—	0

+: activating effect; —: inhibiting effect; 0: no effect.

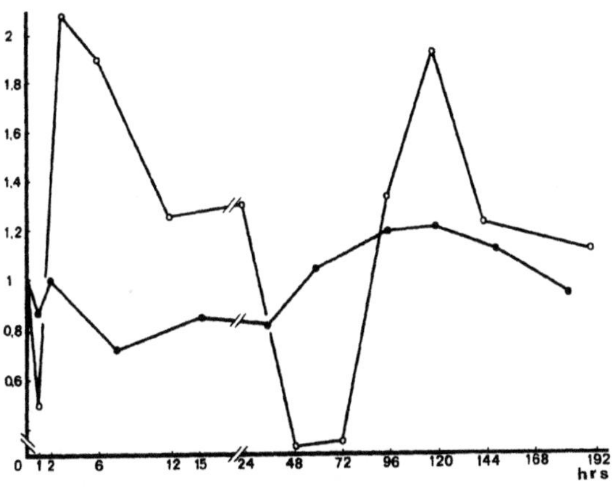

Fig. 2. Variations of the rate of adenylate cyclase/phosphodiesterase and cyclic AMP content during regeneration. Abscissa: duration of regeneration (h) Ordinate: rate of adenylate cyclase/phosphodiesterase activities (●), cyclic AMP (○), 1 = 1.42 nM/g fresh weight.

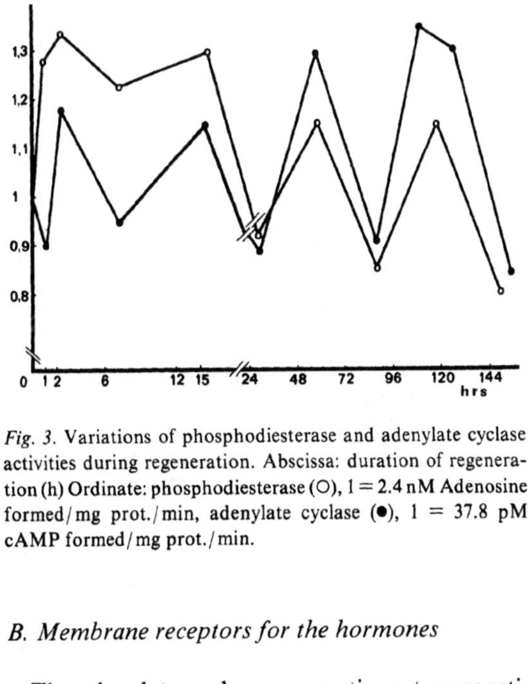

Fig. 3. Variations of phosphodiesterase and adenylate cyclase activities during regeneration. Abscissa: duration of regeneration (h) Ordinate: phosphodiesterase (○), 1 = 2.4 nM Adenosine formed/mg prot./min, adenylate cyclase (●), 1 = 37.8 pM cAMP formed/mg prot./min.

B. Membrane receptors for the hormones

The adenylate cyclase enzymatic system constitutes the membrane receptor to the hormones. Several arguments support this assertion. The cyclase system exists in planarians, as demonstrated by Franquinet *et al.* (1976, 1978), and its activity has been localised by Moraczewski & Duma (1978) using an ultrastructural-histochemical technique.

Quantitative variations of cAMP and enzyme activities (adenylate cyclase and phosphodiesterase). The variations of concentration of cAMP during planarian regeneration have been established (Fig. 2). These are the result of the activities of adenylate cyclase and phosphodiesterase which degrades the cAMP (Fig. 3).

Effect of the hormones on cyclase activity. As shown in Table 1 hormones activate the adenylate cyclase activity while their antagonists have an inhibitory effect.

It may be concluded from these data that the adenylate cyclase, localised in the cellular membrane, presents receptors to serotonin and catecholamines which modulate its activity.

C. Consequences of the hormone action on the cellular metabolism of DNA and RNA

DNA synthesis and mitotic activity. For reasons which are not clear, and which probably involve a deficiency of the permeability system for thymidine, the planarians do not incorporate exogenous ^{3}H thymidine into their DNA. Therefore, ^{32}P was used as a marker for DNA synthesis after DNA purification. The variations in the rates of DNA labelling during planarian regeneration are shown in Fig. 4. Between 10 and 12 h after sectioning a large number of cells enter into the S phase. After a minimum of DNA synthetic activity around 24 h, DNA syn-

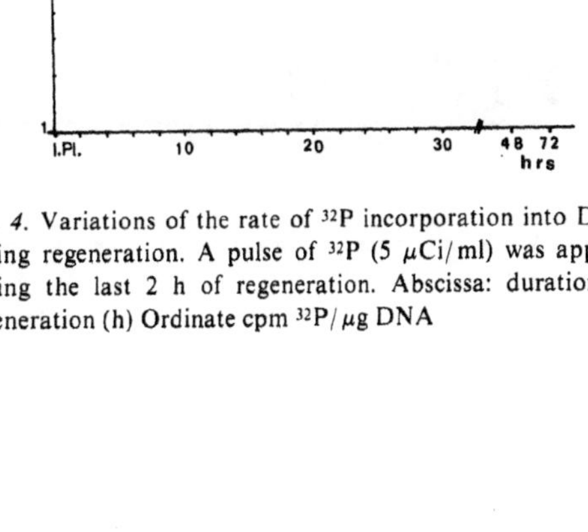

Fig. 4. Variations of the rate of ^{32}P incorporation into DNA during regeneration. A pulse of ^{32}P (5 µCi/ml) was applied during the last 2 h of regeneration. Abscissa: duration of regeneration (h) Ordinate cpm ^{32}P/µg DNA

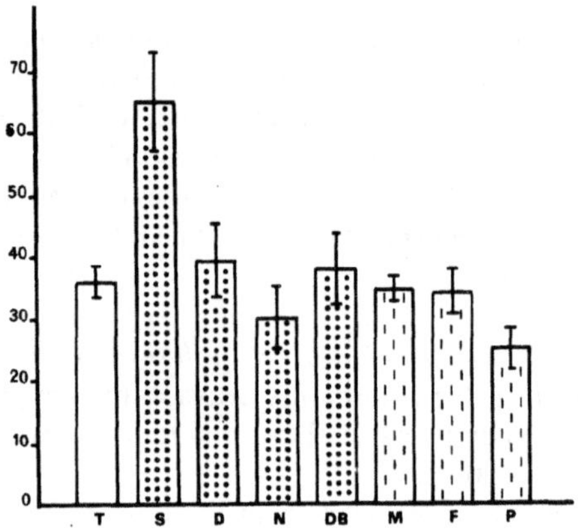

Fig. 5. Effect of neurohormones and their antagonists on ^{32}P incorporation into DNA of cultured cells. T: Control cultures, (S) Serotonin, (D) Dopamine, (N) Noradrenaline (DB) Dibutyryl cyclic AMP, (M) Methiothepine (F) Fluphenazine, (P) Propranolol treated cultures. The different products were added to the culture medium at the moment of the cellular dissociation. The cultivation lasted 18 h. Ordinate: ^{32}P cpm DNA/mg protein.

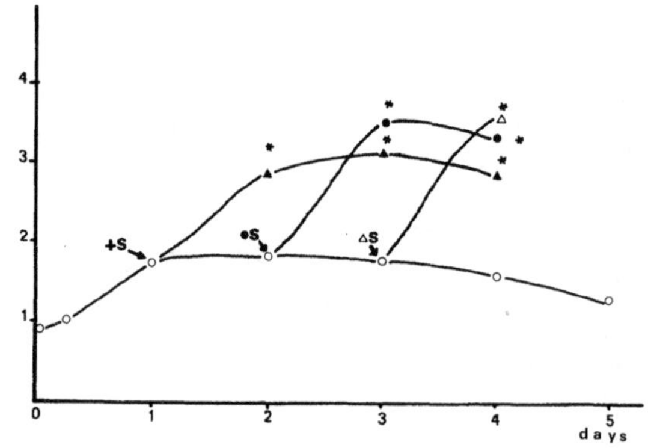

Fig. 6. Effect of the addition of Serotonin (S→) on the mitotic rate of cultured cells. Abscissa: duration of cultures (days) ordinate mitotic rate (%).

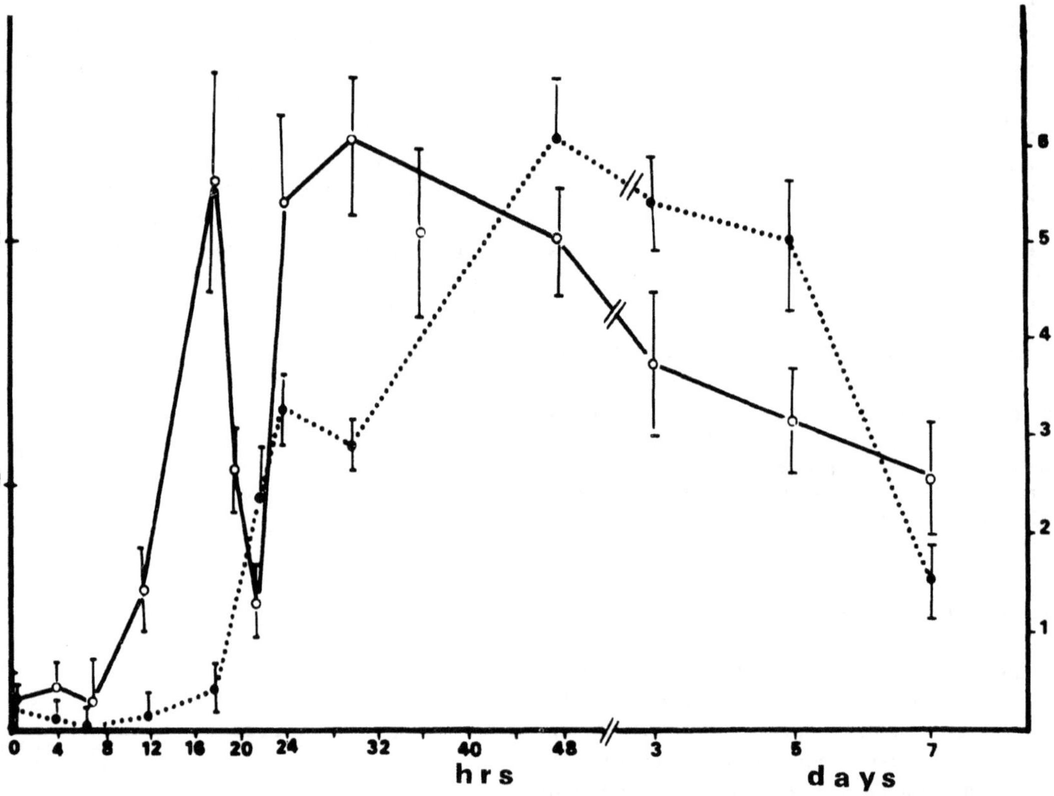

Fig. 7. Variations of the rate of RNA and protein labelling during regeneration. ^{3}H orotic acid or ^{14}C amino acids were added to regenerating fragment bath during the last two hrs of regeneration. Abscissa: duration of regeneration (h and days) Ordinate: left 10^{-2} cpm $^{3}H/\mu g$ RNA (O) right 10^{-3} cpm $^{14}C/\mu g$ protein (•).

thesis slowly increased again until 72 h of regeneration had elapsed. It is important to remember that Coward *et al.* (1970) have shown that the thymidine kinase activity measured during planarian regeneration attained a first maximum at 12 h and a second peak at 72 h after sectioning. This enzymatic activity is known to be linked with the replicative phase of DNA in the cell cycle (Plageman & Erbe 1972). The resulting DNA accumulated in the planarians reaches a maximum at 18 h, when it represents an increase of 80% compared to control DNA content. The DNA synthesizing cells have been identified indirectly by a histochemical method that detects DNA polymerase activity (Chevaillier & Philippe, 1976, 1977). The cells which have such activity are located, at 48 h, in the peripheral zone of the blastema. As observed by Gabriel (1968), and confirmed by our own observations, mitoses are stimulated only after the end of the first 24 h of regeneration.

Effect of neurohormones on DNA synthesis and on mitotic rate in cultured cells. When serotonin is added to the culture medium of planarian cells a

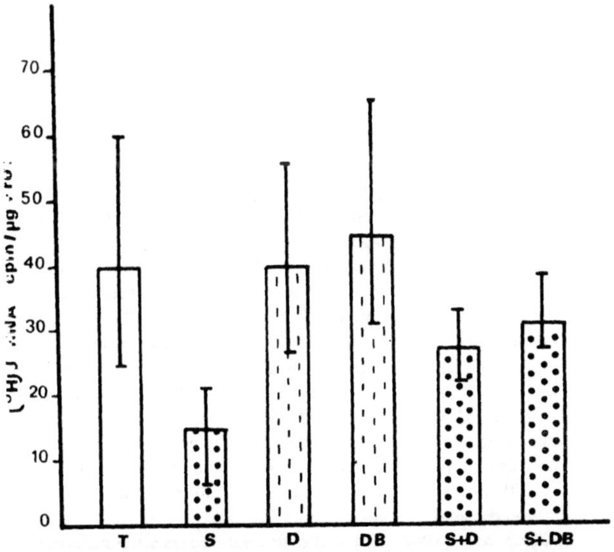

Fig. 8. Effect of neurohormones on ^3H uridine incorporation into RNA of cultured cells. (T) Control cultures, (S) Serotonin, (D) Dopamine, (DB) Dibutyryl cyclic AMP, (S + D) Serotonin + Dopamine, (S + DB) Serotonin + Dibutyryl cAMP treated cultures. The different products were added to the culture medium at the moment of the cellular dissociation. The cultivation lasted 18 h. Ordinate: ^3H cpm RNA/μg protein.

stimulation of DNA synthesis is observed (Fig. 5). Moreover, the rate of mitosis also is stimulated to the same extent at each serotonin addition (Fig. 6). On the other hand the catecholamines have no effect on DNA synthesis nor on mitotic rate, and cAMP does not seem to interfere directly with DNA synthesis.

RNA synthesis. The pattern of RNA synthesis also presents a biphasic aspect with a minimum at 21 h after sectioning (Fig. 7). It has been shown that during the first phase principally ribosomal RNA was synthesized while after 24 h of regeneration heterogenous RNA was also produced in greater amounts (Martelly & Le Moigne 1980).

Effect of neurohormones and cAMP on RNA synthesis. The effects of neurohormones on RNA synthesis have been tested either on regenerating fragments or *in vitro* on cultured cells. In both cases serotonin has an inhibitory effect on RNA synthesis measured either with ^3H orotic acid or with ^3H uridine incorporation. Dopamine, noradrenaline and cAMP on its own have no apparent effect but when serotonin was added to the culture medium with other neurohormones, or with cAMP, the inhibitory effect of the serotonin was suppressed, and the level of RNA synthesis rose to the control level (Fig. 8). This shows that dopamine has an activatory effect on RNA synthesis in cells which are in an inhibited state with respect to the production of RNA. The effect of dopamine seems to take place directly through cAMP (unpublished data).

From all these results it may be concluded that some of the neurohormones which have been detected in planarians have either direct or indirect effects on DNA and RNA synthesis. The role of noradrenaline is not yet fully understood. Protein synthesis, which is stimulated after RNA synthesis, (Fig. 7) has not yet been studied in relation to the neurohormones.

Discussion and conclusions

These results establish a chronology for the different biochemical events which occur during planarian regeneration. Moreover, causal relations between some of them can now be assumed.

Soon after section serotonin is the first hormone

involved in cellular activation. It stimulates the activity of a Ca^{++}-independent adenylate cyclase, and hence the concentration of cAMP increases and then later decreases when the ratio of the activity of adenylate cyclase to phosphodiesterase is in favour of the latter. After the decrease of cAMP content DNA synthesis is induced in accord with what happens in other cellular systems (Whitfield, *et al.* 1979). The steps which take place between the cAMP and DNA peaks are not fully known. We have evidence, however, that Ca^{++} has an important role at this period and, in fact, it stimulates the activity of Ca^{++}-dependent protein kinases (Moraczewski & Franquinet 1980). These protein kinases may directly induce DNA synthesis. Subsequently the rise in dopamine concentration may be responsible for the induction of RNA synthesis, its effect being possibly mediated by cAMP-dependent protein kinases, which present a peak of activity at the same time (18 h). Since we found that protein synthesis is stimulated after RNA synthesis (Martelly & Le Moigne 1981) and reaches a maximum at 48 h when cellular differentiation has been wholly initiated, it follows that the increase in activities of adenylate cyclase, phosphodiesterase, protein kinases, and other enzymes may not come from *de novo* synthesis but rather from activation of preformed enzymes. This could explain the rapidity of the response to wounding. After 24 h of regeneration, at a time when all the activities are reduced, cellular proliferation and differentiation take place, and these are under the control of other regulatory mechanisms.

References

Chevaillier, P. & Philippe, M., 1976. Activité DNA-polymérase nucléaire dans les spermatozoïdes de souris et évolution de cette activité au cours de la spermatogenèse. Exp. Cell Res., 99: 237–244.

Chevaillier, P. & Philippe, M., 1977. In situ detection and characterization of DNA polymerase activities in the nucleus of Eukaryotic cells. Chromosoma, 63: 385–399.

Coward, S. J., Hirsh, F. M. & Taylor, J. H., 1970. Thymidine kinase activity during regeneration of planarian. J. Exp. Zool., 173: 269–278.

Franquinet, R., 1979. Rôle de la sérotonine et des catécholamines dans la régénération de la Planaire Polycelis tenuis. J. Embryol. Exp. Morphol., 51: 85–95.

Franquinet, R., Stengel, D. & Hanoune, J., 1976. The adenylate cyclase system in fresh water Planarian (Polycelis tenuis Iijima) Comp. Biochem. Physiol., 53B: 329–333.

Franquinet, R., Le Moigne, A. & Hanoune, J., 1978. The adenylate cyclase system of Planarian Polycelis tenuis. Activation by serotonin and guanine nucleotides. Biochim. Biophys. Acta, 538: 88–97.

Franquinet, R. & Le Moigne, A., 1979. Relations entre les variations des taux de sérotonine et d'AMP cyclique au cours de la régénération d'une Planaire. Biol. Cell., 34: 71–76.

Franquinet, R. & Catania, R., 1979. Localisation et étude microspectrofluorométrique de la sérotonine et des catécholamines chez la Planaire normale et en régénération. C.R. Acad. Sc., 289: 339–342.

Gabriel, A., 1968. Influence du moment de l'irradiation aux rayons X sur la régénération et le taux d'incorporation de l'uridine tritiée chez la Planaire Dugesia gonocephala (Turbellarie Triclade) C.R. Acad. Sc., 266: 820–822.

Gabriel, A., 1969. Etude morphologique et évolution biochimique des néoblastes au cours des premières phases de la régénération des Planaires d'eau douce. Ann. Embryol. Morphol., 3: 49–69.

Le Moigne, A. & Martelly, I., 1974. Effet de l'emploi préventif ou différé de l'actinomycine D sur la régénération des Planaires jeunes et adultes. C.R. Acad. Sc., 279: 367–369.

Lender, R. & Klein, N., 1961. Mise en évidence de cellules sécretrices dans le cerveau de la Planaire Polycelis nigra. Variation de leur nombre au cours de la régénération postérieure. C.R. Acad. Sc., 253: 331–333.

Martelly, I., Borney, C. & Le Moigne, A., 1976. Mise en évidence de deux phases dans les synthèses d'ARN au cours de la régénération de la Planaire (Polycelis tenuis) C.R. Acad. Sc., 282: 1805–1808.

Martelly, I. & Le Moigne, A., 1979. Comparaison des effets de l'actinomycine D sur les synthèses d'ARN des Planaires jeunes et adultes en régénération C.R. Soc. Biol., 173: 1023–1030.

Martelly, I. & Le Moigne, A., 1980. Ribonucleic acid metabolism during Planarian regeneration. Reprod. Nutr. Develop., 20 (5A): 1527–1537.

Martelly, I. & Le Moigne, A., 1981. Regeneration of Planarians: quantitative differences of RNA and protein synthesis depending on the age, submitted.

Moraczewski, J. & Duma, A., 1978. Ultrastructural localisation of adenyl-cyclase activity in neoblasts of Turbellarian Dugesia lugubris (O. Schmidt). J. Exp. Zool., 203: 491–496.

Moraczewski, J. & Franquinet, R., 1980. Mise en évidence et activités des protéines kinases AMP cyclique dépendantes et AMP cyclique indépendantes chez la Planaire Polycelis tenuis. C.R. Acad. Sci., 290: 1567–1570.

Plageman, P. C. W. & Erbe, J., 1972. Thymidine transport by cultured Novikoff hepatoma cells and uptake by simple diffusion and relationship to incorporation into deoxyribonucleic acid. J. Cell. Biol., 55: 161–178.

Sauzin, M. J., 1967a. Etude ultrastructurale de la différenciation du néoblaste au cours de la régénération de la Planaire Dugesia gonocephala I: Différenciation en cellule nerveuse. Bull. Soc. Zool. Fr., 92: 313–318.

Sauzin, M. J., 1967b. Etude ultrastructurale de la différenciation du néoblaste au cours de la régénération de la Planaire Dugesia gonocephala II: Différenciation musculaire. Bull. Soc. Zool. Fr., 92: 613–616.

Sauzin-Monnot, M. J., 1973. Etude ultrastructurale des néoblastes de Dendrocoelum lacteum au cours de la régénération. J. Ultrastruc. Res., 45: 206–222.
Sauzin-Monnot, M. J., 1976. Action de broyats de blastèmes de régénération sur l'activité synthétique de fragments postérieurs de Planaires Dendrocoelum lacteum sectionnées en arrière du pharynx. C.R. Acad. Sci. Paris, 282: 1885–1888.
Spiegelman, M. & Dudley, P. L., 1973. Morphological stages of regeneration in the planarian Dugesia tigrina: A light and electron microscopic study. J. Morphol., 139: 155–184.
Welsh, J. H., 1972. Catecholamines in the invertebrates. 4 – Platelminthes (flat worms). In Catecholamines, Blaschko, H. and Muschall, F. edit 87–88 Springer Verlag, Berlin.
Whitfield, J. F., Boynton, A. L., Mac Manus, T. P., Sikorska, M. & Tsang, B. K., 1979. The regulation of cell proliferation by calcium and cyclic AMP. Mol. Cell. Biochem., 27: 155–179.

Cell activation during regeneration of planarians

Jerzy Moraczewski
Institute of Zoology, University of Warsaw, 00927 Warsaw, Poland

Keywords: Turbellaria, regeneration, neurohormones, adenylate cyclase

Abstract

Planarians are a good subject for studies of cell differentiation. Each multicellular organism is maintained by continuous production, differentiation and ageing of cellular elements. Each cell has a specific position defined by specific regional boundaries.

After amputating a part of the body this positional information changes, involving, probably, the first range of cellular activation, the activation of membrane receptors. At the same time in an injured organism the level of neurohormones, which can be now coupled with activated receptors, increases. In the opinion of many authors neurohormones act on the regenerative cells through the medium of adenylate cyclase. This enzyme converts ATP to cAMP and by means of this cyclic nucleotide the second range of cellular activation is initiated i.e. changes of activation of cAMP dependent protein kinases.

The sequence of these processes plays the principal role in the ensuing cell differentiation.

Introduction

Regeneration has long been one of the basic problems of modern biology. Turbellarians, and especially planarians, are extremely suitable for carrying out investigations in this field. In the mid-sixties at the symposium in Athens (Kiortris & Trampusch (Eds.), 1965) a number of problems concerning regeneration were presented and at the end of the nineteen-sixties some comprehensive publications on the regeneration of Turbellarians appeared (Sengel 1967, Brøndsted 1969, Goss 1969). Regeneration has been studied as a morphogenetic phenomenon and morphological processes occurring during regeneration have been known for a long time. Little is known, however, about this process at the cellular level and almost nothing at the molecular level. The most significant phenomena seem to occur 2 or 3 h after sectioning or wounding. Afterwards the usual proliferation of cells and their differentiation takes place. This problem has been investigated using biochemical methods (see Martelly, et al. 1981, this volume) and by means of ultrastructural cytochemistry (see Moraczewski, *et al.* 1981).

Discussion

The basic problems to be solved when investigating regeneration are as follows: How are the cells activated for proliferation, migration or differentiation? How are signals from the outside recognised by the cell and transmitted inside by means of a cell membrane? How can the structures of intracellular molecules modify molecules or the structure of cell surfaces?

The same questions may be put for all processes of differentiation but these are especially significant in the case of regeneration because they concern changes occurring in an already developed organism, after embryogenesis. In an adult the pro-

Hydrobiologia 84, 203–207 (1981). 0018-8158/81/0843-0203/$01.00.

cess of cell exchange (renewal) is limited to old cells only, their limited number participating in the process. During regeneration a great number of cells become activated and thus able to proliferate. This is why the total reconstruction of amputated parts of the body becomes a very good experimental system suitable for biochemical studies because of the great number of cells engaged. Morphogenesis occurring during regeneration involves the cellular cycle and thus cell differentiation. It differs from phenomena of embryonic organogenesis by the participation of differentiated or resting cells which have limited development possibilities. Cellular activation during regeneration differs also from that during carcinogenesis, in which, after ther proliferation stage, transformed cells lose their abilities for regulation, and there is no differentiation stage. On the contrary during regeneration the proliferation stage is limited to some number of descendant cells, and they always undergo differentiation.

A live organism is an integrated unity and each region (unit) is determined by a system of coordinates related to positional information. Between extreme (boundary) regions of each unit a gradient of positional information is formed. During growth all cells within a determined unit are provided with information depending on their position and they interpret it by adequate differentiation. This theory of positional information (Wolpert 1971) is useful for regeneration as explained by Wolpert in experiments on *Hydra*.

I have also tried to prove this using the example of turbellarians (Moraczewski 1977). Amputation of a part of a turbellarian, amphibian's limb or the column of a *Hydra* disorders the whole system of co-ordinates deciding the fate of a cell. In the stump this system forms again. If among the remaining cells there are undifferentiated cells in the resting stage, or not fully differentiated, they can be stimulated to differentiate or to dedifferentiate and proliferate. In the injured fragment of a tricladid there are differentiated cells which can not dedifferentiate, undifferentiated cells and cells that can dedifferentiate. The last two categories of cells can obtain new information and reproduce the lost part of the body. This information is transmitted by hormones, probably the aminenergic neurohormones serotonin and catecholamines (Lender & Klein 1961, Lenicque 1973, Franquinet 1979, Franquinet & Catania 1979).

Cellular activation requires a specific interaction of cell membrane and extracellular information. The functioning effect of hormones on target cells is quite well known. In this system the enzyme adenylate cyclase (AC) is very important. The hormone coupling the receptor in cell membrane activates AC which catalyzes the ATP transformation to 3′5′ cAMP. cAMP is very important in controlling the cell metabolism. It may, for example, phosphorilate proteins by means of cAMP dependent protein kinases (PK). The level of cAMP in the cell depends on the activity of another enzyme, 3′5′ cAMP phosphodiesterase (PDE), the activity of which is controlled through the calcium concentration in the cell by means of calmodulin. Some of the mentioned links of the chain of molecular changes connected with cellular activation have been investigated.

Serotonin level (Franquinet 1979) in the organism of injured planarian drops in the first hour following transection (Fig. 1) and rises in the second hour but remains on a constant level during the entire preliminary period of regeneration, i.e. during six hours.

Adenylate cyclase activity initially changes according to changes of serotonin level (Franquinet *et al.* 1976, Franquinet *et al.* 1978) but after the second hour of regeneration the level drops dis-

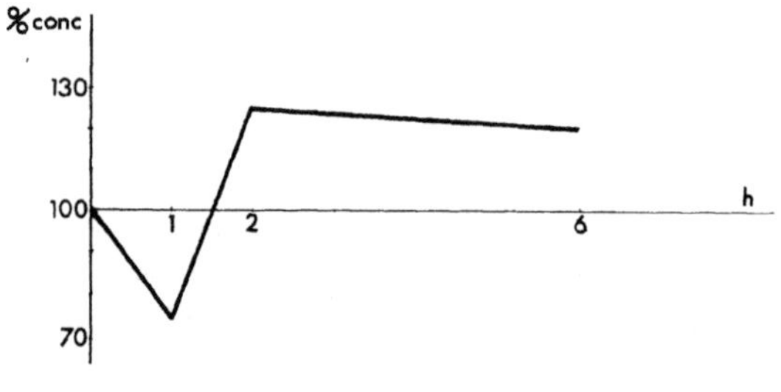

Fig. 1. Changes of concentration of serotonine during the 6 h of regeneration. The concentration in intact planarian was considered as 100%.

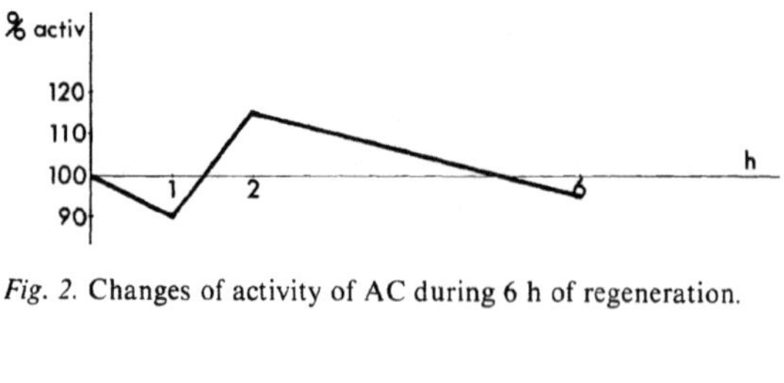

Fig. 2. Changes of activity of AC during 6 h of regeneration.

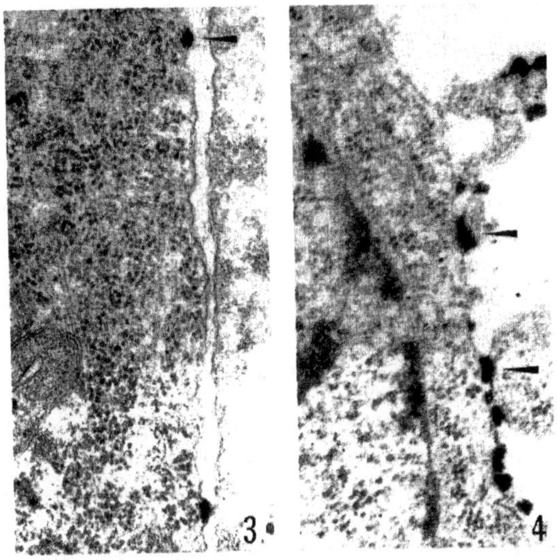

Fig. 3 & 4. Localization of an AC activity (arrows) in non-regenerating (3) and regenerating (4) planarian tissue × 33 500.

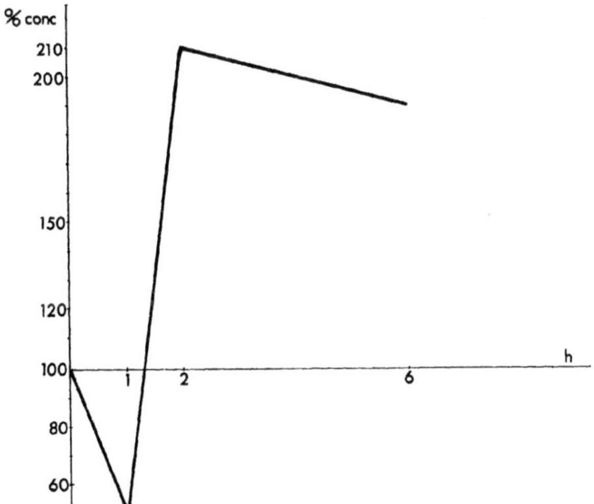

Fig. 5. Changes of a concentration of cAMP during the 6 h of regeneration.

tinctly (Fig. 2). Using ultracytochemical methods (Morawska, et al. 1981) it has been observed that the highest AC activity occurs in the second and third hour of regeneration (Fig. 3; Fig. 4).

Consistent with the change of AC activity the cAMP level drops in the first hour of regeneration (Le Moigne *et al.* 1976, Franquinet & Le Moigne 1979) to increase six-fold in the second hour (Fig. 5).

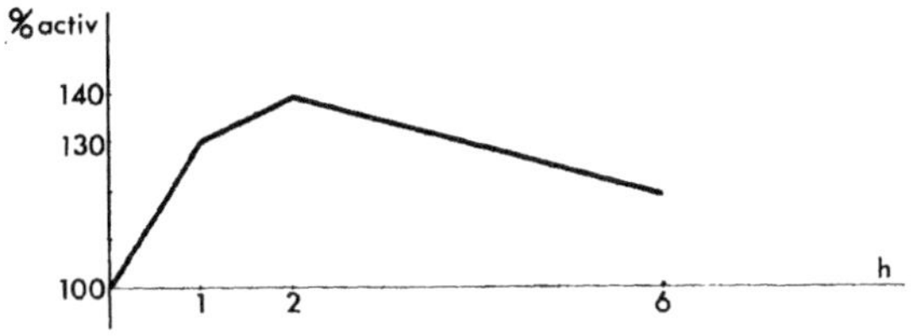

Fig. 6. Changes of an activity of PDE during the 6 h of regeneration.

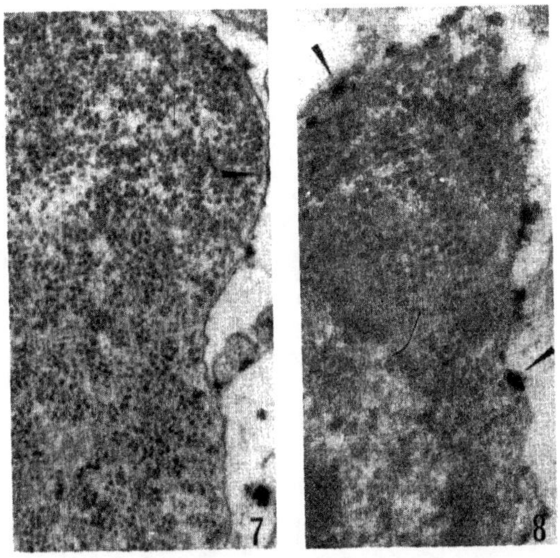

Fig. 7 & 8. Localization of PDE activity (arrows) in non-regenerating (7) and regenerating (8) planarian tissue × 33 500.

PDE activity (Franquinet & Coullon 1981) increases in the first two hours after the operation (Fig. 6) and afterwards decreases. This increase of activity together with AC activity decrease in the first hour of regeneration breaks the curve of cAMP level in regenerating Tricladida. Higher AC activity in the second hour even at high PDE activity is sufficient for a marked increase of cAMP which is in the regenerating organism. And again the biochemical results have been confirmed by our ultrastructural studies (Duma & Moraczewski 1980, Duma 1981) (Fig. 7, 8). Photographs show the activity in the third hour of regeneration as compared with the tissue of a non-regenerating turbellarian.

Recently (Moraczewski & Franquinet 1980), we have observed correlation between the cAMP level in tissues of turbellarian and the cAMP dependent PK activity (Fig. 9). Rapid increase of cAMP in regenerating planarian causes distinct increase of

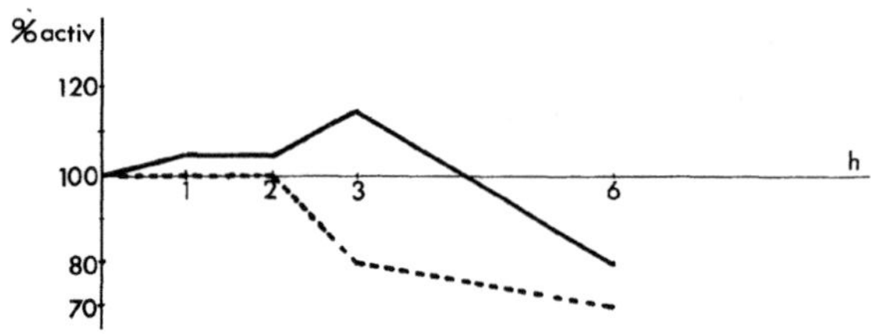

Fig. 9. Changes of an activity of PK cAMP dependent, and PK Ca^{+2} dependent, during the 6 h of regeneration.

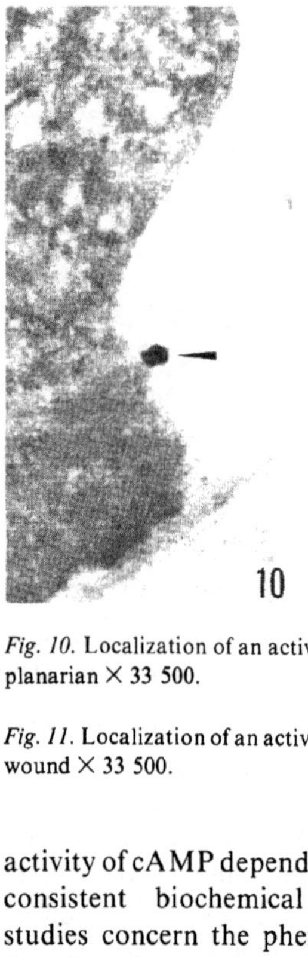

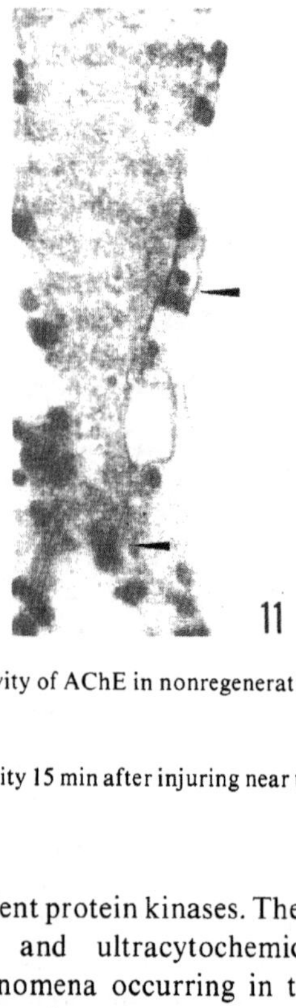

Fig. 10. Localization of an activity of AChE in nonregenerating planarian × 33 500.

Fig. 11. Localization of an activity 15 min after injuring near the wound × 33 500.

activity of cAMP dependent protein kinases. These consistent biochemical and ultracytochemical studies concern the phenomena occurring in the first three hours of regeneration.

On the basis of these results and assumptions it can be assumed that in the first three hours the sequence of phenomena is following:
Amputation of a part of the body results in a local osmotic shock near the wound and in a vast depolarization of nervous cells. Fifteen minutes after the operation mass excretion of ACHE to the intercellular space may be observed (Figs. 10, 11). The reaction of nervous cells is violent because in the first hour they produce a large amount of neurosecretion. Rapid depolarization of nervous cells probably causes the release of a great amount of free Ca^{++}ions. Calcium activates PDE and decreases the cAMP level in the cells. The appearance of serotonin and catecholamines in the medium activates AC and increases the amount of cAMP in the cells. It seems that this process is on a broad scale, i.e., the majority or a great number of cells undergo the changes in the activity of enzymes because of the disordered positional information. After 3 or 6 h the amount of substances examined and the activity of enzymes returns to normal. But in undifferentiated cells the process of differentiation begins. It occurs probably due to changes in cAMP level and phosphorylation of the proper proteins. This is analogous to changes during the transformation of cells into malignant ones. But in this case after an aggregation to undifferentiated cells is attained the blastema becomes a self-regulating system where the reproduced system of co-ordinates of positional information begins to function and the lost part of the body is harmoniously reproduced. Syntheses of nucleic acids and cell proliferation are the subject of the publication by J. Martelly *et al.* (1981) presented at the same symposium.

Acknowledgements

I acknowledge my debt to Doctor R. Franquinet for his biochemical data. I thank Drs. A. Le Moigne, I. Martelly and R. Franquinet for the helpful discussions during development of this report. This work was supported in part by a grant of the Polish Academy of Sciences.

References

Brøndsted, H. V., 1969. Planarian regeneration. Pergamon Press, Oxford.

Duma, A., 1981. Activity in cAMP phosphodiesterase in the early regeneration stage of planarian Dugesia lugubris (O. Schmidt). Ultracytochemical studies. Acta Medica Polona (in press).

Duma, A. & Moraczewski, J., 1980. Ultracytochemistry of cyclic 3′5′ nucleotide phosphodiesterase activity in the planarian Dugesia lugubris (O. Schmidt). Histrochemistry 66: 211–220.

Franquinet, R., Stengel, O. & Hanoune, J., 1976. The adenylate cyclase system in a freshwater planarian (Polycelis tenuis Iijima). Comp. Biochem. Physio. 53B: 329–333.

Franquinet, R., Le Moigne, A. & Hanoune, J., 1978. The adenylate cyclase system of Planarian Polycelis tenuis. Activation by serotonin and guanine nucleotides. Biochim. Biophys. Acta 539: 88–97.

Franquinet, R., 1979. Rôle de la sérotine et des catécholamines dans la régéneration de la Planaire polycelis tenuis. J. Embryol. Exp. Morphol. 51: 85–95.

Franquinet, R. & Le Moigne, A., 1979. Relations entre les variations des taux de sérotine et d'AMP cyclique au cours de la régéneration d'une Planaire. Biol. Cell. 34: 71–76.

Franquinet, R. & Catania, R., 1979. Localisation et étude microspectrofluorimetrique de la serotonine et des catecholamines chez la Planaire normale et en regeneration. C. R. Acad. Paris (in press).

Franquinet, R. & Coullon, J., 1981. Regeneration et nucleotides cycliques. Ann. Biol. (in press).

Goss, R. J., 1969. Principles of regeneration. Academic Press, New York.

Kiortsis, V. & Trampusch, H. A. L., 1965. Regeneration in animals and related problems, North-Holl. Publ. Comp., Amsterdam.

Le Moigne, A., Martelly, I. & Franquinet, R., 1976., Variation des syntheses d'ARN et de nucleotides cyclique au cours de la regeneration des Planaries. Bull. Soc. Zool. France, 101, suppl. 3: 29–33.

Lender, R. & Klein, N., 1961. Mise en evidence de cellules secretri ces dans le cerveau de la Planaire polycelis nigra. Variation de leurs nombres au cours de la regeneration posterieur. C. R. Acad. Sci. Paris 253: 331–333.

Lenicque, P. M., 1973. Morphogenetic action of 5 hydroxytryptamine and some analogues substances on the regeneration of the planarian worm Dugesia tigrina. Acta Zool. 54: 131–137.

Martelly, I., Franquinet, R. & Le Moigne, A., 1981. Relationship between variations of cAMP, neuromediators and the stimulation of nucleic acid synthesis during planarian regeneration (Polycelis tenuis). In: Schockaert, E. R. & Ball, I. R. Turbellaria. Proc. Third Int. Symp. Hydrobiologia. This volume.

Moraczewski, J., 1977. Asexual reproduction and regeneration of Catenula (Turbellaria, Archoophora). Zoomorphologie 88: 65–80.

Moraczewski, J. & Franquinet, R., 1980. Mise en evidence et activite des proteines kinasses AMP cyclique dependantes et AMP cyclique independantes chez la Planaire Polycelis tenuis. C. R. Acad. Paris 290: 1567–1570.

Morawska, E., Moraczewski, J., Malczewska, M. & Duma, A., 1981. Adenylate cyclase in regenerating tissues of the Planarian Fugesia lugubris (O. Schmidt). In: Schockaert, E. R. & Ball, I. R. Turbellaria. Proc. Third Int. Symp. Hydrobiologia.

Sengel, P., 1967. Aspects recents de la morphologenese chez les Planaires. Chemistry of Learning, red. W. C. Courning. S. C. Ratner. Plenum Press, New York, pp. 73–115.

Wolpert, L., 1971. Positional information and pattern formation. Curr. Top. Dev. Biol. 6: 183–224.

Adenylate cyclase in regenerating tissues of the planarian Dugesia lugubris (O. Schmidt)

E. Morawska, J. Moraczewski, M. Malczewska & A. Duma
Institute of Zoology, University of Warsaw, Warsaw, Poland

Keywords: Turbellaria, *Dugesia lugubris*, regeneration, ultracytochemistry, adenylate cyclase

Abstract

Adenylate cyclase (AC) was localized ultracytochemically in certain tissues of the regenerating planarian *Dugesia lugubris*. Studies were carried out from one hour after injury up to the 5th day of regeneration. It was found that the greatest amount of active AC appears during the initial hours of regeneration in the membranes of the muscle cells near the wound, in the epithelial cells surrounding the wound, and in rhabdite-forming cells and neoblasts.

Introduction

A considerable amount of attention has been given in recent years to the role of adenylate cyclase (AC) and cyclic 3',5' adenosinomonophosphate (cAMP) in controlling cell processes in both animals and plants. Biochemical studies (Franquinet *et al.* 1976, Le Moigne *et al.* 1976, Coulon *et al.* 1978, Franquinet & Le Moigne 1979) have suggested that the AC, and the cAMP phosphodiesterase system connected with it, play an important role in regeneration.

Moraczewski & Duma (1978) studied the localization of AC activity in neoblasts of regenerating planarians. The present study is an extension of that work and was aimed at tracing possible variations in AC activity connected with changes taking place in the cells of regenerating planarians at the ultrastructural level.

Material and methods

The species used in these studies was *Dugesia lugubris* (O. Schmidt) which was maintained in a laboratory culture. The head and prepharyngeal part of a recently transsected planarian was prepared for these studies, as follows:

1. Fixation in 2.5% glutaraldehyde in 0.05 M cacodylate buffer with pH=7,4 for 30 minutes at room temperature.
2. Two buffer washes at room temperature (23 °C). After washing the fixed fragments in 0.05 M cacodylate buffer, they were left in this buffer for 12 h in a refrigerator at 4 °C.
3. Incubation for histochemistry of the fixed parts of the planarian in a medium, prepared after Moraczewski & Duma (1981).
4. Washing:
   a) in 80 mM Tris-maleate acid buffer pH 7.4 for 1 min.
   b) in 0.05M cacodylate buffer pH 7.4. for 2 h on ice.
5. Fixation in 2% OsO_4 in 0,05M cacodylate buffer (pH 7.4) for 2 at 0 °C.
6. Dehydration in increasing concentrations of acetone.
7. Embedding in Epon.

Hydrobiologia 84, 209–212 (1981). 0018-8158/81/0843-0209/$00.80.

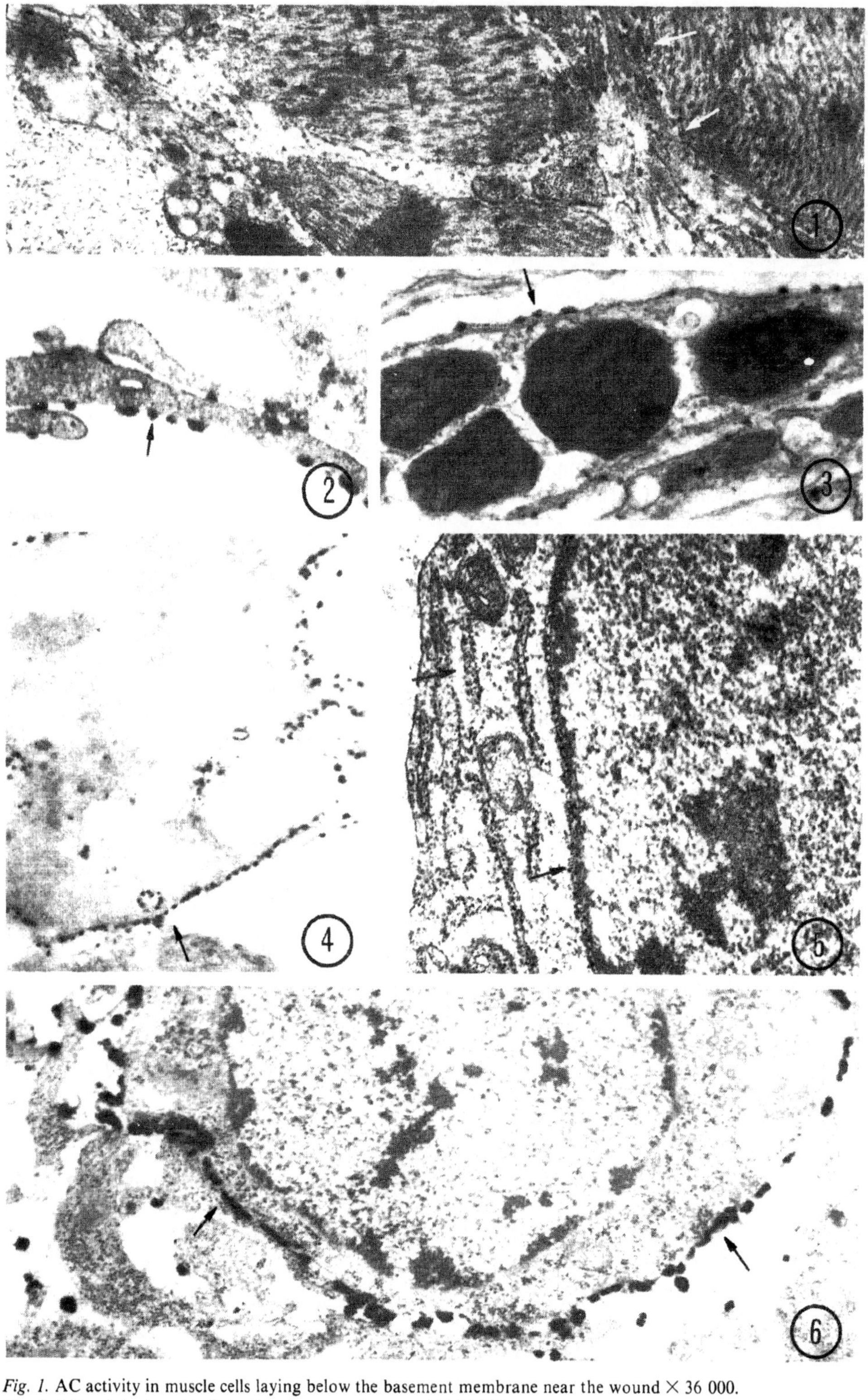

Fig. 1. AC activity in muscle cells laying below the basement membrane near the wound × 36 000.
Fig. 2. AC activity in epithelial cell laying near the wound × 36 000.
Fig. 3. AC activity in secretory cell × 36 000.
Fig. 4. AC activity in degenerating cell close the wound × 36 000.
Fig. 5. AC activity in regenerating cell of one day blastema × 36 000.
Fig. 6. AC activity in differentiating cell in three days blastema. × 36 000.

Results and discussion

A distinct increase in AC activity was observed one hour after transection and this increase was even more distinct during the third hour of regeneration. Biochemical studies have revealed a distinct increase in cAMP level during the second hour of regeneration, followed by a decrease lasting up to the 12th (Le Moigne *et al.* 1976). Our observations refer only to the region round the wound and variations in AC activity in definite types of cells. Increase in adenylate cyclase activity is the response to information created as the results of injury. This information is probably transmitted by means of hormones (Franquinet *et al.* 1978, Franquinet 1979).

During the initial three hours of regeneration a distinct increase was observed in AC activity in the membranes of the muscle cells which lay below the basement membrane near the wound (Fig. 1). The presence of AC in muscle cell membranes was also found in a non-regenerating planarian (Moraczewski & Duma 1981), but this enzyme did not occur with such great intensity and was not limited to a spatially defined group of cells.

The increase in AC activity in the epithelial cells surrounding the wound is interesting. During the initial hours of regeneration the epithelial cells surrounding the wound became flattened and stretched by extending processes in the direction of the wound. Analogous changes have been described by many authors (Betchaku 1979; Spiegelman *et al.* 1973, Morita *et al.* 1974, Pedersen, 1975). Cells forming the processes are characterized by a large nucleus and increased number of filaments in the cytoplasm (Fig. 2). On the basis of these facts it may be suggested that the occurrence of AC activity in the membranes of epithelial cells surrounding the wound is connected with the attainment by these cells of the capacity for movement.

The secretory cells we observed as occurring numerously in the wound itself and in its surroundings, frequently had considerable AC activity (Fig. 3). Precipitate grains were observed in cells in different stages of differentiation. In cells exhibiting the morphological characteristics of the early stages of differentiation precipitates were localized chiefly in the mitochondria.

During the initial hours of regeneration large amounts of fine precipitate were observed to occur in cytoplasmic membranes of degenerating cells (Fig. 4). These cells contained a large number of lysosomes in the cytoplasm. Coward & Bennett (1974) found much acid phosphatase activity connected with these cells. They suggested that it is a question not so much of degeneration and necrosis, as dedifferentiation of tissues. This assumption is borne out by the fact that we did not observe any nuclei with distinct symptoms of necrosis in any of the stages examined.

We suggest that both undifferentiated cells and dedifferentiating cells of certain tissues in the vicinity of the wound participate in regeneration. It may be that transmission of information to cells and AC activation are based on co-action of at least two or more substances. In *Polycelis tenuis* serotonin acts synergically with guanine nucleotids (Franquinet *et al.* 1979).

The processes taking place in the wound region in this early stage of regeneration do not appear to be connected with neoblasts, since they rarely occur near the wound. The few neoblasts which we observed during this period exhibited no signs of AC activity, which appears in later stages (Moraczewski & Duma 1978). The results obtained indicate that adenylate cyclase activity is very low between the 6th and 12th h of regeneration. Moreover, during these specific precipitates are to be observed almost exclusively on the epithelial cells covering the wound, and neoblasts begin to accumulate in the wound area. Similar results have been obtained by other authors (Morita & Best 1969, Spiegelman & Dudley 1973).

Fairly high adenylate cyclase activity appears suddenly in the regenerating cells at the first day (Fig. 5). At this stage of regeneration it is possible to speak for the first time of the true blastema containing distinct aggregations of neoblasts. It may consequently be suggested that the high AC activity observed on the first day of regeneration is connected with accumulation of neoblasts.

It may be that in turbellarians the accumulation of cells connected with the formation of the blastema in some way affects increase in AC activity. Adenyl cyclase is active on the first day of regeneration; on the second day its activity decreases, to rise again on the third day when it attains a very high level, after which it again decreases on the 4th and 5th day.

Although adenyl cyclase activity is very high on

the third day of regeneration it can be clearly seen that more of the active enzyme occurs in cells morphologically weakly differentiated and that these cells are mainly more deep localized (Fig. 6). It may be suggested, therefore, that AC is active only in early and middle stages of cell differentiation. It would also seem justifiable to speak of a spatial differentiation of distribution of the blastema cells.

Increase in AC activity during the initial hours of regeneration would thus appear to be connected with activation of all cells potentially capable of participating in the regeneration process. The second stage concerns only those cells which directly participate in the regeneration processes, i.e. cells involved in migration and differentiation.

References

Betchaku, T., 1970. The cellular mechanism of the formation of a regeneration blastema of fresh water planarian Dugesia dorotocephala. I. The behaviour of cells in a tiny fragment isolated in vitro. J. exp. Zool., 174: 253-280.

Coulon, J., 1978. Fluctuation of adenylate cyclase activity during anterior regeneration in Owenia fusiformis. J. Embryol. exp. Morph., 48: 73-78.

Coward, S. J. & Bennett, C. E., 1974. Lysosomes and liposomal enzyme activity in the regenerating planarian. Evidence in support of dedifferentiation. J. exp. Zool., 189: 133-137.

Franquinet, R., Stengel, D. & Hanoune, J., 1976. The adenylate cyclase system in a freshwater planarian (Polycelis tenuis Iijima), Comp. Biochem. Physiol., 53: 329-333.

Franquinet, R., Le Moigne, A. & Hanoune, J., 1978. The adenylate cyclase system of Planaria (Polycelis tenuis). Activation by serotonine and guanine nucleotides, B.B.A., 539: 88-97.

Franquinet, R. & Le Moigne, A., 1979. Rélations entre les variations des taux de sérotonine et d'AMP cyclique au cours de la régénération d'une Planaire. Biol. Cell., 39: 71-76.

Le Moigne, A., Martelly, J. & Franquinet, R., 1976. Variations des syntheses d'ARN et nucleotides cycliques au cours de la régénération des planaires. Bull. Soc. zool. Fr. Suppl., 101: 29-34.

Moraczewski, J. & Duma, A., 1978. Ultrastructural localization of adenyl cyclase activity in neoblasts of turbellarian Dugesia lugubris (O. Schmidt), J. exp. Zool., 203: 491-496.

Moraczewski, J. & Duma, A., 1981. Localization of adenylate cyclase activity in the tissues of an intact planarian Dugesia lugubris (O. Schmidt). Histochemistry (in press).

Morita, M. & Best, J.B., 1974. Electron microscopic studies of planarian regeneration II. Changes in epidermis during regeneration. J. exp. Zool., 187: 345-358.

Pedersen, K. J., 1976. Scaning electron microscopical observation on epidermal wound healing in the planarian Dugesia tigrina. W. Roux Arch., 179: 251-273.

Spiegelman, M. & Dudley, P. L., 1973. Morphological stages of regeneration in the planarian Dugesia tigrina: a light and electron microscopic study. J. Morph., 139: 155-184.

Morphology of the Turbellaria at the ultrastructural level

Reinhard M. Rieger
Dept. of Zoology, Univ. of North Carolina at Chapel Hill, Wilson Hall 046 A, Chapel Hill, NC 27514, U.S.A.

Keywords: Turbellaria, morphology, ultrastructure, body wall, nervous system, protonephridia, parenchyma, digestive system, reproductive system

Abstract

The paper reviews the following systems: body wall, nervous system, protonephridia, parenchyma, digestive system, reproductive system, and includes a summary of the literature. New information is presented for the catenulid and neoophoran body wall-construction, the kalyptorhynch proboscis, the catenulid and haplopharyngid protonephridial construction, and the prolecithophoran spermatozoon and female germ cell. Examples of new features, as well as examples of how electron microscopy has clarified the relative position of structures and their substructures are given from the subcellular level to the organization of whole organs. Fine structural features linking different turbellarian orders are summarized. They apparently support Karling's (1974) latest assessment of the affinities between the turbellarian orders which is based primarily on light histological data, they add the recognition of a special link between the Macrostomida and Haplopharyngida and they suggest the existence of three main evolutionary lines within the Turbellaria.

Introduction

Over 270 papers on various aspects of turbellarian fine structure have been published since the first published electron micrograph of turbellarian tissue, namely that of a planarian eye by Wolken (1958). The vast majority of these papers have used transmission electron microscopy. Only recently has scanning electron microscopy been applied increasingly to the Turbellaria (e.g. Reuter 1978; Smales & Blankespoor 1978; Williams 1978). The order Tricladida figured greatly in the early phase of Turbellarian fine structural research (e.g., Klima 1959, 1961; Pedersen 1959a, b, 1961a; Skaer 1961) but the interest soon expanded to include other orders, especially the Acoela (e.g. Pedersen 1964; Dorey 1965; Klima 1967). The turbellarian photoreceptor, cyrtocyte, rhabdite and spermatozoon became the first ultrastructural features used for investigating various systematic relationships (Eakin 1962; Kümmel 1962; Reisinger & Kelbetz 1964; Hendelberg 1965). Although fine structural information was used early in combination with light microscopy for a species description (Reisinger 1968), it is only recently that this approach has been used more frequently (e.g. Sterrer & Rieger 1974; Rieger & Sterrer 1975; Crezee 1975; Crezee & Tyler 1976; Lanfranchi 1978). Today about one third of the ultrastructural information deals with some aspect of the body wall, about one fifth with gametes and another fifth with regeneration. The last subject and related aspects have been excluded from this present review, since they are summarized by Gremigni (1981). Relatively little is still known about the comparative fine structure of the intestine, the parenchyma and especially about the gonoducts and associated structures.

A complete listing of ultrastructural literature was not intended in this paper, but the given references should provide a first summary of the

Hydrobiologia 84, 213–229 (1981). 0018-8158/81/0843-0213/$03.40.

available literature. In several instances a new synthesis of the data is presented to stimulate the search for further ultrastructural information. Finally, an attempt is made to relate our present knowledge of turbellarian ultrastructure to the understanding of turbellarian systematics as revealed by the light microscope.

Body wall

The fine structure of the *epidermis* was the subject of one of the first truly comparative ultrastructural investigations of free-living flatworms by Bedini & Papi (1974). Twenty-three species of ten different families and eight different orders have been studied by these authors. From these data and various subsequent studies on turbellarian ultrastructure, four basic types of epidermis can be discerned (Figs. 1–4).

The first type can be called the neoophoran epidermis because it is characteristic for all neoophoran orders (Lecithoepitheliata, Prolecithophora, Proseriata, Tricladida, Rhabdocoela) and the Polycladida (Fig. 4). The main characteristic feature of this type of epidermal organization is the presence of a well-developed basement membrane (see Bowen & Ryder 1974; Hori 1979) with a 20–40 nm thick limiting layer of granular fibrous appearance at the epidermal base and a microfibrillar layer of varying thickness (0.2–4 μm).[1] Often additional extracellular materials are deposited between basal foldings of the epidermis and the outer limiting layer of the basement membrane (e.g. coagulated fluid in *Gieysztoria* and *Dalyellia*, see Bedini & Papi 1974; electron-lucent zone in *Dugesia japonica*, see Hori 1979).

The microfilaments in the deeper layer average 80–120 nm in diameter; they are most likely collagenous in chemical composition and they may exhibit complex three dimensional arrangements (e.g. twisted fiber orientations, Fig. 6, see Bouligand 1972, for discussion). With some exceptions (e.g. *Florianella bipolaris*, see Rieger & Sterrer 1975) the microfibrillar layer lacks an inner limiting layer towards the body wall musculature; instead the

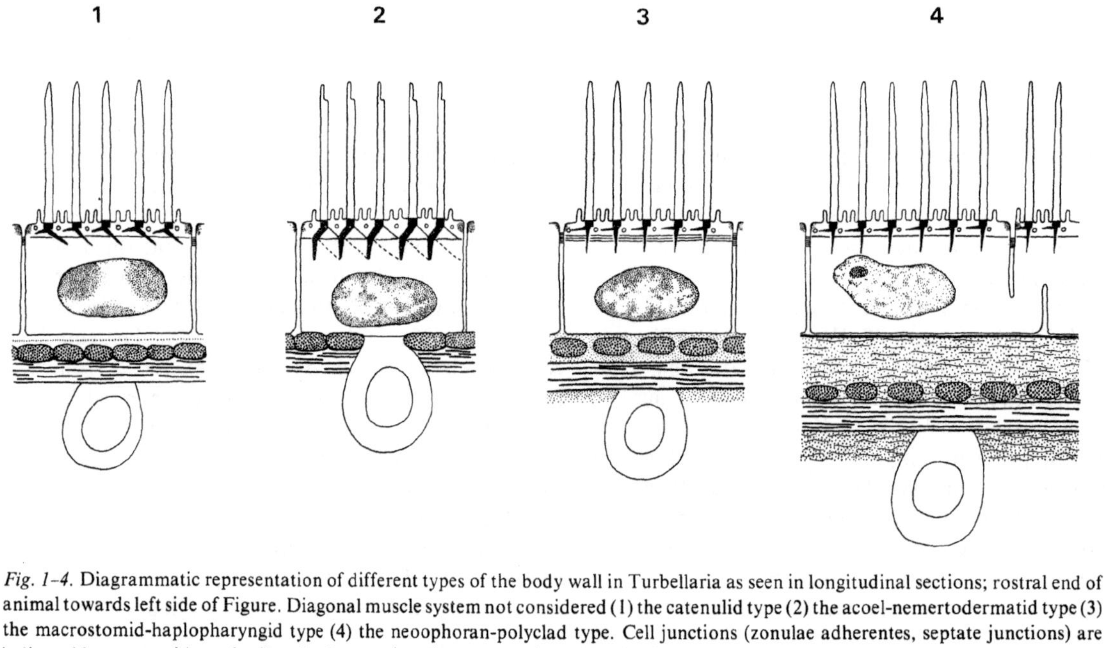

Fig. 1–4. Diagrammatic representation of different types of the body wall in Turbellaria as seen in longitudinal sections; rostral end of animal towards left side of Figure. Diagonal muscle system not considered (1) the catenulid type (2) the acoel-nemertodermatid type (3) the macrostomid-haplopharyngid type (4) the neoophoran-polyclad type. Cell junctions (zonulae adherentes, septate junctions) are indicated between epidermal cells only. Septate junctions are not illustrated in (2), since septa, if present, are often indistinct (Bedini & Papi 1974; Tyler & Rieger 1977). Terminal web and basement membrane are illustrated in relative thickness between the types. Small circles next to ciliary rootlets represent ultrarhabdites. Cell body below muscle layers symbolize sunken epidermal cells, which are seen most commonly in Acoela. The neoophoran-polyclad epidermal cells are shown partly in a transition towards a syncytium, which is common in several lines of the neoophorans.

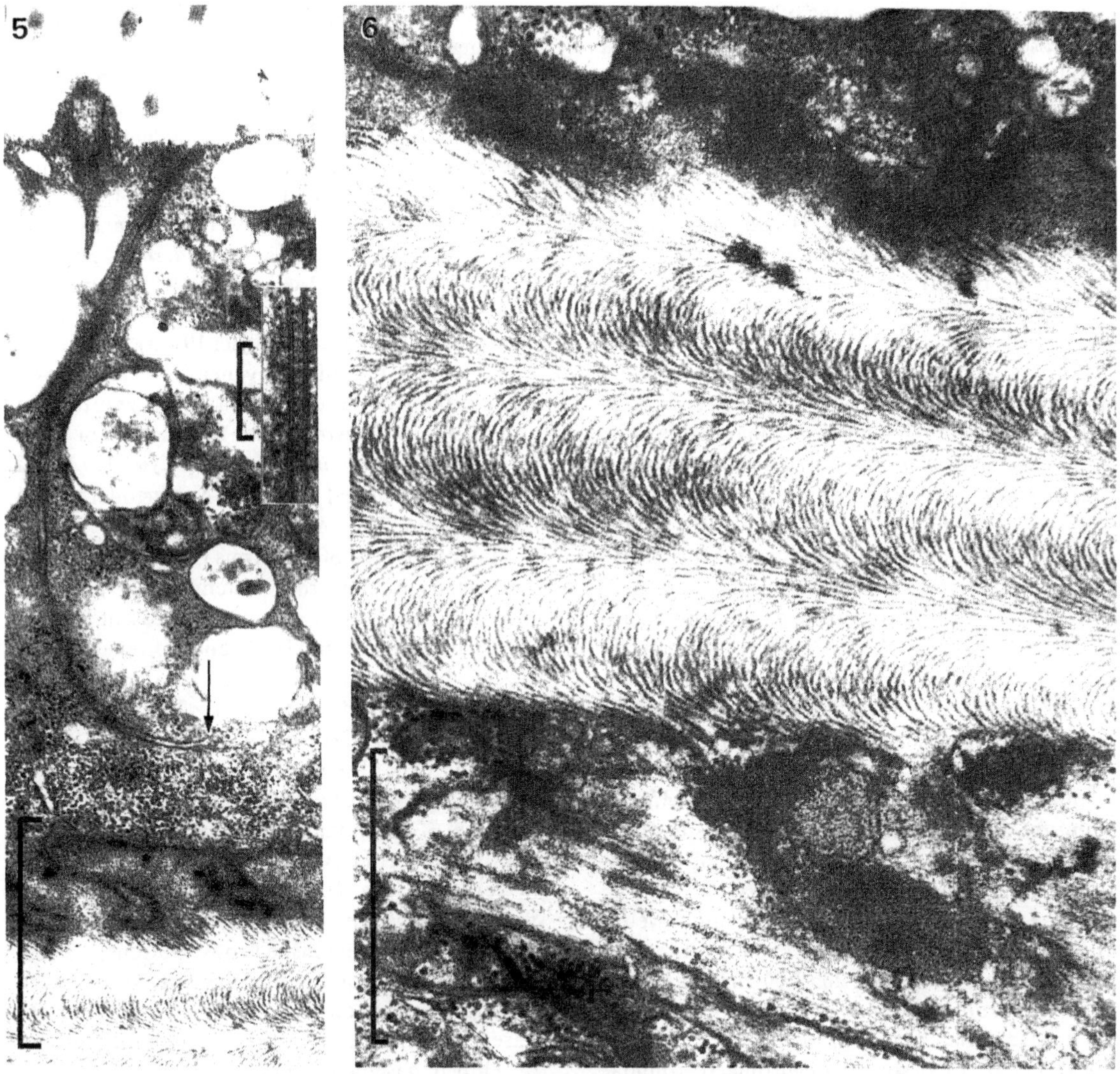

Fig. 5–6. Fine structure of epidermis (5) and basement membrane (6) of *Kytorhynchella* c.f. *meixneri*, from cross-sections. Note in (5) the incomplete separation of epidermal cells (see arrow) and the septate junction (see inset). Note in (5) and in (6) the presence of a fine granular limiting layer at the epidermal face of the basement membrane and the lack of such a layer at the muscle face of the basement membrane in (6). Scale for (5) 0.1 μm; for inset and (5) 0.5 μm; and for (6) 0.5 μm. Standard 2–5% glutaraldehyde and 1–2% Osmiumtetroxyde double fixation, each for 1–2 h. Dehydration with standard ethanol series; Epon-Araldite embedment, section stained with uranyl acetate-lead citrate and studied with a Zeis EM electron microscope.

layer is often (especially in larger forms such as in Tricladida, Polycladida and Lecithoepitheliata) continuous with the fibrous matrix between muscle cells and parenchymal elements.

The epidermis itself shows two characteristic trends. First, a tendency toward truly syncytial organization (e.g. *Kytorhynchella*, Fig. 5; *Xenoprorhynchus*, see Reisinger 1968; *Gyratrix*, see Reuter 1975; *Florianella bipolaris*, see Rieger & Sterrer 1975). This tendency is especially common in the suborder Kalyptorhynchia. Secondly, a tendency towards weakly developed terminal webs in the apical cytoplasm of the cells. Exceptions to this second trend are seen in some members of the

Dalyellidae (Bedini & Papi 1974), some members of the Proseriata (e.g. *Notocaryoplanella*, see Ehlers & Ehlers 1977c) and some members of the Rhabdocoela (e.g. *Litucivis serpens*, see Ehlers & Ehlers 1977c). Occasionally the nuclear portion of the epidermal cells are sunken below the level of the basement membrane (e.g. *Xenoprorhynchus*, see Reisinger 1968; *Prorhynchus, Otoplana*, see Bedini & Papi 1974). The cilia are either provided with a long single rostral rootlet (e.g. Otoplanidae, see footnote 7 in Rieger 1976, p. 222) or display a rostral and vertical rootlet.

The second type can be called the macrostomid epidermis, since it occurs only in the closely related orders Macrostomida and Haplopharyngida (Fig. 3). A basement membrane as described above is lacking in these orders. Fine granular or sometimes fine fibrous material is found deposited around the subepidermal muscle cells and in some forms also between other cell elements which fill the space between gut and body wall (e.g. Dolichomacrostomidae, see Rieger & Tyler 1974). This matrix forms occasionally a 50–200 nm thick irregular layer between epidermal cells and body wall musculature (e.g. especially visible in *Haplopharynx, Microstomum, Myozona*, see Tyler 1976) or it may be present only incompletely (Doe 1981 for *Macrostomum*). In the epidermal cells there is a strong tendency among most species of this group towards a thickening of the terminal web. In *Haplopharynx, Myozona* and *Microstomum*, the terminal web is about 50–100 nm thick (e.g. Tyler 1976), it reaches medium values of about 200–500 nm in *Macrostomum, Promacrostomum* and *Psammomacrostomum* (e.g. Bedini & Papi 1974; Tyler 1976), it is thickest (1–3 μm) in some Dolichromacrostomidae (e.g. *Paramyozonaria simplex*, see Doe 1981), and especially in *Bradynectes*, (see Tyler 1976). In the last species, the microfilaments in the terminal web exhibit a 3-dimensional arrangement similar to that seen in the basement membrane of the neoophoran epidermis. Occasionally, the nuclear portions of some epidermal cells are sunken below or into the subepithelial musculature (e.g. anchor cells of adhesive organs in *Myozona, Macrostomum, Microstomum*, see Tyler 1976; epidermis around mouth opening in *Paramyozonaria simplex* and *Microstomum* sp., see Doe 1981). Two rootlets are associated with epidermal cilia: a horizontal rostral rootlet and a vertical, sometimes (e.g. *Myozona, Psammomacrostomum*, see Tyler 1976) caudally-pointing rootlet.

The third type of epidermis is only found in the orders Acoela and Nemertodermatida (Fig. 2). Its main characteristic feature is a complete lack of an intercellular matrix at the base of the epidermal cells (Ferrero 1973; Bedini & Papi 1974; Tyler & Rieger 1977; but see Smith 1981 for remarks on an undescribed species of *Nemertoderma*) and the presence of a special ciliary rootlet system with interconnecting striated fibers as well as the presence of special ciliary tips (Tyler 1979a). A terminal web is usually present, but generally weakly developed in the Acoela (e.g. 20 nm in *Mecynostomum* and 60 nm in *Convoluta*, see Bedini & Papi 1974; 100 nm in *Paratomella rubra*, after Tyler 1973) and strongly developed in the Nemertodermatida (250–600 nm, Tyler & Rieger 1977).

The fourth type of epidermis is least well known and occurs in the order Catenulida (Fig. 1). In *Catenula*, a basement membrane is absent according to Bedini & Papi (1974) and Soltynska *et al.* (1976), but present as a very thin intercellular membrane according to Figs. 2, 4, and 5 in Moraczewski & Czubaj (1974). Sterrer & Rieger (1974) note the absence of a basement membrane in *Retronectes*, but reexamination establishes now the presence of a thin matrix between the muscle cells and between epidermis and muscle cells in *Retronectes thalia* (Fig. 7). No intercellular matrix could yet be discerned in *Retronectes atypica* (see Doe & Rieger 1977; unpublished observations). These electron microscopical findings parallel the conflicting light histological observations about the basement membrane in the Catenulida (e.g. well developed basement membrane in *Stenostomum virginianum*, see Pullen 1957, and lack of basement membrane in *Rynchoscolex*, see Reisinger 1924). In all cases studied so far (with the possible exception of *Stenostomum virginianum*) the intercellular matrix at the base of the epithelium – if present – is extremely weakly developed. The nuclear portion of the epidermal cells is occasionally sunken below the musculature (Doe 1981). A distinct terminal web is described for some Catenulida (see Bedini & Papi 1974; Soltynska *et al.* 1976) but appears to be lacking in *Stenostomum* (see Soltynska *et al.* 1976) and in the Retronectidae (Sterrer & Rieger 1974; own unpublished observations). The striated ciliary rootlets consist of a rostrally and a caudally oriented fibre.

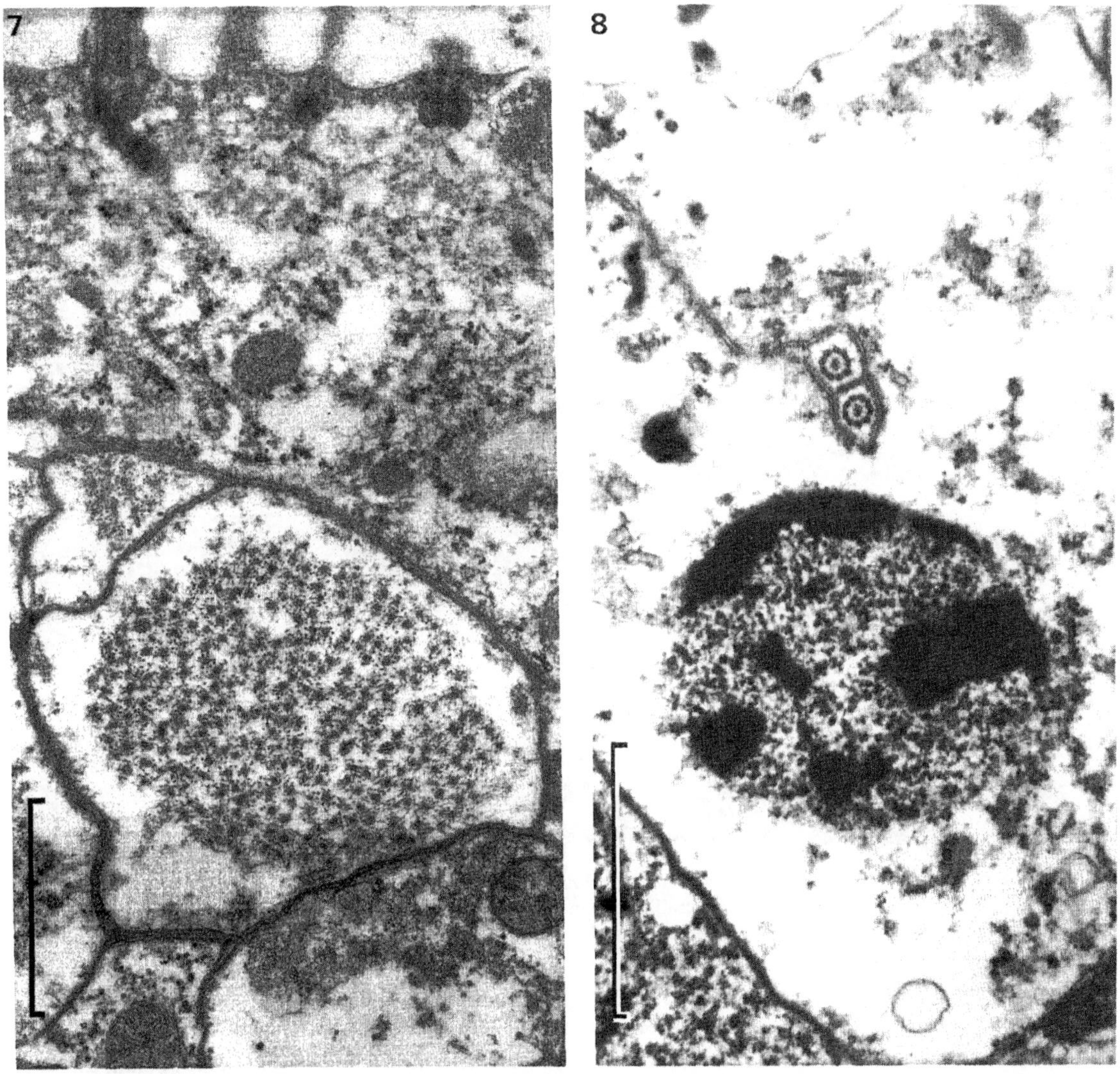

Figs. 7–8. Fine structure of body wall (7) and protonephridial canal (8) in *Retronectes* c.f. *thalia* as seen in cross-sections. In (7) note the matrix between epidermal cell and longitudinal muscle cells and around the large profile of a longitudinal muscle cell. The protonephridial canal containing the two cilia in (8) is formed similarly to the canal of the caudal sensory organ in Macrostomida (see Rieger & Tyler 1974, Fig. 7). Note also in (8) the junction of the ciliary membrane with the wrapping canal cell. Scale for (7) 0.25 μm. Scale for (8) 0.5 μm. For Methods see Sterrer & Rieger (1974).

From their studies on 'cuticular' hard structures associated with special epidermal differentiations (e.g. the kalyptorhynch proboscis) Rieger & Doe (1975) suggest that supportive and hard structures in the phylum Plathelminthes have not evolved through 'true' cuticularizations,[2] but through elaborations of the intracellular fibrous system (cell web, terminal web) or the intercellular fibrous system at the base of the epidermis (basement membrane). The macrostomid type of epidermis represents a trend towards a domination of the cell web for mechanical support, whereas the neoophoran type of epidermis represents a trend towards a domination of the basement membrane system for the same function. The unique interconnecting striated rootlet system of the Acoela and

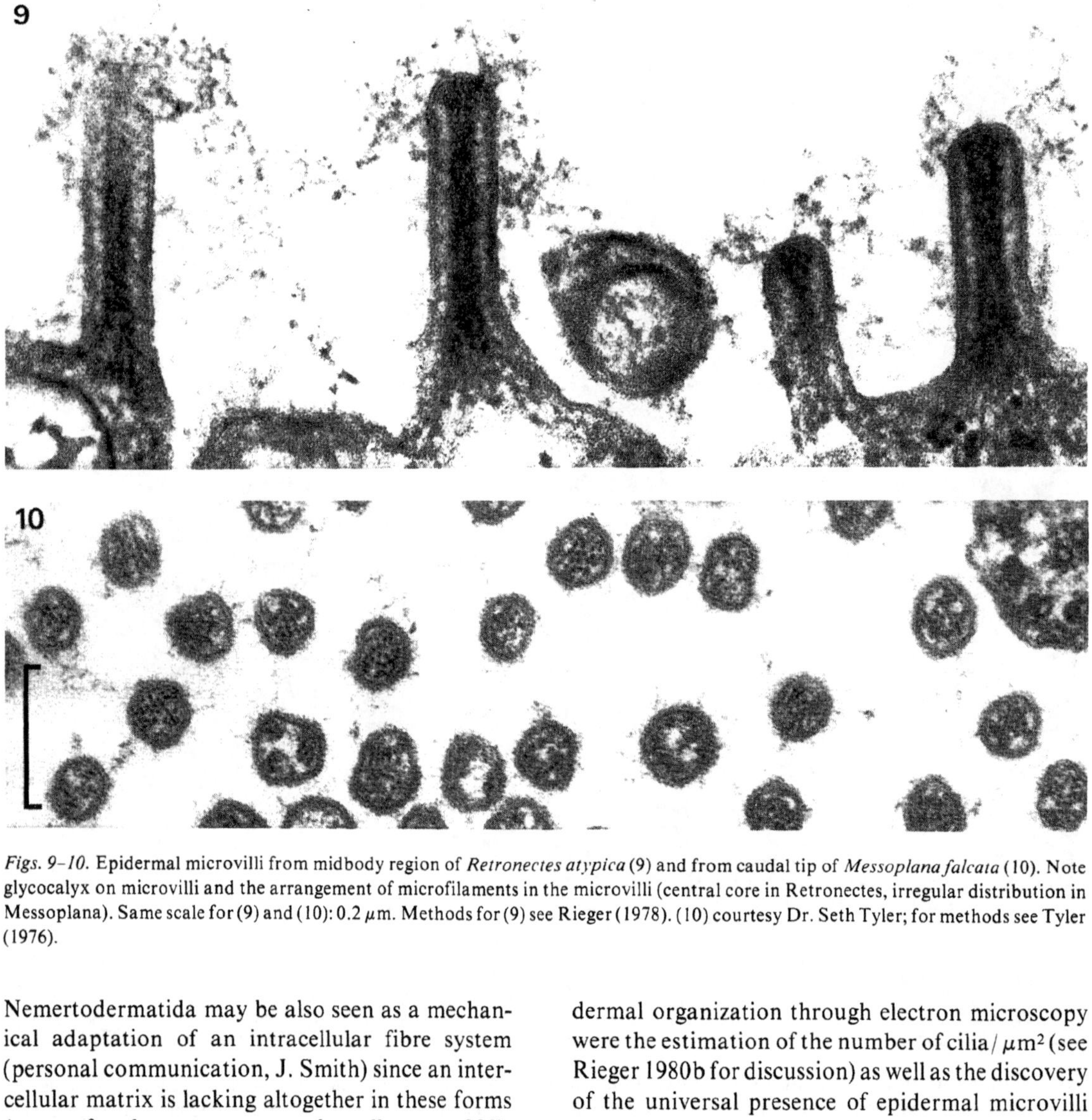

Figs. 9–10. Epidermal microvilli from midbody region of *Retronectes atypica* (9) and from caudal tip of *Messoplana falcata* (10). Note glycocalyx on microvilli and the arrangement of microfilaments in the microvilli (central core in Retronectes, irregular distribution in Messoplana). Same scale for (9) and (10): 0.2 μm. Methods for (9) see Rieger (1978). (10) courtesy Dr. Seth Tyler; for methods see Tyler (1976).

Nemertodermatida may be also seen as a mechanical adaptation of an intracellular fibre system (personal communication, J. Smith) since an intercellular matrix is lacking altogether in these forms (except for the statocyst capsule wall, see p. 220). Finally, it should be noted that the well know fragility of all catenulids (Sterrer & Rieger 1974) is now explained by their extremely weak development of both the intra- and intercellular fibrous systems.

As shown above, ultrastructural studies on the turbellarian epidermis revealed for the first time the patterns of structural variability and functional importance of the cell web and various ciliary structures; (Tyler 1979a; Hendelberg 1981). Other important additions and clarifications of the epidermal organization through electron microscopy were the estimation of the number of cilia/μm^2 (see Rieger 1980b for discussion) as well as the discovery of the universal presence of epidermal microvilli and some kind of vesicle system in the apical cytoplasm of the epidermal cells (= ultrarhabdites, Bedini & Papi 1974). But the patterns of structural variation and the functional significance of these structures are less well understood. Rieger & Rieger (1976) suggested the importance of increased microvillar glycocalyx specializations for uptake of dissolved organics. There are also noticeable differences in the arrangement of the microfilaments in epidermal microvilli (Figs. 9 & 10), which might have phylogenetic and functional significance. Recently, Martin (1978) suggested a new function

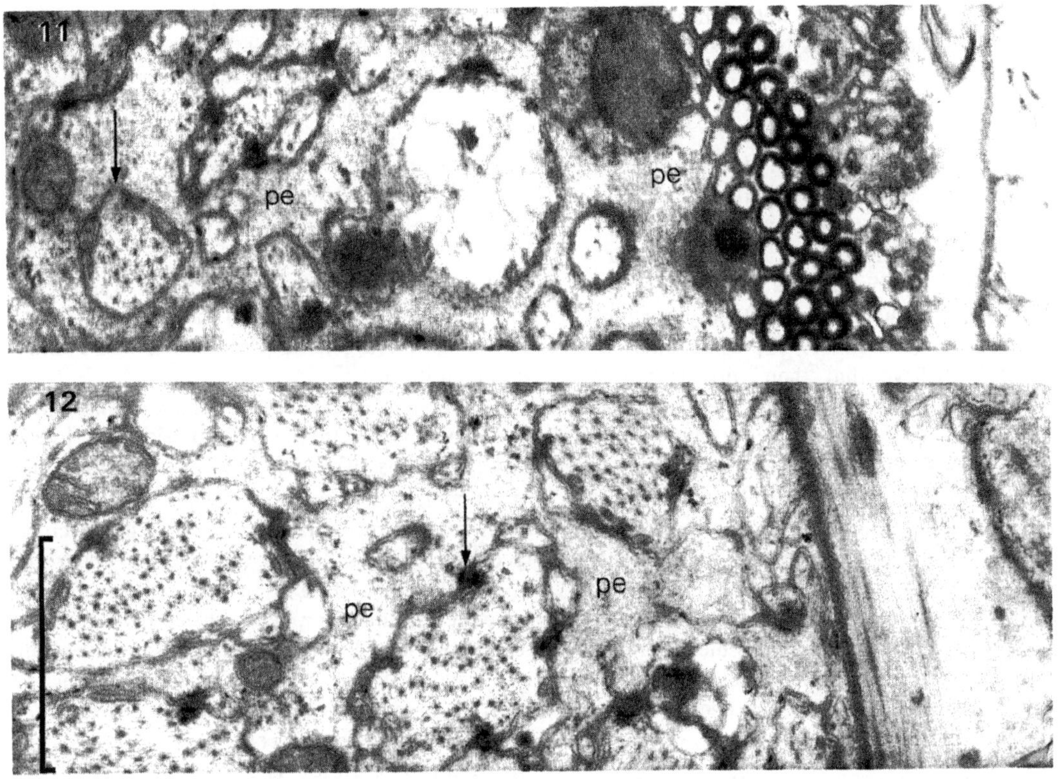

Figs. 11–12. Parts of cross sections just above (11) and just below (12) the junction of the proboscis in *Florianella bipolaris* (see Rieger & Sterrer 1975; Fig. 10a). In (11) note the cross sections of large microvilli of the cone-epithelium and the smaller microvilli of the proboscis sheath-epithelium in right side of picture; note also the extension of proboscis cone-epithelium (pe) between necks of gland cells and muscle cells (→). In (12) note part of circular muscle surrounding the bulb below junction on right side, the matrix between this circular muscle and extensions of cone-epithelium (pe), the lack of matrix between cone-epithelium and the profiles of the central muscles and the presence of special junctions between the cone-epithelium (pe) and cone muscle cells (→). Same scale for (11) and (12) 1.0 μm. For Methods see Rieger and Sterrer (1974).

for rhabdites in the mucous production for ciliary gliding. However, he did not discuss the possibility that such a function may actually be carried out by the ultrarhabdites, common in regular epidermal cells of Turbellaria. From the work of Bedini & Papi (1974), Doe (1981) & Tyler (1976) considerable structural diversity is suggested for the ultrarhabdite system.

Of the well developed gland cell systems in the turbellarian epidermis, two have received special attention: the adhesive organs (see Rieger & Tyler 1979; Sopott-Ehlers 1979, for summary) and the rhabdite glands (see Smith *et al.* MS). Tyler's (1977) study on the turbellarian adhesive system is seen as a major step in the application of ultrastructural features towards a better understanding of turbellarian phylogeny (e.g. the support of the separation of the Acoela-Nemertodermatida from the other orders, as well as the better recognition of the dichotomy between the Macrostomida-Haplopharyngida on one hand and the Polycladida-Neoophora on the other hand). New data on the organization of the rhabdites in Turbellaria support the earlier conclusions based on adhesive organs (e.g. that the occurrence of true rhabdites is limited to the orders of the Macrostomida and the Neoophoran lines; Acoela have a different kind of rod-shaped secretion). However, Smith *et al.* (MS), stress the need for further studies especially in the Acoela, Proseriata and Tricladida.

Studies on the *body wall musculature* are few. They have demonstrated the presence of the usual thick and thin myofilaments, as well as the presence of dense bodies (homologous to the Z-system in striated muscle) and various specializations of the sarcoplasmic reticulum (Moraczewski & Czubaj 1974; DeEquileor & Valvassori 1975; see also Reuter 1977). The suggestion of the presence of epitheliomuscular cells in Catenulida (Moraczewski & Czubaj 1974, p. 37) must be rejected.

Studies of specialized areas of the body wall are even rarer [e.g. the unciliated epidermis in Otoplanidae (Ehlers 1977); the proboscis of Kalyptorhynchia (Schockaert & Bedini 1977)]. Schockaert & Bedini (1977) believe that their ultrastructural findings do not add to our understanding of the evolution of the eukalyptorhynch conorhynch. But this surely is only the case because we lack yet comparative data from other Kalyptorhynchia. That one will find a considerable degree of variation in the structure of the eukalyptorhynch conorhynch can be illustrated by the organization of the proboscis cone eipthelium of *Florianella bipolaris,* which lacks any clearly visible basement membrane and has ramifying processes between the longitudinal muscle fibers of the cone itself (Figs. 11 & 12).

Nervous system

Our present knowledge on the fine structure of the central and peripheral nervous system in Turbellaria has been recently discussed by Reuter & Lindroos (1979a, b). Besides several members of the Tricladida and Polycladida only one genus of the Rhabdocoela and two genera of the Catenulida have been investigated so far. Up to four different types of neurons have been identified in the neoophoran orders, but only a single type of neuron was seen in the Catenulida. Reuter & Lindroos (1979a, b) do not mention the occurrence of neuroglia in the rhabdocoel *Gyratrix.* Moraczewski *et al.* (1977) mention the lack of neuroglia in the Catenulida. Neuroglia is, however, known from light and electron microscopy at least in the Polycladida (e.g. type E and type G cells in *Leptoplana,* see Turner 1946) and glial elements have been seen by electron microscopy in *Notoplana's* peripheral system (Koopowitz & Chien 1974). The peripheral nervous system in *Notoplana* consists of three plexus: an infra-epithelial, a subepithelial, and a submuscular. Nerve cell bodies are especially associated with the infraepithelial and the submuscular plexus. In *Gyratrix* only a subepithelial and submuscular plexus has been found, both of which lack any nerve cell bodies. The lack of nerve cell bodies in the peripheral nervous system and the occurrence of distinct neuronal cell types in *Gyratrix* is interpreted as a derived condition within the Turbellaria. The Turbellarian peripheral nervous plexus has both primitive and advanced features according to Koopowitz & Chien (1974) and does not seem to show close phylogenetic relationships to the cnidarian condition.

Ciliated epidermal receptors have been studied more extensively in the Turbellaria (e.g. Ehlers 1977; Williams 1977; Doe 1981). Two main groups of monociliated sensory cells, each with two subtypes can be recognized. (1) monociliated collar receptors (Ehlers 1977), which either bear the microvillar collar free at the epithelial surface or have the collar sunken below the epithelial surface (e.g. Proseriata, Rhabdocoela). These collar receptors have a regular 9 + 2 axoneme and function most likely in mechanoreception (see Ehlers 1977); (2) monociliated sensory cells, which lack a collar of microvilli. The cilium is usually exposed to the epidermal surface (but see Fig. 20, MacRae 1967) and either provided with a 9 + 2 axoneme or the cilia are shorter, showing various changes in the axonemal configuration (e.g. see Reuter 1975, type I, III and IV receptors). The functional significance of these structural variations in the ciliary receptors are not well understood (see Storch 1973). Rheo-, tango- and chemoreceptive functions have been most often suggested for this latter group of ciliary receptors (see Reuter 1975). Multiciliated receptors are less common, but have been noted in the Otoplanidae (see Bedini *et al.* 1975) and in the pharynx of *Haplopharynx* and various macrostomids (see Doe 1981).

The fine structure of the *statocyst* has been studied only in the Acoela (Ivanov *et al.* 1972; Ferrero 1973). Because of the lack of cilia and special sensory cells the acoel statocyst appears different from most other static sense organs in the lower metazoa. Ferrero (1973) describes the capsule wall as an intercellular structure, whereas Ivanov *et al.* (1972) indicate the same structure to be intracellular. Unpublished observations of the statocyst in a species of *Nemertoderma* and one of *Convoluta* (courtesy J. Smith) support Ferrero's interpretation.

Photoreceptors have been studied with the electron microscope in the Macrostomida, Proseriata, Tricladida, Rhabdocoela, Prolecithophora and Polycladida (Palmberg *et al.* 1980; Durand & Gourboult 1977; Bedini *et al.* 1977; Ehlers & Ehlers 1977a, b; Ruppert 1978; Lanfranchi *et al.* 1981; Burt & Bance 1981). The organization of the rhabdomeric cerebral pigment cup eye (e.g. see Fig. 3 in MacRae 1966) is rather uniform in these orders,[3] except for the cerebral eyes of the Müller's larva, where microvillar and cilia-bearing sensory cells have been noted (Ruppert 1978). The first clearly ciliarytype eye in Turbellaria was described from the epidermal eye of polyclad larvae by Ruppert (1978). Possible ciliary photoreceptors have been noted before near the brain in the Otoplanidae by Ehlers & Ehlers (1977a, b).

Some information is also available on more complex glandular-sensory organs in Turbellaria (Rieger & Tyler 1974; Crezee & Tyler 1976) but very little is still known about the structure of the complex frontal organ, which according to Ax (1963) and Karling (1974) is a characteristic feature of the Turbellarian archetype (see Ruppert 1978, for fine structure of frontal organ in polyclad larvae).

Protonephridia

Fine structural data on the flame-bulb cell and sometimes also on other regions of the protonephridial system have been obtained in the orders Catenulida, Macrostomida, Lecithoepitheliata, Tricladida and Polycladida (see Ishii 1980a, b; Ruppert 1978; Brandenburg 1974; Wilson & Webster 1974 for summaries). The cyrtocyte of the catenulid *Stenostomum* is unique because of its weir and because it only contains two flagella. That the number of flagella might be characteristic for the whole order is suggested by data on the protonephridial system of *Retronectes thalia* (Fig. 8). In the other orders usually well over ten flagella are reported in the cyrtocytes.[4] Striking is the similarity in the parallel arrangements of two types of cytoplasmic rods in the cyrtocyte weir of *Macrostomum* and that of the parasitic flatworms (see Figs. 13–15). Wilson & Webster (1974) suggested that this configuration is brought about by interdigitation of microvilli-like processes from the cyrtocyte and the adjacent canal cell. Since such a weir system is not common outside the platyhelminthes (found in *Priapulus*, and possibly in *Urnatella*), the similarities between *Macrostomum* and the parasitic flatworms may reflect a homology. Contrary to the claim of Brandenburg (1974, p. 6) the weir of the flame-bulb cell in *Xenoprorhynchus* consists of only one type of rods. This has been verified also for the closely related genus *Prorhynchus* (unpublished observation, courtesy Dr. S. Tyler). In *Xenoprorhynchus* the cross-section through the cyrtocyte-weir shows at one place a special thick cytoplasmic process (Fig. 15). A somewhat similar disruption of the regular rods in the weir is also seen in *Macrostomum* and in *Fasciola* (see Figs. 13 & 14). Especially in *Macrostomum* and *Fasciola*, the pictures suggest that the weir is not formed as a cylindrical outgrowth, but more likely as a sheath which folds back on itself to form the tube of the weir. Further studies are needed to clarify these complex threedimensional relationships between cyrtocyte and adjacent canal cells. This is also true for the cyrtocytes of the Tricladida, which appear, at first, quite different in their organization from the forms discussed above (see especially McKanna 1968; Ishii 1980a).

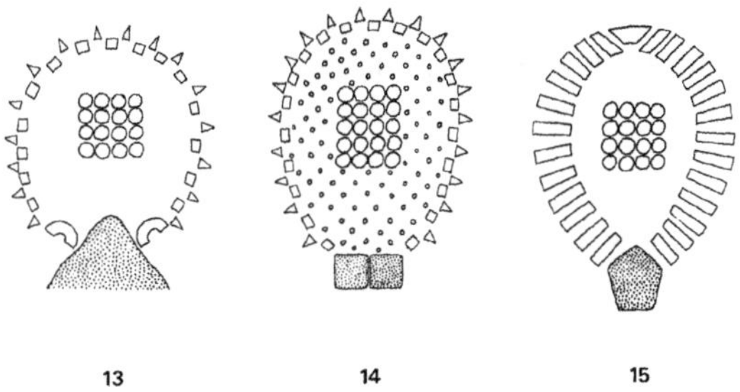

Figs. 13–15. Diagrammatic representations of cross sections through the weir of the cyrtocytes in *Macrostomum tuba* (13) in *Fasciola hepatica* (14) and in *Xenoprorhynchus steinböcki* (15). (13) and (15) after Reisinger (1968); (14) after Brandenburg (1974). Connections between rods excluded in diagram.

Parenchyma

In an attempt to summarize our recent ultrastructural advances in recognizing the cells filling the body cavity (space between body wall and other main organ systems) Rieger (1980a) distinguished three classes of cells: (1) sunken gland cell bodies of epidermal, pharyngeal or genital glands; (2) stem cells, neoblasts or replacement cells; and (3) one to several kinds of so-called parenchymal cells. The first class should be expanded here to include generally epidermal or gastrodermal cell-elements which are deeply sunken below the epithelial levels. This first, not truly parenchymal class of cell bodies apparently represents the numerical majority of cell bodies seen between body wall and the other organ systems in the Haplopharyngida and especially in most Macrostomida (unpublished observations). Epidermal stem cells containing numerous centrioles have been seen in the Catenulida and the Haplopharyngida (see Rieger 1980a for literature) and in the Tricladida (see Hori 1978). Neoblast-type cells are described for the Acoela, Polycladida and Tricladida (see Rieger 1980a for literature). The third class of so-called parenchymal cells is the most heterogeneous of Rieger's (1980a) groups. The ramified or fixed parenchymal cells in the Acoela, Polycladida and Tricladida, the glycogen-storage cells found in the Tricladida, and the fluid-filled cells in *Myozona* (Macrostomida) and *Archimonotresis* (Prolecithophora) may represent all various specializations of the same cell type. Different specialization may have led to the vacuole cells which are known as chordoid tissue (Ax 1966) in some Acoela, Catenulida and Macrostomida, to the granular strands of retronectid catenulids (Doe & Rieger 1977), and to the pigment cells in the Tricladida (see Palladini *et al.* 1979)[5].

An intercellular matrix between these parenchymal cells is completely lacking only in the Acoela, Nemertodermatida, and some Catenulida[6] (see Rieger 1980a, for summary). A matrix is only weakly developed and often lacking in the Proseriata, Rhabdocoela, and Prolecithophora, but usually well developed in the Polycladida, Tricladida, and Lecithoepitheliata (in the latter three orders, the matrix is likely to be collagenous) and in the Macrostomida and Haplopharyngida (in these two orders, the matrix is granular to finly fibrous). In general, the matrix is usually well developed around muscle cells (see Reuter 1977), and smaller representatives have less intercellular matrix developed in the parenchyma (exception: the Nemertodermatida, Acoela; some Macrostomida, e.g. *Paramyozonaria simplex,* see Rieger & Tyler 1974).

Digestive system

Recently Doe (1981) undertook a comparative fine-structural study of the pharynx simplex in the Catenulida, Acoela, Haplopharyngida, and Macrostomida. Based on the presence of a pharyngeal nerve ring, of a single postoral commissure, of sensory cells with variable numbers of cilia and without microvillar collars and because of a distinct gland ring with a special epithelial transition zone, Doe (1981) was able to define the new pharynx simplex coronatus, which is confined to the orders Haplopharyngida and Macrostomida. His data further support a possible multiple origin of the pharynx simplex in the Acoela from a simple connection of the gastrodermis and epidermis as is seen in the Nemertodermatida (own unpublished observation). The pharynx simplex of the Catenulida and the Acoela studied by Doe (1981) illustrates considerable within-group variability. The similarities of the ciliated pharynx simplex of the catenulid *Retronectes atypica* and the pharynx simplex coronatus are presently seen as functional convergences according to Doe (1981). His data show also that the pharynx simplex is not a simple invagination of the body wall but that on the caudal and lateral mouth-margins of a ventrally located pharynx simplex, reorientation of the epidermal cells must occur if the inward beat of the cilia is to be established for the whole pharynx tube. Within the level of the pharynx simplex, the pharynx simplex coronatus should be considered of primary importance in the evolution of the higher pharynx-types. However, electron microscopical studies of higher pharynx-types are generally few and restricted to the pharynx plicatus of some Tricladida (e.g. Bowen & Ryder 1973; Farnesi & Pascolini 1973; Pascolini & Gargiulo 1974; Kishida & Asai 1977) and the pharynx bulbosus of a few Rhabdocoela (Reuter 1975; Holt & Mettrick 1975).

As early as 1967, ultrastructural studies have supported the hypothesis of a multinucleated organization of the digestive parenchyma of the Acoela, but these data have not been widely discussed until recently (see Smith 1981). A cellular organization of the intestine has been recently confirmed at the ultrastructural level for the closely related Nemertodermatida (Tyler & Rieger 1977); but the tendency of the loss of gut lumen (see Tyler & Rieger 1977, Fig. 8) and of the occurrence of complex cell interdigitation has been stressed in this group.

Little ultrastructural data are yet available on the intestinal structure of the other flatworm orders (e.g. see Doe, in review, for Catenulida, Haplopharyngida and Macrostomida; Bowen *et al.* 1974, for Tricladida; Holt & Metrick 1975, for a Rhabdocoel).

Whereas Doe (1981) mentions that in the Haplopharyngida and Macrostomida the regular gastrodermal cells bear, in addition to the cilia, microvilli which are more numerous and longer (1–2 μm) than the microvilli on the epidermis (8–28/μm²), Bowen *et al.* (1974) state that intestinal microvilli in *Polycelis* are restricted to the apical cell margins and Holt & Mettrick (1975) confirm the total lack of microvilli and cilia in *Syndesmis*. The rather uniform occurrence of microvilli on the epidermis, therefore appears in sharp contrast to the much more variable occurrence and distribution of microvilli on the gastrodermal cells.

Reproductive system

The fine structure of the male gametes have been investigated in all orders, except for the Lecithoepitheliata and Haplopharyngida (see Hendelberg 1977; Tyler & Rieger 1977; Rieger 1978). The biflagellate spermatozoa of the neoophoran orders Proseriata, Tricladida and Rhabdocoela show a characteristic '9 + 1' organization of the flagellar axoneme which agrees well with the 9 + 1 axonemal pattern in the order Polycladida. No indications of flagella or basal bodies have yet been found in the two prolecithophoran species studied by electron microscopy (see also Ehlers 1981). This is further substantiated by electron micrographs of spermatozoa in the bursal organ of the peculiar prolecithophoran *Acanthiella* (Fig. 16). Characteristic of all three prolecithophoran species is the presence of elaborated membrane-bounded cisternae or alveoli in the middle part of the spermatozoon. Flagella are also lacking in spermatozoa of the orders Macrostomida and Haplopharyngida (Hendelberg 1977; unpublished EM data on various Macrostomida and *Haplopharynx*, courtesy Dr. S. Tyler and Dr. D. Doe). There is no evidence yet that the peculiar lateral bristles seen in some freshwater macrostomids (Bedini & Papi 1969) are actually derived flagella. The spermatozoa of the Acoela are biflagellate but their axonemal structure is variable (9 + 2, 9 + 0, 9 + 1). At least in some species the 9 + 1 axonemal pattern of the Acoela does not correspond to the 9 + 1 pattern found in the neoophoran orders and the Polycladida. Negative staining techniques to reveal more clearly the 9 + 1 axonemal pattern in the Acoela would be desirable for better comparison (e.g. see Thomas 1975). The most primitive spermatozoon in the Turbellaria is found in the order Nemertodermatida, which have a uniflagellate spermatozoon divided into the characteristic regions of head, middle piece and tail, considered to be a primitive feature of metazoan spermatozoa (Tyler & Rieger 1977). The most aberrant spermatozoa in Turbellaria are those of the order Catenulida, where multiple, short ciliary structures are formed during spermatogenesis in at least one species of the genus *Retronectes* (Rieger 1978).

The fine structure of the female gametes has been studied in greater depth only in the Polycladida (Boyer 1972; Domenici *et al.* 1975) and the Tricladida (Gremigni 1974, 1979), but some preliminary data are available for immature oocytes of the Acoela (Henley 1974) and the eggshell in the Rhabdocoela (Domenici & Gremigni 1977). In the endolecithal egg of the Polycladida the formation of cortical granules (eggshell globules)[7] preceeds the yolk synthesis. This is similar to the light microscopical findings in the entolecithal eggs of the Macrostomida (Rieger 1971). The cortical granules of the Polycladida consist of a dense center, which is asymmetrically surrounded by a diffuse outer rim, again a feature similar to the 'eggshell globules' in the Macrostomida (unpublished light histological data). Very similar inclusions have been noticed in the ooplasma of the fertilized ectolecithal egg of *Acanthiella* (Fig. 17): On the basis of the presence of the male pronucleus, they are here interpreted as eggshell globules. Cortical granules are also found in the ectolecithal eggs of the triclad families Planariidae and Dendrocoeliidae while autosynthetic yolk is present – probably as a derived character – in the Dugesiidae (Gremigni 1979).

Our knowledge of the ultrastructure of the other parts of the reproductive system is poor. Only few papers have been published on the structure of the wall of the gonads and the gonoducts in the Tricladida (e.g. Franquinet & Lender 1973; Schilt

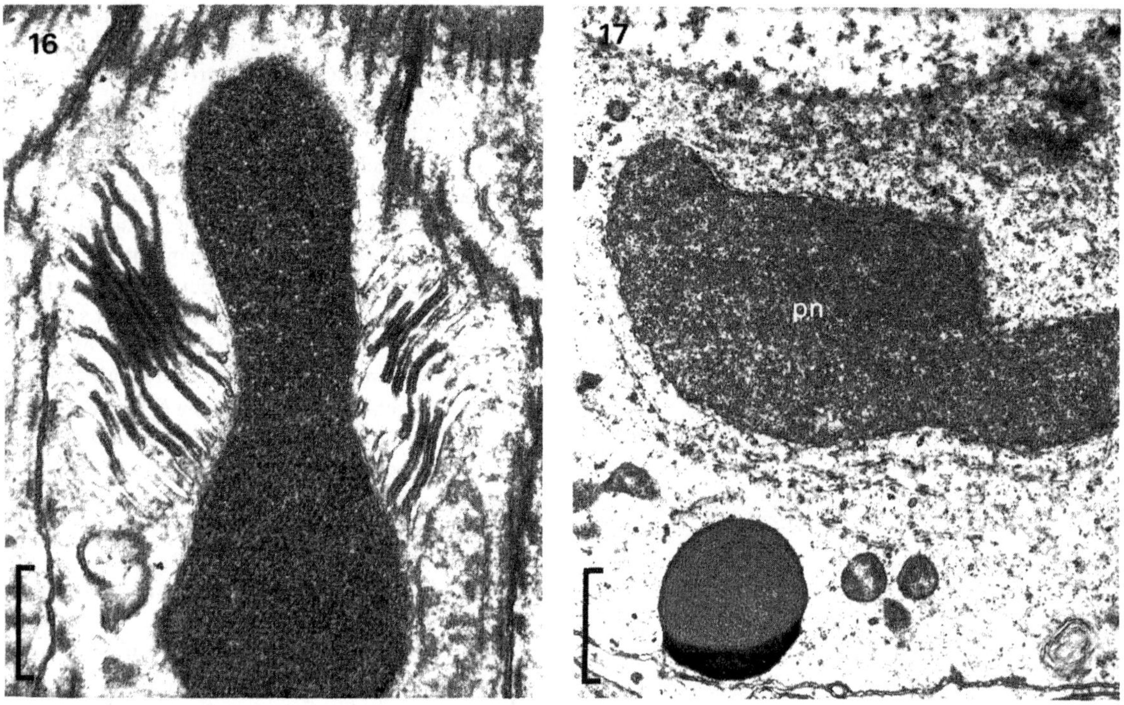

Figs. 16–17. Spermatozoon (16) and portion of oocyte with male pronucleus (pn) and an eggshell globule below the germinal vesicle (17) of *Acanthiella chaetonotoides.* The spermatozoon is seen in an oblique cross section through the middle region (see Rieger and Sterrer, 1975, Fig. 20a). Scale for (16): 0.5 μm. Scale for (17): 1.0 μm. For Methods see Rieger & Sterrer (1975).

1976; but see Martens & Schockaert 1981 for Proseriata). Studies on the 'cuticular' hard structures of the male and female genital organs confirm so far the lack of true cuticular structures (see Doe 1977). The penis stylets both in the Acoela and Proseriata appear to be formed intracellulary (Mainitz 1977; Ehlers & Ehlers 1980; see Lanfranchi 1978, Fig. 4).[8] This seems also to be the case for the bursal mouthpiece in the Acoela (see Henley 1974, Fig. 4; unpublished observations, courtesy J. Smith).

Conclusions

Electron microscopy has changed our understanding of turbellarian morphology generally in two ways: either through revealing new, yet unseen structures, or through clarifying the relative position of structures and their substructures to a greater satisfaction than light microscopy (Tyler 1979b). Examples of first discoveries of new, yet unseen structures occur at a variety of organizational levels: *cell organelles* [shelfed cilia (Tyler 1973); ultrarhabdites (Bedini & Papi 1974); microvillar circle of epidermal receptors (Lyons 1973; Bedini *et al.* 1973); 9 + 1 axonemal pattern in sperm flagella (Klima 1961)]; *complex cell specializations* [weir of the cyrtocyte (Kümmel 1962)]. *new cell types* [epidermal stem cells in Catenulida (Moraczewski 1977), Haplopharyngida (Doe MS), and Tricladida (Hori 1978)]; *microorgans* [duogland adhesive system (Tyler 1976)].

Examples of the first clarification of the relative position of structures and substructures likewise range from the subcellular level all the way to the level of organs: *cell organelles* [rhabdite substructure (Kelbetz 1962); structure and spatial relationship of ciliary rootlets (Klima 1961; Dorey 1965); bilayered structure of the epidermal basement membrane in Neoophora (Török & Röhlich 1959; Pedersen 1961)]; *cells and tissues* [special structure and spatial relationship of epidermal cells (Bedini & Papi 1974), of the parenchymal cells (Pedersen

1961, 1966), and of the digestive parenchyma of Acoela (Klima 1967; Kozloff 1972)]; *organs* [special structure of the pharynx simplex (Doe 1981); caudal glandular-sensory organ in Macrostomida (Rieger & Tyler 1974)].

In applying the new information towards a better understanding of turbellarian systematics, special attention is needed for identifying convergences and parallel evolution, since the number of possible structural solutions to specific functional problems appears to be lower at the ultrastructural level than at the light microscopical level (see Rieger & Tyler 1979, for discussion). However, with the use of the appropriate systematic methodology, fine structural features can be successfully used – together with all other characteristics – to reconstruct the phylogeny and improve the presently used classification of the Turbellaria. Newly discovered ultrastructural ties between the turbellarian orders are seen between the neoophoran orders Proseriata, Tricladida, Rhabdocoela and the order Polycladida [axonemal pattern of sperm flagella (Hendelberg 1965); construction of adhesive organs (Tyler 1976); construction of the epidermis[9] (this paper)]. The fine structural features of the adhesive system (Tyler, 1976), of the pharynx simplex coronatus (Doe 1981) and of the epidermis (this paper) are seen as a special link between the Haplopharyngida and the Macrostomida. The structure of the collar receptors are interpreted as linking the Proseriata and Rhabdocoela with the Prolecithophora (Ehlers 1977). All neoophoran orders (except the Lecithoepitheliata and the Prolecithophora) appear linked with the Haplopharyngida and the Macrostomida through the organization of the duo gland adhesive system (Tyler 1976). This is supported by the rhabdite substructure which links the Macrostomida with the Polycladida, Proseriata, Tricladida, Rhabdocoela and Temnocephalida (Smith *et al.* MS) and the organization of the weir in the cyrtocyte, which appears to link the Macrostomida with the Trematoda (this paper), a group which is clearly derived from the rhabdocoel evolution. New ultrastructural ties have also been found between the order Acoela and Nemertodermatida (shelfed cilia and rootlet system, Tyler & Rieger 1977). Finally, our knowledge of the fine structure of the Catenulida concerning the protonephridial structure (this paper) and the sperm morphology (Rieger 1978) supports a rather distant relationship of the Catenulida with the rest of the orders.

These ultrastructural findings agree well with Karling's (1974) assessment of the affinities between the turbellarian orders based primarily on light histological data. They support the position of the Polycladida close to the neoophoran orders and they favour Karling's (1974) second alternative, which expresses the early separation of the Catenulida as well as the separation of the Acoela and Nemertodermatida from the rest of the orders.

The new ultrastructural information adds to this the recognition of special ties between the Haplopharyngida and Macrostomida and suggests the existence of three major evolutionary lines within the Turbellaria: (1) the catenulid line, (2) the nemertodermatid-acoel line and (3) the line leading to the remaining orders. Within the third group, a dichotomy leading on one hand to the Haplopharyngida-Macrostomida and on the other hand to the neoophoran orders and the Polycladida is indicated. Among the neoophoran orders the Prolecitophora and especially the Lecithoepitheliata remain of uncertain systematic position because we know at present of only few special ties which unite these two groups with the other orders.

Acknowledgment

I would like to acknowledge first the free and open exchange of ideas and material with Dr. Seth Tyler, Dr. D. Doe, Dr. E. Ruppert, and especially Mr. J. Smith, III. I thank my wife for technical assistance and encouragement. Dr. Seth Tyler, Mr. J. Smith, Dr. M. Fransen and B. Jolley, I thank for various improvements in the final preparations of my manuscript. The line drawings have been executed after my designs by Mrs. Linda Hillowe. The original electron microscopy reported in this paper was carried out in the D.P. Costello Electron Microscope Laboratory of the Department of Zoology, UNC, Chapel Hill. The work was supported by NSF grant GB 42211 (R. M. Rieger, P.I.).

References

Ax, P., 1963. Relationships and Phylogeny of the Turbellaria. In: (E. C. Dougherty, ed.) The Lower Metazoa. Berkeley: Univ. California Press.

Ax, P., 1966. Das chordoide Gewebe als histologisches Lebensformsmerkmal der Sandlückenfauna des Meeres. Naturwiss. Rundsch. 19: 282–289.

Bedini, C., Ferrero, E. & Lanfranchi, A., 1973. The Ultrastructure of Ciliary Sensory Cells in Two Turbellaria Acoela. Tissue & Cell 5: 359–372.

Bedini, C., Ferrero, E. & Lanfranchi, A., 1975. Fine Structural Observations on the Ciliary Receptors in the Epidermis of Three Otoplanid Species (Turbellaria Proseriata). Tissue & Cell 7: 253–266.

Bedini, C., Ferrero, E. & Lanfranchi, A., 1977. Fine-Structural Changes Induced by Circadian Light-Dark Cycles in Photoreceptors of Dalyelliidae (Turbellaria: Rhabdocoela). J. Ultrastruct. Res. 58: 66–77.

Bedini, C. & Lanfranchi, A., 1974. The Fine Structure of Photoreceptors in Two Otoplanid Species (Turbellaria, Proseriata). Z. Morphol. Tiere 77: 175–186.

Bedini, C. & Papi, F., 1969. Peculiar Patterns of Microtubular Organization in Spermatozoa of Lower Turbellaria. In: (B. Baccetti, ed.) Comparative Spermatology, Proc. 1st Int. Symp. New York: Acad. Press.

Bedini, C. & Papi, F., 1974. Fine Structure of the Turbellarian Epidermis. In: (N. W. Riser & M. P. Morse, ed.) Biology of the Turbellaria. New York: McGraw-Hill.

Bouligand, Y., 1972. Twisted Fibrous Arrangements in Biological Materials and Cholesteric Mesophases. Tissue & Cell 4: 189–217.

Bowen, I. D. & Ryder, T. A., 1973. The Fine Structure of the Planarian Polycelis tenuis (Ijima) I. The Pharynx. Protoplasma 78: 223–241.

Bowen, I. D. & Ryder, T. A., 1974. The Fine Structure of the Planarian Polycelis tenuis (Ijima). III. The Epidermis and External Features. Protoplasma 80: 381–392.

Bowen, I. D., Ryder, T. A. & Thompson, J. A., 1974. The Fine Structure of the Planarian Polycelis tenuis Iijima. II. The Intestine and Gastrodermal Phagocytosis. Protoplasma 79: 1–17.

Boyer, B. C., 1972. Ultrastructural Studies of Differentiation in the Oocyte of the Polyclad Turbellarian, Prostheceraeus floridanus. J. Morphol. 136: 273–294.

Brandenburg, J., 1974. The Morphology of the Protonephridia. Fortschr. Zool. 23: 1–15.

Burt, M. D. B. & Bance, G. N., 1981. Ultrastructure of the eye of Urastoma cyprinae (Turbellaria, Alloeocoela). (Abstract) In: Schockaert, E. R. & Ball, I. R., 'Turbellaria', Proc. Third Int. Symp. Hydrobiologia.

Crezee, M., 1975. Monograph of the Solenofilomorphidae (Turbellaria: Acoela). Int. Rev. Gesamten Hydrobiol. 60: 769–845.

Crezee, M. & Tyler, S., 1976. Hesiolicium gen. n. (Turbellaria, Acoela) and Observations on its Ultrastructure. Zool. Scr. 5: 207–216.

de Eguileor, M. & Valvassori, R., 1975. Fine Structure of Flatworm Muscles: Dugesia tigrina (Girard) and Dugesia lugubris (Schmidt). Monit. Zool. Ital. 9: 37–50.

Doe, D., 1977. Fine Structure of 'Cuticular' Structures in Platyhelminthes. Am. Zool. 17: 970.

Doe, D., 1981. Comparative Ultrastructure of the Pharynx Simplex in Turbellaria. Zoomorphologie. 97: 133–193.

Doe, D. A. & Rieger, R. M., 1977. A New Species of the Genus Retronectes (Turbellaria, Catenulida) from the Coast of North Carolina, U.S.A. Mikrofauna des Meeresbodens 66: 1–10.

Domenici, L., Galleni, L. & Gremigni, V., 1975. Electron Microscopical and Cytochemical Study of Egg-Shell Globules in Notoplana alcinoi (Turbellaria, Polycladida). J. Submicrosc. Cytol. 7: 239–247.

Domenici, L. & Gremigni, V., 1977. Fine Structure and Functional Role of the Coverings of the Eggs in Mesostoma ehrenbergii (Focke) (Turbellaria, Neorhabdocoela). Zoomorphologie 88: 247–257.

Dorey, A. E., 1965. The Organization and Replacement of the Epidermis in Acoelous Turbellarians. Q. J. Microsc. Sci. 106: 147–172.

Durand, J. P. & Gourbault, N., 1977. Étude Cytologique des Organes Photorecepteurs de la Planaire Australienne Cura pinguis. Can. J. Zool. 55: 381–390.

Eakin, R. M., 1962. Lines of Evolution of Photoreceptors. J. Gen. Physiol. 46: 359A–360A.

Ehlers, B., 1977. 'Trematoden-artige' Epidermisstrukturen bei einem freilebenden proseriaten Strudelwurm. Acta Zool. Fenn 154: 129–136.

Ehlers, U., 1977. Vergleichende Untersuchungen über Collar-Rezeptoren bei Turbellarian. Acta Zool. Fenn. 154: 137–148.

Ehlers, U., 1981. Fine structure of the giant aflagellate spermatozoon in Pseudostomum quadrioculatum (Leuckart) (Plathelminthes, Prolecithophora). In: Schockaert, E. R. & Ball, I. R., 'Turbellaria', Proc. Third Int. Symp. Hydrobiologia. This volume.

Ehlers, B. & Ehlers, U., 1977a. Die Feinstruktur eines ciliaren Lamellarkörpers bei Parotoplanina geminoducta Ax (Turbellaria; Proseriata). Zoomorphologie 87: 65–72.

Ehlers, B. & Ehlers, U., 1977b. Ultrastruktur pericerebraler Cilienaggregate bei Dicoelandropora atriopapillata Ax und Notocaryoplanella glandulosa Ax (Turbellaria, Proseriata). Zoomorphologie 88: 163–174.

Ehlers, B. & Ehlers, E., 1980. Struktur und Differenzierung penialer Hartgebilde von Carenscoilia bidentata Sopott (Turbellaria, Proseriata). Zoomorphologie 95: 159–167.

Ehlers, U. & Ehlers, B., 1977c. Monociliary Receptors in Interstitial Proseriata and Neorhabdocoela (Turbellaria, Neoophora). Zoomorphologie 86: 197–222.

Farnesi, R. M. & Pascolini, R., (1973). Investigations on the Ultrastructure of the Pharyngeal Glands in Dugesia lugubris S.L. Boll. Zool. 40: 137–147.

Ferrero, E., 1973. A Fine Structure Analysis of the Statocyst in Turbellaria Acoela. Zool. Scr. 2: 5–16.

Franquinet, R. & Lender, T., 1973. Etude Ultrastructurale des Testicules de Polycelis tenuis et Polycelis nigra (Planaires). Evolution des Cellules Germinales males Avant la Spermiogenese. Z. Mikrosk. Anat. Forsch. (Leipz.) 87: 4–22.

Graff, L. von, 1889. Enantia spinifera, der Repräsentant einer neuen Polycladen-Familie. Mitt. Naturwiss. Ver. Stiermark (1889): 1–16.

Gremigni, V., 1974. The Origin and Cytodifferentiation of Germ Cells in Planarians. Boll. Zool. 41: 359–377.

Gremigni, V., 1979. An Ultrastructural Approach to Planarian Taxonomy. Syst. Zool. 28: 345–355.

Gremigni, V., 1981. The problem of cell totipotency, dedifferentiation and transdifferentiation in Turbellaria. In: Schockaert, E. R. & Ball, I. R. 'Turbellaria', Proc. Third Int. Symp. Hydrobiologia. This volume.

Hendelberg, J., 1965. On Different Types of Spermatozoa in Polycladida, Turbellaria. Ark. Zool. 18: 267–304.

Hendelberg, J., 1977. Comparative Morphology of Turbellarian Spermatozoa Studied by Electron Microscopy. Acta. Zool. Fenn. 154: 149–162.

Hendelberg, J., 1981. The system of epidermal ciliary rootles in Turbellaria. (Abstract) In: Schockaert, E. R. & Ball, I. R., 'Turbellaria'. Proc. Third Int. Symp. hydrobiologia. This volume.

Henley, C., 1974. Platyhelminthes (Turbellaria). In: (A. C. Giese & J. S. Pearse, ed.) Reproduction of Marine Invertebrates v.1. Acoelomate and Pseudocoelomate Metazoans. New York: Academic Press.

Holt, P. A. & Mettrick, D. F., 1975. Ultrastructural Studies of the Epidermis and Gastrodermis of Syndesmis franciscana (Turbellaria: Rhabdocoela). Can. J. Zool. 53: 536–649.

Hori, I., 1978. Possible Role of Rhabdite-Forming Cells in Cellular Succession of the Planarian Epidermis. J. Electron Micros. 27: 89–102.

Hori, I., 1979. Structure and Regeneration of the Planarian Basal Lamina: An Ultrastructural Study. Tissue & Cell. 11: 611–621.

Ishii, S., 1980a. The Ultrastructure of the Protonephridial Flame Cell of the Freshwater Planarian Bdellocephala brunnea. Cell Tissue Res. 206: 441–449.

Ishii, S., 1980b. The Ultrastructure of the Protonephridial Tubules of the Freshwater Planarian Bdellocephala brunnea. Cell Tissue Res. 206: 451–458.

Ivanov, V. P., Mamkaev, Y. & Pevzner, ■., 1972. Electron microscopic study of the statocyst in the Acoelan Turbellarian Convoluta convoluta. Eng. Transl. Zh. Evol. Biokhim. Fiziol. 8: 162–168.

Karling, T. G., 1940. Zur Morphologie and Systematik der Alleocoela cumulata and Rhabdocoela lecithophora. Acta. Zool. Fenn. 26: 1–260.

Karling, T., 1974. On the Anatomy and Affinities of the Turbellarian Orders. In: (N. W. Riser & M. P. Morse, ed.) Biol. of the Turbellaria. New York: McGraw-Hill.

Kelbetz, S., 1962. Zum Feinbau der Rhabditen. Naturwissenschaften: 49: 348.

Kishida, Y. & Asai, E., 1977. Les Cristeaux Intranuclearies des Cellules Pharyngiennes chez la Planaire, Dugesia japonica Ichikawa et Kawakatsu. J. Electron Micros. 26: 145–147.

Klima, J., 1959. Das Bild des endoplasmatischen Reticulums von Planaria alpina in Abhängigkeit vom pH-Wert des Fixierungsmittels (I). Protoplasma 51: 415–435.

Klima, J., 1961. Elektronenmikroskopische Studien über die Tricladen (Turbellaria). Protoplasma 54: 101–162.

Klima, J., 1967. Zur Feinstruktur des acoelen Süsswasserturbellars Oligochoerus limnophilus Ax & Dörjes. Ber. Nat-wiss.-Med. Ver. Innsb. 55: 107–124.

Køie, M. & Bresciani, J., 1973. On the Ultrastructure of the Larva of Kronborgia amphipodicola Christensen & Kanneworff 1964 (Tubellaria, Neorhabdocoela) Ophelia 12: 171–203.

Koopowitz, H. & Chien, P., 1974. Ultrastructure of Nerve Plexus in Flatworms: I. Peripheral Organization. Cell Tiss. Res. 155: 337–351.

Kozloff, E., 1972. Selection of Food, Feeding, and Physical Aspects of Digestion in the Acoel Turbellarian Otocelis luteola. Trans. Amer. Micros. Soc. 91: 556–565.

Kümmel, G., 1962. Zwei neue Formen von Cyrtocyten. Vergleich der bisher bekannten Cyrtocyten und Erörterung des Begriffes 'Zelltyp'. Z. Zellforsch. Mikrosk. Anat. 57: 172–201.

Lanfranchi, A., 1978. Morphology and Taxonomy of Two New Otoplanids (Turbellaria, Proseriata) from the Ligurian Sea. Zool. Scr. 7: 249–254.

Lanfranchi, A., Bedini, B. & Ferrero, E., 1981. The Ultrastructure of the eyes in larval and adult Polyclads (Turbellaria). In: Schockaert, E. R. & Ball, I. R. 'Turbellaria', Proc. Third Int. Symp. Hydrobiologia. This volume.

Lyons, K. M., 1973. The Epidermis and Sense Organs of the Monogenea and Some Related Groups. Adv. Parasitol. 11: 193–232.

MacRae, E. K., 1966. Fine Structure of the Photoreceptors in a Marine Flatworm. Z. Zellforsch. Mikrosk. Ahat. 75: 469–484.

MacRae, E. K., 1967. The Fine Structure of Sensory Receptor Processes in the Auricular Epithelium of the Planarian, Dugesia tigrina. Z. Zellforsch. Mikrosk. Anat. 82: 479–494.

McKanna, J. A., 1968. Fine Structure of the Protonephridial System in Planaria. I. Flame Cells. Z. Zellforsch. Mikrosk. Anat. 92: 509–523.

Mainitz, M., 1977. The Fine Structure of the Stylet Apparatus in Gnathostomulida Scleroperalia and its Relationship to Turbellarian Stylets. Acta Zool. Fenn. 154: 163–174.

Martens, E. E. & Schockaert, E. R., 1981. Observations on the ultrastructure of the copulatory organ of Archilopsis unipunctata (Fabricius, 1826). In: Schockaert, E. R. & Ball, I. R., 'Turbellaria', Proc. Third Int. Symp. Hydrobiologia. This volume.

Martin, G., 1978. A New Function of Rhabdites: Mucous Production for Ciliary Gliding. Zoomorphologie 91: 235–248.

Moraczewski, J., 1977. Asexual Reproduction and Regeneration of Catenula (Turbellaria, Archoophora). Zoomorphologie 88: 65–80.

Moraczewski, J. & Czubaj, A., 1974. Muscle Ultrastructure in Catenula lemnae (Dug.) (Turbellaria, Archoophora). J. Submicrosc. Cytol. 6: 29–38.

Moraczewski, J., Czubaj, H. & Bakowska, J., 1977. Organization and Ultrastructure of the Nervous System in Catenulida (Turbellaria). Zoomorphologie 87: 87–95.

Nicaise, G., 1973. The Gliointerstitial System of Molluscs. Int. Rev. Cytol. 34: 251–332.

Palladini, G., Medolago-Albani, L., Margotta, V., Conforti, A. & Carolei, A., 1979. The Pigmentary System of Planaria. I. Morphology. Cell Tissue Res. 199: 197–202.

Palmberg, I., Reuter, M. & Wikgren, M., 1980. Ultrastructure of Epidermal Eyespots of Microstomum lineare (Turbellaria, Microstomida). Cell Tissue Res. 210: 21–32.

Pascolini, R. & Gargiulo, A., 1974. Relationships between Neurosecretion and Glandular Secretion on the Pharynx of Dugesia lugubris S.L. Boll. Zool. 41: 43–46.

Pedersen, K. J., 1959a. Some Features of the Fine Structure and Histochemistry of Planarian Subepidermal Gland Cells. Z. Zellforsch. Mikrosk. Anat. 50: 121–142.

Pedersen, K. J., 1959b. Cytological Studies on the Planarian Neoblast. Z. Zellforsch. Mikrosk. Anat. 50: 799–817.

Pedersen, K. J., 1961a. Studies over ferskvandstricladernes cytologi (with an English summary). Copenhagen: Munksgaard.

Pedersen, K. J., 1961b. Studies on the Nature of Planarian Connective Tissue. Z. Zellforsch. Mikrosk. Anat. 53: 569–608.

Pedersen, K. J., 1964. The Cellular Organization of Convoluta convulata, an Acoel Turbellarian: a Cytological, Histochemical and Fine Structural Study. Z. Zellforsch. Mikrosk. Anat. 64: 655–687.

Pedersen, K. J., 1966. The Organization of the Connective Tissue of Discocoelides langi (Turbellaria, Polyclada). Z. Zellforsch. Mikroskop. Anat. 71: 94–117.

Pullen, E. W., 1957. A Histological Study of Stenostomum virginianum. J. Morphol. 101: 579–621.

Reisinger, E., 1924. Die Gattung Rhynchoscolex. Z. Morphol. Oekol. Tiere 1: 1–37.

Reisinger, E., 1968. Xenoprorhynchus, ein Modellfall für progressiven Funktionswechsel. Z. Zool. Syst. Evolutionsforsch. 6: 1–55.

Reisinger, E. & Kelbetz, S., 1964. Feinbau und Entladungsmechanismus der Rhabditen. Z. Wiss. Mikrosk. Mikrosk. Tech. 65: 472–508.

Reuter, M., 1975. Ultrastructure of the Epithelium and the Sensory Receptors in the Body Wall, the Proboscis, and the Pharynx of Gyratrix hermaphroditus (Turbellaria, Rhabdoceola). Zool. Scripta 4: 191–204.

Reuter, M., 1977. Ultrastructure of the Stylet Protractor Muscle in Gyratrix hermaphroditus (Turbellaria, Rhabdocoela). Acta. Zool. (Stockh.) 58: 179–184.

Reuter, M., 1978. Scanning and Transmission Electron Microscopic Observations on Surface Structures of Three Turbellarian Species. Zool. Scr. 7: 5–11.

Reuter, M. & Lindroos, P., 1979a. The Ultrastructure of the Nervous System of Gyratrix hermaphroditus (Turbellaria, Rhabdocoela) I. The Brain. Acta Zool. (Stockh.) 60: 139–152.

Reuter, M. & Lindroos, P., 1979b. The Ultrastructure of the Nervous System of Gyratrix hermaphroditus (Turbellaria, Rhabdocoela) II. The Peripheral Nervous System and the Synapses. Acta Zool. (Stockh.) 60: 153–161.

Rieger, R. M., 1971. Die Turbellarienfamilie Dolichomacrostomidae nov. fam. (Macrostomidae) I. Teil. Vorbemerkungen und Karlingiiae nov. Subfam. I. Zool. Jahrb. Abt. Syst. Oekol. Geogr. Tiere 98: 236–314.

Rieger, R. M., 1976. Monociliated Epidermal Cells in Gastrotrichia: Significance for Early Metazoan Evolution. Z. Zool. Syst. Evolutionsforsch. 14: 198–226.

Rieger, R. M., 1978. Multiple Ciliary Structures in Developing Spermatozoa of Marine Catenulida (Turbellaria). Zoomorphologie 89: 229–236.

Rieger, R. M., 1980a. A new Group of Interstitial Worms, Lobatocerebridae nov. fam. (Annelida) and its Significance for Metazoan Phylogeny. Zoomorphologie 95: 41–84.

Rieger, R. M., 1980b. Fine structure of the Body Wall, the Nervous System, and the Digestive System of the Lobatocerebridae Rieger (Annelida) and Remarks to the Organization of the Glio-interstitial Systems in Annelida. J. Morphol. (in press).

Rieger, R. & Doe, D., 1975. The Proboscis Armature of the Turbellaria Kalyptorhynchia. A Derivative of the Basement Lamina? Zool. Scr. 4: 25–32.

Rieger, R. & Rieger, G., 1976. Fine Structure of the Archiannelid Cuticle, and Remarks on the Evolution of the Cuticle within the Spiralia. Acta Zool. (Stockh.) 57: 53–68.

Rieger, R. & Sterrer, W., 1975. New Spicular Skeletons in Turbellaria, and the Occurrence of Spicules in Marine Meiofauna. Z. Zool. Syst. Evolutionsforsch. 13: 207–248.

Rieger, R. M. & Tyler, S., 1974. A New Glandular Sensory Organ in Interstitial Macrostomida I. Ultrastructure, Mikrofauna des Meeresbodens 42: 1–41.

Rieger, R. & Tyler, S., 1979. The Homology Theorem in Ultrastructural Research. Am. Zool. 19: 655–664.

Ruppert, E. E., 1978. A Review of Metamorphosis of Turbellarian Larvae. In (F. S. Chia & M. Rice, ed.) Settlement and Metamorphosis of Marine Invertebrate Larvae. New York: Elsevier.

Schilt, J., 1976. Aspect Ultrastructural de l'Ovovitelloducte de Dugesia lugubris O. Schmidt (Turbellarie, Triclade). Bull. Soc. Zool. Fr. 101: 527–533.

Schockaert, E. R. & Bedini, C., 1977. Observations on the Ultrastructure of the Proboscis Epithelia in Polycystis naegelii Kolliker (Turbellaria Eukalyptorhynchia) and Some Associated Structures. Acta Zool. Fenn. 154: 175–191.

Smales, I. R. & Blankespoor, H. D., 1978. The Epidermis and Sensory Organs of Dugesia tigrina (Turbellaria: Tricladida) A Scanning Electron Microscopic Study. Cell Tissue Res. 193: 35–40.

Skaer, R. J., 1961. Some Aspects of the Cytology of Polycelis nigra. Q.J. Microsc. Sci. 102: 295–317.

Smith, J. P. S., 1981. Fine-structural observations on the central parenchyma in Convoluta sp. In: Schockaert, E. R. & Ball, I. R., 'Turbellaria', Proc. Third Int. Symp. Hydrobiologia. This volume.

Smith, J., Tyler, S., Thomas, M. B. & Rieger, R. M. The Morphology of Turbellarian Rhabdites: Phylogenetic Implications. Trans. Am. Microsc. Soc. (MS submitted)

Soltynska, M. S., Mroczka, B. & Moraczewski, J., 1976. Ultrastructure of Epidermis in Turbellaria from the Family Catenulida (Turbellaria). J. Submicrosc. Cytol. 8: 293–301.

Sopott-Ehlers, B., 1979. Ultrastruktur der Haftapparate von Nematoplana coelogynoporoides (Turbellaria, Proseriata). Helgol. Wiss. Meeresunters. 32: 365–373.

Sterrer, W. & Rieger, R., 1974. Retronectidae- a New Cosmopolitan Marine Family of Catenulida (Turbellaria). In: (N. W. Riser & M. P. Morse, ed.) Biology of the Turbellaria. New York. McGraw-Hill.

Storch, V., 1973. Vergleichende elektronenmikroskopische Untersuchungen über Receptoren von Wirbellosen (Nemertinen, Turbellarien, Molluscen, Anneliden, Aschelminthen). Verh. Dtsch. Zool. Ges. 66: 61–65.

Thomas, M. B., 1975. The Structure of the 9 + 1 Axonemal Core as Revealed by Treatment with Trypsin. J. Ultrast. Res. 52: 409–422.

Török, L. J. & Röhlich, P., 1959. Contributions to the Fine Structure of the Epidermis of Dugesia lugubris. O. Schm. Acta. Biol. Acad. Sci. Hung. 10: 23–48.

Turner, R. S., 1946. Observations on the Central Nervous System of Leptoplana acticola. J. Comp. Neurol. 85: 53–65.

Tyler, S., 1973. An Adhesive Function for Modified Cilia in an Interstitial Turbellarian. Acta Zoologica 54: 139–151.

Tyler, S., 1976. Comparative Ultrastructure of Adhesive Systems in the Turbellaria. Zoomorphologie 84: 1–76.

Tyler, S., 1977. Ultrastructure and Systematics: an Example from Turbellarian Adhesive Organs. Mikrofauna des Meeresbodens 61: 271–286.

Tyler, S., 1979a. Distinctive Features of Cilia in Metazoans and their Significance for Systematics. Tissue & Cell 11: 385–400.

Tyler, S., 1979b. Contributions of Electron Microscopy to Systematics and Phylogeny: Introduction to the Symposium. Am. Zool. 19: 541–543.

Tyler, S. & Rieger, R., 1977. Ultrastructural Evidence for the Systematic Position of the Nemertodermatida (Turbellaria). Acta Zool. Fenn. 154: 193–207.

Williams, J. B., 1977. Studies on the Epidermis of Temnocephala II. Epidermal Sensory Receptors of Temnocephala novae-zelandiae. Aust. J. Zool. 25: 187–191.

Williams, J. B., 1978. Studies on the Epidermis of Temnocephala III. Scanning Electron Microscope Study of the Epidermal Surface of Temnocephala dendyi. Aust. J. Zool. 26: 217–224.

Wilson, R. A. & Webster, A., 1974. Protonephridia. Biol. Rev. Camb. Philos. Soc. 49: 127–160.

Wolken, J., 1958. Studies of Photoreceptor Structures. Ann. N.Y. Acad. Sci. 74: 164–181.

FOOTNOTES

[1] An outer limiting layer often appears to be missing in members of the Proseriata (e.g. *Otoplana truncaspina, Notocayoturbella bigermaria, Parotoplana truncaspina,* see Bedini & Papi 1974, Figs. 11a, b & 15; Ehlers & Ehlers 1977c, Fig. 5a–c).

The presence of two distinct layers in the basement membrane has been noted by light microscopists (see Karling 1940, p. 139–141). Karling's assertion of a possible dual origin of the basement membrane has not yet been confirmed by electron microscopical studies, but recent ultrastructural studies of basement membrane regeneration point to the importance of the interactions between epidermal cells and muscel cells in basement membrane formation (e.g. Hori 1979). Ultrastructural data of Polyclad larvae (Ruppert 1978) and of the larva of *Kronborgia* (Køi & Bresciani 1973) show the basement membrane as an undevided, homogeneous structure early in development. Such a simple, uniform basement membrane has also been seen in adult animals of a few smaller neoophoran species (e.g. *Acmostomum dioicum,* see Karling 1940, *Messoplana falcata,* see Tyler 1976, *Litucivis serpens,* see Ehlers & Ehlers 1977c).

[2] The only possible epidermal hard structure which appears to be derived from true cuticularization is seen in the marginal spines of *Enantia spinifera,* Graff, 1889.

[3] The peculiar cerebral photoreceptors in the Otoplanidae are possibly not correctly interpreted by Bedini & Lanfranchi, (1974, Fig. 3). From their Figs. 8 & 9 it appears that the highly lamellated photoreceptor part of the sensory cell is surrounded (like a tightly fitting cap) by a second cell type (nuclei to be seen in upper part of Fig. 8). This cell would correspond in position to the pigment cells in the other known cerebral pigment cup eyes.

[4] Ruppert (1978, Fig. 5d) illustrates a single cross section through the flame bulb of a polyclad larva with four cilia. Only four cilia are seen also in protonephridial ducts of Haplopharynx (unpublished data, courtesy Dr. Seth Tyler).

[5] Rieger (1980a, p. 66) mentions the occurrence of cells with numerous electron-dense granules immediately below the basement membrane in *Adenorhynchus, Plagiostomum* and Otoplanids. These cells appear similar to the epidermal gliointerstitial cell system of *Lobatocerebrum.* They also could be interpreted as pigment cells. The similarity of pigment cells with cells of the gliointerstitial system has been discussed earlier (Nicaise 1973).

[6] Rieger (1980a, p. 67) states incorrectly that Catenulida also lack a basement membrane (see p. 216).

[7] The functional significance of these granules is not clear because of conflicting evidence from *Prostheceraeus* (Boyer 1972) and *Notoplana* (Domenici *et al.* 1975).

[8] The nature of the stylet sheath in *Gyratrix* is not obvious in Fig. 1 of Reuter (1977).

[9] The epidermal construction also ties the Lecithoepitheliata and Prolecithophora with the other neoopheran orders.

Development of cilia in embryos of the turbellarian Macrostomum

Seth Tyler
Electron Microscopy Center and Dept. of Zoology, Washington State Univ., Pullman, and Dept. of Anatomy Dalhousie Univ., Halifax, Nova Scotia, Canada

Keywords: Turbellaria, *Macrostomum*, ultrastructure, ciliogenesis

Abstract

Electron microscopy of *Macrostomum hystricinum* raised in culture shows that ciliogenesis in the worm's epidermal blastomeres begins in embryos 39–41 h old with kinetosomal and *de novo* genesis of presumptive basal bodies, which are morphologically distinguishable from centrioles of the mitotic apparatus, and proceeds by the migration of basal bodies to the apical plasma membrane of the cells and their production there of ciliary axonemes by an age of 51–53 h when the bastomeres emerge between yolk cells on the embryo's surface. Ciliogenesis continues throughout development with the addition of cilia virtually one by one to the expanding epidermal cells' surfaces. At no time in ciliogenesis are stages seen that might show derivation of these multiciliated cells from the primitive monociliated cell type presumably present in the ancestors of the Turbellaria.

Introduction

The form of ciliation on epidermal cells has recently been cited as a character that may hold a key to our understanding of metazoan evolution. The initial spur to this consideration came from Nørrevang (1964) and Nørrevang & Wingstrand's (1970) studies of monociliated cells in epithelia of an enteropneust and of echinoderms and their comparison of these with choanocytes of sponges. Lyons (1973a,b) later pointed to the occurrence of 'collar cells' in a cnidarian and of a 'possible collar cell' (actually probably a sensory cell) in a turbellarian and discussed the significance of these findings for the planuloid-acoeloid theory of metazoan evolution. The definitive study on the phylogenetic significance of monociliated cells is that of Rieger (1976) whose extensive comparative studies give strong support to the theory that a monociliated epidermal cell type is ancestral for the Metazoa. Rieger wisely limits consideration in this theory to epidermal cells that are used in locomotion or in creating water currents, and excludes sensory cells, protonephridia, and other monociliated cell types as being specialized types whose relation to epidermal cells is uncertain. According to Rieger, monociliated epidermal cells are found in the Placozoa, Porifera, Cnidaria, Gnathostomulida, Gastrotricha, and the archicoelomates, whereas a derived multiciliated cell type is to be found in at least some gastrotrichs, ectoprocts, entoprocts, and bogonophorans, and is especially characteristic of the spiralian phyla.

Although the monociliated cell type is clearly seen as being ancestral to the multiciliated cell type in the Metazoa, the course of evolution that transformed that monociliated cell into the multiciliated cell is difficult to envision. If the Metazoa is monophyletic, there must be such a transformation. One of the prime means by which this transformation might be investigated is through studies of the development of these cell types, to see whether, for instance, the development of multiciliated cells passes through stages that resemble the

Hydrobiologia 84, 231–239 (1981). 0018-8158/81/0843-0231/$01.80.

monociliated condition. Such a development would not necessarily be interpreted as a recapitulation in the Haeckelian sense, but simply as a source of morphological characters that could be cited in a transformation sequence for demonstrating homology of the two cell types (homology criterion two of Rieger & Tyler 1979).

In a sense, such a transformation has already been found through studies of ciliogenesis in vertebrate epithelia. The same epithelial cell in chicken trachea (Kalnins & Porter 1969) or in monkey oviduct (Anderson & Brenner 1971), for instance, produces a single rudimentary cilium directly from the cell's diplosome early in stages of ciliogenesis and then later becomes multiciliated. Some other vertebrate epithelia may show similar transitions during ciliogenesis, though in these epithelia, mono- and multiciliated conditions usually occur separately in two distinct populations of cells of the same cell type (Sorokin 1968; Ludwig & Metzger 1977). Ciliogenesis in still other vertebrate epithelia gives rise to strictly multiciliated cells (Steinman 1968; Dirksen 1971; Sorokin 1968; Chang *et al.* 1979). The resemblance of the early monociliated stage in vertebrate epithelia to the presumably primitive monociliated cell type goes farther than simply being a matter of having a single cilium; that cilium usually arises in a cup-shaped depression in the cell surface and has a diplosomal basal body with or without a rootlet something like epithelial cells in a placozoan or echinoderm.

If a similar plasticity in ciliogenesis could be found in multiciliated cells of a spiralian, the evolutionary step to the derived multiciliated epithelial cell type characteristic of the major phylogenetic line of the Spiralia could be better explained. To this end, I have studied the process of ciliogenesis in embryos of two species of turbellarians, one an acoel, *Archaphanostoma*, and the other a macrostomid, *Macrostomum*. Either the Acoela (and closely related Nemertodermatida) or the Macrostomida have been considered to be most like a turbellarian archetype by various speculators, or even to be most like the ancestral bilaterian (see the most recent discussions of these positions by Ax 1963; Karling 1974; Salvini-Plawen 1978), and the conservative nature of their morphology makes them likely candidates to show links with a monociliated ancestor. Ciliogenesis in the acoel is described in a separate paper (Tyler, in preparation); this paper concentrates on the major events of ciliogenesis in *Macrostomum*.

Materials and methods

All of the embryos used in this study were raised in culture following methods of Apelt (1969). The cultures were started from a stock of *Macrostomum hystricinum* that was collected at the mouth of the Lawrencetown River, Nova Scotia, in surface sand at the edges of small tidal pools. Animals were placed in 60-mm diameter petri dishes with a culture of an unidentified species of the diatom *Nitzschia* isolated from sand at the Lawrencetown collection site and grown in Guillard's f/2 medium (Guillard & Ryther 1962). The culture dishes were held at 18–20 °C on a 16 h-light/8 h-dark cycle.

Eggs laid by the adult worms were monitored for appropriate stages of development and fixed at specific stages for electron microscopy. Because the shell of the egg proved to be virtually impermeable to many of the reagents used in processing, the eggs were punctured with an electrochemically sharpened tungsten needle immediately after they were placed in the primary fixative which was 2% paraformaldehyde + 2.5% glutaraldehyde in 0.1 M cacodylate buffer (pH 7.4). Following a fixation period of 2–3 h, the eggs were rinsed in buffer 2 h or overnight, fixed 1 1/2 – 2 h in cacodylate-buffered 1.5% OsO_4, stained *en bloc* with uranyl acetate, dehydrated in an ethanol series, and embedded in Epon 812 or an Epon/Araldite mixture using propylene oxide for infiltration. Serial sections from the embedment blocks were mounted on slot grids with a composite parlodion/carbon film or, in some instances, on standard mesh grids, and stained with aqueous uranyl acetate and lead citrate.

The stages of the ciliogenesis reported on here were found in embryos of age 32–35 h, 39–41 h, 51–53 h, and in hatchlings of approximately 100 h age (h after laying). Two specimens from each age group were examined.

Results

32-to-35-h embryos

Up to the age of at least 32–35 h, the embryos completely lack cilia and lack even structures that could be interpreted as precursors to cilia formation. The entire embryo is clothed in a 'yolk mantle' ('Dottermantel': Seilern-Aspang 1958), a 1–5 μm thick monolayer of 4–8 flattened yolk-rich blastomeres. Blastomeres that will eventually give rise to the epidermis lie beneath this mantle. Cells of the yolk mantle have a pale (i.e., relatively ribosome-poor) cytoplasm and relatively larger nuclei than other blastomeres of the embryo, and they contain relatively more of the embryo's spherical 0.5–3.3 μm diameter yolk granules and lipid droplets. The external surface of the yolk cells, that facing the egg shell, is smooth, devoid of microvilli and other such projections.

39-to-41-h embryos – presumptive basal body formation

By 39–41 h, the blastomeres that will eventually give rise to epidermis are recognizable by the presence in their cytoplasm of numerous centrioles (Fig. 1A,C). These blastomeres lie immediately beneath the yolk mantle, which has thinned over them. They are not epithelial at this stage, that is they are not arranged in a discrete layer and show no differentation of their outermost sides as they do in older embryos. There are indications that such a differentiation is in progress, however, in that small bundles of tonofiliaments are present, as well as dense granules near dictyosomes of the Golgi apparatus, which are presumably developing ultrarhabdites. Rather than completely encompassing the embryo, the epidermal blastomeres are grouped in what appear to be two blastema-like masses on one side of the embryo, presumably the ventral side as Bresslau (1904) has described for *Mesostoma.*

The numerous centrioles in the epidermal blastomeres are clearly presumptive basal bodies, morphologically distinguishable from centrioles of the mitotic apparatus. In comparison with mitotic centrioles, presumptive basal bodies are shorter (ca. 0.2 μm long as compared to ca. 0.3 μm) long, and appear in longitudinal section to have sides of unequal length and a slightly narrower distal end than proximal end, whereas centrioles in longitudinal section have parallel sides of equal length (compare Fig. 1B with centriole c_2 in 1C). Differences in the diameter between proximal and distal ends of the presumptive basal bodies may be related to a change in tilt of the axial plane of triplet microtubules in the body as can be seen in comparing Fig. 1E with 1F. In Fig. 1E the angle subtended from the tangential by individual triplets is less than the angle in triplets of Fig. 1F, so in all likelihood the triplets are twisted. Such a twisting in triplets of basal bodies in monkey oviduct also gives rise to differences in proximal *vs.* distal diameters (Anderson 1972). Further characteristics that distinguish presumptive basal bodies from mitotic centrioles in *Macrostomum* are the presence in basal bodies of a dense material filling the lumen of subfiber A (Fig. 1E,F) and the presence of a thin dense plate-like structure at the proximal end (arrow in Fig. 1B). The distal ends of basal bodies in addition appear to have an incomplete plate-like structure that closes that end. Centrioles of the mitotic apparatus have a cartwheel-like structure at their proximal ends as is typical of centrioles; presumptive basal bodies apparently lack this cartwheel.

Though presumptive basal bodies are morphologically distinguishable from centrioles of the mitotic apparatus, I will still refer to them as centrioles as this is the convention in the literature. All centrioles in Fig. 1 are presumptive basal bodies with the probable exception of that one labeled c_2 in Fig. 1C.

Centrioles appear in clusters in the epidermal blastomeres (Fig. 1C) evidently as a result of generation of new centrioles near fully formed ones. The clusters contain both fully formed centrioles (presumptive basal bodies and, more rarely, mitotic centrioles) and procentrioles which are small centrioles with a raspberry-like profile as they are seen in longitudinal section. The association with mitotic centrioles (Fig. 1C, 2a) indicates that the generative process may start with these forms; procentrioles are more frequently seen on presumptive basal bodies. Many of the procentrioles are generated by the kinetosomal mode of genesis (as defined by Pitelka 1974), that is they form in direct association with a parent centriole (e.g., pc_1 and pc_2 on c_3 in Fig. 1C).

One or several procentrioles appear at the same time on a single parent centriole; they may appear at various angles with respect to the axis of the parent centriole and at either the proximal or distal end of that parent (Fig. 2c). Procentrioles also arise *de novo* in these cells, i.e., without a parent centriole nearby; for example serial sections through the procentrioles in Figure 1D showed there were

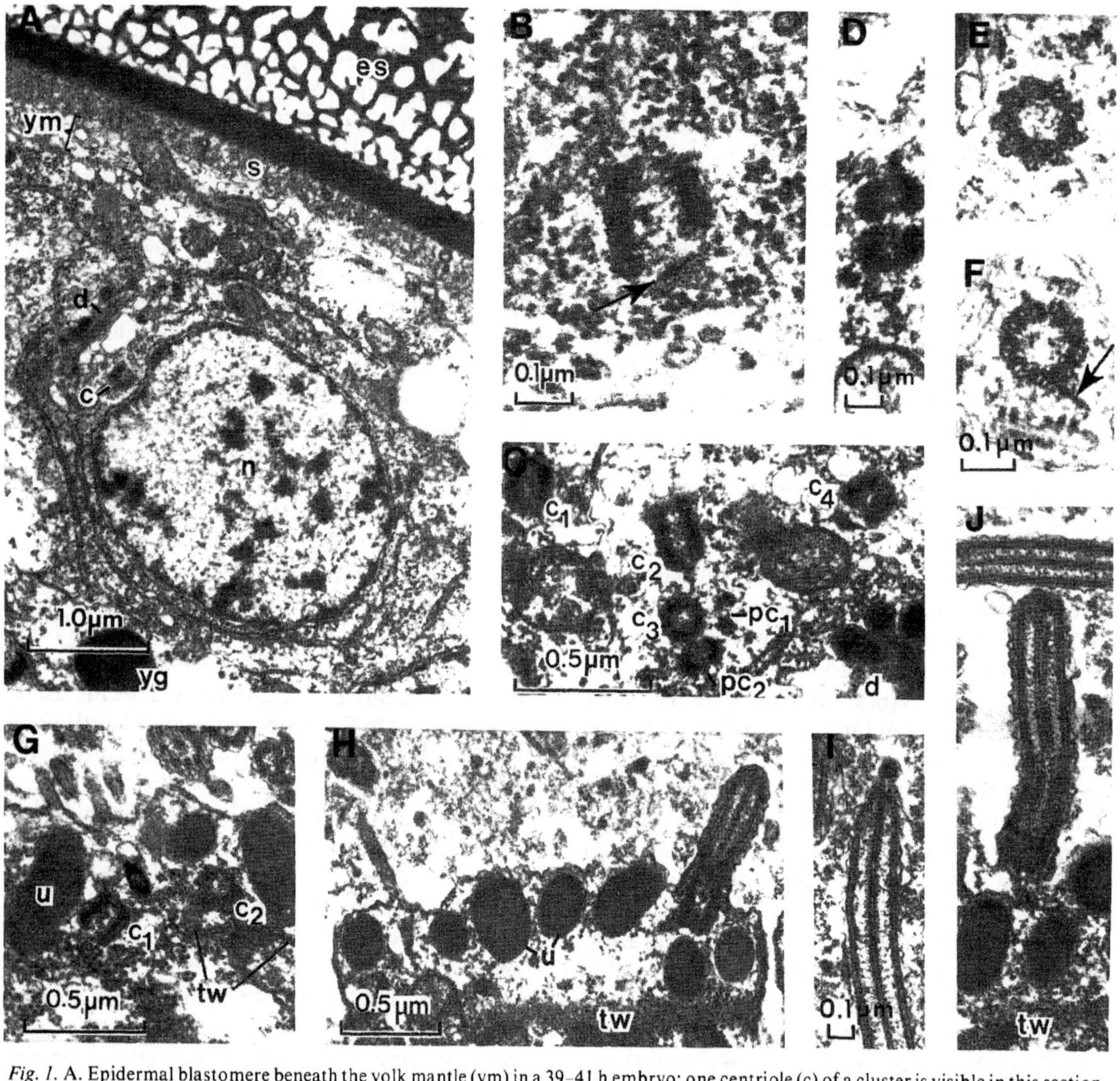

Fig. 1. A. Epidermal blastomere beneath the yolk mantle (ym) in a 39–41 h embryo; one centriole (c) of a cluster is visible in this section.
B. Presumptive basal body deep in the cytoplasm of an epidermal blastomere; arrow marks plate-like structure at basal end.
C. Kinetosomal genesis of centrioles: cluster of four centrioles (c_1–c_4) and two procentrioles (pc_1 & pc_2) in a 39–41 h embryo. C_2 is probably a centriole of the mitotic apparatus, whereas the other centrioles are presumptive basal bodies.
D. *De novo* genesis of two procentrioles; serial sections showed no other centriole within at least 0.3 μm of these two.
E. & F. Presumptive basal bodies in transverse section; arrow in F marks basal foot; this centriole lies beneath the terminal web.
G. Centrioles migrating into the apical cytoplasm through discontinuity in the terminal web of a 51–53-h embryo.
H. Newly developing cilium with nascent rootlet.
I. Tip of fully developed cilium.
J. Developing cilium of intermediate length with bluntly rounded tip.

no other centrioles in their vicinity. Such *de novo* procentrioles more frequently occur in pairs, and their orientation relative to one another may be perpendicular, in tandem, or at oblique angles; some procentrioles appear in isolation from all other centriolar forms (Fig. 2d).

There are usually several clusters of centrioles in a given blastomere. Most frequently the larger clusters are seen in the vicinity of a dictyosome of the Golgi complex.

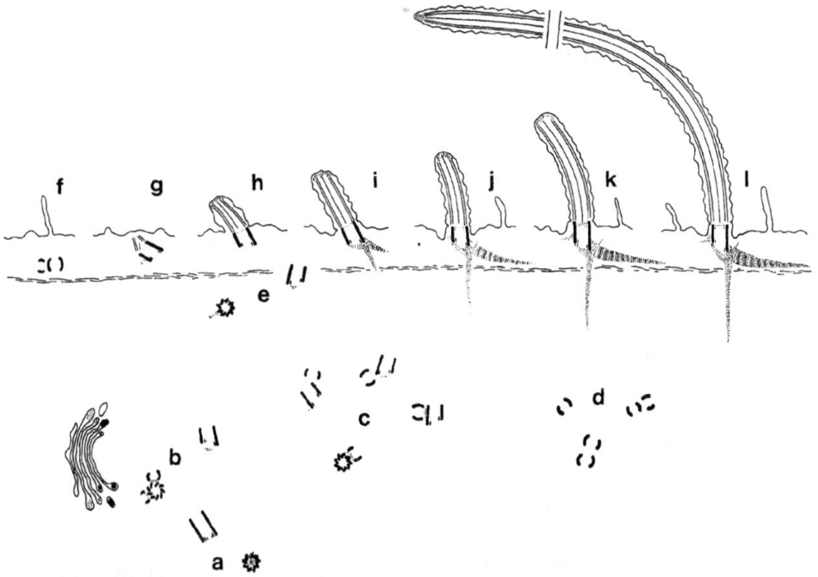

Fig. 2. Reconstruction of ciliogenesis in *Macrostomum*. Each section of the figure depicts actual structures seen in electron micrographs, and they are arranged here in an order believed to represent the course of events. (a) centrioles of the mitotic apparatus in association with kinetosomal genesis of presumptive basal bodies (b) near a Golgi apparatus; (c) form of kinetosomal genesis of procentrioles; (d) form of *de novo* genesis of procentrioles; (e) migration of presumptive basal bodies through terminal web – a basal foot may appear at this stage; (f) *de novo* genesis of procentrioles above terminal web; (g) – (k) initiation and development of axoneme and rootlets; (l) fully formed cilium; rootlet directed toward right is rostral rootlet.

51-to-53-h embryos – cilia formation

By 51–53 h, the epidermal blastomeres have emerged through the yolk mantle to the surface of the embryo and a few cilia have developed on their surface. The blastomeres apparently emerge between cells of the yolk mantle, pushing the yolk cells aside. At this stage, they are clearly epithelial, with a well differentiated terminal web, ultrarhabdites, microvilli, and cell junctions that are characteristic of the epidermis in a fully developed worm (Fig. 1H). They emerge in two distinct areas, probably corresponding to the two areas of epidermal blastema directly overlying the brain blastema which Bresslau (1904) identified in developing *Mesostoma.* I could not find in my material what Seilern-Aspang (1958) refers to in *Macrostomum appendiculatum* as a 'provisional epithelium', an epithelium that he says temporarily covers the dorsocentral cells, later to be overgrown by definitive epidermis; this may be because I have not sectioned completely through any one embryo.

Genesis of centrioles, as was evident in the 39–41-h embryos, continues in this older embryo. However, relatively fewer procentrioles are to be found in each blastomere, and there are not as many of the large clusters of centrioles as were visible in the younger embryos, but rather more widely separated pairs consisting of a centriole plus a procentriole. Centrioles can apparently be generated even in the apical portion of a cell distal to its terminal web as is evident from the finding of a *de novo* pair of procentrioles close to the terminal web in a 51-h embryo (Fig. 2f).

Cilia develop on the surface of the newly emergent epidermal blastomeres one at a time from presumptive basal bodies that reach the surface of a cell through discontinuities in the terminal web (Fig. 1G,2e). The presumptive basal bodies migrate (or are moved) individually through the cytoplasm without any evident preferred orientation during that migration. A basal foot may develop on the basal body even before it passes through the terminal web (Fig. 1F, 2e). At the apical plasma membrane where axoneme formation is initiated (Fig. 2g), basal body orientation again is not consistent; some orient at a right angle to the membrane, others can be seen to have short microtubules of the axoneme develop on them even as they lie oblique or even nearly parallel to the membrane. These first few axonemal microtubules are only loosely organized and of irregular lengths (Fig. 2g,h) so that transverse sections of these complexes show relatively disorganized axonemes often with less than nine doublets. By the time an axoneme is 0.5 μm long, it has a standard 9+2 pattern and is oriented more or less perpendicular to the cell surface (Fig. 1H,2i). In the course of this positioning, alar sheets are added, linking each microtubular fiber at the distal end of the basal body to the plasma membrane, and ciliary rootlets begin to form at the proximal end. The basal foot is the first of the rootlet-like fibers to be added, often even before the basal body is positioned at the plasma membrane, and the rostral rootlet and then vertical rootlet are added next. The rostral and vertical rootlets continue to elongate as the axoneme elongates. A given blastomere may have only one or a few fully developed cilia before basal bodies of future cilia have even positioned themselves at the cell surface.

Growing cilia have tips that are bluntly rounded (Fig. 1H,J, 2h–k). Only longer cilia, of a length of at least 5–7 μm have the tapered tip shape characteristic of cilia in fully developed animals (Fig. 1I, 2l; see also Tyler 1979). Steinman (1968) describes a similar differentiation of ciliary tips in *Xenopus.*

There is no pattern in the way in which cilia are added to the surface of the cells; they are not added in rows, nor do the newly positioning basal bodies orient to already established bodies as do basal bodies in acoel embryos (Tyler, in prep.). At first, the orientation of individual cilia is relatively disorganized; their rostral rootlets, rather than aligning in the same direction, point in directions as much as 55° apart from one another. As the ciliary field in a cell grows, the rootlets align with one another and lie parallel by the time the embryo hatches. Lyons (1973c) has described a similar transitory disorientation of ciliary rootlets in developing epidermis of the monogenean *Entobdella soleae.* The spacing of the cilia in developing blastomeres of *Macrostomum* is relatively irregular; nearest neighbors are between 0.4 and nearly 0.9 μm apart (measured center-to-center). By the hatchling stage more of the cilia are closer to their nearest neighbors; the range in center-to-center spacing between nearest neighbors is 0.5–0.65 μm. At no time are the rootlets of adjacent cilia in contact with one another.

Hatchling

The young worm that emerges from the egg is completely covered with a ciliated epidermis. The cells in this epidermis are broader and less deep than those of the 51–53-h embryo, having a surface area on the order of 790 μm^2, for instance (in cells measured in the posterior third of the body), in contrast to a surface area on the order of 60 μm^2 for cells of the newly forming epidermis. As there is no evidence of cell division in the epidermal blastomeres of the 51–53-h embryos, nor for immigration of cells from the parenchyma, I assume that the epidermis spreads solely by flattening and spreading of its component cells. Bresslau (1904) assumed the same for epidermal spreading in *Mesostoma;* Seilern-Aspang (1958), on the other hand, claims expansion takes place in *Macrostomum appendiculatum* by cell division.

Cilia are added to the expanding surface of the cells throughout embryogenesis. Ciliogenesis apparently continues even in the hatchling's epidermal cells; presumptive basal bodies can be seen in their apical cytoplasm. Besides ciliated epidermal cells, the hatchling's epidermis also has processes from a variety of gland and sensory cells, cells that were not to be seen in earlier embryonic stages examined.

Discussion

Many of the events of ciliogenesis in *Macrostomum* have parallels with ciliogenetic events in other animals. At the same time, however, the course of that ciliogenesis taken as a whole is unique, at least unlike that known in the vertebrates or the ciliated protozoans, groups in which ciliogenesis has been studied most thoroughly. A kinetosomal mode of genesis for basal bodies, in principle like that found here in *Macrostomum,* occurs, for example, on the diplosome of primary cilia in some vertebrate respiratory epithelia (Sorokin 1968; Kalnins & Porter 1969), in monkey oviduct (Anderson & Brenner 1971), and in lamprey nephron (Yousen 1979), and it occurs on walls of both presumptive and cilium-bearing basal bodies in chicken trachea, although production of most basal bodies in these tissues and other vertebrate epithelia (see e.g., Chang *et al.* 1979, for references) proceeds by a deuterosomal mode. In ciliated protozoans, both kinetosomal genesis (Dippel 1968; Allen 1969) and *de novo* genesis (Grimes 1973; Ruffolo 1976) of basal bodies occurs as is seen in *Macrostomum,* but kinetosomal genesis in ciliates takes place on cilium-bearing basal bodies rather than only on deeply lying presumptive basal bodies as is the case in *Macrostomum.* The fact that kinetosomal genesis takes place in *Macrostomum* often without a strict perpendicular orientation between daughter and parent centrioles is paralleled in ciliates (Grimes 1973), vertebrate epithelia (Kalnins *et al.* 1972), and in the prototrochal cells of an annelid (Kalnins 1967).In animals other than *Macrostomum* the formation of an accessory structure on the basal body such as the basal foot usually takes place only as the basal body positions itself at the cell surface, in contrast to the earlier growth seen in *Macrostomum,* but rootlets may form on migrating presumptive basal bodies in chicken trachea for example (Kalnins *et al.* 1972). Production of basal bodies and addition of cilia to the cell surface is a continuous process in *Macrostomum,* proceeding after the initial major wave of basal body production and even after the first cilia are fully formed; such a continuous production is seen as well in other ciliating cells (see e.g., Kalnins *et al.* 1972; Dirksen 1974).

Ciliogenesis in the acoel turbellarian *Archaphanostoma* (Tyler, in prep.) shares many features with that in *Macrostomum.* This turbellarian also produces at least some of its basal bodies by a kinetosomal mode and adds cilia to its surface in a one-by-one manner. Like *Macrostumum, Archaphanostoma* has basal bodies that are morphologically distinguishable from centrioles of its mitotic apparatuses. In contrast to *Macrostomum,* however, the cilia in the acoel are added in an orderly fashion at specific focal points, and from the beginning of the growth of the ciliary field the cilia are strictly oriented in parallel with one another through a system of interconnecting rootlets. (See Tyler, in prep., for further comparison).

The phylogenetic significance of ciliogenesis in the turbellarians is difficult to gauge, especially since it has parallels with ciliogenesis in even as widely separated a group as the Vertebrata. Evidently the processes of ciliogenesis are as conservative and universally occurring as is the 9+2 ciliary axonemal pattern. Such a result could have been expected at the outset. The question of prime importance to be answered here, though, as was pointed out in the introduction, is whether ciliogenesis in these lower spiralians, presumably more conservative members of the spiralian line of multiciliated animals, would reflect an ancestry in the monociliated line of evolution. The answer is 'no' for although there is, in essence, a monociliated stage in *Macrostomum's* epidermal cells this is a transitory state resulting simply from the fact that cilia are added one at a time to the cell's surface. The cells are clearly set to be multiciliated from the beginning as they have a mass production of basal bodies taking place even before they emerge into an epithelial position through the embryo's yolk mantle. The transitory monociliated condition has no other characteristics of the ancestral monociliated type, such as the presence of diplosome, or a collar of microvilli. The acoel turbellarian similarly has epidermal cells that are destined to be multiciliated from the outset (Tyler, in prep.).

Besides the two turbellarians *Macrostomum* and *Archaphonostoma,* the only spiralians in which ciliogenesis has been studied are the mollusc *Aequipecten* and the annelid *Nereis,* and only brief reports exist on these. Kalnin's (1967) abstract on ciliating prototrochal cells of the trochophore of *Nereis* describes the production of basal bodies (by kinetosomal mode) and their positioning at the cell's surface, and both of these processes could be quite similar to such processes in *Macrostomum.* In what would be striking contrast to this, Ash & Stephens (1975) claim in their description of ciliogenesis in ctenidia of *Aequipecten* that ciliogenesis is like that of the vertebrates, proceeding by a deuterosomal production of basal bodies; however, their micrographs do not convincingly support this claim. A process resembling vertebrate ciliogenesis is also claimed by Soltynska *et al.* (1976) to take place in epidermal replacement cells of a catenulid, a member of a group whose present classification in the Turbellaria is questioned; these authors make only passing reference, however, and do not show deuterosomal basal-body production in their figures.

Lyons (1973a) has already pointed to the fact that the differences between epidermal cells in cnidarian planulae, which are monociliated, and turbellarians, which are multiciliated, is an obstacle to acceptance of the planuloid-acoeloid theory of the origin of the Bilateria. My own findings that a fundamental difference remains even when one looks at the earliest stages of development of multiciliated epidermal cells in turbellarians leaves the planuloid-acoeloid theory still in question. Lyons (1973b) thought that the mono-versus-multiciliated obstacle could be overcome with the discovery of any monociliated cells in the Turbellaria such as sensory cells (and collar-cell-like sensory cells are now known in this group – see Ehlers 1977, for references). Salvini-Plawen (1978) in the latest discussion on theories of the origin of the Metaozoa evidently follows such reasoning and supports the view that the Turbellaria is the most conservative of the bilaterians; the remaining bilaterians then would be derived from a turbellarian-like form. This scheme ignores the major dichotomy between those phyla with monociliated epidermis and those with multiciliated epidermis, and would derive the primitively monociliated groups such as the Gnathostomulida or echinoderms from a multiciliated ancestor. Since the Turbellaria is a fundamentally multiciliated group, with epidermal cells that are multiciliated in all adults (Bedini & Papi 1974; Tyler, unpublished observations) and that are, as can be seen now at least for *Macrostomum* and *Archaphonostoma,* even strictly multiciliated throughout development, such a scheme loses its validity. Collar-cell-like sensory cells or even protonephridial cells are not good candidates for bridging the gap; by virtue of their specialized nature it does not seem legitimate to compare them to true epidermal cells (Rieger 1976). Alternative phylogenetic schemes such as that proposed by Heider (discussed by Ax 1963) may be more acceptable on these grounds if they keep all the spiralian groups in a single derived branch and do not derive monociliated groups from multiciliated ancestors.

At present, the best developmental model for derivation of multiciliated cells from the ancestral monociliated cell type lies in those vertebrate epithelia that have a single cilium with an accessory centriole in cells that later become multiciliated, as was explained in the introduction. The vertebrates, however, are members of the line of evolution for which the monociliated cell type is basic (Rieger 1976). My finding that turbellarians do not show such a developmental link to the monociliated epidermal cell type underscores the fundamental distinctness of the spiralian line of evolution.

Acknowledgements

This research was begun under tenure of a Killam Fellowship at Dalhousie University, and I thank the Killam Trust and the Department of Anatomy there for their support. Dr. David Chapman generously provided lab space, equipment, and advice. I thank David Bentley and Joyce Davis for their untiring help with the figures and Mary Plummer for her diligence in typing the manuscript.

References

Allen, R. D., 1969. The morphogenesis of basal bodies and accessory structures of the cortex of the ciliated protozoan Tetrahymena pyriformis. J. Cell Biol. 40: 716–733.

Anderson, G. W., 1972. The three-dimensional structure of the basal body from the rhesus monkey oviduct. J. Cell Biol. 54: 246–265.

Anderson, R. G. W. & Brenner, R. M., 1971. The formation of basal bodies (centrioles) in the rhesus monkey oviduct. J. Cell Biol. 50: 10–34.

Apelt, G., 1969. Fortpflanzungsbiologie, Entwicklungszyklen und vergleichende Fruhentwicklung acoeler Turbellarien. Mar. Biol. 4: 267–325.

Ash, B. M. & Stephens, R. E., 1975. Ciliogenesis during the sequential formation of molluscan gill filaments. Dev. Biol. 43: 340–347.

Ax, P., 1963. Relationships and phylogeny of the Turbellaria. In Dougherty, E. C. (Ed.), The Lower Metazoa. Univ. California Press, Berkeley. pp. 191–224.

Bedini, C. & Papi, F., 1974. Fine structure of the turbellarian epidermis. In Riser, N. W. & Morse, M. P. (Eds.), Biology of the Turbellaria. New York: McGraw-Hill, pp. 108–147.

Bresslau, E., 1904. Beitrag zur entwicklungsgeschichte der turbellarien. I. Die Entwicklung der Rhabdocölen und Alloiocölen. Z. wiss. Zool. 76: 213–332.

Chang, P., *et al.*, 1979. An ultrastructural study of morphogenesis of fibrogranular complex and centriole in ductuli efferentes of chinese hamster. Tiss. Cell 11: 401–412.

Dippell, R. V. 1968. The development of basal bodies in paramecium. Proc. Nat. Acad. Sci., USA 61: 461–468.

Dirksen, E. R. 1971. Centriole morphogenesis in developing ciliated epithelium of the mouse oviduct. J. Cell Biol. 51: 286–302.

Dirksen, E. R., 1974. Ciliogenesis in the mouse oviduct. A scanning electron microscope study. J. Cell Biol. 62: 899–904.

Ehlers, U., 1977. Vergleichende untersuchungen uber collarrezeptoren bei turbellarien. Acta Zool. fenn. 154: 137–148.

Grimes, G. W., 1973. Origin and development of kinetosomes in Oxytricha fallax. J. Cell Sci. 13: 43–53.
Guillard, R. R. L. & Ryther, J. H., 1962. Studies on marine planktonic diatoms. I. Cyclotella nana Hustedt and Detonula confervacea (Cleve) Gran. Can. J. Microbiol. 8: 229–239.
Kalnins, V., 1967. Ultrastructural observations on the development of cilia in the trochophore of Nereis limbata. Biol. Bull. 133: 472–473.
Kalnins, V. I., Chung, C. K. & Turnbull, C., 1972. Procentrioles in ciliating and ciliated cells of chick trachea. Z. Zellforsch. 135: 461–471.
Kalnins, V. I. & Porter, K. R., 1969. Centriole replication during ciliogenesis in the chick tracheal epithelium. Z. Zellforsch. 100: 1–30.
Karling, T. G., 1974. On the anatomy and affinities of the turbellarian orders. In: Eds. Riser, N. W. & Morse, P. W. New York: McGraw-Hill. pp. 1–16.
Ludwig, H. & Metzger, H., 1977. Ciliogenesis in human fetal oviduct. Scanning Electron Microscopy, 1977(II): 361–366.
Lyons, K. M., 1973a. Evolutionary implications of collar cell ectoderm in a coral planula. Nature 245: 50–51.
Lyons, K. M., 1973b. Epidermal fine structure and development in the oncomiracidium larva of Entobdella soleae (Monogenea). Parasitology 66: 321–333.
Lyons, K. M., 1973c. Collar cells in planula and adult tentacle ectoderm of the solitary coral Balanophyllia regia (Anthozoa Eupsammiidae). Z. Zellforsch., 145: 57–74.
Nørrevang, A., 1964. Choanocytes in the skin of an enterpneust. Nature (London) 204: 398–399.
Nørrevang, A. & Wingstrand, K. G., 1970. On the occurrence and structure of choanocyte-like cells in some echinoderms. Acta Zool. (Stockh.) 51: 249–270.
Pitelka, D. R., 1974. Basal bodies and root structures. In: Ed. Sleigh, M. A. Cilia and Flagella. New York: Academic Press, pp. 437–469.
Rieger, R. M., 1976. Monociliated epidermal cells in Gastrotricha: significance for concepts of early metazoan evolution. Z. zool. Syst. Evolutionsf. 14: 198–226.
Rieger, R. & Tyler, S., 1979. The homology theorem in ultrastructural research. Am. Zool 19: 655–665.
Ruffolo, J. J., Jr., 1976. Cortical morphogenesis during the cell division cycle in Euplotes: An integrated study using light optical, scanning electron and transmission electron microscopy. J. Morph. 148: 489–528.
Salvini-Plawen, L. v., 1978. On the origin and evolution of the lower Metazoa. Z. zool. Evolutionsf. 16: 40–88.
Seilern-Aspang, F., 1958. Die Entwicklung von Macrostomum appendiculatum (Fabricius). Zool. Jb. Anat. 76: 311–330.
Soltynska, M. S., Mroczka, B. & Moraczewski, J., 1976. Ultrastructure of epidermis in Turbellaria from the family Catenulida (Archoophora). J. Submicr. Cytol. 8: 293–301.
Sorokin, S. P., 1968. Reconstructions of centriole formation and ciliogenesis in mammalian lungs. 1968. J. Cell Sci. 3: 207–230.
Steinman, R. M., 1968. An electron microscopic study of ciliogenesis in developing epidermis and trachea in the embryo of Xenopus laevis. Am. J. Anat. 122: 19–56.
Tyler, S., 1979. Distinctive features of cilia in metazoans and their significance for systematics. Tiss. & Cell. 11: 385–400.
Tyler, S., (in prep.) Ciliogenesis in embryos of the acoel turbellarian Archaphanostoma.
Yousen, J. H., 1979. Centriole duplication and ciliogenesis in differentiation of the lamprey tubular nephron. Anat. Rec. 193: 729.

The system of epidermal ciliary rootlets in Turbellaria

Jan Hendelberg
Dept. of Zoology, Univ. of Göteborg, Box 25059, S-400 31 Göteborg, Sweden

Keywords: Turbellaria, Acoela, ultrastructure, ciliary rootlets, glycogen

Summary

The rootlets of the kinetic cilia form patterns of different types in the different turbellarian subgroups (cf. Rieger 1981). In the Acoela a rather complex system of ciliary rootlets is found in the epidermis (Dorey 1965; Hendelberg & Hedlund 1973; Bedini & Papi 1974). In the acoel *Childia groenlandica* (Levinsen) the four rootlets of each cilium make contact with those of adjacent cilia at two levels (Hendelberg & Hedlund 1974). Distinct granules are found in the interior of the main rootlets (Hendelberg & Hedlund 1974; Bedini & Papi 1974, Fig. 16) and basal bodies (Silveira 1972; Hendelberg & Hedlund 1974) of the epidermal kinetic cilia of acoels. Similar granules, probably of identical structure, can be seen in nemertodermatids, in the same positions (Tyler & Rieger 1977, Figs. 3 & 6). Such granules were studied in *C. groenlandica* with histochemical methods adapted for electron microscopy. Like Silveira (1972) I found the granules of the basal bodies to be Thiéry-positive, and thus evidently to be made up of or at least to contain polysaccharide material. The granules of the main rootlets were also found to be Thiéry-positive (Hendelberg 1976). Digestion experiments (Hendelberg & Hellmén 1978 and unpublished results) strongly support the concept that the granules are glycogen beta-particles.

We know that cilia can function as kinetic organelles without any rootlets. But we are still uncertain about the function of the rootlets when occurring. Most probably they form an anchorage, a function which may be favoured by branching rootlets making contact with each other. Another function which has been discussed is the transmittance of impulses regulating the ciliary beat. Glycogen granules represent an energy deposit. The functional implication of these granules in the interior of the ciliary rootlets and basal bodies is not clear. However, the observations raise the question of how energy is transmitted to the cilia. Are the ciliary rootlets, when occurring, involved? This question will be further discussed, with references, in a future full report on the digestion experiments (to be published elsewhere).

References

Bedini, C. & Papi, F., 1974. Fine structure of the turbellarian epidermis. In: Eds. Riser, N. W. & Morse, M. P., Biology of the Turbellaria, New York: MacGraw-Hill, pp. 108–147.

Dorey, A. E., 1965. The organization and replacement of the epidermis in acoelous turbellarians. Q. J. microsc. Sci. 106: 147–172.

Hendelberg, J., 1976. Granules of glycogen beta-particle type demonstrated in epidermal ciliary rootlets of acoelous turbellarians. J. Ultrastruct. Res. 54: 491.

Hendelberg, J. & Hedlund, K. O., 1973. Electron microscope studies of the junctions between ciliary rootlets and the orientation of ciliary axonemes in the epidermis of acoelous turbellarians. J. Ultrastruct. Res. 44: 440.

Hendelberg, J. & Hedlund, K. O., 1974. On the morphology of the epidermal ciliary rootlet system of the acoelous turbellarian Childia groenlandica. Zoon 2: 13–24.

Hendelberg, J. & Hellmén, E., 1978. Electron microscope studies of the effects of amylase digestion on granules occurring in basal bodies and ciliary rootlets. J. Ultrastruct. Res. 63: 106–107.

Rieger, R. M., 1981. Morphology of the Turbellaria at the ultrastructure level. In: Eds. Schockaert, E. R. & Ball, I. R. Turbellaria. Proc. of Third Int. Symp. Hydrobiologia. This volume.

Silveira, M., 1972. Association of polysaccharide material with certain flagellar and ciliary structures in turbellarian flatworms. Revue Microsc. Electron. 1:96.

Tyler, S. & Rieger, R. M., 1977. Ultrastructural evidence for the position of the Nemertodermatida (Turbellaria). Acta Zool. fenn. 154: 193–207.

Hydrobiologia 84, 240 (1981). 0018-8158/81/0843-0240/$00.20.

Ultrastructure of the epidermis of adult and embryonic Paravortex species (Turbellaria, Eulecithophora)

Barbara M. MacKinnon[1], Michael D. B. Burt[1] & Alan W. Pike[2]
[1] *Dept. of Biology, Univ. of New Brunswick Fredericton, N.B. Canada*
[2] *Dept. of Zoology, Univ. of Aberdeen, Aberdeen, Scotland*

Keywords: Turbellaria, Eulecithophora, *Paravortex, Cerastoderma,* parasites, ultrastructure, epidermis, embryo capsule

Abstract

The epidermis and associated structures of adult and embryonic *Paravortex cardii* and *Paravortex karlingi,* internal parasites of *Cerastoderma edule,* have been examined using scanning and transmission electron microscopy. The cellular epidermis of adult *Paravortex* bears cilia and microvilli which differ in number and distribution between *P. karlingi* and *P. cardii.* Cellular organelles include mitochondria, lipid bodies, Golgi bodies, and ultrarhabdites. Epidermal nuclei are located in the proximal portion of the cells. The development of the tegument of embryo *Paravortex* has been described and a possible origin for the embryo capsule is suggested. These findings are discussed in relation to the phylogenetic status of the Turbellaria in relation to other Platyhelminthes and in the functional adaptation of the epidermis for a parasitic mode of life.

Introduction

There exist in the literature numerous reports of ultrastructural investigations of the epidermis of various Turbellaria. Several of these have used ultrastructural characteristics, including cilia, as evolutionary or taxonomic features (Rieger 1976; Tyler 1977, 1979; Tyler & Rieger 1977) while others have been concerned with the structure of cilia and ciliary rootlets and their locomotory or sensory functions (Bedini, Ferrero & Lanfranchi 1973, 1975; Hendelberg & Hedlund 1974; Olson & Rattner 1975; Hendelberg 1976; Ehlers, B. 1977; Ehlers, U. 1977; Martin 1978a).

Few ultrastructural studies have been conducted on commensal or parasitic Turbellaria (Burt & Phillips 1969; Bresciani & Køie 1970; Køie & Bresciani 1973; Nichols 1975; Burt & Fleming 1978; Fleming 1979). We have examined two species of the genus *Paravortex: P. karlingi* Pike & Burt 1981 and *P. cardii* Hallez 1909. Both species are parasitic as adults in *Cerastoderma* (= *Cardium*) *edule,* the common cockle in Great Britain. *P. cardii* is found in the digestive gland and *P. karlingi* in the intestine. For a more details: see Pike & Burt (1981).

This ultrastructural investigation of the epidermis and associated structures of adult and embryonic *P. karlingi* and *P. cardii* was undertaken in hopes of contributing towards our understanding of the evolutionary position of the Turbellaria in relation to other Platyhelminthes and of the functional adaptations of the tegument of the Turbellaria which are concommitant with a parasitic mode of existence.

Materials and methods

Cerastoderma edule, the molluscan host of both *P. karlingi* and *P. cardii,* was collected from the Eden estuary in St. Andrews and the Ythan estuary near Aberdeen in Scotland. Animals were maintained in running sea water until dissected. Cockles

Hydrobiologia 84, 241–252 (1981). 0018–8158/81/0843–0241/$02.40.

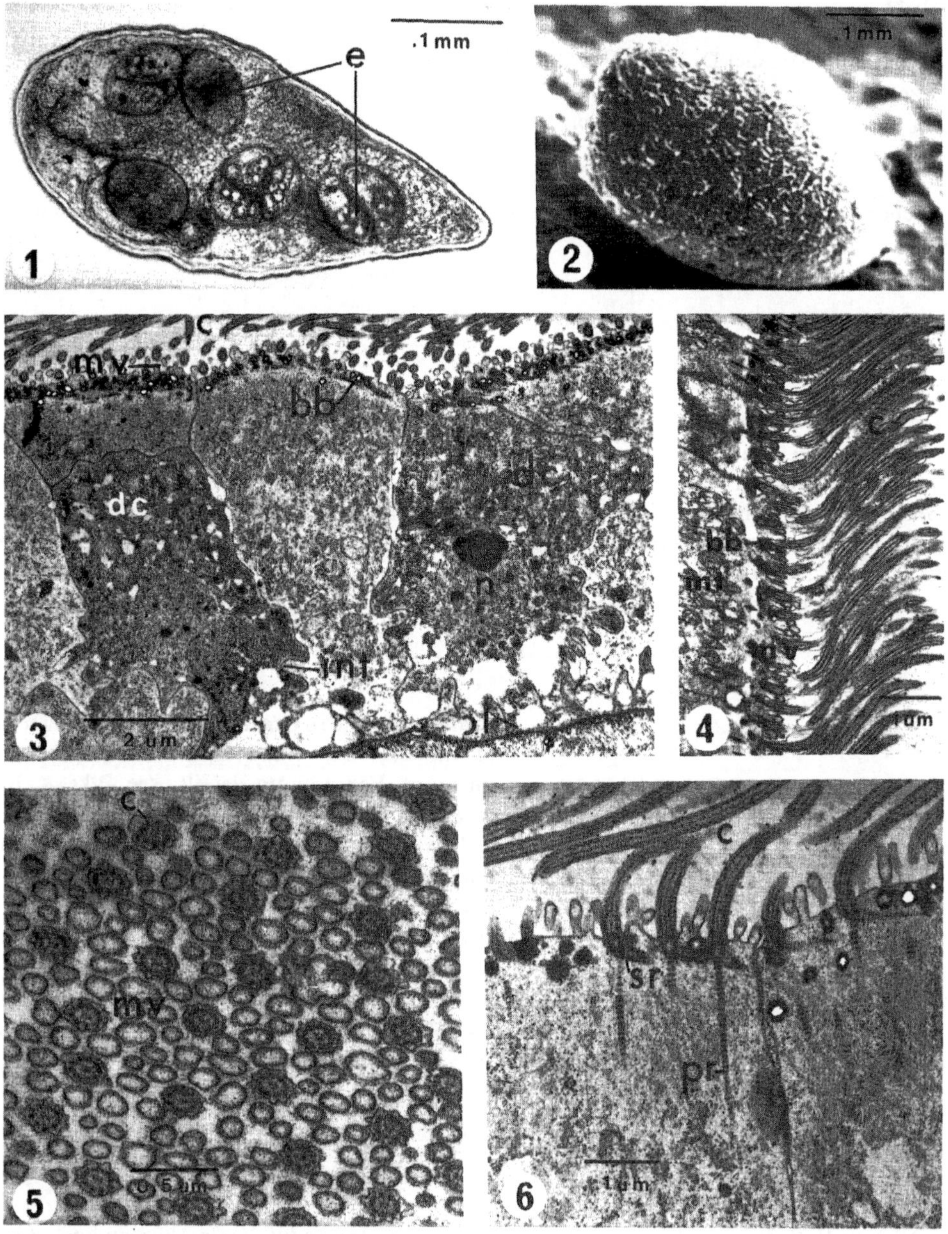

Fig. 1. Free-swimming gravid *P. karlingi* showing embryos.
Fig. 2. Scanning electron micrograph of *P. karlingi.*
Fig. 3. Transmission electron micrograph of *P. karlingi* showing the cilia and microvilli of the surface, basal bodies of the cilia, darkly staining cells, interdigitation of these cells, nucleus, mitochondria, Golgi bodies and basal lamella.
Fig. 4. TEM of the epidermis of *P. karlingi* to show cilia, microvilli, basal bodies and mitochondria.
Fig. 5. Transverse section through the cilia and microvilli of *P. karlingi.*
Fig. 6. TEM of *P. karlingi* showing primary and secondary rootlets, ultrarhabdites, basal bodies, cilia and microvilli.

were carefully opened in sea water and the Turbellaria were removed from the digestive gland (*P. cardii*) and from the intestine (*P. karlingi*) and some were fixed immediately for light and electron microscopy. Others were examined alive and photographed. Although free-swimming larvae were released from the adults, they were not investigated in this study.

Worms for examination with the transmission electron microscope were fixed in 3–5% glutaraldehyde in Dulbecco phosphate buffer, post-fixed in 1% osmium tetroxide and embedded in either EMix (EMScope, London, UK) or Epon Araldite. Thin sections were mounted on formvar coated copper grids and stained with uranyl acetate and lead citrate. Sections were examined and photographed on an EM6B or on a Phillips EM200. Specimens for scanning electron microscopy were fixed in the same way, dehydrated, critical point dried, coated with gold palladium and examined on a Cambridge Stereoscan.

Results

Paravortex karlingi

P. karlingi (Fig. 1) from the intestine of *C. edule* is 0.47 mm long by 0.23 mm wide (see Pike & Burt 1981) and contains up to 20 embryo capsules distributed in the parenchyma of the worm. Each capsule usually contains two embryos. Scanning electron microscopy reveals that the epidermis of *P. karlingi* is uniformly ciliated (Fig. 2) with a few raised bumps located at random over its surface. These protrusions are covered with cilia similar to those over the rest of the epidermis. Although microvilli are present as evidenced by transmission electron microscopy studies, they cannot be seen using scanning electron microscopy as they are occluded by the long and numerous cilia.

Transmission electron microscopy

The epidermis of *P. karlingi* (Fig. 3) is approximately 7–8 μm thick and is composed of both cuboidal cells and irregularly shaped cells. These cells form a monolayer although in many places there appears to be more than a single cell layer due to the overlapping of the irregularly-shaped cells in any one section. These irregularly shaped cells show a large surface area amplification in that their surface membranes are convoluted. These cells interdigitate with neighbouring cells especially in the proximal portion of the cells. Some of the cells within the epidermis have a cytoplasm which is more electron-dense than that of others. There seems to be a range in the staining properties of these cells from very dark to lightly granular. These cells have no other striking characteristics which suggest that they are different.

Cilia, approximately 2–3 μm long, and microvilli, about 0.25 μm long, cover the distal surface of the epidermis (Fig. 4). There are about 12 cilia per μm^2 and the microvilli are seen to be well dispersed. The cilia have rounded or tapered tips. The axonemal structure is the typical nine doublets with two singlets in the centre of the cilium. In transverse section (Fig. 5), the outer membranes of the cilia appear crenulated but this is believed to be a fixation artefact. No sensory cilia were seen. From tangential sections taken close to the epidermis, the ratio of cilia to microvilli can be seen to be approximately 1:6. Typical ciliary basal bodies lie just underneath the external cell membrane (Figs. 3 and 4). The rootlets have two branches, a primary rootlet which runs perpendicular to the surface and a secondary rootlet which lies almost parallel to the surface (Fig. 6). These caudally pointed secondary rootlets lie more or less in the same direction but are not strictly parallel to each other. No glycocalyx was visible in any of the sections examined.

The epidermal cells do not contain a cell web. Many cells contain numerous mitochondria, usually situated in the distal portion of the cells (Fig. 4), ultrarhabdites close to the surface, rhabditoid bodies distributed throughout the cells, Golgi bodies, vacuoles and lipid droplets. Other cells have a granular appearance and lack these cellular inclusions other than a few Golgi bodies and pale, poorly defined mitochondria. The nuclei of all these cells are positioned in the proximal portion.

Occasionally, cells were seen protruding beyond the normal layer of epidermal cells (Fig. 7 and 8). These cells have granular cytoplasm in their distal portions and contain numerous vesicles of varying electron density in the more proximal part of the cells. Some of these protruding cells were entirely devoid of cilia and microvilli while others were covered with them in the same way as the normal

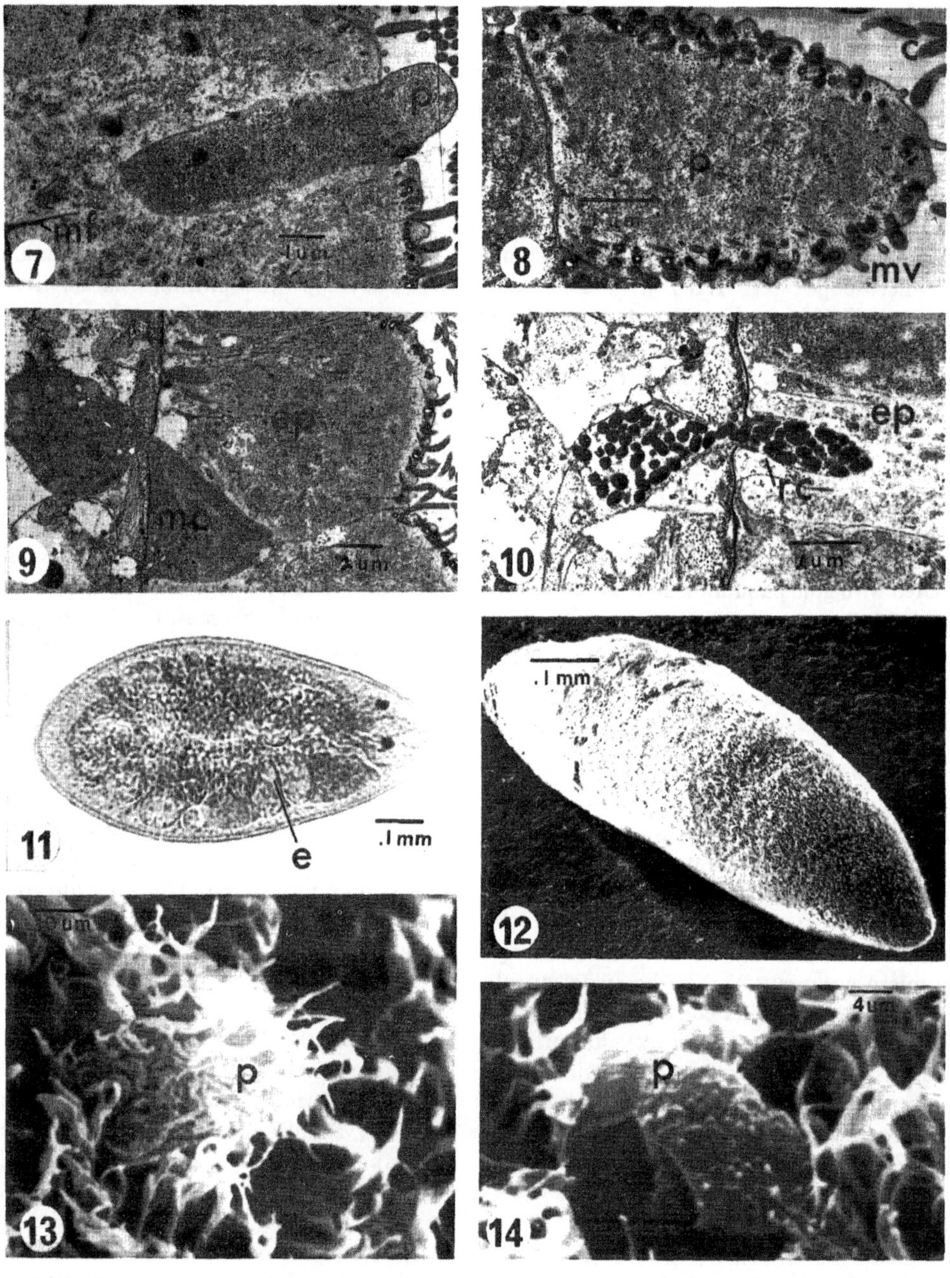

Fig. 7. Transmission electron micrograph of *P. karlingi* showing a protruded non-ciliated cell. A fold of the proximal membrane can be seen.

Fig. 8. TEM of a protruding cell of *P. karlingi* which bears cilia and microvilli.

Fig. 9. TEM of *P. karlingi* showing a cell which is partially in the epidermis and partially in the sub epidermal area.

Fig. 10. TEM of a rhabdite-forming cell entering the epidermis of *P. karlingi.*

Fig. 11. Whole mount of a fixed and stained specimen of *P. cardii.*

Fig. 12. Scanning electron micrograph of *P. cardii.*

Fig. 13. SEM of *P. cardii* showing ciliated bump on epidermis.

Fig. 14. SEM of *P. cardii* showing non-ciliated protrusion on epidermis.

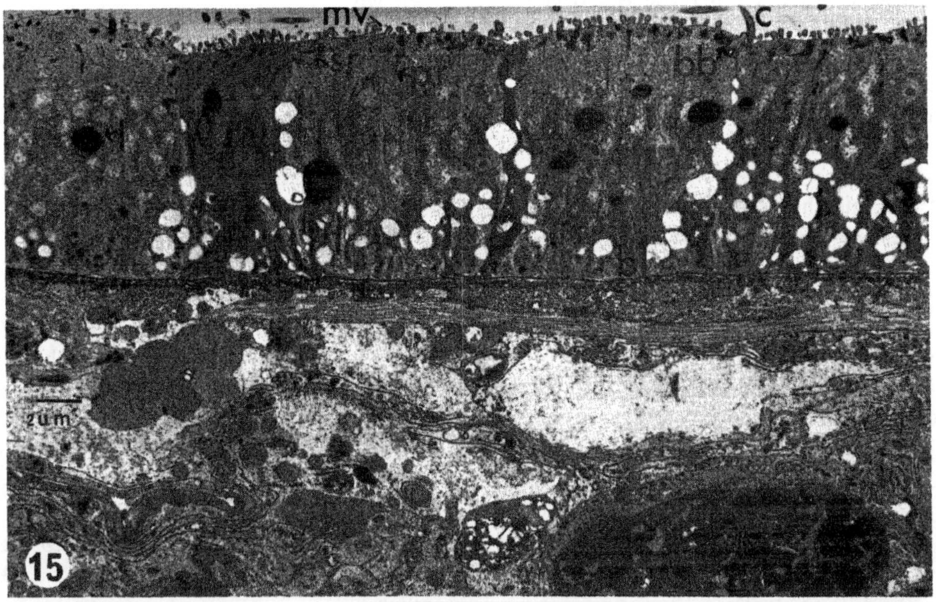

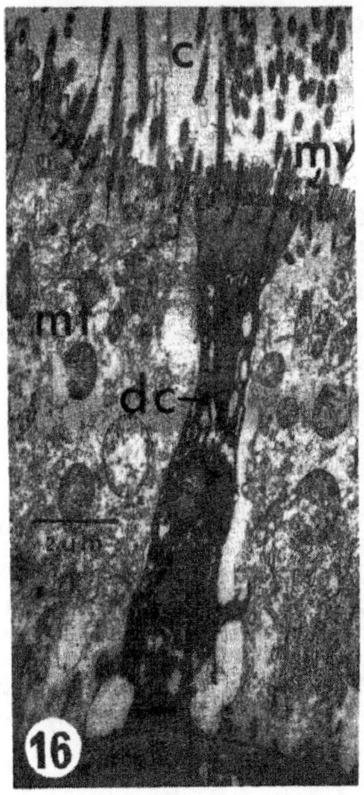

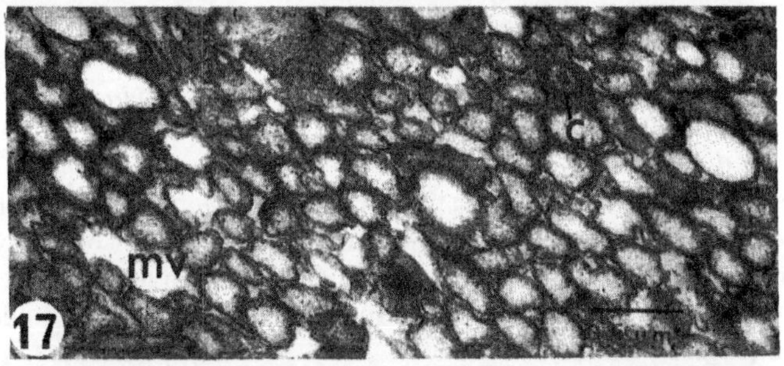

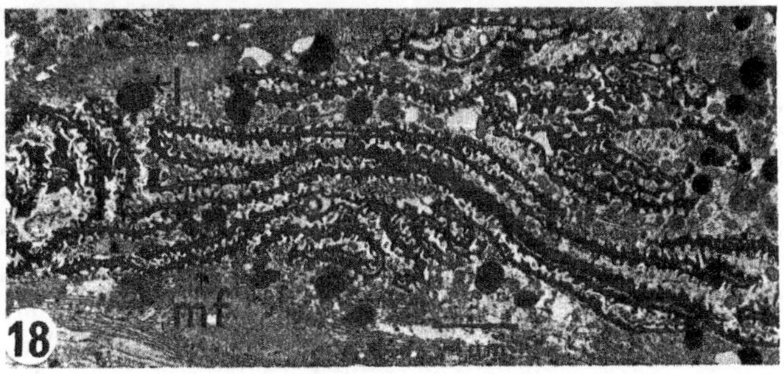

Fig. 15. Transmission electron micrograph of the epidermis of *P. cardii*, showing cilia and microvilli on the surface, basal bodies, ciliary rootlets, mitochondria, basal lamella and the variation in electron density of the epidermal cells.

Fig. 16. TEM of the epidermis of *P. cardii* showing electron-dense cell, cilia, microvilli, and mitochondria within the cells.

Fig. 17. TEM of a transverse section through the cilia and microvilli of *P. cardii*.

Fig. 18. TEM of the electron-dense lamellar folds seen subepidermally in both *Paravortex* species.

epidermal cells. It is possible that those seen to have a smooth surface could have had a single cilium which was missed in sectioning. Septate desmosomes were seen between neighbouring cells.

Although the epidermis is clearly divided into separate cells, there are numerous places where membrane folds can be seen extending from the proximal cell membrane into the base of the epidermal cells (see Fig. 7). These do not enclose separate cells but extend only about one third of the distance from the basal lamella to the outer surface of the epidermal cells. The entire epidermis rests on a thin fibrous basal lamella. Subepidermally, circular and longitudinal muscles are evident. In some places, cells can be seen with part of the cell in the epidermis and part underneath the basal lamella (Figs. 9 and 10). These are presumed to be migrating into the epidermis through the muscle layers and may be either replacement epidermal cells (Fig. 9) or glandular cells containing rhabdites (Fig. 10).

Paravortex cardii

P. cardii from the digestive gland of *C. edule* (Fig. 11) is 0.79 × 0.33 mm and contains up to 60 capsules each with two embryos. Scanning electron microscopy shows the adult worm to be uniformly ciliated (Fig. 12), like *P. karlingi*. Occasional ciliated bumps and nonciliated protrusions are found (Figs. 13 and 14).

Transmission electron microscopy

The ultrastructure of the epidermis of *P. cardii* is almost identical to that of *P. karlingi* (Fig. 15) with the following exceptional differences: the epidermis is approximately 5 μm thick; the cilia are about 4–5 μm long and the microvilli about 0.5 μm in length (Fig. 16). Thus the epidermis is apparently thinner and the cilia and microvilli longer than in *P. karlingi*, the ratio of cilia to microvilli being about 1:30 (Fig. 17). There are about 3 cilia/μm^2 and the microvilli are so densely distributed in places that their bases are often contiguous. The cellular inclusions, types of epidermal cells, basal lamella and subepidermal migrating cells are much the same as in *P. karlingi*.

Embryos

The ultrastructure of the embryos and the development and formation of the capsules of both *P. karlingi* and *P. cardii* are apparently identical and thus will be treated together.

In the mesenchyme of the adult worms, just below the epidermis, can be seen numerous clusters of folded lamellar bodies. At the ultrastructural level these lamellae are electron-dense, smooth on one surface and raised into either ridges or finger-like projections on the other surface (Fig. 18). There are often lipid droplets and mitochondria surrounding these lamellae. The lamella or sheath forming the capsule around the embryos has an identical structure to these electron-dense folds (Fig. 19), although, when surrounding an embryo, they are usually stretched quite thin. The capsule surrounding young embryos has been seen to be continuous with the folded lamella in the parenchyma.

The actual age of the embryos examined could not be determined since adult worms were not kept *in vitro* and watched for the beginning of development. The youngest embryo seen (Fig. 20) was an encapsulated ovum which had not begun cellular division for the production of the epidermis and other organ systems. The few cells comprising the ovum had a granular cytoplasm, numerous mitochondria and a few Golgi bodies. Their outer membranes were smooth. Surrounding the ovum was a layer of vitelline cells containing yolk droplets, mitochondria and empty vacuoles. The capsule appeared thick and electron-dense.

When cell division begins to form epidermal cells, there are still vitelline cells surrounding the embryo although some yolk material can be seen within the embryo (Fig. 21). No sign of cilia or microvilli could be seen at the external membrane of the embryo. The capsule remains thick and electron-dense. As cell division and maturation proceed, distinct cells can be seen forming the epidermal layer of the embryo and these cells now bear both cilia and microvilli (Fig. 22). These epidermal cells contain electron-dense material around the bases of the cilia (possibly material used in forming new ciliary basal bodies and rootlets) mitochondria, microtubules, Golgi bodies and proximally situated nuclei. As the embryos grow, the capsule wall or lamella becomes thinner.

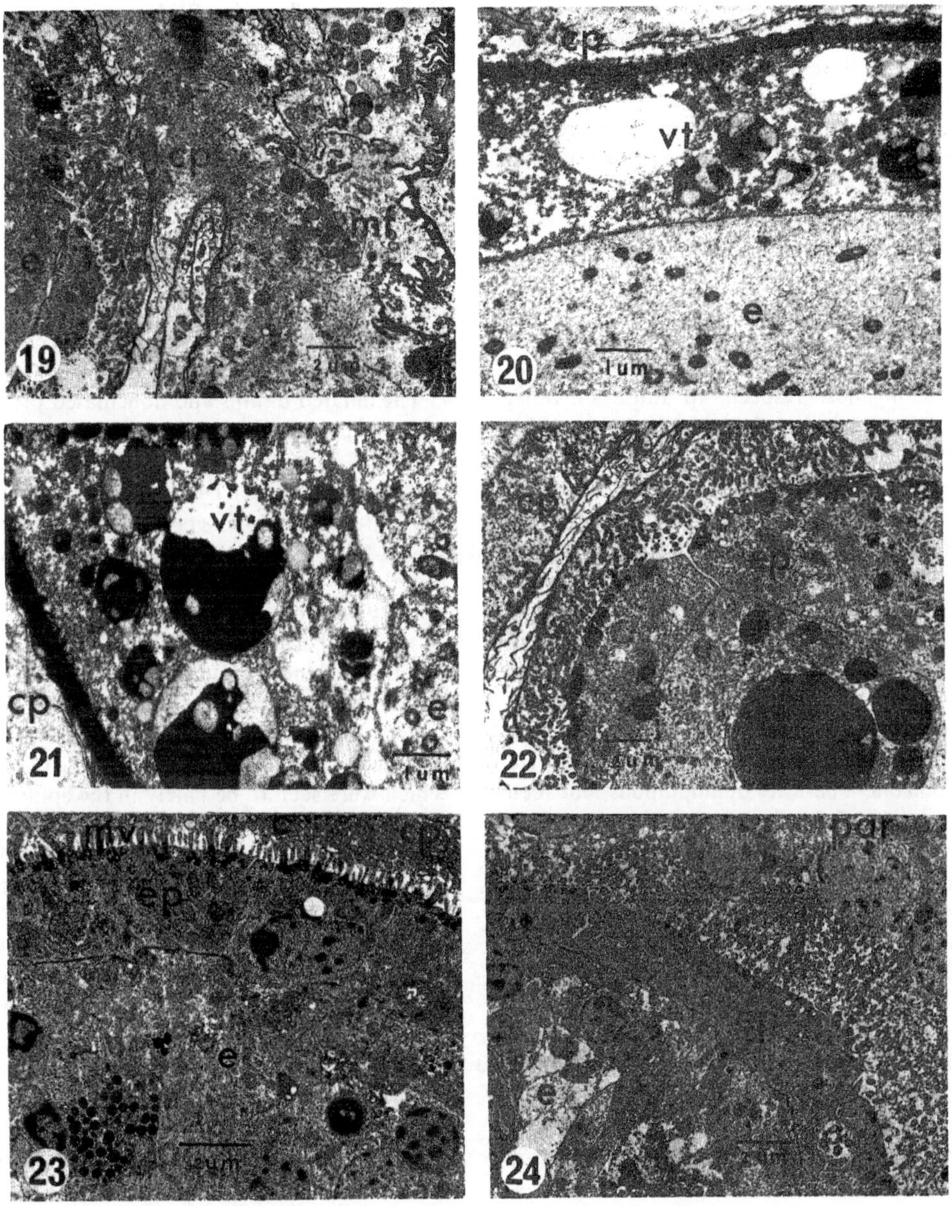

Fig. 19. Transmission electron micrograph of lamellar folds in the parenchyma of *Paravortex* in close contact with structurally similar embryo capsule.
Fig. 20. TEM of a young embryo of *Paravortex* surrounded by vitelline material and capsule.
Fig. 21. TEM of developing embryo of *Paravortex.*
Fig. 22. TEM of an embryo with epidermal cells bearing cilia and microvilli. The capsule appears thinner than in previous micrographs.
Fig. 23. TEM of a mature embryo still enclosed by the capsule.
Fig. 24. TEM of a mature embryo, freed from its capsule, and migrating through the parent tissue.

As the epidermal cells develop further and take on an adult appearance (Fig. 23), the inside of the embryo rapidly develops numerous neoblasts and other organ systems become evident. The capsule at this stage is quite thin. When the embryo is freed of its capsule and wanders freely in the parenchyma of the adult worm (Fig. 24), the epidermis with its cilia and microvilli appears fully functional.

Discussion

The epidermis of adult *P. karlingi* and *P. cardii* is similar in most respects to those of other Turbellaria described to date (see Pederson 1976; Bautz 1977; Ehlers, B. 1977) differing only in a few minor characteristics. As can be seen from the results, all the cilia seen in micrographs have either rounded or tapered tips. Cilia with shelf-like structures, as described by Tyler and Rieger (1977) and Tyler (1979) were not seen in either species examined here. Similarly, sensory cilia as described by Bedini *et al.* (1973, 1975) were not seen although it is suspected that they existed but were missed in sectioning. Bresciani & Køie (1970) describe cilia with swollen tips containing vesicles in *Kronborgia amphipodicola,* a parasite of amphipods, but such were not seen in *P. karlingi* or *P. cardii.*

Although a glycocalyx was not identified in either *P. karlingi* or *P. cardii,* it is possible that a thin layer covering the epidermis did exist but was stripped off due to the fixation procedures. Bresciani & Køie (1970) describe a glycocalyx in *K. amphipodicola* but it is difficult to identify in their micrographs. The same may be true in this case. The epidermis of *P. cardii* and *P. karlingi* lacks a cell web which is common to most Turbellaria. This is not a unique feature to *Paravortex,* however, since cell webs are lacking in *Prorhynchus, Otoplana, Dendrocoelum, Gyratrix,* and *Typhloplanidae* (except *Opistomum*) (Bedini & Papi 1974), and in *Kronborgia amphipodicola* (Bresciani & Køie 1970). As the absence of cell web occurs in both free-living and parasitic Turbellaria, the loss of this structure is not considered to reflect any adaptation to a parasitic mode of life in *P. karlingi* and *P. cardii.*

Glandular cells, other than rhabdite-forming cells, found within the epidermis of numerous Turbellaria (Bedini & Papi 1974; Bresciani & Køie 1970; Bowen & Ryder 1974) and often associated with an adhesive function (Tyler 1977; Martin 1978b; Sopott-Ehlers 1979) were not seen in the *Paravortex* species examined, even in horizontal sections through the epidermis. Such gland cells are easily seen using light microscopy and it is interesting to note that Hallez (1909) and more recently Pike & Burt (1981) mention the absence of gland cells often considered to be so characteristic of the Turbellaria. The only glandular type cells seen in the present study of *P. karlingi* and *P. cardii* are those involved, presumably, in the production of rhabdites or ultrarhabdites.

The ciliated bumps and nonciliated protrusions of the epidermis identified on *P. karlingi* and *P. cardii* are unlike epidermal protrusions described from other Turbellaria (Bresciani & Køie 1970; Bautz 1977; Ehlers B. 1977; Sopott-Ehlers 1979). Although transmission electron microscopy reveals no indication that these bumps have a sensory or adhesive function, further investigation is needed before eliminating these as possible functions of these protrusions.

The darkly staining cells seen in the epidermis seem to be similar to the other epidermal cells in every character other than their density. The cells which were seen to be partly in the epidermis and partially in the subepidermal area, were darkly stained also, in terms of electron density. These are suspected to be migrating into the epidermis, as happens in other Turbellaria (Dorey 1965; Skaer 1965; Soltynska, Mroczka & Moraczewski 1976). It could be that the dark cells observed in the epidermis were newly-formed cells which had recently been added to the epidermis.

Embryos

The development of the epidermis of the embryos of *P. cardii* and *P. karlingi* seems to be a relatively straightforward division, and differentiation, of cells at the external surface of the embryo. No 'primary' and 'secondary' epidermal layers, as described by Skaer (1965) were seen, although in mature worms, cells from the parenchyma probably replenish the epidermal layer. The first components of the epidermal cells, once multiplication of the cells at the periphery of the embryo takes place, seem to be the cilia and microvilli and thus it can be suggested that their respective functions of movement and nutrient absorption are of primary importance to the developing embryo.

Our observation that the capsule sheath surrounding the embryos is structurally similar and continuous with the folded lamella found within the parenchyma of adult worms, leads us to the tentative suggestion that the capsules of the embryos may be formed from this lamella in the parenchyma. Hallez (1909) in his original description of *P. cardii* was unable to follow or describe the formation of the capsule satisfactorily although he suggested that it originated in the shell glands and was transported to the egg via the female atrium and oviduct. He suggested, as did Linton (1910) that the shrivelled 'membranes' seen in the parenchyma were discarded embryo capsules. Wahl (1906, in Hallez 1909) suggested that the cocoon (= capsule) of *Paravortex scrobicularia* was formed in the mesenchyme, while Linton (1910) admitted that the origin of the capsule of *P. cardii* was uncertain. Patterson (1912) suggested that the embryo capsules of *Graffila* (= *Paravortex*) *gemellipara* were formed in the mesenchyme and that the shell gland, a portion of the poorly developed and possibly rudimentary female genital system, was nonfunctional in these parasitic Turbellaria. While it seems obvious from whole mount specimens of *P. cardii* and *P. karlingi* that these folded lamella are capsules evacuated by mature embryos (they being present in pockets around the entire worm and not all located near the ovary) it could be possible that these lamellae are re-used by newly formed embryos. Obviously this is a purely speculative suggestion based on our observation that the capsules of immature embryos were continuous with these lamellae.

Adaptations to a parasitic mode of life

Bresciani & Køie (1970) showed that the microvilli of *K. amphipodicola* were longer and more numerous than those of other Turbellaria and suggested that this was a reflection of their parasitic mode of existence and an adaptation for increased nutriment absorption through the epidermis as *K. amphipodicola* has no gut. Counting the number of cilia and microvilli in *K. amphipodicola*, from their micrographs, we found the ratio to be about 1:10, slightly higher than that of *P. karlingi* (1:6) and much lower than *P. cardii* (1:30). Unfortunately other authors working on the ultrastructure of the epidermis of free-living Turbellaria have not mentioned the ratio of cilia to microvilli, nor in many cases have they shown a tangential section through both cilia and microvilli from which such a ratio could be established. Based on our interpretation of published micrographs of various free-living Turbellaria (e.g. Bedini & Papi 1974) it seems that several of these show a proportion of cilia to microvilli of between 1:2 and 1:4 in any one plane, which in two dimensions could give a ratio of about 1:4 to 1:16. *K. amphipodicola* has long microvilli and this is presumably a function of increased absorption. Although *P. cardii* and *P. karlingi* both have relatively short microvilli, they have an intestine as well. There are, however, significant differences between these two species. In *P. karlingi*, the cilia are not only shorter than those of *P. cardii* but there are more than four times as many cilia per unit area. The microvilli of *P. karlingi* are also shorter than those of *P. cardii* but there are fewer per unit area. As can be seen from Fig. 5, the microvilli of *P. karlingi* are well dispersed amongst the cilia whereas the microvilli of *P. cardii* (Fig. 16) are so densely distributed that their bases are contiguous with each other. While there is some variation from one part of the worm to another, and also variation based on fixation, the number and size of microvilli in relation to unit area is greater in *P. cardii* than in *P. karlingi* and hence the surface area available for absorption and/or excretion is similarly greater. This could represent an adaptation related to the different environments of the two species. The requirement for motility in *P. karlingi*, presumably having to maintain its position with respect to flow through the gut, is reflected in the larger number of cilia than in *P. cardii* which, living in the digestive gland, has no such requirement. On the other hand, the greater surface area presented by *P. cardii* with respect to its microvilli possibly reflects a greater dependence on direct absorption of nutriment through its outer surface. *P. karlingi*, living in the intestine of the cockle, may have more particulate food available to it for intestinal feeding than would *P. cardii* in the digestive gland.

It is perplexing to understand where the Acoels would fit into this scheme, with their short microvilli and absence of a well-defined gut. Although we agree with Bresciani & Køie (1970) that an increase in surface area, through development of microvilli in *Kromborgia,* is probably an adaptation to a parasitic mode of life, the diversity seen in the Turbellaria as a whole, with respect to number and size of cilia and microvilli, makes it difficult to draw any general conclusions with respect to development of parasitism *per se.*

The interdigitation of the cells comprising the epidermis of *P. cardii* and *P. karlingi* may be reflecting a considerable metabolic interaction between these cells, possibly in transfer of nutrient material. The absence of glandular cells in the epidermis of these worms is a rather rare characteristic within the Turbellaria. It is tempting to suggest that if the host (cockle) reaction to these parasites is only slight, these worms may have little need for extensive protective or adhesive secretions. However, in most characteristics of the epidermis of *P. cardii* and *P. karlingi,* they are strikingly similar to many free-living Turbellaria.

Phylogenetic considerations

It is now generally accepted that the ancestral progenitor of the Monogenea, Digenea and Cestoda was a rhabdocoele-like animal (Llewellyn 1965; Lyons 1973; Lumsden 1975). The adult forms of these Platyhelminthes have a syncytial tegument, with subtegumental cells containing the nuclei. Projections on the outer surface are in the form of spines, finger-like projections or microtriches, not cilia and microvilli as in the Turbellaria. However, the epidermis of some larval forms of many parasitic Platyhelminthes show striking similarities to the Turbellarian epidermis. Fournier (1979) has shown that the oncomiracidium of *Polystoma* (Monogenea) has ciliated cells within its epidermis. These cells have nuclei within the epidermal layer. The syncytial tegument of the adult *Polystoma* bears microvilli similar to the epidermis of Turbellaria. The miracidium of *Schistosoma mansoni* (Digenea) bears rows of ciliated cells in its epidermis (Meuleman *et al.* 1978). These cells also bear microvilli but differ from the epidermal cell of Neorhabdocoela in having insunk nuclei.

Many Platyhelminthes with syncytial teguments have membranous folds extending from the proximal cell membrane of the tegument into the cytoplasm of the syncytium, as we have seen here in *Paravortex* and which are found in *Gyratrix* which has a syncytial epidermis (Bedini & Papi 1974). It seems likely that these folds in the syncytial epidermis could be the remnants of complex cell boundaries. The folds in the epidermis of *Paravortex,* which has an otherwise cellular epidermis, could indicate a predisposition towards a syncytial tegument as can also be seen in *Urastoma cyprinae* (Burt, unpublished).

Thus, especially when investigating ontogenetic forms of Monogenea and Digenea in particular, we can see striking similarities in the epidermis of these groups with that of the Turbellaria. Within the Turbellaria itself, we can see the presence of different epidermal organizations that are syncytial or have insunk nuclei (a common feature of the Platyhelminthes) as well as the more characteristic, cellular epidermis.

Acknowledgements

We are most grateful to Mr. Andy Lucas, Department of Zoology, University of Aberdeen, and Mr. Paul Nitishin, of the University of New Brunswick for assistance with the preparation of the micrographs and particularly to Mr. Gerry Bance, of the University of New Brunswick, for invaluable help with the transmission electron microscopy material. The help of Dr. R. B. Johnson, Department of Botany, University of Aberdeen is gratefully acknowledged in relation to the SEM work.

We also gratefully appreciate the provision of space and facilities, for part of this work, by Prof. M. S. Laverack in the Gatty Marine Laboratory, St. Andrews University, and help in collecting cockles from Sheila and David Burt and from Fiona, David John, and Graeme Jamieson. Financial assistance (to M.D.B.B.) from NATO (through a Sen. Sci. Research Grant No. B (109)) and from NSERC of Canada (through Operating Grant A-2358) is gratefully acknowledged.

Abbreviations

bb - basal bodies
bl - basal lamella
c - cilia
cp - capsule
dc - dark cells
e - embryos

ep – epidermis
g – Golgi bodies
int – interdigitation (of cells)
l – lipid
lf – lamellar fold
mc – migrating cell
mf – membranous folds
mt – mitochondria
mv – microvilli
n – nucleus
nb – neoblasts
p – projections of epidermis
par – parenchyma of mother
pr – primary rootlet
rc – rhabditogen cells
sr – secondary rootlet
ur – ultrarhabdites
vt – vitelline material

References

Bautz, A., 1977. Structure fine de l'épiderme chez des planaires triclades terrestres et paludicoles. Arch. Zool. Exp. gén. 118: 155–172.

Bedini, C., Ferrero, E. & Lanfranchi, A., 1973. The ultrastructure of ciliary sensory cells in two Turbellaria Acoela. Tissue Cell 5: 359–372.

– 1975. Fine structural observations on the ciliary receptors in the epidermis of three otoplanid species (Turbellaria, Proseriata). Tissue Cell 7: 253–266.

Bedini, C. & Papi, F., 1974. Fine structure of the turbellarian epidermis. In: Ed. Riser, N. W. and Morse, M. P. Biology of the Turbellaria. New York: McGraw Hill.

Bowen, I. D. & Ryder, T. A., 1974. The fine structure of the planarian Polycelis tenuis (Iijima) iii: The epidermis and external features. Protoplasma 80: 381–392.

Burt, M. D. B. & Fleming, L. C., 1978. Phylogenetic clues in the ultrastructure of Turbellaria and Cestoda teguments. Parasitology 77: xxx–xxxi.

Burt, M. D. B. & Phillips, W. L., 1969. The fine structure of Urastoma cyprinae, a microturbellarian found in oysters. Am. Zool. 9: 517.

Bresciani, J. & Køie, M., 1970. On the ultrastructure of the epidermis of the adult female of Kronborgia amphipodicola Christensen and Kanneworff, 1964 (Turbellaria, Neorhabdocoela). Ophelia 8: 209–230.

Dorey, A. E., 1965. The organization and replacement of the epidermis in acoelous turbellarians. Quart. J. Microsc. Sci. 106: 147–172.

Ehlers, B., 1977. 'Trematoden-artige' Epidermisstrukturen bei einem freilebenden proseriaten Strudelwurm (Turbellaria, Proseriata). Acta Zool. fenn. 154: 129–136.

Ehlers, U., 1977. Vergleichende untersuchungen über collarrezeptoren bei Turbellarien. Acta Zool. fenn. 154: 137–148.

Fleming, L. C., 1979. On the biology of the turbellaria associated with the queen crab, Chionoecetes opilio (O. fabricius). PhD. Thesis, Univ. of New Brunswick.

Fournier, A., 1979. Evolution du tégument des Polystoma (Monogènes, Polystomatidae) au cours du cycle. Z. Parasitenkde. 59: 169–185.

Hallez, P., 1909. Biologie, organisation, histologie et embryologie d'un rhabdocoele parasite du Cardium edule L. Paravortex cardii n.sp. Arch. Zool. Exp. gén. 9: 429–544.

Hendelberg, J., 1976. Granules of glycogen beta-particle type demonstrated in epidermal ciliary rootlets of acoelous turbellarians. J. Ultrastr. Res. 54: 491.

Hendelberg, J. & Hedlund, K. O., 1974. On the morphology of the epidermal ciliary rootlet system of the acoelous turbellarian Childia groenlandica. Zoon. 2: 13–24.

Køie, M. & Bresciani, J., 1973. On the ultrastructure of the larva of Kronborgia amphipodicola Christensen and Kanneworff, 1964 (Turbellaria, Neorhabdocoela). Ophelia 12: 171–203.

Linton, E., 1910. On a new rhabdocoele commensal with Modiolus plicatulus. J. Exp. Zool. 9: 371–386.

Llewellyn, J., 1965. Evolution of Parasites. Symposia of the British Society for Parasitology 2: 47–78.

Lumsden, R. D., 1975. Surface ultrastructure and cytochemistry of parasitic helminths. Exp. Parasitol. 37: 267–339.

Lyons, K., 1973. The epidermis and sense organs of the Monogenea and some related groups. Advances in Parasitology 11: 193–232.

Martin, G. G., 1978a. Ciliary gliding in lower invertebrates. Zoomorphologie 91: 249–261.

Martin, G. G., 1978b. The duo-gland adhesive system of the archiannelids Protodrilus and Saccocirrus and the turbellarian Monocelis. Zoomorphologie 91: 63–75.

Meuleman, E. A., Lyarus, D. M., Khan, M. A., Holzmann, P. J. & Sminia, T., 1978. Ultrastructural changes in the body wall of Schistosoma mansoni during the transformation of the miracidium into the mother sporocyst in the snail host Biomphalaria pfeifferi. Z. Parasitenkd. 56: 227–242.

Nichols, K. C., 1975. Observations on lesser-known flatworms: Temnocephala. Int. J. Parasitol. 5: 245–252.

Olson, G. E. & Rattner, J. B., 1975. Observations on the substructure of ciliary rootlets. J. Ultrastr. Res. 51: 409–417.

Patterson, J. T., 1912. Early developments of Graffila gemellipara – a supposed case of polyembryony. Biol. Bull. 22: 173–205.

Pederson, K. J., 1976. Scanning electron microscopical observations on epidermal wound healing in the planarian Dugesia tigrina. Wilhelm Roux's Archives 179: 251–273.

Pike, A. W. & Burt, M. D. B., 1981. Paravortex karlingi. sp. nov. from Cerastoderma edule L., in Britain. In: Schockaert, E. R. & Ball, I. R. Turbellaria. Proc. Third Int. Symp. Hydrobiologia. This volume.

Rieger, R. M., 1976. Monociliated epidermal cells in Gastrotricha: significance for concepts of early metazoan evolution. Z. Zool. Syst. Evol. 14(B): 198–226.

Skaer, R. J., 1965. The origin and continuous replacement of the epidermal cells in the planarian Polycelis tenuis (Iijima). J. Embryol. exp. Morph. 13: 129–239.

Soltynska, M. S., Mroczka, B. & Moraczewski, J., 1976. Ultrastructure of epidermis in turbellaria from the family Catenulida (Archoophora). J. Submicrosc. Cytol. 8: 293–301.

Sopott-Ehlers, B., 1979. Ultrastruktur der Haftapparate von Nematoplana coelogynoporoides (Turbellaria, Proseriata). Helgoländer wiss. Meeresunters 32: 365–373.

Tyler, S., 1977. Ultrastructure and systematics: an example from Turbellarian adhesive organs. Mikrofauna Meeresboden 61: 271–286.

Tyler, S., 1979. Distinctive features of cilia in metazoans and their significance for systematics. Tissue Cell 11: 385–400.

Tyler, S. & Rieger, R. M., 1977. Ultrastructural evidence for the systematic position of the Nemertodermatida (Turbellaria). Acta Zool. fenn. 154: 193–207.

Ultrastructural observations on paracnids. I: Coelogynopora axi Sopott (Turbellaria, Proseriata)

Beate Sopott-Ehlers
II. Zoologisches Institut und Museum der Univ. Göttingen, Berliner Strasse 28, D-3400 Göttingen, F.R.G.

Keywords: Turbellaria, Proseriata, Coelogynoporidae, ultrastructure, paracnids

Abstract

The 'Schlauchdrüsen' or paracnids of *Coelogynopora axi* Sopott, 1972 consist of two components: a muscle cell and a secretory cell.

The secretory cell is provided with a tube, which bears a border of microvilli. In the normal position the tube is situated in the interior of the secretory cell, and the microvilli stand at the inner side of the tube. After expulsion of the tube the microvilli are situated at its free surface.

The evagination takes place in response to chemical stimuli and is effected by the contraction of the myofibrils of the muscle cell.

The paracnids are supposed to be mechanisms of defense.

However, conformities with nematocysts and spirocysts of the cnidarians do not exist.

The paracnids in other species of the Coelogynoporidae, for example in *Invenusta paracnida* (Karling, 1966) and *Carenscoilia bidentata* Sopott, 1972 differ from those of *C. axi* in many details.

Introduction

Many turbellarians, for example species of the genus *Microstomum* and *Archimonocelis* have the ability to store cnids, which are derived from the food.

Within the family Coelogynoporidae conspicuous refractive glands are very common. These refractive glands have a certain similarity to cnids and are called paracnids (Karling 1966).

Among these paracnids there are some types, which show a clearly developed, eversible tube, for example, the glands of *Invenusta paracnida* (Karling, 1966), *Coelogynopora axi* Sopott, 1972 and *Coelogynopora gallica* Sopott-Ehlers, 1976. In the paracnids of other species, such as *Coelogynopora tenuis* Meixner, 1938 and *Carenscoilia bidentata* Sopott, 1972 these tubes are absent.

Lightmicroscopical observations lead to the conclusion that the glands of the Coelogynoporidae are autochthonous formations (cf. Karling 1966), but a more exact analysis of the fine structure of the different types of glands is impossible at the lower limits of light microscopy.

Materials and methods

The animals were extracted from sand samples taken from a beach at the eastern side of the island of Sylt (North Sea).

Before fixation for TEM the animals were kept in sea water diluted with distilled water (Sopott-Ehlers 1979). For further steps of TEM-preparation see Ehlers & Ehlers (1977.)

The methods for SEM-preparation are described in Ehlers (1977.)

Hydrobiologia 84, 253–257 (1981). 0018-8158/81/0843-0253/$01.00.

Results

The paracnids of *Coelogynopora axi* are situated in the peripheral parenchyme of the ventrolateral sides of the animals and have a diameter of about 10 μm. In live individuals they are round in the top view and heartshaped in profile. The tube measures 8 μm (Sopott-Ehlers 1972, p. 26f.).

As electronmicroscopy reveals, the paracnids of *Coelogynopora axi* consist of two cells: a muscle cell with myofibrils and a cell, which contains secretion.

The muscle cell surrounds the secretory cell in a cup-like manner, forming an envelope. Within the myofibrils no cell boundaries were observed, and apparently a single cell forms the muscle envelope of the secretory cell. Cell organelles, such as mitochondria, rough ER, and glycogen particles, can be found in peripheral sections of the proximal region of the muscle cell. A nucleus could not be clearly identified. A layer of basement lamina, which separates from the subepidermal basement lamina (Fig. 1B), runs between the muscle envelope and the surrounding tissue, as well as between muscle cell and secretory cell. The enveloping cell and the innermost layer of the basement lamina are linked to each other by hemidesmosomes (Fig. 1E). Branches of nerve cells run closely to the muscle envelope.

In the normal position, which means in the state of rest, the paracnids of *C. axi* are droplet-shaped (Fig. 1A). The cell containing secretion stretches distally to the body surface and has its opening – with microvilli – between epidermal cells. The cytoplasm surrounding this orifice (Fig. 1C) contains many microtubules. Epidermal cells and secretory cell are linked by zonulae adhaerentes and septate desmosomes. In the distal region the cell membrane is pleated several times in the longitudinal axis of the gland cell.

The epidermal cells of many turbellarians are characterized by the existence of a layer of tonofilaments, the cell web, (Bedini & Papi 1974). Such tonofilaments, are also contained in the distal part of the gland cell but more closely packed than in the cell web of the epidermal cells. This layer of tonofilaments continues into the interior of the gland cell (Fig. 1A, B) and builds up the eversible tube. The membrane, which bounds the layer of tonofilaments, forms a border of tightly arranged microvilli. In the proximal part of the cell the tube touches a plasm area, which is enclosed by a membrane and contains secretion granules (Fig. 1D). This area is linked to adjacent cell vacuoles by septate desmosome-like structures. The hemispherically shaped nucleus is located a short distance beneath this bowl containing secretion. Cell organelles- dictyosomes, rough ER, glycogen particles, free ribosomes and mitochondria (which have a conspicuous dark matrix) are found in the vicinity of the nucleus. Further away from the nucleus cell organelles occur in a marginal position. The largest part of cell volume contains vacuoles. Their contents have a fibrous to granulated texture, which may be coagulated fluid.

The process of eversion is an evagination. It results from contraction of the myofibrils of the muscle cell. During evagination the paracnid becomes lance-shaped (Fig. 2E). The nucleus of the secretory cell moves to the level of the epidermis. The tube is turned inside out, so that the microvilli are on the free surface (Fig. 2A-E). The tube extends beyond the epidermis like a chimney. The substances of the gland cell can be discharged through the distal opening of the tube.

Discussion

The evagination of the tube can be caused by addition of a few droplets of fixatives like picric acid or glutaraldehyde to the sea water. The stimulus is probably transmitted by the branches of the nerve cells in the vicinity of the enveloping cell to this cell. This causes to contraction of the myofibrils and evaginates the tube. Thus a reaction to chemical stimuli occurs. One can conclude from this, that the paracnids of *C. axi* have a function in protection against enemies; a function in prey capturing is also possible.

Many extrusomes in the protozoa (Hausmann 1970), as well as nematocysts and spirocysts in the cnidarians have a similar significance. Beyond this the spirocysts are supposed to have a function in substrate attachment (Mariscal *et al.* 1977). Substrate attachment is surely not a function of the paracnids, because the animals have well-developed adhesive organs. Beyond these functional aspects as mechanisms in protection or capturing prey, extrusomes, nematocysts and spirocysts are not com-

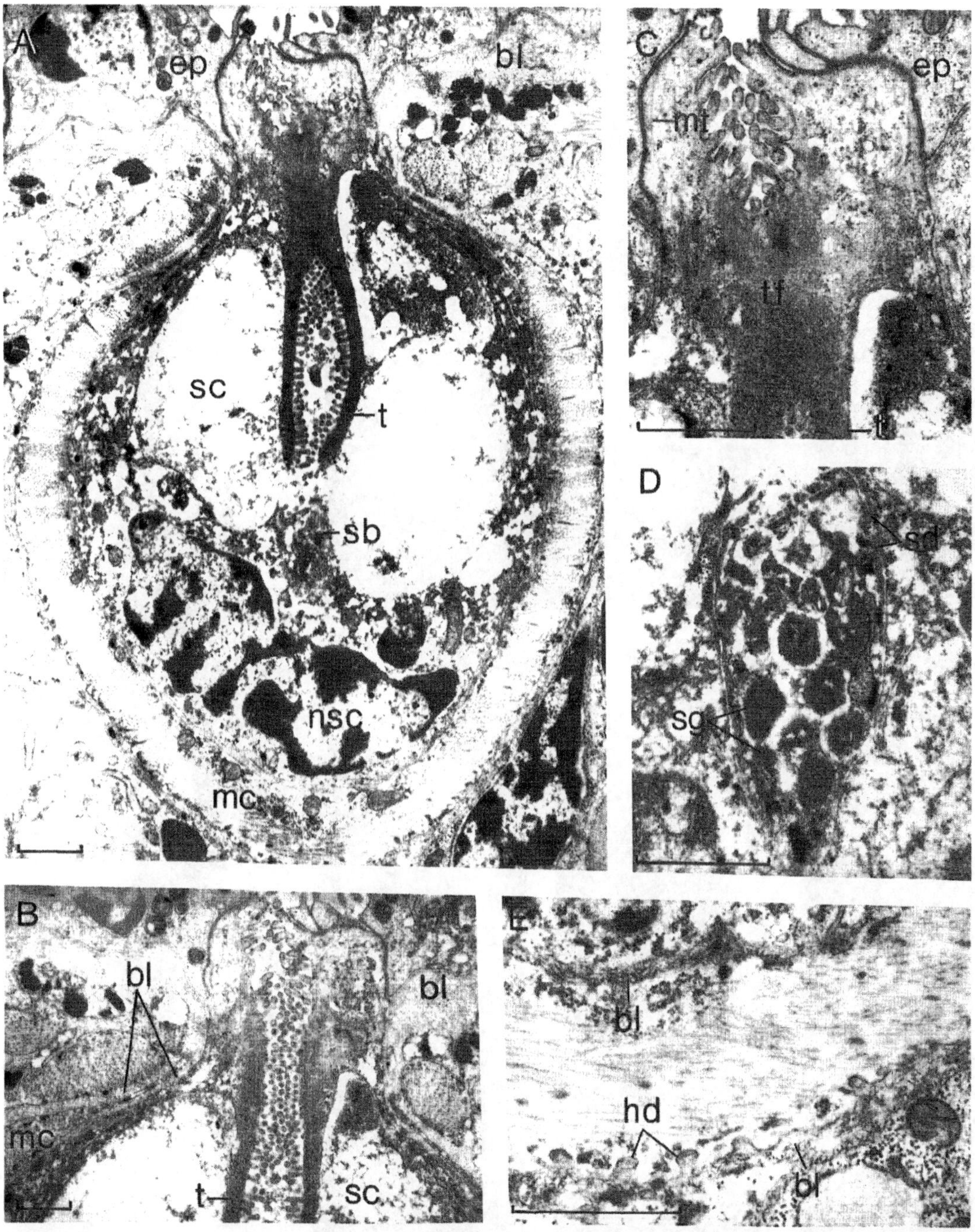

Fig. 1. Coelogynopora axi. A-E. TEM-micrographs. A. Paracnid in longitudinal section. B. Distal part of a paracnid. C. Microtubules in the vicinity of the orifice of the secretory cell. D. Bowl with secretion granules. E. The two layers of basement lamina. In the upper part the basement lamina between the secretory cell and the muscle cell; in the lower part the basement lamina between the muscle cell and the adjacent tissue. (Scale in all figures = 1 μm)

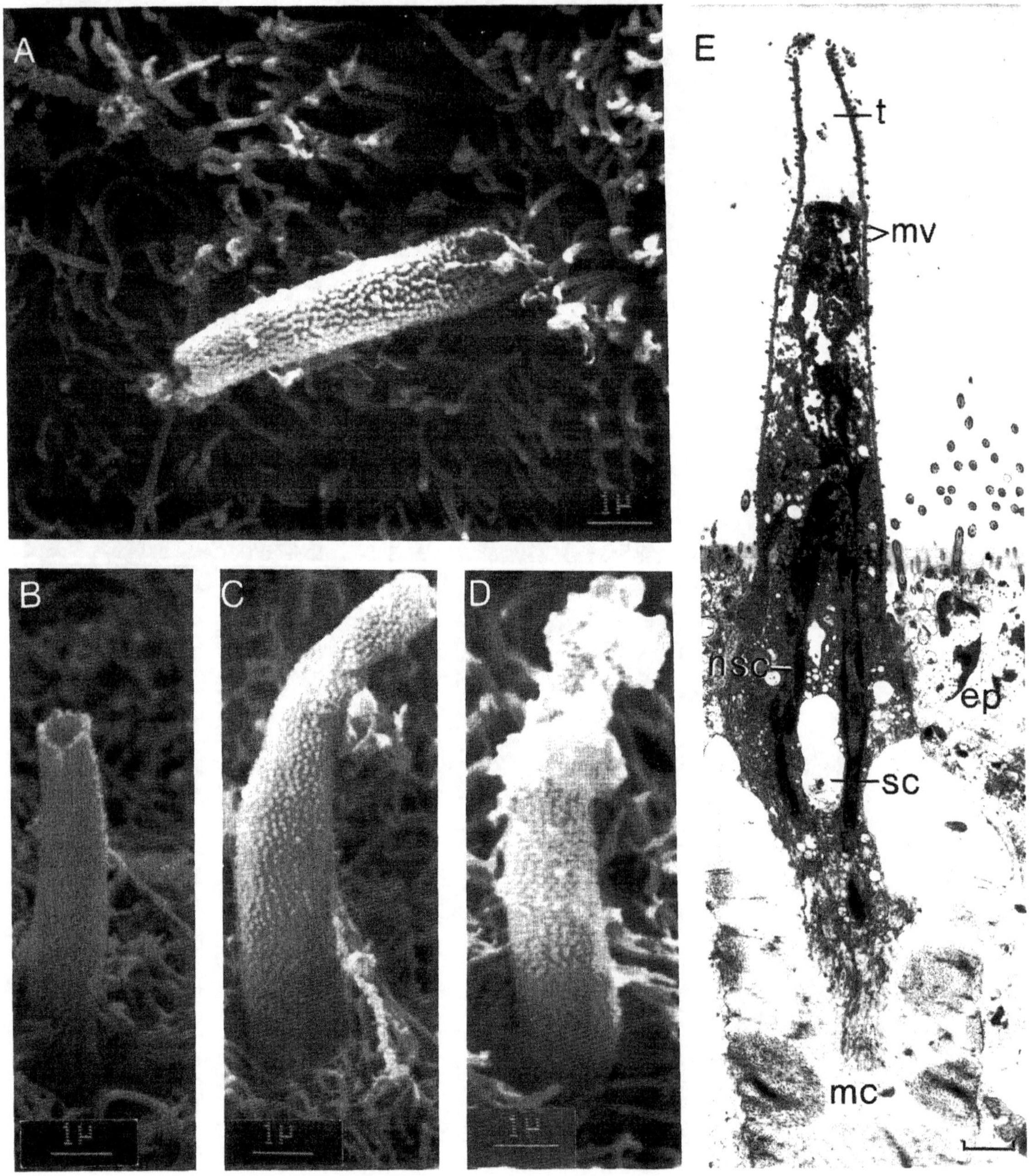

Fig. 2. Coelogynopora axi. A-D. SEM-micrographs. A and C. Completely evaginated tube. B. Partly evaginated tube. D. Completely evaginated tube with secretion. E. TEM-micrograph of an evaginated tube in longitudinal section.
(Scale in all figures = 1 μm)

parable with the paracnids of *C. axi*. Extrusomes develop from specialized regions of the cytoplasm, i.e., they are organelles. Nematocysts and spirocysts are differentiated within special cells, so they are also organelles (e.g. Mariscal 1974; Mariscal *et al.* 1976, 1977; Mariscal & McLean 1976; Ivester 1977; Satterlie & Case 1978). In contrast the paracnids of *C. axi* consist of two cells and consequently they are not organelles. Furthermore, no relations to the colloblasts of the ctenophora (Franc 1978) exist.

Karling's lightmicroscopical observations and the present electronmicroscopical finding make it very evident, that paracnids are genuine structures of the Coelogynoporidae. In addition, the ability of *C. axi* to regenerate the paracnids after injuries, e.g., damage of the tail, shows that paracnids are autochthonous formations.

Due to their similar lightmicroscopical structure, it is probable that the paracnids of *Coelogynopora gallica* correspond in their ultrastructure to those of *C. axi*. Therefore paracnids are probably not species-specific differentiations. In the species *Invenusta paracnida* and *Carenscoilia bidentata* two totally different fine-structured types of paracnids occur (Sopott-Ehlers in preparation).

Acknowledgements

Financial support and support in SEM-facilities was provided by the Akademie der Wissenschaften und der Literatur, Mainz.

Thanks are due to Dipl.-Biol. Daniel F. Whybrew for correcting the English in the manuscript and to Miss E. Hildenhagen for technical assistence.

Abbreviations

bl – basement lamina
ep – epidermis
hd – hemidesmosomes
mc – muscle cell
mt – microtubules
mv – microvilli
nsc – nucleus of the secretory cell
sb – bowl containing secretion granules
sc – secretory cell
sd – septate desmosome-like structures
sg – secretion granules
t – tube
tf – tonofilaments

References

Bedini, C. & Papi, F., 1974. Fine structure of the turbellarian epidermis. In: Biology of the Turbellaria. Riser, N. W. & Morse, M. P., Eds., New York: McGraw-Hill, pp. 108–147.

Ehlers, B., 1977. 'Trematodenartige' Epidermisstrukturen bei einem freilebenden proseriaten Strudelwurm (Turbellaria, Proseriata). In: Karling, T. G. & Meinander, M. Eds. The Alex. Luther Centennial Symposium on Turbellaria. Acta Zool. fenn. 154: 129–136.

Ehlers, U. & Ehlers, B., 1977. Monociliary receptors in interstitial Proseriata and Neorhabdocoela (Turbellaria, Neoophora). Zoomorph. 86: 197–222.

Franc, J. M., 1978. Organization and function of ctenophore colloblasts: an ultrastructural study. Biol. Bull. 155: 527–541.

Hausmann, K., 1978. Extrusive organelles in protists. Int. Rev. Cytology 52: 197–276.

Ivester, M. S., 1977. Nematocyst differentiation in the anthozoan Renilla reniformis (Pallas). Trans. Am. Microsc. Soc. 96: 238–247.

Karling, T. G., 1966. On nematocysts and similar structures in turbellarians. Acta Zool. fenn. 116: 1–28.

Mariscal, R. N., 1974. Nematocysts. In: Muscatine, L. & Lenhoff, H. M., Eds. Coelenterate Biology, reviews and new perspectives, pp. 129–178. New York, San Francisco, London: Academic Press.

Mariscal, R. N., Bigger, Ch. H. & McLean, R. B., 1976. The form and function of cnidarian spirocysts. 1. Ultrastructure of the capsule exterior and relationship to the tentacle sensory surface. Cell Tiss. Res. 168: 465–474.

Mariscal, R. N. & McLean, R. B., 1976. The form and function of the cnidarian spirocysts. 2. Ultrastructure of the capsule tip and wall and mechanism of discharge. Cell Tiss. Res. 169: 313–321.

Mariscal, R. N., McLean, R. B. & Hand, C., 1977. The form and function of cnidarian spirocysts. 3. Ultrastructure of the thread and the function of spirocysts. Cell Tiss. Res. 178: 427–433.

Satterlie, R. A. & Case, J. F., 1978. Neurobiology of the gorgonian coelenterates, Muricea californica and Lophogorgia chilensis. II. Morphology. Cell Tiss. Res. 187: 379–396.

Sopott, B., 1972. Systematik und Ökologie von Proseriaten (Turbellaria) der deutschen Nordseeküste. Mikrofauna Meeresboden 13: 1–72.

Sopott-Ehlers, B., 1979. Ultrastruktur der Haftapparate von Nematoplana coelogynoporoides (Turbellaria, Proseriata). Helgoländer wiss. Meeresunters. 32: 365–373.

Fine-structural observations on the central parenchyma in Convoluta sp.

Julian P. S. Smith III
Dept. of Zoology, Univ. of North Carolina at Chapel Hill, Chapel Hill, NC 27514, U.S.A.

Keywords: Turbellaria Acoela *Convoluta* sp., ultrastructure, central parenchyma

Abstract

Although the turbellarian order Acoela occupies a significant position in many theories on the origin of the Metazoa, detailed ultrastructural observations on members of this group are few in number. The acoelan central parenchyma has been recently described as at least partly syncytial on the basis of electron-microscopical data. These data, however, are insufficient for a complete understanding of the anatomical construction of the acoelan central parenchyma. In the present paper, a description of the micro-anatomy of the central parenchyma in juveniles of *Convoluta* sp. is presented. This region of the body consists of a central syncytium and certain specialized 'wrapping cells' which are associated with the syncytium. The wrapping cells are tentatively placed in three classes based on their positions and morphologies. On the grounds that the wrapping cells and the central syncytium form a closely-knit anatomical unit, it is proposed that all of these elements together should be considered homologous to the epithelial intestine of the other turbellarian orders. A brief consideration of the phylogenetic implications of these results suggests that the construction of the central parenchyma in *Convoluta* sp. is a derived feature in relation to the other Turbellarian orders, and perhaps in relation to certain other Acoela.

Introduction

Because of the significant position occupied by the turbellarian order Acoela in many theories of Metazoan evolution (Salvini-Plawen 1978, with references), an in-depth understanding of the phylogenetic position of the Acoela *vis-à-vis* the rest of the Turbellaria is necessary before these theories can be evaluated. Although there is accumulating evidence that the Acoela and Nemertodermatida are closely related phylogenetically, and most probably respectively derived and primitive (see Tyler & Rieger 1977), the question of from which of these two orders the higher turbellarians are to be derived is apparently still open (see Hendelberg 1977). These problems are due in no small part to the still rather incomplete knowledge of the ultrastructural anatomy of the forms concerned.

Since the time of the first published ultrastructural studies of acoels by Pedersen (1964) and Dorey (1965), which largely dispelled the notion that acoel tissues were syncytial, ultrastructural studies by Klima (1967), Kozloff (1972) and several Russian authors (see Ivanov & Mamkaev 1977) have all suggested that at least a portion of the central parenchyma in certain acoels is syncytial. None of these studies have given a sufficiently detailed account of the morphology of the central parenchyma to allow a decision as to what exactly should be considered homologous to the epithelial gut found in other turbellarian orders. The results given in the present paper are intended as a preliminary basis for such a decision.

Hydrobiologia 84, 259–265 (1981). 0018-8158/81/0843-0259/$01.40.

Materials and methods

Specimens of an undescribed species of the genus *Convoluta* were collected from mid-tide-level surface sediment of low-energy beaches at Swansboro and at Piver's Island, North Carolina. Sexually-mature specimens of this species measure approximately 500 μm to 600 μm in length and have separate male and female gonopores, a bursa seminalis with a 'cuticular' mouthpiece, and an unciliated, muscular penis invaginated into a muscular vesicula seminalis. A more complete description of this new species will be presented in a future paper.

Sexually-mature specimens of this *Convoluta* sp. collected in the months of June and July were allowed to copulate and lay eggs in seawater-filled fingerbowls kept in the dark. Embryonic stages just prior to hatching as well as juveniles and adults obtained from the sediment were processed for transmission electron microscopy using the glutaraldehyde and osmium tetroxide double fixation described in Tyler & Rieger (1977), with the addition of 1% alcian blue 8GX or 1% tannic acid to the glutaraldehyde primary fixative in some cases (noted below as 'AB', or 'TA', respectively). For the present study, data were taken from: serial thin cross sections of one 133 μm. long (fixed length) juvenile (AB), serial thin cross sections of one 95 μm. long (fixed length) pre-hatchling, sagittal thin sections of one 104 μm. long (fixed length) juvenile (TA), and sagittal thin sections of one 490 μm. long (fixed length) adult (TA). The thin sections were collected on copper grids, doubly stained in uranyl acetate and lead citrate, and examined and photographed using a Zeiss EM9S-2 electron microscope.

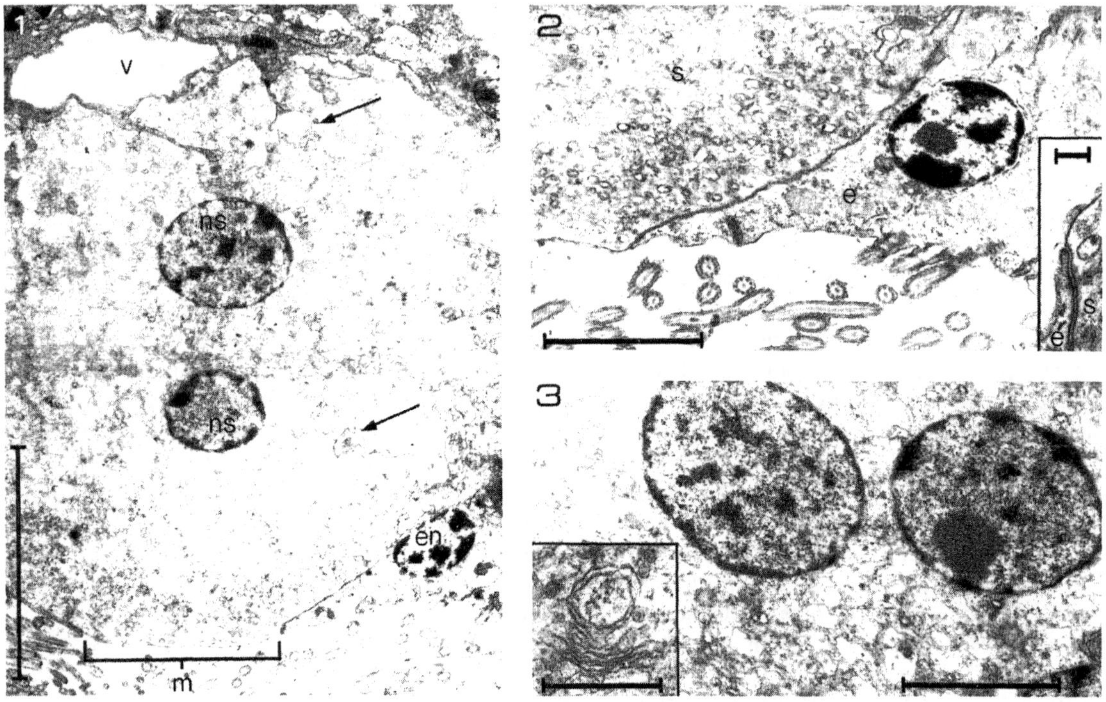

Fig. 1. Cross-section of a juvenile *Convoluta* sp. through the mouth, showing the general organization of the central syncytium. The arrows indicate the processes of wrapping cells invaginated into the central syncytium. Scale = 5 μm.
Fig. 2. Edge of the mouth opening, showing an epidermal cell and the central syncytium at the mouth. Scale = 2 μm.
Inset: Juncture of the central syncytium and an epidermal cell. Scale = 200 nm.
Fig. 3. Nuclei of the central syncytium. Scale = 2 μm.
Inset: A golgi apparatus of the central syncytium. Scale = 500 nm.

Results

Figure 1 shows a cross-section through the mouth region of a juvenile *Convoluta* sp. The central or digestive parenchyma occupies approximately one third of the cross-sectional area of the body at the level of the mouth. The major volume of the central parenchyma consists of a highly-vacuolated syncytium (Figs. 1–3).* At the mouth, this syncytium reaches the ventral surface of the body, and here it is bordered by epidermal cells (Fig. 2). The membrane surrounding the central syncytium is continuous across the mouth and peripherally, where the syncytium is in contact with other cells. The thin and extended nature of the epidermal cells bordering the mouth, where the central syncytium is somewhat protruded, suggests that there is some sort of attachment of the central syncytium to the epidermis here, although no typical junction is visible by transmission electron microscopy (Fig. 2, inset). The nuclei of the central syncytium are roughly spherical, of about 3 μm diameter, with single, rounded nucleoli (Fig. 3). The heterochromatin of these nuclei is only slightly condensed, with small patches scattered through the nucleus and a thin rim at the periphery (Fig. 3). The cytoplasm of the central syncytium contains an extremely large number of irregular vacuoles, 200–400 nm. in size. These give way to smaller rounded vesicles of about 75 nm. diameter in the immediate region of the mouth (Fig. 2). Mitochondria are abundant in the central syncytium, as are short segments of rough endoplasmic reticulum. The numerous Golgi bodies found in the central syncytium appear to be producing mostly small, dense cored vesicles of about 50 nm. diameter (Fig. 3, inset). Preliminary studies of a pre-hatchling of *Convoluta* sp. show that a well-defined, membrane-bounded central syncytium is already present at this stage in development (Fig. 4).

The central syncytium of *Convoluta* sp. is completely surrounded (except at the mouth) by the processes of certain specialized cells, which are here called 'wrapping cells'. Extensions of these wrapping cells are also invaginated into the central syncytium (Figs. 1 & 5–8). In all cases so far

* The term 'syncytium' will be used in this paper to denote a polyenergid state. At present, it is not possible to say whether the central portion of the digestive parenchyma in *Convoluta* sp. is 'syncytial' or 'plasmodial', *sensu stricto*.

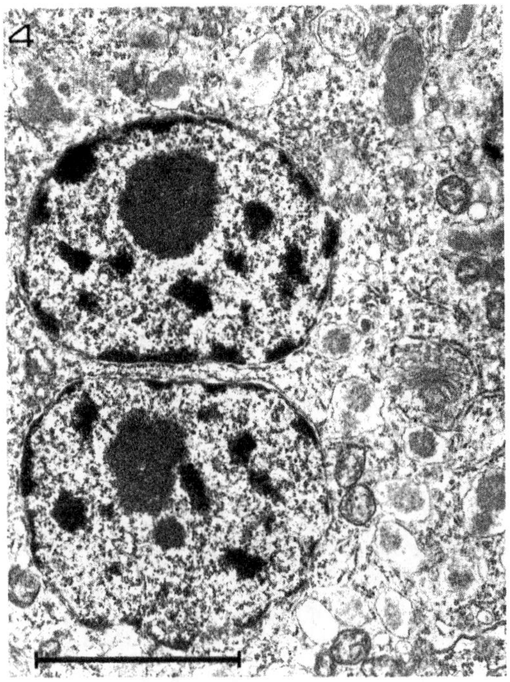

Fig. 4. Central portion of the digestive parenchyma in a pre-hatchling of *Convoluta* sp., showing the syncytial nature of this tissue. Also note the marked synthetic activity in the cytoplasm. Scale = 2 μm.

examined, the membranes of both the wrapping cells and of the central syncytium are unbroken, and maintain the nearly constant, small intercellular space characteristic of acoel tissues. No junctions have been seen between the wrapping cells or between the wrapping cells and the central syncytium. The nuclei of the wrapping cells are mostly located around the periphery of the central syncytium; rarely, the nucleus of a wrapping cell is invaginated into the central syncytium (Figs. 5 & 8). The nuclei of the wrapping cells are similar to those of the central syncytium in their heterochromatin pattern and their posession of a prominent, rounded, nucleolus. The wrapping cells have been assigned to three classes, based on their ultrastructure and position. This assignment is not meant to exclude the possibility, however, that two, or even all three types of wrapping cell may be different stages of differentiation of the same cell type.

Type I wrapping cells have a large cell body, and relatively few processes, and are found in the ventral and lateral regions surrounding the central syncytium (Figs. 5 & 6). In some cases, the cell body

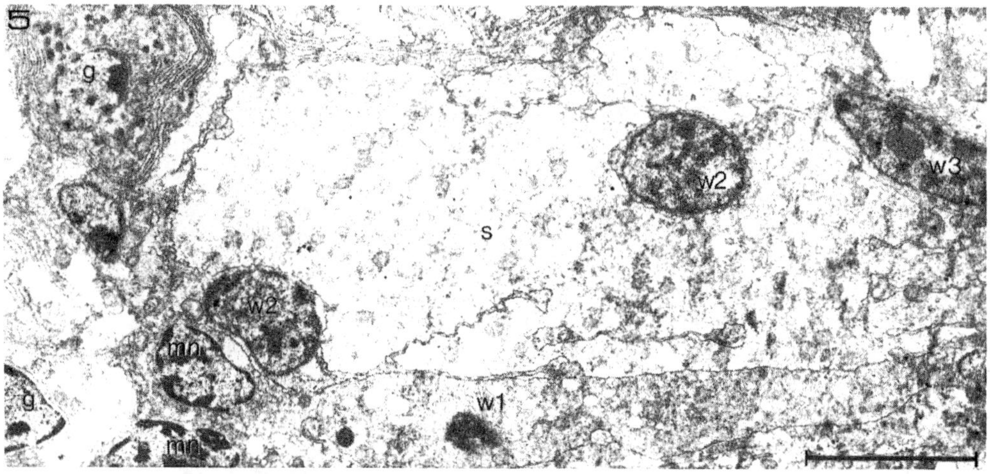

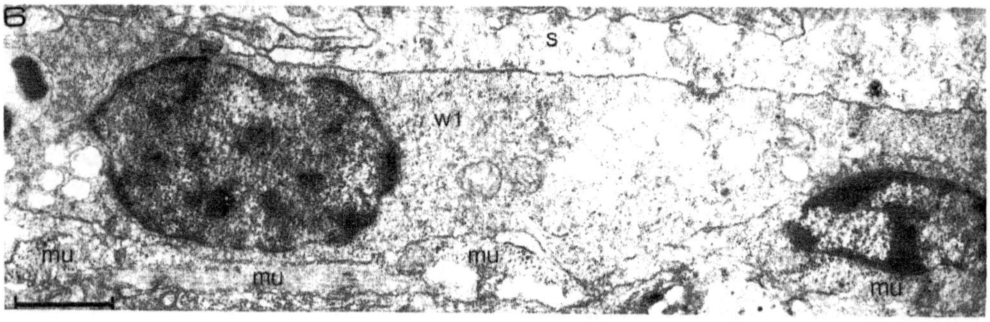

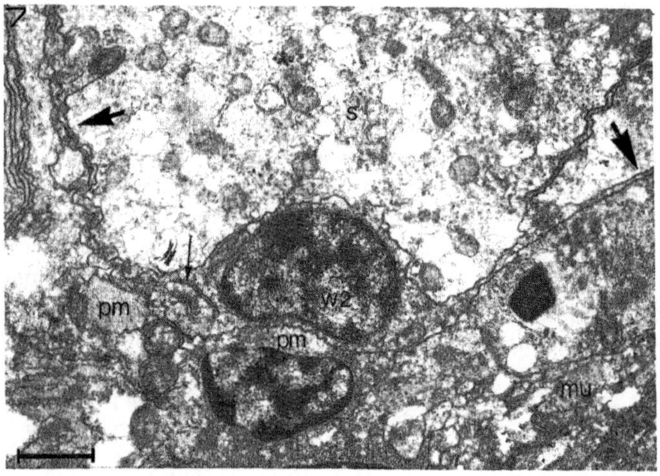

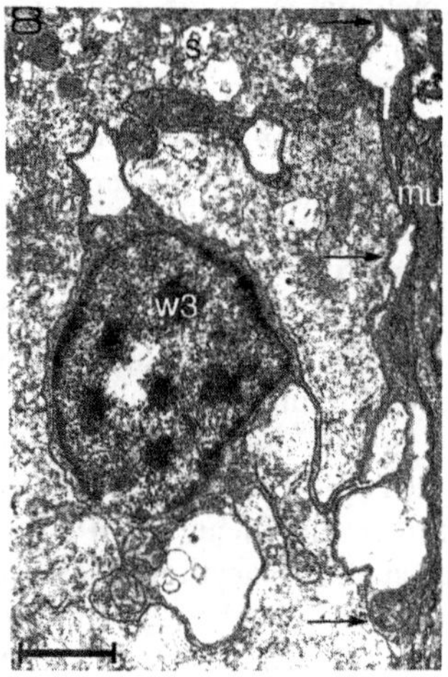

Fig. 5. Cross-section of a juvenile *Convoluta* sp., anterior to the mouth, showing the spatial relationships of the three types of wrapping cells, the central syncytium, and nearby gland and muscle cells. The dorsal surface is at the top of the figure. Scale = 3 μm.
Fig. 6. Type I wrapping cell, lying between the muscles of the ventral body wall and the central syncytium. Scale = 1 μm.
Fig. 7. Type II wrapping cell. The small arrow indicates the Golgi apparatus and a segment of rough ER. Note that the processes of the type II cell separate the parenchymal muscle fibers from the central syncytium.

Large arrows: processes of the type II cell overlap the processes of other wrapping cells (left) and partly cover the cell body of a type I wrapping cell (right). Scale = 1 μm.
Fig. 8. Type III wrapping cell. Arrows: Vacuole-containing processes of the type III wrapping cell separate the central syncytium from the muscle fibers of the dorsal body wall. Scale = 1 μm.

is slightly removed from contact with the central syncytium. In these cases, however, a process from the cell reaches the central syncytium. The rather homogeneous cytoplasm of the type I cells contains rough endoplasmic reticulum, several Golgi bodies, mitochondria, and numerous rounded vacuoles of about 200–400 nm. diameter (Fig. 6). Type II wrapping cells have a small cell body and numerous extended processes (Fig. 7). The processes of the type II cells, as thin as 35 nm. in some places, overlap one another and the processes of the other wrapping cells at the periphery of the central syncytium (Fig. 7). Because sagittal sections show that the type II cells are also extremely extended in the anterio-posterior direction, and because no figures are seen which could be interpreted as fingerlike protrusions of these cells cut in cross-section, it is probable that the processes of the type II cells have the form of thin sheets, both peripherally to the central syncytium, and where they are invaginated into it. Although no quantitative measurements have been made, it seems likely that the type II cells are responsible for covering the major proportion of the outer surface of the central syncytium. The cytoplasm of the type II cells contains relatively fewer vacuoles than that of the type I cells, and those vacuoles that are present in the type II cells are much more irregular than those of the type I cells. The other cytoplasmic components of the type II cells are similar to those of the type I cells. The type III wrapping cells appear to be confined to the dorsal side of the central syncytium (Figs. 1, 5 & 8). They are characterized by an extremely vacuolated cytoplasm, both in the region near the nucleus, and in their extended processes. In all other aspects are similar to the type II cells.

In the juvenile *Convoluta* sp., the wrapping cells or their processes are often the only elements interposed between the body wall and the central syncytium (Figs. 6 & 8). Laterally, the processes of the wrapping cells do not diverge from the immediate region of the central syncytium, and do not ramify into the rest of the body. Furthermore, all of the data at present indicate that it is only the wrapping cells described above that make contact with the central syncytium. Although there are longitudinally and dorsoventrally-oriented muscle fibers that lie near the central syncytium, they are always separated from the central syncytium by at least one wrapping cell process (Fig. 7).

Discussion

The question of what should properly be called 'central parenchyma' in the Acoela, and therefore, what should be considered homologous to the endodermal digestive epithelium of other Turbellarians, cannot be completely settled on the basis of the limited data available at present. However, the anatomical results of the present study, as well as recent functional observations on digestion (Kozloff 1972; Ivanov & Mamkaev 1977), argue strongly for a new understanding of what constitutes the 'central parenchyma' of the Acoela.

It should be clear from the results of this study that the wrapping cells in *Convoluta* sp. effectively isolate the central syncytium from the rest of the body, inasmuch as the wrapping cells are the only cells which make contact with the central syncytium (excepting, of course, the epidermal cells bordering the mouth). Furthermore, the processes of the wrapping cells either enclose the central syncytium or they are invaginated into it; they do not ramify into the rest of the body. Thus, the wrapping cells and the central syncytium, taken together, form a well-defined anatomical unit.

Certainly, the most direct evidence bearing on whether or not this 'anatomical unit' is the functional and phylogenetic equivalent of an epithelial digestive tract should come from functional studies. The studies of feeding and digestion in acoel turbellarians which have so far appeared in the literature, however, have been directed mainly towards an elucidation of the role of the central syncytium in these processes, and not towards an understanding of the functions performed by other cell types which may be involved. The syncytial state of the central parenchyma in the Acoela has been regarded as a specialization for the intracellular digestion of food items which are too large to be digested intracellularly in a single cell (Mamkaev & Seravin 1963, cited in Boguta & Mamkaev 1972; see also Mamkaev 1979; Kozloff 1972). Although no data have emerged to contradict this hypothesis, it is the ontogenetic origin and the maintainence of the central syncytium that is the prime concern here. The suggestion that the central part of the digestive parenchyma is rendered syncytial by the movements of cells with respect to one another in response to feeding (Boguta & Mamkaev 1972) seems very unlikely in the light of the observation that the pre-hatchling of *Convoluta* sp. has a well-developed central syncytium. A more likely hypothesis is that the syncytial portion of the central parenchyma in some acoels originates and is maintained through the 'holocrine' or 'morphonecrotic' secretion of cells which lie peripherally and release digestive enzymes (see Ivanov & Mamkaev 1977; Mamkaev 1979). If this theory can be extended to the digestive process in the *Convoluta* sp. of the present study, it seems that the most likely candidate for the 'holocrine-secretory' maintainence of the central syncytium is one of the wrapping cell types, perhaps the type I cells. The assertion of Markosova (1972; see also Mamkaev 1979) that the cells responsible for the 'holocrine' or 'morphonecrotic' secretion in *Convoluta convoluta* belong to the 'peripheral parenchyma' is a problem in terminology. Any cell type in the acoel Turbellaria whose primary task is the synthesis of digestive enzymes belongs to the digestive tract, and therefore cannot be considered a part of the 'peripheral parenchyma', according to Pedersen's (1964) definition of the term. For the present, then, the view that the central (digestive) parenchyma of *Convoluta* comprises the central syncytium as well as the three types of wrapping cells is a reasonable one.

The type of construction described above for the central parenchyma of *Convoluta* sp. may be rather widely distributed among the Acoela, considering the small, but diverse group of acoel species in which syncytia have been observed in the central parenchyma with the electron microscope (*Oligochoerus limnophilus*-Klima 1967; *Dopisthoporus* sp.-Doe 1976; *Otocelis luteola*-Kozloff 1972; *Anaperus biaculeatus, Convoluta convoluta* and *Aphanostoma diversicolor*-Ivanov & Mamkaev 1977). This notwithstanding, there are some acoels for which preliminary data indicate that the digestive tract is not constructed in a fashion similar to that of *Convoluta* sp. The digestive parenchyma of *Oxyposthia praedator* is described as consisting of 'amoeboid phagocytes' (Ivanov & Mamkaev 1977). The digestive tract of *Proporus minimus* has been described as epithelial on the basis of light microscopy (Dörjes 1971; see also Karling 1974). The importance of what is apparently a completely cellular digestive parenchyma in these two acoel species lies in their intermediate position, with respect to the digestive tract, between the partially-syncytial construction of the digestive tract seen in the forms cited above, and the epithelial digestive tract nearly always seen by electron microscopy in non-acoel Turbellaria (Nemertodermatida-references below; Tricladida-Bowen, Ryder & Thompson 1974, with references to earlier work; Macrostomida-Doe 1976; Rhabdocoela-Holt & Mettrick 1975).

Ivanov & Mamkaev (1977; see also Mamkaev 1979) have suggested that the primitive state of the digestive tract in the Turbellaria was an 'amorphous digestive parenchyma with wandering phagocytes' (exemplified by *Oxyposthia praedator*), which secondarily gave rise to a digestive epithelium in non-acoel Turbellaria. However, there is fairly good evidence that the order Nemertodermatida, the sister group to the Acoela, is primitive to the latter, based on the ultrastructure of the sperm and of the infraciliature (Tyler & Rieger 1977). Additional support for this idea has come from the recent discovery of a small amount of intercellular matrix between the epidermis and the muscles of the body wall in *Nemertoderma* sp. (author's unpublished data), which is a feature unknown for any member of the Acoela studied so far, but common to most other Turbellaria (see Rieger 1981). In view of this apparent primitive position of the Nemertodermatida in relation to the Acoela, the observation that the intestine in the two nemertodermatid species so far studied by EM is not merely cellular, but possibly epithelial, is highly significant (*Flagellophora* sp. - Tyler & Rieger, 1977; *Nemertoderma* sp. - author, unpublished). Current ultrastructural evidence thus suggests that the type of construction of the central parenchyma described for *Convoluta* sp. must be viewed as a derived condition in relation to a primitive epithelial condition of the digestive tract in the Turbellaria.

Acknowledgements

The author wishes to thank Dr. R. M. Rieger, who has provided much helpful discussion and comment concerning the thoughts contained herein, and Dr. David A. Doe, who provided unpublished electron micrographs of *Diopisthoporus* sp. The author's thanks are also due to Elena St. Jean for translation of the Russian papers. This study was supported by NSF grant #GB42211 (R. M. Rieger, P.I.) and by a grant from the Sigma Xi Scientific Society (S. L. Gardiner, P.I.).

Abbreviations

e - epidermal cell
en - nucleus of epidermal cell
g - nucleus of gland cell
m - mouth
mn - nucleus of muscle cell
mu - muscle fiber of body wall
ns - nucleus of central syncytium
pm - parenchymal muscle fiber
s - central syncytium
v - cytoplasmic vacuole of type III wrapping cell
w1 - type I wrapping cell
w2 - type II wrapping cell
w3 - type III wrapping cell

References

Boguta, K. K. & Mamkaev, Yu. V., 1972. Structure of the parenchyma of acoelous turbellarians (in Russian). Vestn. Leningr. Univ. Biol. 27: 15–29.

Bowen, I. D., Ryder, T. A. & Thompson, J. A., 1974. The fine structure of the planarian Polycelis tenuis (Ijima) II. The intestine and gastrodermal phagocytosis. Protoplasma 79: 1–17.

Doe, D., 1976. Comparative ultrastructure of the pharynx simplex in Turbellaria. Doctoral dissertation, Dept. of Zoology, Univ. of North Carolina at Chapel Hill.

Dorey, A. E., 1965. The organization and replacement of the epidermis in acoelous turbellarians. Q. J. Microsc. Sci. 106: 147–172.

Dörjes, J., 1971. Monographie der Proporidae und Solenofilomorphidae (Turbellaria, Acoela). Senckenb. Biol. 52: 113–137.

Hendelberg, J., 1977. Comparative morphology of turbellarian spermatozoa studied by electron microscopy. Acta Zool. Fenn. 154: 149–162.

Holt, P. A. & Mettrick, D. F., 1975. Ultrastructural studies of the epidermis and gastrodermis of Syndesmis franciscana (Turbellaria: Rhabdocoela). Can. J. Zool. 53: 536–549.

Ivanov, A. V. & Mamkaev, Yu. V., 1977. Über die Struktur des Digestionsparenchyms bei Turbellaria Acoela. Acta Zool. Fenn. 154: 59–61.

Karling, T. G., 1974. On the anatomy and affinities of the turbellarian orders. In: Eds. Riser, N. W. & Morse, M. P., Biology of the Turbellaria, New York: McGraw-Hill.

Klima, J., 1967. Zur Feinstruktur des acoelen Süsswasserturbellars Oligochoerus limnophilus (Ax & Dörjes). Ber. Naturwiss.-Med. Ver. Innsb. 55: 107–124.

Kozloff, E., 1972. Selection of food, feeding and physical aspects of digestion in the acoel turbellarian Otocelis luteola. Trans. Am. Microsc. Soc. 91: 556–565.

Mamkaev, Yu. V., 1979. On the histological organization of the turbellarian digestive system (in Russian). Tr. Zool. Inst. Akad. Nauk USSR 84: 13–24.

Mamkaev, Yu. V. & Seravin, L. N., 1963. Feeding of the acoelous turbellarian Convoluta convoluta (in Russian, cited in Boguta & Mamkaev 1972). Zool. Zh. 47: 197–205.

Markosova, T. G., 1976. Acid phosphatase activity during digestion in the anintestinal turbellarian Convoluta convoluta. J. Evol. Biochem. Physiol. (Engl. Transl. Zh. Evol. Biokhim. Fiziol.) 12: 183–184.

Pedersen, K. J., 1964. The cellular organization of Convoluta convoluta, an acoel turbellarian. A cytological, histochemical, and fine-structural study. Z. Zellforsch. Mikrosk. Anat. 64: 655–687.

Rieger, R. M., 1981. Morphology of the Turbellaria at the ultrastructure level. In: Schockaert, E. R. & Ball, I. R. Turbellaria. Proc. Third Int. Symp. Hydrobiologia. This volume.

Salvini-Plawen, L. von, 1978. On the origin and evolution of the lower Metazoa. Z. Zool. Syst. Evolutionsforsch. 16: 40–88.

Tyler, S. & Rieger, R., 1977. Ultrastructural evidence for the systematic position of the Nemertodermatida (Turbellaria). Acta Zool. Fenn. 154: 193–207.

The ultrastructure of the eyes in larval and adult polyclads (Turbellaria)

Alberto Lanfranchi, Celina Bedini & Enrico Ferrero[1]
Istituto di Biologia Generale, Univ. di Pisa, Via Volta 6, I-56100 Pisa, Italia
[1] *Present address: Istituto di Zoologia e Anatomia Comparata, Univ. di Trieste, Italia*

Keywords: Turbellaria, Polycladida, epidermal eye, cerebral eye, ultrastructure

Abstract

Examination of the larvae of *Thysanozoon brocchii* and *Stylochus mediterraneus* shows that they have both epidermal and cerebral eyes, while the young worms of *Notoplana alcinoi* have only cerebral eyes. A description is given of the ultrastructure of both kinds of eyes. The epidermal eye consists of one cup-shaped pigmented cell, whose cavity is filled with lamellae of ciliary origin. A small covering cell is located over the cup-opening. The cerebral eye is made up of three cells: one pigmented cell with ciliary projections and two rhabdomeric-type photoreceptor cells. The cerebral eye in the adult is formed of a pigmented cup without cilia and at least three rhabdomeric-type photoreceptors. A number of remarks of a morpho-onthological nature are presented.

Introduction

The morphology at the ultrastructural level of turbellarian photoreceptors has been reported for the adults of several groups (Wolken 1958; Press 1959; Röhlich Török 1961; Eakin 1963; Mac Rae 1964, 1966; Stewart 1966; Kishida 1967; Burt & Phillips 1969; Bedini *et al.* 1973; Bedini & Lanfranchi 1974; Carpenter *et al.* 1974; Durand & Gourbault 1975; Ehlers & Ehlers 1977a, b). The ultrastructure of the eyes of turbellarian larvae has been briefly described by Ruppert (1978) for the polyclads.

Trocophore-type larvae are known indeed to occur in polyclads, catenulids and perhaps nemertodermatids. Such larvae are named Müller's larva, Götte's larva and Luther's larva.

The aim of the present investigation is to obtain further information on the morphology of the photoreceptors of polyclads during larval and direct development, so as to throw light on possible ultrastructural differences occuring in the larvae, the young worms and the adults.

Materials and methods

Specimens of the polyclads *Thysanozoon brocchii* (Risso, 1818), *Stylochus mediterraneus* Galleni, 1976 and *Notoplana alcinoi* (O. Schmidt, 1861) collected in the Ligurian Sea near Leghorn, were maintained separately in culture dishes. Müller's larvae of *T. brochii* and Götte's larvae of *S. mediterraneus* were examined at the three-eyed stage, in which there were one asymmetrically located epidermal eye and two symmetrically located cerebral eyes. The young worms of *N. alcinoi* were studied at the stage with four cerebral eyes Fixation was carried out in 1% osmium tetroxide in phosphate buffer 0,1 M, followed by dehydration and epon-araldite mixture embedding using routine methods. Thin sections stained with both solutions of uranyl acetate and lead citrate were examined through a Siemens Elmiskop 1A electron microscope.

Hydrobiologia 84, 267–275 (1981). 0018-8158/81/0843-0267/$01.40.

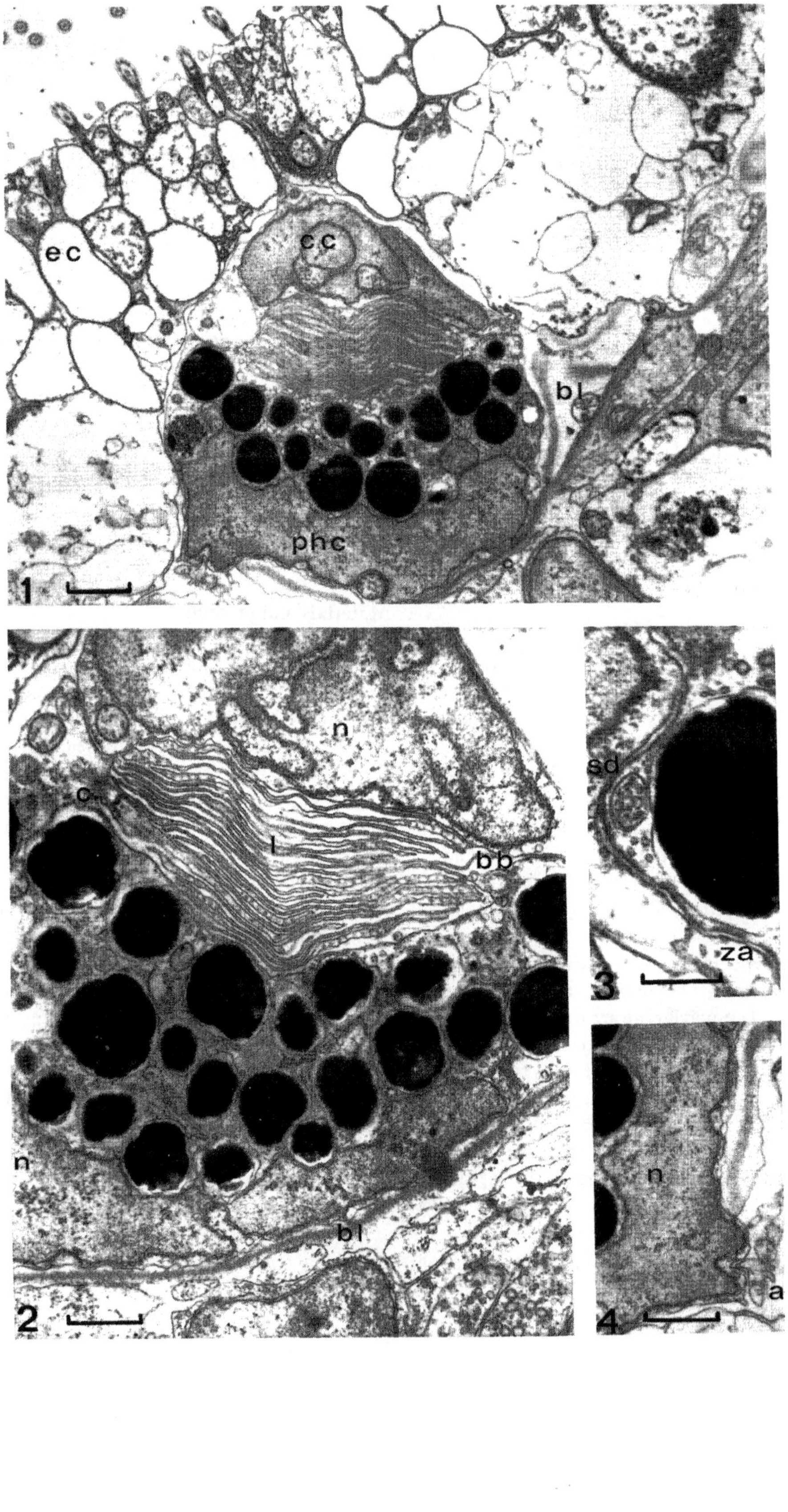
ec
cc
bl
phc
1
n
c
l
bb
n
bl
2
sd
za
3
n
a
4

Results

Epidermal eye A single epidermal eye has been found in Müller's larvae of *T. brochii* and in Götte's larvae of *S. mediterraneus*. It is lacking in the young worms of *N. alcinoi*.

This epidermal eye is located in an antero-lateral position and lies between the epithelial cells and the basal lamina (Fig. 1). It consists of one cup-shaped epithelial cell (maximum diameter: 10–13 μm), whose nucleus is situated at the base. The nucleus surface is indented. Pigment granules and mitochondria are located in the indentations. The cytoplasm is occupied almost entirely by pigment granules, generally arranged in a double row. The granules are about 1,5 μm in diameter; they are bounded by a membrane, and their content is homogeneous and strongly electrondense. The Golgi apparatus is present near the cup cavity. The elements of the smooth endoplasmic reticulum are scanty. The cup cavity is filled with stacked laminae (Fig. 2), which are caused by the flattening of cilia of the $9 \times 2 + 2$ pattern, and whose basal bodies are situated inside the lateral wall bounding the cavity. The laminae are about 350 Å thick, and contain the microtubular doublets. In the regions devoid of microtubules they are thinner (150 Å).

A small sphere-cap shaped cell, occupied almost entirely by its nucleus, is located over the cup-opening. Its cytoplasm is limited to a thin layer, which increases in volume in the lateral regions in contact with the pigmented cell. In this cytoplasm, mitochondria, vesicles and Golgi complexes are present. In the contact regions between both cells zonulae adhaerentes and septate desmosomes may be noted (Fig. 3). A nervous projection runs just under the cup-shaped photoreceptor cell and above the basal lamina and appears to be in contact with this cell (Fig. 4).

Cerebral eye. Two cerebral eyes are present in the Müller's and Götte's larvae we examined (Fig. 5). The young worms of *Notoplana alcinoi* were studied at the four-cerebral-eye stage (Fig. 6).

The cerebral eye is located antero-laterally under the basal lamina near the cerebral ganglion (Fig. 7). It is made up of three cells: one pigmented cell and two photoreceptor cells. Together, the three cells form a globose mass (maximum diameter: 12–20 μm), the pigmented region of which is oriented toward the interior of the body (Fig. 8).

Pigmented cup. This is a crescent-shaped nucleus on the bottom with pigment granules generally arranged in a double row over it. The cytoplasmic regions devoid of granules contain mitochondria, vesicles and Golgi complexes. The diameter of the granules is about 1 μm; they are bounded by a membrane and their content is homogeneous and highly electrondense. The pigmented cup borders continue in slender cytoplasmic projections, bounding a space not much larger that the space bounded by the cup. In the region which is diametrically opposite the base of the cup, there are cytoplasmic projections with mitochondria and numerous basal bodies, which produce cilia (about 500), partially filling the cup cavity. The cilia (10 μm in length), with $9 \times 2 + 2$ pattern, and devoid of dynein arms, are closely packed, and, in some regions, joined by septate desmosome-like structures (Fig. 9). There are no ciliary rootlets. There is an axon bundle in contact with the ciliated projections.

Ganglionar cells. Two ganglionar cells are located near the pigmented cup borders, in diametrically opposite positions. They are made up of a cell body, a dendritic process with a photosensitive portion and an axon (Fig. 10). The cell body, situated immediately outside the cup, contains an irregularly-shaped nucleus with chromatin condensed into aggregates; Golgi complexes, small vesicles, free ribosomes and glycogen granules are to be found in the cytoplasm. The dendritic process (Fig. 11) contains numerous mitochondria and vesicles in the proximal region. The intermediate region, at its point of entrance into the cytoplasmic projections of the cup, is constricted (1.5 μm wide) and contains vesicles and microtubules. Inside the cup the distal dendritic

Fig. 1. Epidermal eye. Note the pigmented photoreceptor cell and the covering cell (scale = 2 μm).

Fig. 2. Epidermal eye. Note the closely stacked laminae produced by the cilia (scale = 1 μm).

Fig. 3. Epidermal eye. Note the zonulae adhaerentes and septate desmosomes between the pigmented cell and the covering cell (scale = 0.5μm).

Fig. 4. Epidermal eye. Note the nervous projections in contact with the photoreceptor cell (scale = 1 μm).

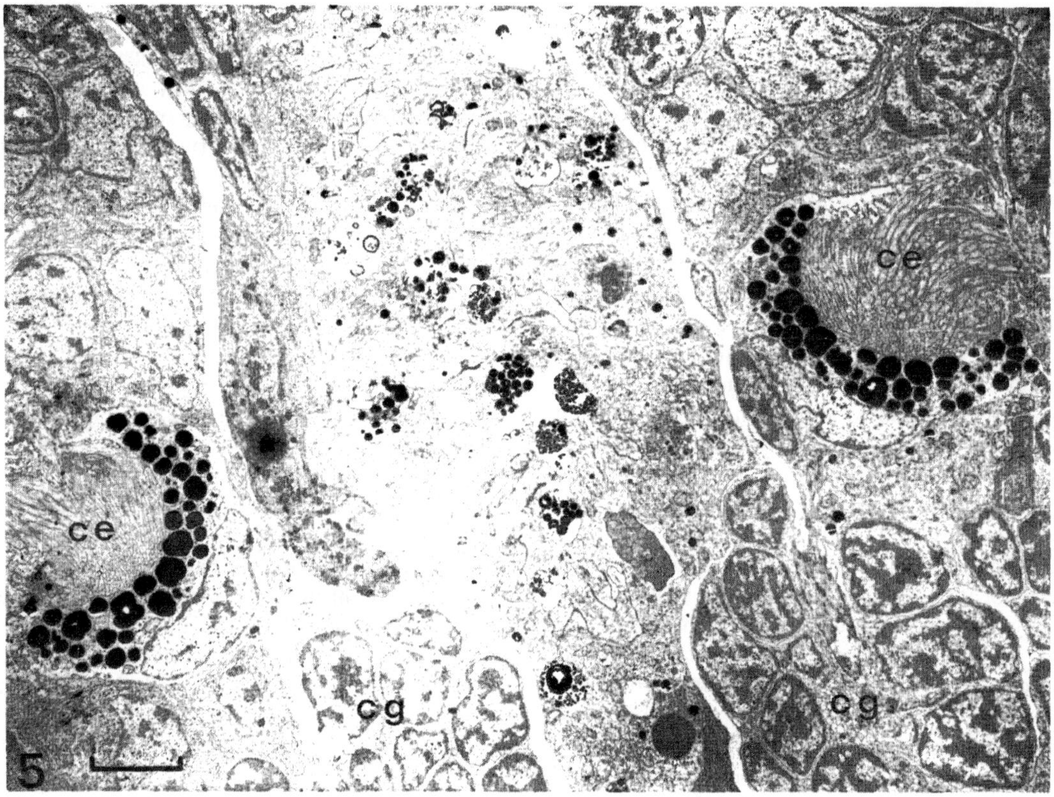

Fig. 5. Cerebral eyes. Note their localization in the mesoderm (scale = 5 μm).

Fig. 6. Cerebral eyes. Note the face to face orientation of the cerebral eyes in *Notoplana alcinoi*.

region spreads out like a fan and gives rise to rhabdomeric microvilli (2.5 μm in length), which extend as far as the base of the cup. At the base of the microvilli, numerous vesicles and a few mitochondria are to be found. The pigment cell is joined to the dendritic process by zonulae adhaerentes and septate desmosomes.

The cerebral eye in the adult. Only cerebral eyes are to be found in the adult. They are made up of a pigmented cup without cilia, which contains the dendritic processes of at least three rhabdomeric-type photoreceptor cells (Fig. 12).

Discussion

Contrary to the conclusions drawn by Ruppert (1978), the epidermal and cerebral eyes of Müller's and Götte's larvae, as well as the cerebral eyes of the

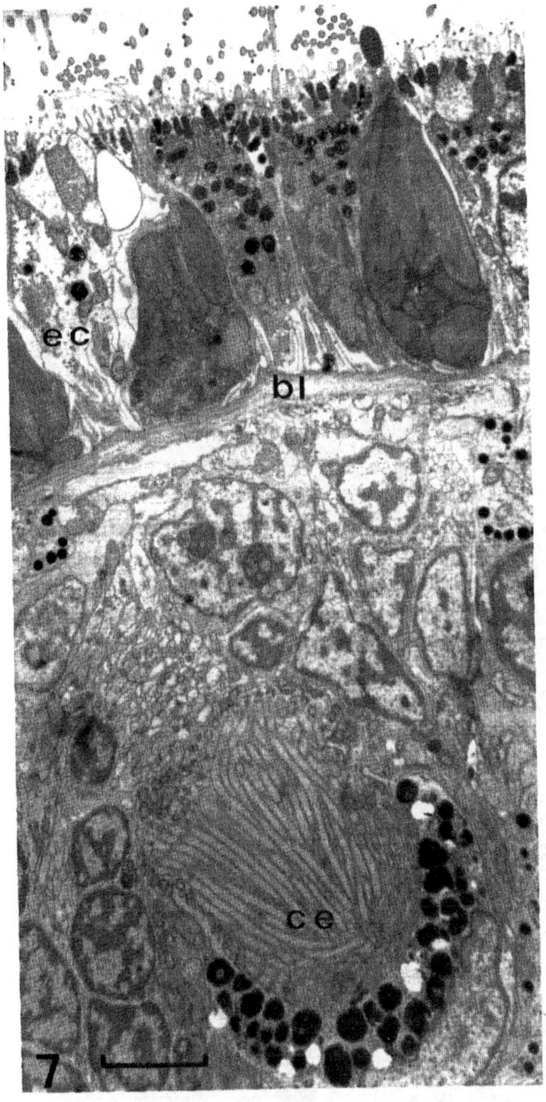

Fig. 7. Cerebral eye. Note its position with respect to the epidermis (scale = 5 μm).

young worms of *Notoplana alcinoi*, are not definitive but transitory organs. According to our ultrastructural observations, the epidermal eye conceivably sinks into the parenchyma and changes into a cerebral eye with the addition of two ganglionar cells. The cilia of the epithelial pigmented cell persist in the larval cerebral eye, while they disappear in the cerebral eye of the adult. In the latter, the number of photoreceptor ganglionar cells increases. With references to the development of polyclads, Kato (1940) initially observed that two eyes form in the epidermis, and subsequently, that one of these sinks into the mesoderm. Before this happens the other eye is divided, giving rise to one eye which remains located in the epidermis, while the other sinks into the mesoderm. Our ultrastructural findings appear to agree with Kato's embryological data, but not with the results obtained by Lang (1884). While not going through a larval-lobed stage, *N. alcinoi* has, in the case of young worms, cerebral eyes whose structure is the same as those of Müller's and Götte's larvae. These eyes will change into the typical cerebral eyes of the adult. Although *N. alcinoi* has been considered as a species with direct development, it must be emphasized that the 'miniature worm' possesses organs which will undergo transformation before becoming the definitive organs of the adult.

The covering cell of the pigmented cup of the epidermal eye and the bundle of cilia of the cerebral eye might function as a lens, according to Ruppert's hypothesis. The cilia of the cerebral eye, however, might still have a photosensitive function, owing to the facts that they are probably derived from those of the epidermal eye, that ciliary rootlets and dynein arms are not to be found, and that septate desmosome-like structures are present among them. Ciliary aggregations (Ehlers & Ehlers 1977b) and bundles of interwoven cilia (Palmberg *et al.* 1980) have been described in turbellarians, otoplanids and macrostomids respectively. Both types of structures have been interpreted as photosensitive organelles.

Two pigmented cup ocelli are present in the trochophore of the polychaete *Harmothoë imbricata* (Holborow & Laverack 1972). Since Müller's and Götte's larvae are of the trochophore type, morphological affinities might exist between their photoreceptors. On the other hand, the epidermal ocellus of the trochophore does not appear to be of the ciliary type, as is the asymmetrical epidermal eye of the polyclads which we examined, and its photosensitive portion does not belong to the pigmented cell. The trochophore ocellus seems more similar to the cerebral eye of the polyclads, even though there is no bundle of cilia and it is situated in the epidermis and not in the mesoderm. In conclusion, the epidermal eye of the larval polyclads appear to be, structurally, more primitive than the ocellus of the polychaete trochophore.

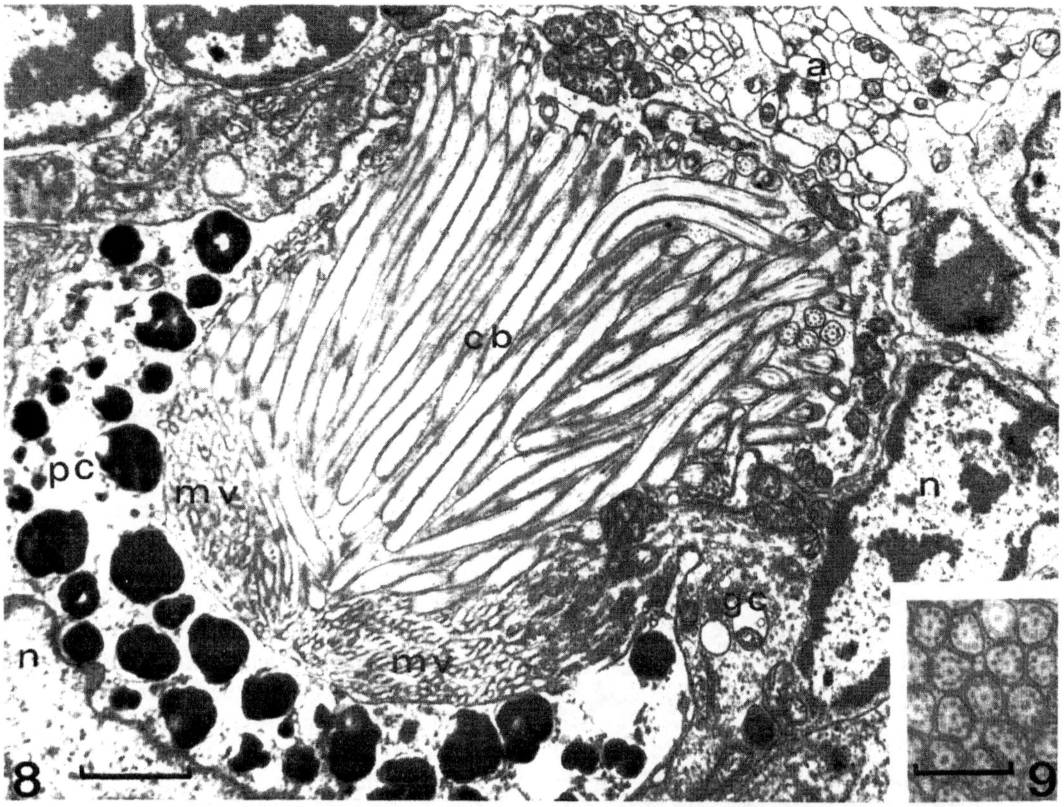
a
cb
pc
mv
n
mv
gc
n
8
9

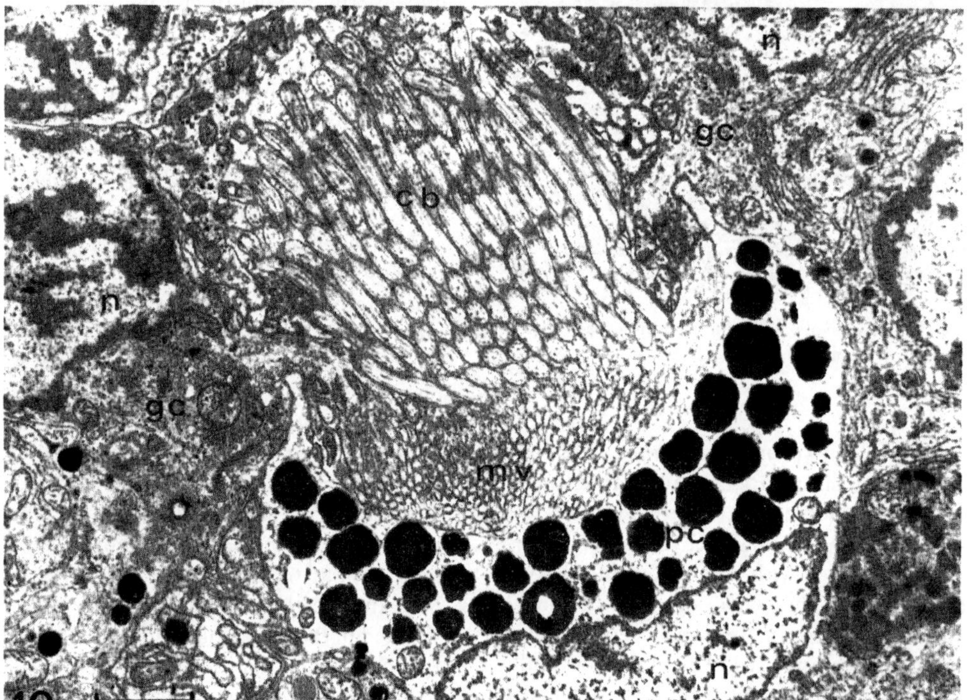
n
gc
cb
n
gc
mv
pc
n
10

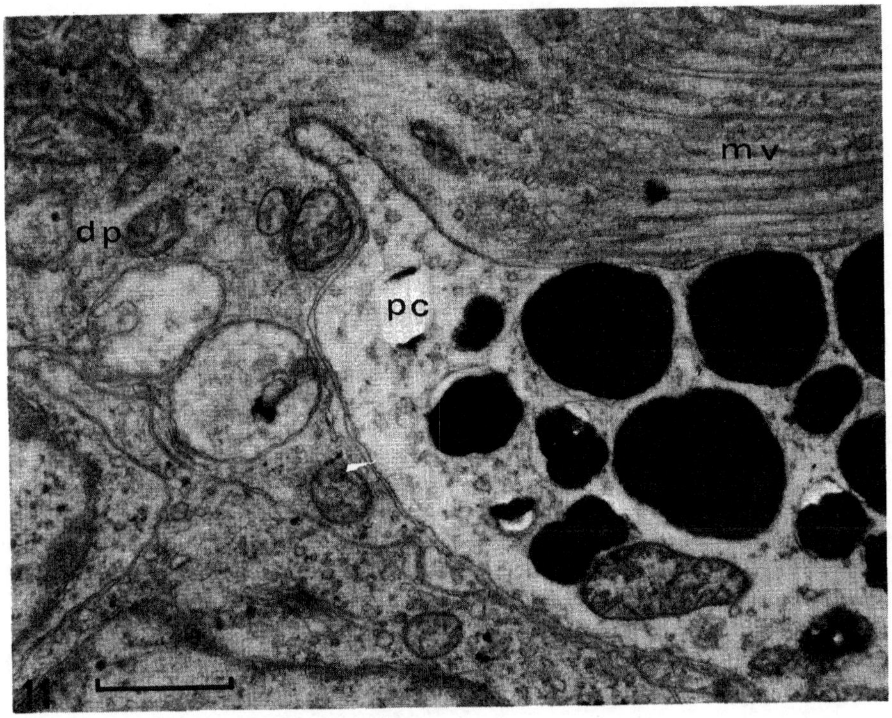

Fig. 11. Cerebral eye. Note the detailed structure of the dendritic process and the presence of zonulae adhaerentes and septate desmosomes between the pigment cell and the photoreceptor (scale = 1 μm).

Acknowledgements

This work was supported by Consiglio Nazionale delle Ricerche, Gruppo Nazionale Scienze Neurologiche.

We should like to thank Dr. L. Galleni for having provided the polyclads.

Abbreviations

a – axons
bb – basal body
bl – basal lamina
c – cilium
cb – cilium bundle
ce – cerebral eye
cg – cerebral ganglion
dp – dendritic process
ec – epithelial cells
gc – ganglionar cell
l – laminae
mv – microvilli
n – nucleus
pc – pigmented cup
ph – photoreceptor
sd – septate desmosomes
za – zonulae adhaerentes

Fig. 8. Cerebral eye. Note the pigmented cell projections giving rise to a ciliary bundle, which partially fills the cup cavity (scale = 2 μm).
Fig. 9. Cerebral eye. Note the septate desmosome-like structures between the cilia (scale = 1 μm).
Fig. 10. Cerebral eye. Note how the rhabdomeric dendritic processes of the two photoreceptors penetrate into the cup cavity (scale = 2 μm).

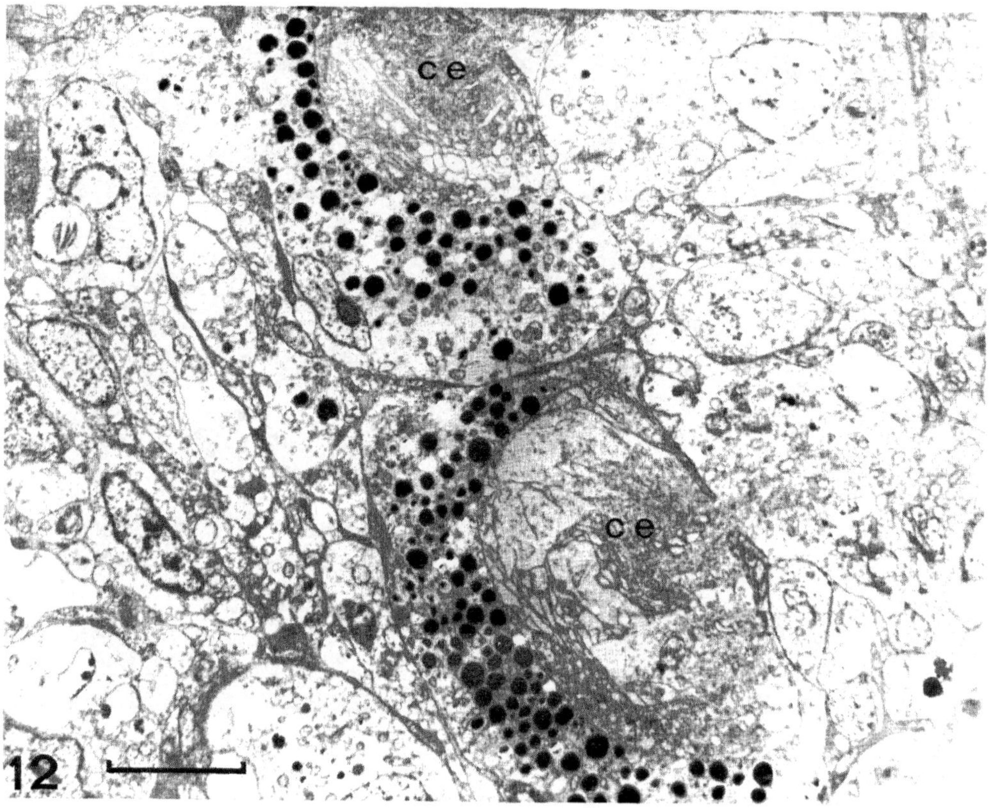

Fig. 12. Cerebral eyes in the adult. Note the pigmented cup without cilia (scale = 10 μm).

References

Bedini, C., Ferrero, E. & Lanfranchi, A., 1973. Fine structure of the eyes in two species of Dalyelliidae (Turbellaria, Rhabdocoela). Monit. Zool. Ital. (N.S.) 7: 51–70.

Bedini, C. & Lanfranchi, A., 1974. The fine structure of photoreceptors in two otoplanids species (Turbellaria, Proseriata). Z. Morph. Tiere 77: 175–186.

Burt, M. D. B. & Phillips, W. L., 1969. The fine structure of Urastoma cyprinae, a microturbellarian found in oysters. Am. Zool. 9: 622.

Carpenter, K. S., Morita, M. & Best, J. B., 1974. Ultrastructure of the photoreceptor of the planarian Dugesia dorotocephala. I. Normal eye. Cell. Tiss. Res. 148: 143–158.

Durand, J. P. & Gourbault, N., 1975. Etude cytologique des organes photorécepteurs de Dendrocoelopsis (Amyadenium) chattoni (Triclade paludicole hypogé). Ann. Spéléol. 30: 129–135.

Eakin, R. M., 1963. Lines of evolution of photoreceptors. In: Eds., Mazia, D. & Tyler, A., General physiology of cell specialization, New York: Mc Graw-Hill, pp. 393–425.

Ehlers, B. & Ehlers, U., 1977a. Die Feinstruktur eines ciliären Lamellarkörpers bei Parotoplanina geminoducta Ax (Turbellaria, Proseriata). Zoomorphologie 87: 65–72.

Ehlers, B. & Ehlers, U., 1977b. Ultrastruktur pericerebraler Cilienaggregate bei Dicoelandropora atriopapillata Ax und Notocaryoplanella glandulosa Ax (Turbellaria, Proseriata). Zoomorphologie 88: 163–174.

Holborow, P. L. & Laverack, M. S., 1972. Presumptive photoreceptor structures of the trochophore of Harmothoë imbricata (Polychaeta). Mar. Behav. Physiol. 1: 139–156.

Kato, K., 1940. On the development of some japanese polyclads. Jap. J. Zool. 8: 537–574.

Kishida, Y., 1967. Electron microscopic studies on the planarian eye. I. Fine structures of the normal eye. Sci. Rep. Kanazawa Univ. 12: 75–110.

Lang, A., 1884. Die Polycladen des Golfes von Neapel. Fauna u. Flora Golfes Neapel XI, 688 pp.

MacRae, E. K., 1964. Observations on the fine structure of photoreceptor cells in the planarian Dugesia tigrina. J. Ultrastruct. Res. 10: 334–349.

MacRae, E. K., 1966. The fine structure of photoreceptors in a marine flatworm. Z. Zellforsch. 75: 469–484.

Palmberg, I. Reuter, M. & Wikgren, M., 1980. Ultrastructure of epidermal eyespots of Microstomum lineare (Turbellaria, Macrostomida). Cell Tiss. Res. 210: 21–32.

Press, N., 1959. Electron microscope study of the distal portion of a planarian retinular cell. Biol. Bull. 117: 511–517.

Röhlich, P. & Török, L. J., 1961. Elektronenmikroskopische Untersuchungen der Augen von Planarien. Z. Zellforsch. 54: 362–381.

Ruppert, E. E., 1978. A review of metamorphosis of turbellarian larvae. In: Chia & Rice, Settlement and Metamorphosis of Marine Invertebrate Larvae, Elsevier North-Holland Biomedical Press, pp. 65–81.

Stewart, A. M., 1966. Ultrastructure of the eyes in three freshwater triclad flatworms. Am. Zool. 6: 615–616.

Wolken, J., 1958. Studies of photoreceptor structures. Ann. N.Y. Acad. Sci. 74: 164–181.

Ultrastructure of the eye of Urastoma cyprinae (Turbellaria, Alloeocoela)

M. D. B. Burt and G. N. Bance
Dept. of Biology, Univ. of New Brunswick Fredericton, N.B., Canada E3B 5A3

Keywords: Turbellaria, ultrastructure, eye, *Urastoma*

Summary

Urastoma cyprinae (Graff) is a microturbellarian which has been recorded both as a free-living organism by Westblad (1955) and Marcus (1951) and as a commensal in various lamellibranch molluscs (see Burt & Drinnan 1968). The material used in this study came from oysters, *Crassostroea virginica*, collected off the coast of Prince Edward Island, in which hosts it occurs in large numbers especially during the summer months when the oysters are spawning (Fleming *et al.* 1981). When *U. cyprinae* is exposed to light as happens, for example, when an oyster is opened, it shows a marked negative phototactic response.

Preliminary work on the fine structure of the photoreceptors in *U. cyprinae* shows that the two eyes each consist of: (1) a single cup cell full of relatively large, electron-dense pigment granules; (2) a tripartite conical lens system; and (3) what appear to be two photosensitive rhabdomes. The pigment cup cell has a single, well defined nucleus situated basally and close to the membrane of the pigment cell furthest away from the rhabdomeres. The lens system consists of a cone made up of three, separate but equal, parts. Each part has two, flat inner surfaces which join at an angle of 120°, an outer rounded surface, and a rounded upper surface. When these three parts fit together, the cone-shaped lens is formed with the apex of the lens within the 'cup' of the pigment cell and the rounded, convex, broad end of the cone lying more or less at the same level as the top of the pigment cup and below the epidermis layer. The rhabdomeres lie between the electron dense lenses and the inside of the pigment cup. They show connections to the visual cells which are bipolar: one extension joining the rhabdomeres; the other constituting the axon which extends into the centrally situated brain or into the longitudinal, lateral nerves. The axons that enter the brain, form connections with other axons from the other eye. The axons that extend posteriorly in a lateral position, presumably play a role in facilitating the avoidance reaction.

The chemical nature of the unusual lens has not yet been determined. This is presently under investigation and will be reported later at which time our work will be discussed in relation to other types of rhabdomeric eyes in the Turbellaria.

References

Burt, M. D. B. & Drinnan, R. E., 1968. A microturbellarian found in oysters off the coast of Prince Edward Island. J. Fish. Res. Bd. Can. 25: 2521–2522.

Fleming, L. C., Burt, M. D. B. & Bacon, G. B., 1981. Some Commensal Turbellaria of the Canadian East Coast. In: Schockaert, E. R. & Ball, I. R. (Eds.). Turbellaria. Proc. Third Int. Symp. Hydrobiologia. This volume.

Marcus, E., 1951. Turbellaria brasileiros (9). Bolm. Fac. Filos. Cienc. Univ. Sao Paulo. (Zool.) 16: 5–215.

Westblad, E., 1955. Marine 'Alloeocoels' (Turbellaria) from North Atlantic and Mediterranean coasts. I. Ark. Zool. 7: 491–526.

Hydrobiologia 84, 276 (1981). 0018-8158/81/0843-0276/$00.20.

Observations on the ultrastructure of the copulatory organ of Archilopsis unipunctata (Fabricius, 1826) (Proseriata, Monocelididae)

Els E. Martens and Ernest R. Schockaert
Dept. SBM, Limburgs Universitair Centrum, B – 3610 Diepenbeek, Belgium.

Keywords: Turbellaria, Proseriata, *Archilopsis unipunctata*, ultrastructure, copulatory organ

Abstract

The copulatory organ in adult specimens of *Archilopsis unipunctata* has been studied by transmission electron microscopy.

This copulatory organ is of the conjuncta-duplex type with eversible cirrus. The seminal vesicle, lined with a nucleate epithelium, is surrounded by spirally arranged muscles. The fibres are enclosed in a sheath that is continuous with the septum of the bulbus and the basement lamina of the male canal epithelium. Distally to the seminal vesicle the bulbus is filled with the secretory cell-necks of the prostate glands. The male canal shows three different parts: seminal duct, ejaculatory duct and eversible cirrus. At the transition of seminal duct and ejaculatory duct two prostate ducts open into the lumen. The structure of the epithelium lining the different parts of the canal is described. The transition into the cirrus may be recognized by an abrupt change in the thickness, the electron density and the stratification in the basement lamina and by the disappearance of the epithelium absent indeed in the cirrus. The material found inside the cirrus-lumen is different according to the zone considered. The origin of this material and of the cirrus teeth is discussed.

Introduction

Though the reproductive apparatus is an important feature used in turbellarian taxonomy, very little attention has been paid to its ultrastructure. To the best of our knowledge, such data are only available for the coelogynoporid *Carenscoilia bidentata* Soppot, 1972 (Ehlers & Ehlers 1980) and for the acoel *Paratomella rubra* Rieger & Ott, 1971 (Mainitz 1977).

In the present study a detailed description is given of the copulatory organ of the monocelidid *Archilopsis unipunctata* (Fabricius, 1926). This copulatory organ is of the so called conjuncta-duplex type, i.e. a copulatory organ with an eversible cirrus enclosed in a bulb delimited by a septum, the prostate vesicle being interposed between the seminal vesicle and the cirrus (for explanation of terminology: see Karling 1956). This type of copulatory organ is of very widespread occurence in many different turbellarian taxa. The main objective of our investigation is to provide a sound basis for later comparison when data on the ultrastructure of the copulatory organ in other Monocelididae (or other turbellaria) become available. We hope that knowledge of the fine structure will provide new and useful data to help elucidate phylogenetic relationships.

Material and methods

Specimens of *Archilopsis unipunctata* were collected in May 1979 in shallow brackish water of the Slack-estuary in Ambleteuse (Pas de Calais, France). The animals were identified from the description by Luther (1960). They were extracted by decantation with an $MgCl_2$-solution isotonic to

Hydrobiologia 84, 277–285 (1981). 0018-8158/81/0843-0277/$01.80.

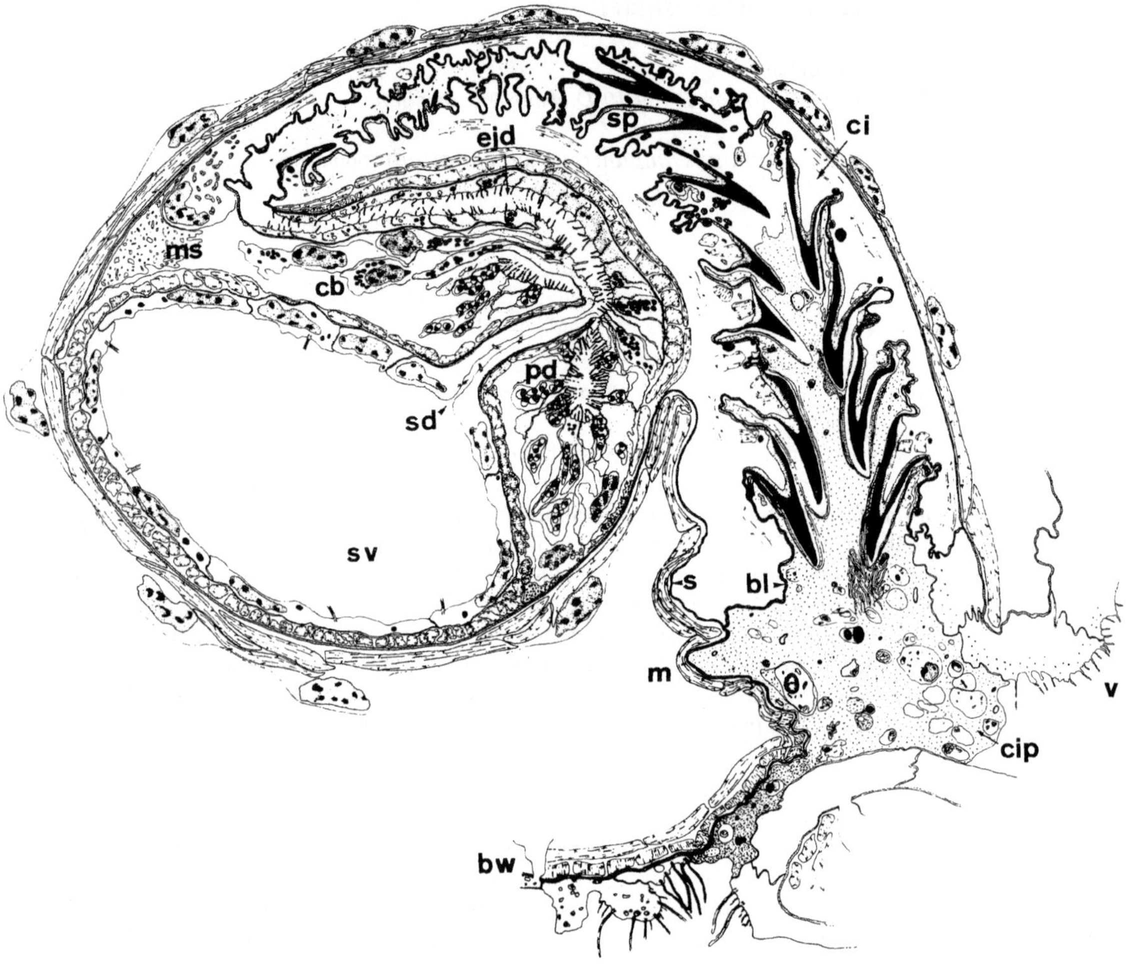

Fig. 1. Reconstruction of the copulatory bulb in *Archilopsis unipunctata*, based on serial 1 μ and ultrathin eponsections.

seawater (Boaden 1963; Sterrer 1968). After relaxation with $MgCl_2$ numerous individuals were fixed in 2% glutaraldehyde in 0.1 M phosphate buffer, with 10% sucrose (pH 7.3), at 4 °C for 2–3 h; rinsed in the same buffer and postfixed in 1% phosphate buffered osmiumtetroxide at 4 °C for 2 h. Dehydration occured in graded ethanols and flat embedding in Epon 812 using propylene oxide as an infiltration solvent. Ultrathin serial sections, alternated with 1 μm sections, were cut with glass knives on a Reichert OMU 3 ultramicrotome and maintained on Pioloform coated cupper grids. Sections were stained with 2% aqueous uranylacetate and 1.2% aqueous lead citrate, and viewed with a Philips EM 400 electron microscope. For light microscopy the 1 μm Epon sections were stained with toluidine blue or with fuchsine-methylene-blue.

Observations

The general organisation of the copulatory organ of *Archilopsis unipunctata* as known from light-microscopy (Fig. 3, 4) is, except for some minor details, confirmed by electronmicroscopy (Fig. 1, 2).

The sperm canal leading from the unpaired intracapsular seminal vesicle to the atrium can be subdivided in three sections: seminal duct surrounded by the prostate vesicle, ejaculatory duct

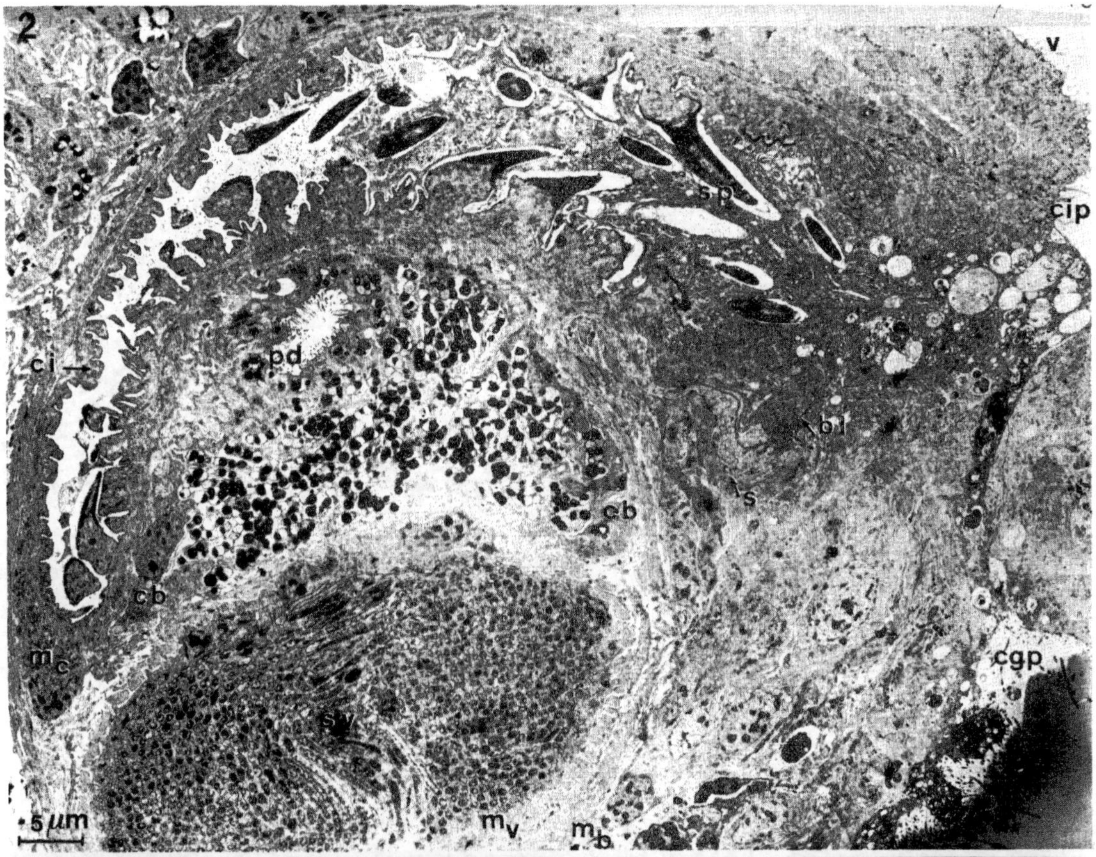

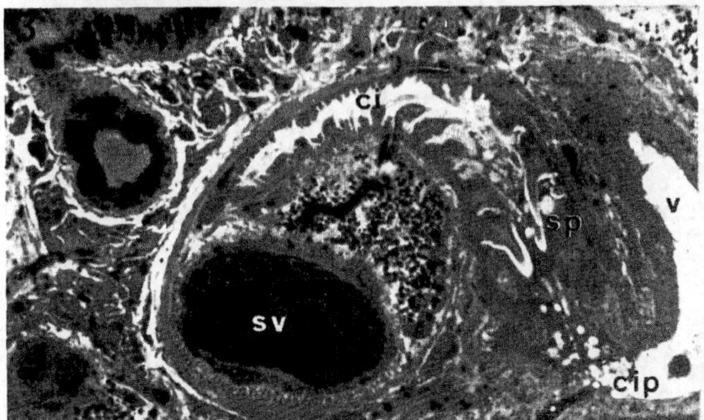

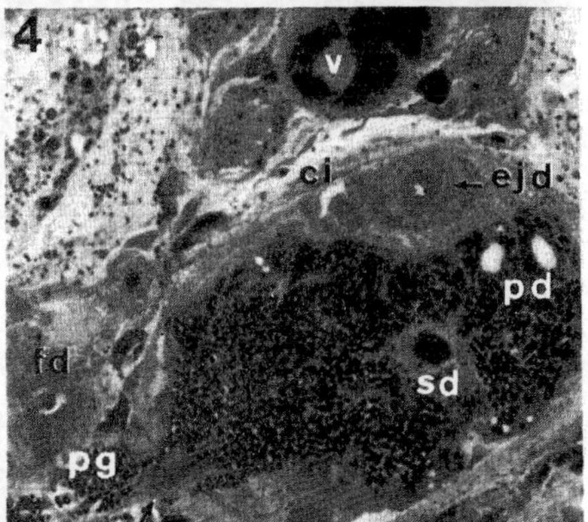

Fig. 2. Electronmicrograph of a sagittal section through the copulatory bulb.

Fig. 3. Micrograph of a 1 µm epon section from the same series as shown in Fig. 2.

Fig. 4. Micrograph of a 1 µm epon section from a crossectioned individual showing the entrance of a vas deferens and prostate glands, seminal duct, both prostate ducts, ejaculatory duct and proximal end of the cirrus. Some elements of the female apparatus are indicated as well.

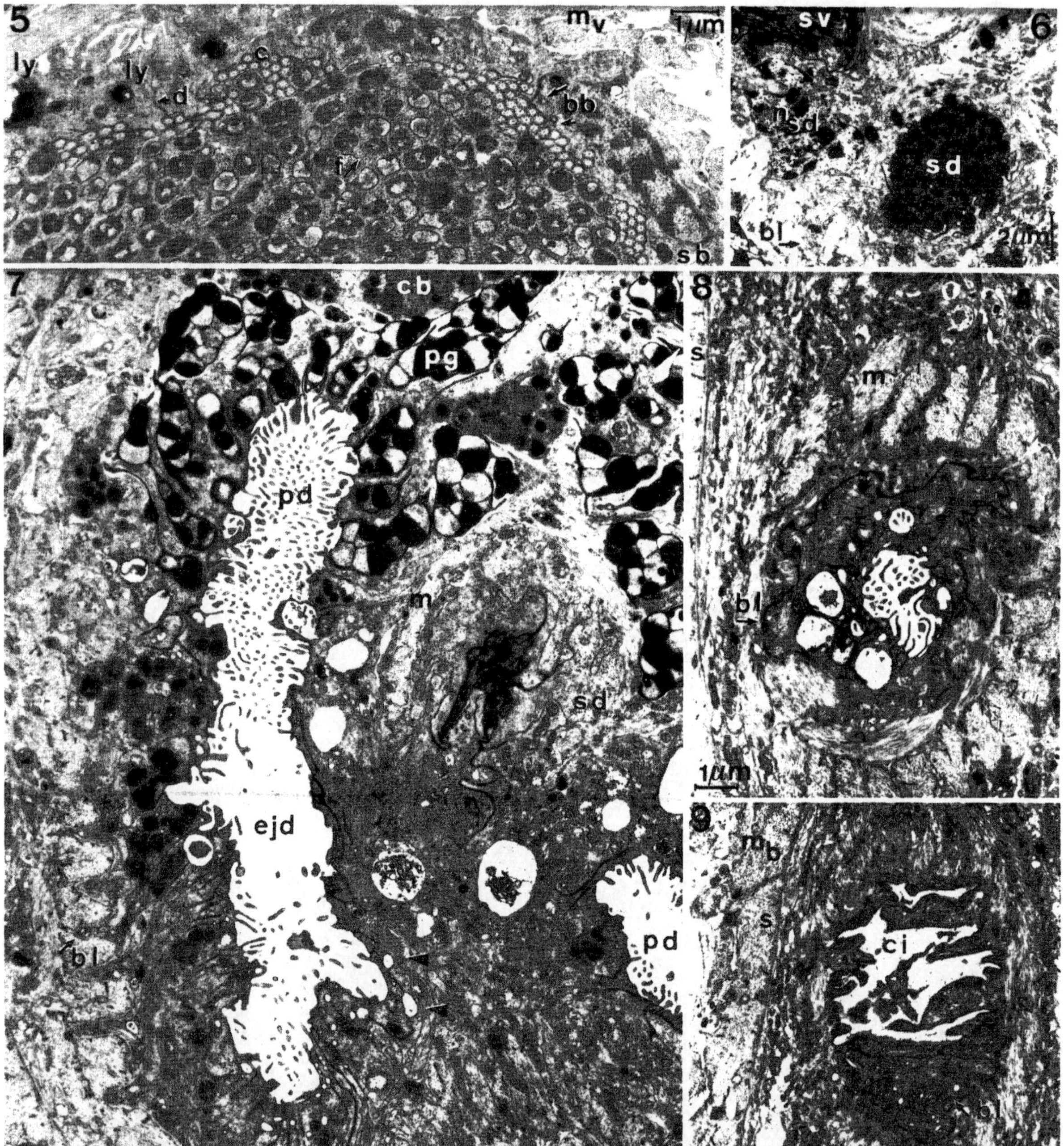
5
ly
ly
d
c
m_v
1 µm
bb
f
sb
6
sv
n
sd
sd
bl
2 µm
7
cb
pg
pd
m
sd
ejd
bl
pd
s
1 µm
8
s
m
bl
1 µm
9
m_b
s
ci
bl
m
1 µm

and cirrus, the whole complex being enclosed in the bulb lined by a septum of fine fibrous material. The prostate vesicle of the light-microscopists actually consists of the cell-necks of the prostate glands of which the cell bodies are situated outside the septum (Fig. 4). The cell-necks enter the bulb at its left and right side, and discharge into two narrow ducts, called here prostate ducts. The latter are tubular diverticles of the sperm canal, evaginated from the sperm canal at the transition between the seminal duct and the ejaculatory duct. The seminal vesicle and the seminal duct are surrounded by a single muscle layer, while the ejaculatory duct and the cirrus are enveloped by inner circular and outer longitudinal muscles. The whole bulb is enclosed in a single layer of muscles, outside the septum, of which the fibres are mainly directed longitudinally and are continuous with the muscles of the atrium and of the body wall.

The muscles are of the striated type with peripherally situated 'dense bodies' (Fig. 7, 8) as described by Reuter (1977) and by Mattison, Nilsson & Fange (1974). Apparently only four cells are involved in the formation of the muscle sheath around the cirrus (and perhaps even around the ejaculatory duct). Their nuclei are situated at the proximal half of the cirrus.

The septum and the basement lamina of the epithelium of the sperm canal consist of a fine fibrous material (Fig. 7). The muscle-fibrils around the seminal vesicle are embedded in the same kind of material (Fig. 2, 5). From its electrondensity this material is probably more packed in the septum. On the other hand an homogeneous electrondense condensation appears in the basement lamina at the most distal part of the ejaculatory duct and of the cirrus as will be described below.

The seminal vesicle (Fig. 5) is lined by a flattened cellular epithelium bearing scattered cilia. These can be recognised by their 9 + 2 configuration between the tails of spermatozoa with 9 + 1 configuration (Hendelberg 1969, 1974, 1977) and by their basal bodies. Juxtanuclear patches of rough endoplasmic reticulum, numerous free ribosomes, a well elaborated Golgi apparatus and some secretion granules indicate secretory activity of the cells. Lysosome-like bodies occur as well. Cell contacts are realised by a zonula adherens and deeper septate desmosomes.

The seminal duct (Fig. 6, 7) has an epithelium formed by four elongated cells the nuclei of which are situated at the transition of the seminal vesicle and the seminal duct. The cell-walls are highly convoluted in transverse section and are held together by deep septate desmosomes. Some cilia occur. In the proximal part of the seminal duct, the cytoplasm of the epithelium can be compared with that of the cells lining the seminal vesicle; in the more distal part the cells are higher, the apical half of the cytoplasm is rather dense and homogeneous with most organelles lying in the basal half.

The prostate ducts (Fig. 2, 7) are devoid of a muscle sheath and no basement lamina occurs under its epithelium. The cells of this epithelium exhibit branching basal expansions which join the nucleiferrous cell bodies situated at the periphery of the copulatory bulb. Some of these cells contain numerous secretory vesicles filled with dense to moderately dense material. Most mitochondria, dictyosomes, and numerous ribosomes are found near the nucleus. The necks of the prostate gland cells interlace with the basal expansions of the epithelium cells. They pierce between (or through?) these cells and discharge their content into the prostate ducts. They are packed with membrane bound granules, about 1 μm long, showing a divi-

Fig. 5. Section through the wall of the seminal vescile showing the flattened ciliated epithelium and the muscles embedded in a fine fibrous matrix. Note the difference between cilia and spermflagella.

Fig. 6. Section through the seminal duct near the seminal vesicle showing the nucleus of a cell of the epithelium lining the seminal duct.

Fig. 7. Section through a part of the copulatory bulb with prostate ducts, seminal duct and initial section of the ejaculatory duct. Compare, the epithelium lining the seminal duct in this micrograph with Fig. 6.
Note the higher electrondensity of the septum compared with the basal lamina of the epithelia and the dense bodies in the muscles.

Fig. 8. Cross sectioned ejaculatory duct near the cirrus. Note the highly vacuolated epithelium and the basal lamina becoming abruptly electrondens.

Fig. 9. Section at the end of the ejaculatory duct and very proximal end of the cirrus (upper part). Note absence of an epithelium in the cirrus.

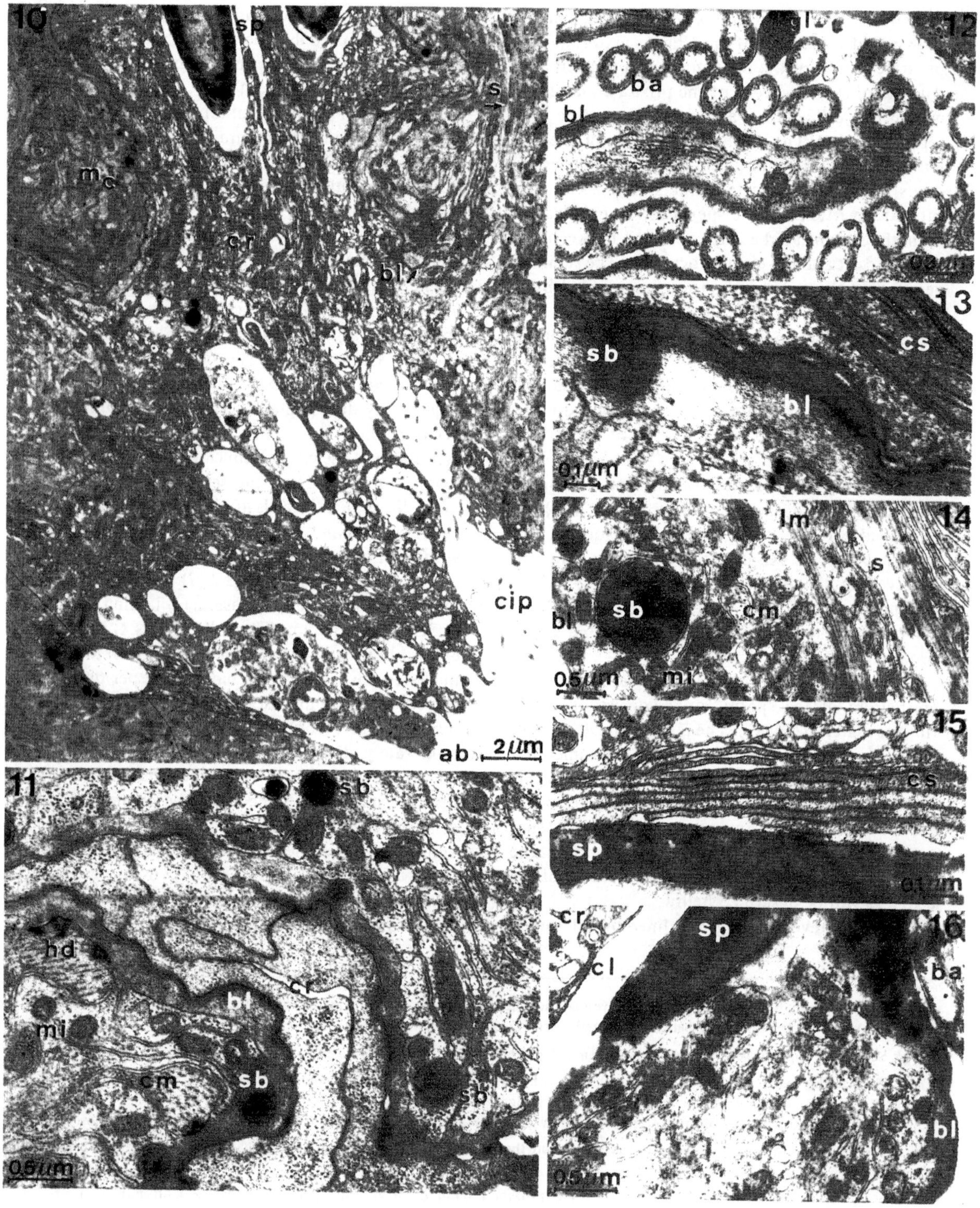
10
sp
s
m_c
cr
bl
cip
ab
2 µm
11
sb
hd
bl
cr
mi
cm
sb
sb
0.5 µm
12
gl
ba
bl
13
sb
cs
bl
0.1 µm
14
lm
s
bl
sb
cm
0.5 µm
mi
15
cs
sp
0.1 µm
16
cr
sp
cl
ba
bl
0.5 µm

sion of their content into an electrondense and an electronlucent zone. Apically all cells are held together by a deep zonula adherens and septate desmosomes. The luminal wall bears numerous microvilli.

The ejaculatory duct (Fig. 7) can easely be recognised by its double muscle sheath and by the uninterrupted basement lamina. No nuclei occur above this basement lamina and the nucleiferrous cellparts are intermigled between those of the prostate ducts. The connections between superficial and nucleiferrous cellparts are concentrated at the transition of the seminal duct and the ejaculatory duct. Since very many cell borders are found in the superficial layer, and only few nuclei, a branching of the cells must be postulated. Between the cell borders, connected to each other by deep septate desmosomes, many intercellular spheric expansions can be seen, while in the most distal part of the ejaculatory duct large vacuoles also occur (Fig. 8). The epithelium is secretory, the secretory granules being most apparent in the proximal part of the duct. When approaching the cirrus, the basement lamina suddenly becomes electrondense (Fig. 8), while the epithelium itself ends where the cirrus begins (Fig. 9).

In the cirrus no epithelium could be found and the basement lamina forms the lining of the lumen (Fig. 2). This basement lamina gradually thickens and becomes stratified (Fig. 11, 15, 16), i.e. with an inner (at the muscle side) finely fibrous layer, 0.2–0.4 μm thick, an outer (at the lumenside) electrondense layer, and on top of it a thin layer of uniform thickness of 0.18 μm (sometimes lacking). The cirrus spines are formed by the extreme thickening of the electrondense layer growing from 0.02 μm – 0.13 μm between the spines to nearly 5 μm at the tip. The tips of the spines seem to be surrounded by an electronlucent zone of nearly 1 μm (Fig. 1–16). We cannot certify if this zone actually consists of some material or if it must be considered as a fixation artifact.

In the most proximal zone of the cirrus, the lumen is filled with dispersed cell debris (Fig. 2). More distally, cell remnants with recognisable organelles are found, often adhering to the basement lamina, as well as numerous bacteria (Fig. 12). The distal part of the lumen is filled with cell material forming a kind of cirrus plug (Fig. 2, 10). It has all the characteristics of a degenerating epithelium that once possibly had a secretory function: clusters of swollen microvilli or intracellular channel systems, rough endoplasmic reticulum, stacks of cytoplasmic sheets, large lysosome-like bodies, glycogenosomes, apoptotic bodies . . . (Kerr *et al.* 1972; Pipan & Rakovec 1980 with references) (Fig. 10, 15). In the sarcoplasma of the surrôunding muscles large vesicles are often observed (Fig. 11, 14). These vesicles are filled with material with the same electrondensity as the dark layer of the basement lamina. Such electrondense spherules also occur in the inner layer of the basement lamina and seem to contact the dark layer where their dense material spreads (Fig. 11, 13).

The muscle fibrils are attached to the basement lamina by hemidesmosomes (Fig. 11).

Discussion

As stated above the electronmicroscopic approach for the most part confirms what is known so far of the anatomy of the copulatory organ of the conjuncta-duplex type. However, two new features can be emphasised: the presence of the prostate ducts and the fact that the epithelium of the ejaculatory duct is of the insunk type and not

---

Fig. 10. Electronmicrograph of the atrium from a sagittally sectioned individual to show the 'cirrusplug' with some characteristics of a degenerating tissue and the tip of a spine.

Fig. 11. An aspect of the cirruswall in the median zone with cellremnants adhering to the basal lamina and showing secretory activity of the surrounding muscles. Note the stratified basal lamina as occuring between the spines.

Fig. 12. Section through part of the cirrus with numerous bacteria.

Fig. 13. Section through the cirruswall with basal lamina invaded by a secretion granule from the muscles and adhering membrane stacks at the luminal side.

Fig. 14. Electronmicrograph of the musclelayer between septum (right) and basal lamina lining the cirruswall (left) to show a large membranebound secretion granule.

Fig. 15. Detail ot tne cirrusplug with membranestacks. The basal lamina with thick electrondense layer is shown as well.

Fig. 16. Electronmicrograph showing the onset of a spine. Note the stratified basal lamina with the narrow clear zone at the lumen side.

ciliated, while that of the seminal duct and seminal vesicle is ciliated and not insunk (i.e. with the same appearence as the vasa deferentia and efferentia – unpublished data). Considering now the cirrus and its spines it has been revealed that hard structures in Turbellaria are either derivatives of the basement lamina, as the proboscis armature of *Carchorodorhynchus* and *Diascorhynchus* (Rieger & Doe 1975), or are intracellular condensations, as the proboscis hooks in *Cheliplana* and *Gnathorhynchus* (Doe 1976), the copulatory needles in *Carenscoilia bidentata* (Ehlers & Ehlers 1980) and the rods of the copulatory organ in *Paratomella rubra* (Mainitz 1977). A true cuticula, i.e. a hardened extracellular layer, forms the copulatory stylet in *Gnathostomulida* (Rieger & Mainitz 1977; Mainitz 1979). In our oppinion the cirrus spines in *Archilopsis* are derivatives of the basement lamina. However, nothing is known about the formation or the origin of this material.

From our observations two possible ways for the formation of the spines can be postulated:

(1) In the subadult animal the cirrus lumen is lined by an epithelium responsible for the secretion of the hard material that is deposited at the basal side of the epithelium upon or in the basement lamina. This epithelium then degenerates eventually to be torn off at copulation, leaving behind the cell debris, the cell remnants and the plug. We must however consider the possibility that the cirrus plug is in fact part of the epithelium of the vagina of the partner torn off during copulation. We are convinced that all animals studied have copulated, as withnessed by the presence of the bacteria and the presence of spermatozoa in the female ducts.

(2). The numerous electrondense vesicles observed in the muscle cells surrounding the cirrus indicate that these cells exhibit a secretory activity. Secretion of proteins or lipoproteins by muscles is not at all exceptional, neither in vertebrates (Young *et al.* 1980 with references; Dadoune *et al.* 1978 with references) nor in invertebrates (Dorsett & Roberts 1980; Økland 1980 with references).

The electronmicrographs Figs. 11 and 13 strongly suggest that the secreted material is transferred to the basement lamina covering the muscles and may be responsible for or at least participate in, the formation of the superficial electrondense layer.

Future investigations on subadult animals and the use of histochemical techniques may help us to elucidate the histogenesis of the cirrus and of other elements in the copulatory organ.

Acknowledgements

We wish to thank Miss E. Plaum, Mrs. and Mr. Withofs and Miss G. Vliegen for their technical assistance, help with photographic work and typing the manuscript.

Abbreviations

ab – apoptotic body
ba – bacteria
bb – basal bodies of cilia
bl – basement lamina
bw – body wall
c – cilia
cb – cell body
cgp – common genital porus
ci – cirrus
cip – cirrus plug
cl – lumen of cirrus
cm – circular muscles
cr – cytoplasmatic remnants
cs – cytoplasmatic sheets
ejd – ejaculatory duct
ep_{ej} – epithelium of ejaculatory duct
d – desmosomes
f – flagella of spermatozoa
fd – female duct
fp – female porus
gc – golgi complex
gl – glycogen particles
hd – hemidesmosomes
lm – longitudinal muscles
ly – lysosome-like body
m – muscles
m_b – muscles of the bulbus
m_c – muscles of the cirrus
m_v – muscles of the seminal vesicle
mi – mitochondria
ml – microvilli
ms – mesenchyme
n_{sd} – nuclei of the seminal duct
pd – prostate duct
pg – prostate glands
ri – ribosomes
s – septum
sb – secretory vesicle
sd – seminal duct
sp – spines
sv – seminal vesicle
v – vagina
vd – vas deferens

References

Boaden, P. J. S., 1963. The interstitial fauna of some N. Wales beaches. J. Mar. Biol. Asssoc. U.K. 43: 79–96.

Dadoune, J. P., Terquem, A. & Alfonsi, M. F., 1978. High resolution radioautographic study of newly formed protein in striated muscle with emphasis on red and white fibres. Cell tissue Res. 193: 269–282.

Doe, D. A., 1976. The proboscis hooks in Karkinorhynchidae and Gnathorhinchidae (Turbellaria, Kalyptorhynchia) as basement membrane or intracellular specializations. Zool. Scripta 5: 105–115.

Dorset, D. A. & Roberts, J. B., 1980. A transverse tubular system and neuromuscular junctions in a molluscan unstriated muscle. Cell Tiss. Res. 206: 251–260.

Ehlers, & Ehlers, U., 1980. Structur and Differenzierung penialer Hartgebilde von Carenscoilia bidentata Soppott (Turbellaria, Proseriata). Zoomorphologie 95: 159–168.

Hendelberg, J., 1969. On the development of different types of spermatozoa from spermatids with two flagella in the Turbellaria, with remarks on the ultrastructure of the flagella. Zool. Bidr. Uppsala 38: 1–50.

– 1974. Spermiogenesis, sperm morphology and biology of fertilization in the Turbellaria. In: Eds. Riser, N. W. & Morse, M. P., Biology of the Turbellaria. New York: McGraw-Hill, 148–169.

– 1977. Comparative morphology of turbellarian spermatozoa studied by electron microscopy. Acta Zool. fenn. 154: 149–162.

Karling, T. G., 1956. Morphologisch-histologische Untersuchungen an den männlichen Atrialorganen der kalyptorhynchia (Turbellaria). Ark. Zool. Ser. 2, 7: 187–279.

Kerr, J. F. R., Wyllie A. H. & Currie, A. R., 1972. Apoptosis: a basic biological phenomenon with wideranging implications in tissue kinetics. Br. J. Cancer 26: 239–256.

Luther, A., 1960. Die Turbellarien Ostfennoskondiens I, Acoela, Catenulida, Macrostomida, Lecithoepitheliata, Prolecithophora und Proseriata. Fauna Fennica 7.

Mainitz, M., 1977. The fine structure of the stylet apparatus in Gnathostomulida Scleroperalia and its relationship to turbellarian In: Eds. Karling, T. G. & Meinander, M., The A. Luther Symposium on Turbellaria, Acta Zool. fenn. 154: 163–174.

– 1979. The fine structure of Gnathostomulid reproductive organs. Zoomorphologie 92: 241–272.

Mattisson, A., Nilsson, S. & Fange, R., 1974. Light microscopical and ultrastructural organisation of muscles of Priapulus caudatus (Priapulida) and their responses to drugs, with phylogenetic remarks. Zool. Scripte 3: 209–218.

Økland, S., 1980. The heart ultrastructure of Lepidopleuris asellus (Spengler) and Tonicella mormorca (Fabricius) (Mollusca: Polyplacophora). Zoomorphologie 96: 1–19.

Reuter, M., 1977. Ultrastructure of the stylet protractor muscle in Gyratrix hemafroditus (Turbellaria, Rhabdocoela). Acta Zool. fenn. 58: 179–184.

Rieger, R. M. & Doe, D. A., 1975. The proboxis armature of Turbellaria – Kalyptorhynchia, a derivative of the basement lamina? Zool. Scripta 4: 25–32.

Rieger, R. M. & Mainitz, M., 1977. Comparative fine structure of the body wall in Gnathostomulida and their phylogenetic position between Platyhelminthes and Aschelminthes. Zeitschr. Zool. Syst. Evol. – forsch. 15: 9–35.

Rieger, R. M. & Ott, J., 1971. Gezeitenbedingte Wanderungen von Turbellarien und Nematoden eines nordadriatische Sandstrangens. Vie Milieu Suppl. 22: 425–447.

Soppott, B., 1972. Systematik und Ökologie von Proseriaten der deutschen Nordseeküste. Mirkofauna Meeresboden 13: 1–72.

Sterrer, W., 1968. Beiträge zur Kenntnis der Gnathostomulida. I. Anatomie und Morphologie des Genus *Pterognathia* Sterrer Ark. Zool. 22: 1–125.

Yunge, L., Benchimol, S. & Contin, M., 1980. Ultrastructural cytochemistry of atrial muscle cells (VIII). Cell Tiss. Res. 207: 1–11.

Fine structure of the giant aflagellate spermatozoon in Pseudostomum quadrioculatum (Leuckart) (Platyhelminthes, Prolecithophora)

Ulrich Ehlers
II. Zoologisches Institut und Museum der Univ. Göttingen, Berliner Strasse 28, D-3400 Göttingen, F.R.G.

Keywords: aflagellate spermatozoon, Platyhelminthes, Pseudostomum, sperm ultrastructure, sperm motility, comparative morphology

Abstract

The giant aflagellate spermatozoa of *P. quadrioculatum* are composed of two different parts: a thicker head piece and a more slender tail piece. In the head there exist a large elongated nucleus and an elongated mitochondrial derivative situated in a groove-like cavity of the nucleus. In mature spermatozoa the nuclear material is arranged in many small membrane bounded areas. Both structures, nucleus and mitochondrial derivative, are spirally coiled. The outer part of the membrane in the mitochondrial derivative forms many loop-like foldings. Both organelles continue to the tail in form of two small, helically coiled ribbons; the nucleus is anchored within the mitochondrial derivative by an electron-opaque process. A sheath of spirally-orientated cortical microtubules starting from the tip of the head runs to the tip of the tail under the cell membrane. In addition, a second sheath of tubules occurs in the tail region, these tubules also run parallel to each other, but in the opposite direction to the microtubules of the outer sheath.

The possible relations between the structures observed and the motility of the spermatozoa are discussed; in addition, some phylogenetic comments are attempted.

Introduction

The Platyhelminthes have strongly divergent spermatozoa (Hendelberg 1969, 1974, 1977a; Henley 1974); the morphological differences between the primitive metazoan spermatozoon described by Franzén (1956, 1977) and the spermatozoa within the Platyhelminthes can be explained as depending on the special mode of platyhelminth propagation in connection with internal fertilization.

Among free-living Platyhelminthes the relatively most primitive spermatozoon seems to be present in the Nemertodermatida (Hendelberg 1977a; Tyler & Rieger 1975, 1977), while all other Platyhelminthes show a more evolved spermatozoon. The occurrence of a biflagellate spermatozoon is common within the Acoela, the Polycladida, and most groups of the Neoophora. Aflagellate, non-axonemal spermatozoa are found as a rule in the Macrostomida and the Prolecithophora, moreover, spermatozoa devoid of flagella have been observed in some single species of the free-living and parasitic Rhabdocoela (Hendelberg 1969, 1970); in the Catenulida, ciliary structures are known only from developing spermatozoa (Rieger 1978).

It is known that the aflagellate spermatozoa of the Macrostomida and the Prolecithophora are motile (Bedini & Papi 1970; Böhmig 1890; v. Graff 1904–08; Hendelberg 1975; Henley 1974; Newton 1979; Thomas & Henley 1971). This is also true for the giant spermatozoon of *Pseudostomum quadrioculatum*. The spermatozoa of this species have been studied by Jensen (1878), v. Graff (1882), and most thoroughly by Karling (1940).

P. quadrioculatum exhibits a caudal bursa often filled with mature spermatozoa (v. Graff 1913). There is no connection, e.g. a vagina, between the

Hydrobiologia 84, 287–300 (1981). 0018–8158/81/0843–0287/$02.80.

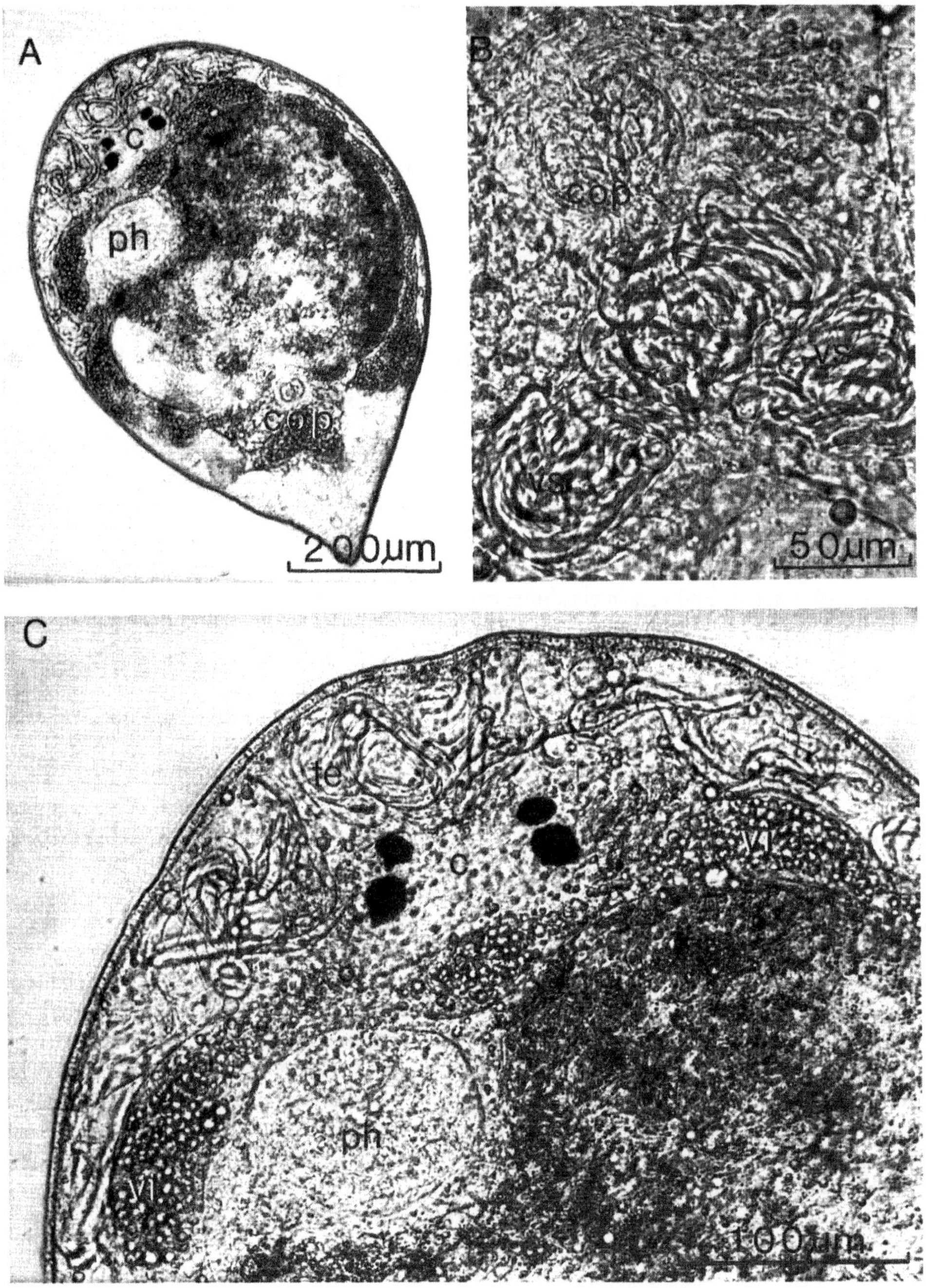

Fig. 1. P. quadrioculatum in male and female maturity. A. Habitus. B. Male genital organs with mature spermatozoa. C. Testes with immature and mature spermatozoa.

bursa and the epidermis; furthermore, this species does not possess any fecundation tract, e.g. ductus spermaticus, between the bursa and the oocytes in the germovitellaries (Karling 1940, 1962; Westblad 1955). The mature spermatozoa, which show helical rotations (v. Graff l.c.), have to wander actively and must penetrate parenchymal tissues on their way towards the bursa or the oocytes.

However, no ultrastructural data are available up to date to demonstrate the morphological basis for this motility. In addition, in only two studies has any ultrastructural investigation of prolecithophoran spermatozoa been attempted (Christensen 1961; Newton 1975). The new findings on the fine structure of the spermatozoon of *Pseudostomum quadrioculatum* may therefore contribute to our knowledge of the prolecithophoran spermatozoon in general.

Materials and methods

The specimens studied were collected in marine and brackish waters on the east coast of the island of Sylt (North Sea). Fixation for transmission electron microscopy was performed at Göttingen one day later; EM processing followed the methods described in Ehlers & Ehlers (1977).

The fine structure of mature spermatozoa was studied on the basis of series of ultrathin sections cut from the testes, the vesiculae seminalis, and the bursa. Within the area of the testes immature spermatozoa were also studied.

Results

The giant spermatozoa of *P. quadrioculatum* are to be found in the voluminous testes situated anteriorly and laterally to the brain (Fig. 1A, C), in the vesiculae seminalis of the male copulatory organ (Fig. 1A, B), and in the bursa lying in the caudal part of animals at female maturity (v. Graff l.c.; Westblad l.c.).

Vital mature spermatozoa are about 120–170 μm long, with a maximum width of about 7–11 μm. Animals at male or female maturity exhibit less than one hundred spermatozoa; this is a relatively low number in comparison with other Platyhelminthes.

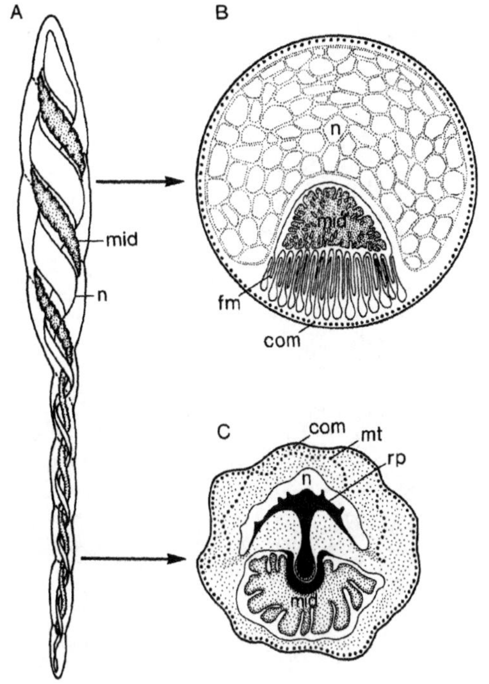

Fig. 2. Schematic diagrams of the mature spermatozoon from the study by light microscopy and electron microscopy. A. Total view. B and C. Cross sections of the head region (B) and of the tail region (C).

From a morphological point of view it may be said that the spermatozoa are composed of two different regions: a head region and a more slender tail region (Fig. 1B, C; 2A). I would like to emphasize that the terms 'head' and 'tail' used in this paper do not correspond with the designations used in the descriptions of less aberrant spermatozoa.

Whereas the immature spermatozoa in the testes are extended (Fig. 1C), the mature spermatozoa in the testes, the vesiculae, or the bursa are always deflected or even coiled. The organelles visible by light microscopy within these mature germ cells always show a helical arrangement with regard to the longitudinal axis of the cells.

Though no flagella are present in the spermatozoa, they move with vigorous spiral rotations.

In the broad head region the elongated nucleus occupies most of the sperm cell. The electron-dense content of the nucleus is seen to alter from a more granular appearance in immature spermatozoa (Fig. 3A) to a densely packed appearance in mature

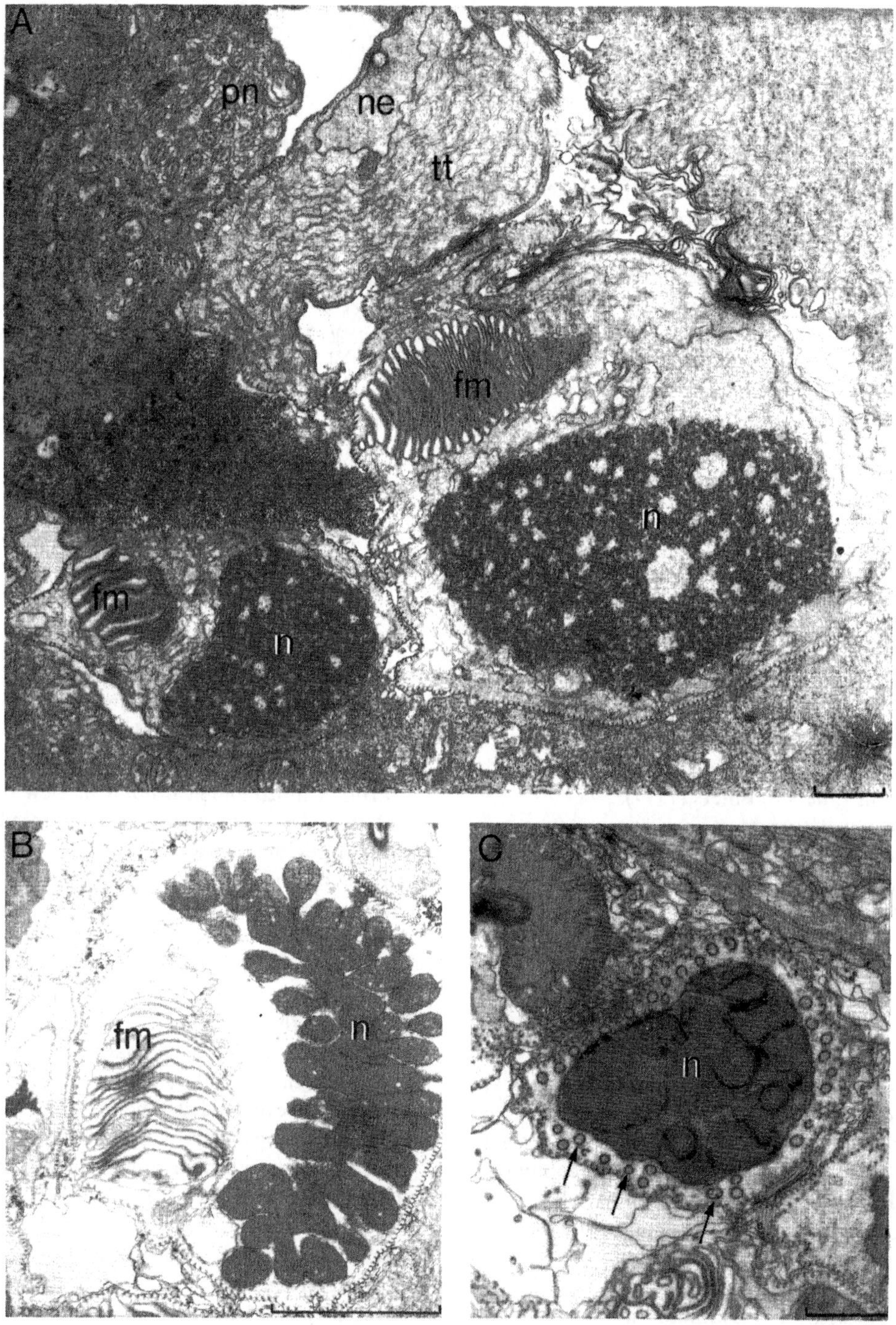

Fig. 3 A. Cross sections of head regions of two immature spermatozoa (lower half) and an oblique section of the tip of the tail of a mature sperm cell (upper half). B and C. Cross sections of head regions of a nearly mature spermatozoon (B) and near the apical part of a fully mature spermatozoon (C), the arrows point to isolated membrane complexes. The scale in all EM-figures (with the exception of Fig. 8B) = 1 μm.

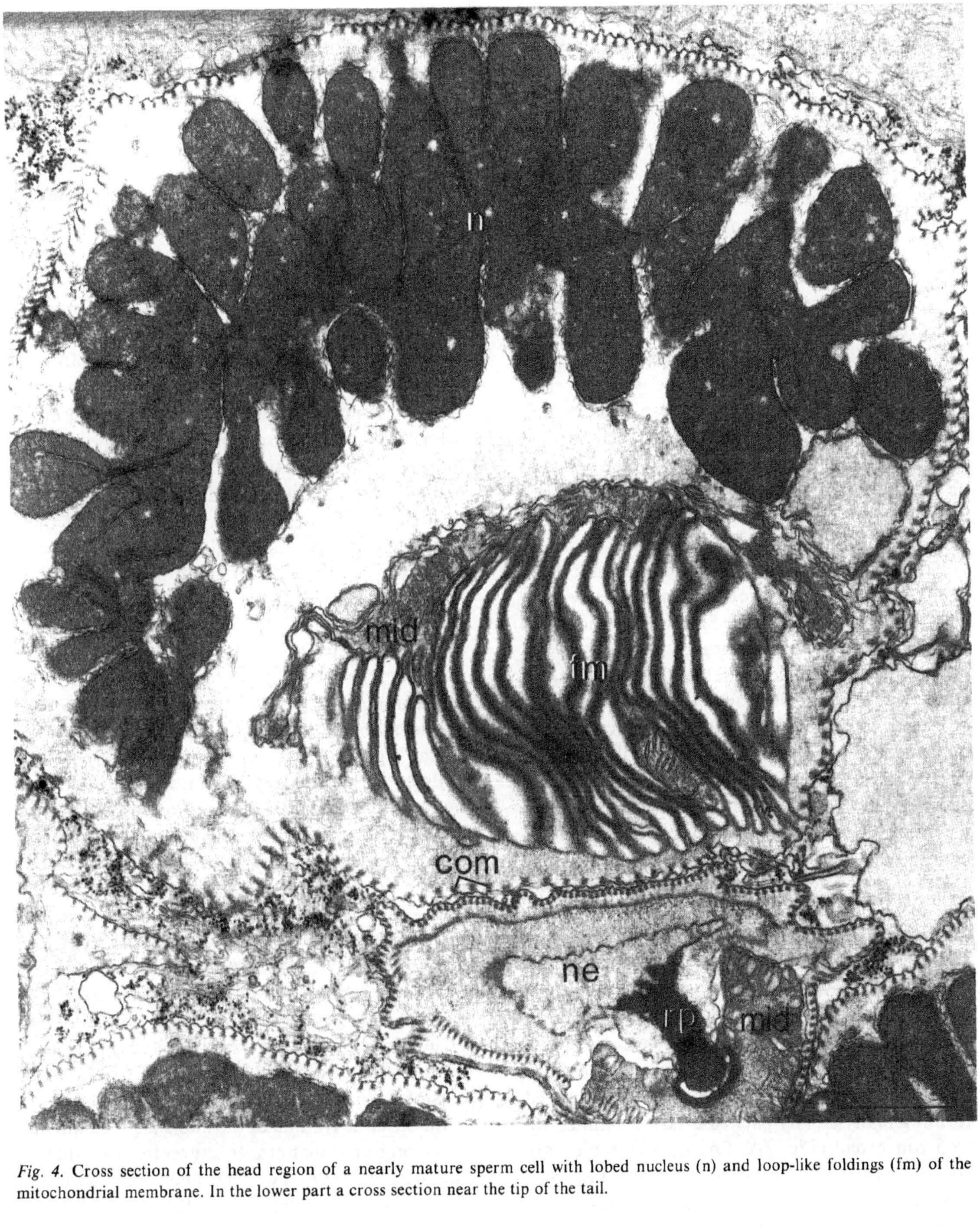

Fig. 4. Cross section of the head region of a nearly mature sperm cell with lobed nucleus (n) and loop-like foldings (fm) of the mitochondrial membrane. In the lower part a cross section near the tip of the tail.

spermatozoa (Fig. 2B, 3C, 4, 6A). During the condensation of the nuclear material the nuclear membrane elongates, sinks into the interior of the nucleus which becomes lobed, and subdivides the opaque material into more or less spherical small areas (Fig. 3B, 4). These areas, which are enveloped by a single membrane initially, are always linked to each other by small electron-dense crossbridges (Fig. 4). In mature spermatozoa nearly all nuclear material is subdivided into hundreds of small areas about 1 μm in diameter (Fig. 6A); these spherical areas are usually surrounded by several membranes (Fig. 5C, D), which result from multiple infoldings of the nuclear envelope. All these areas are situated closely together, and therefore the whole nucleus looks like a compact, electron-dense mass. Nevertheless, between all of these areas there always exists a network of connecting links; the areas do not represent completely isolated compartments. The outer nuclear membrane of mature spermatozoa seems to be devoid of nuclear pores.

Beneath the membrane there is in each nucleus a small spheroidal area, about 0.5 μm across, with an electron-dense lattice (Fig. 5D). This structure is devoid of enveloping membranes.

The nuclear complex extends the full length of the head region, thereby forming a spiral consisting of three complete turns. In cross sections the nucleus of a mature spermatozoon shows a U-shaped outline (Fig. 2A, 6A). A ribbon-like mitochondrial derivative, which extends almost the full length of the head region forming the same helical twists as the nucleus, is situated in the groove-like cavity of the nucleus. This extensive mitochondrial derivative possesses a moderate electron-opaque matrix with many small irregular cristae. The inner and the outer membrane of the mitochondrial derivative are normally widely separated from each other. This may be an artifact. Where the outer membrane of the mitochondrial derivative is turned away from the nucleus, one can see a peculiar structure: the membrane is extremely elongated forming many closely packed disk-like lamina, up to 5 μm broad (Fig. 2A, 5A, B, 6A), which are comparable to extensive dictyosomes or membranous layers of the ER. Very often parts of the mitochondrial matrix with a few cristae can be found between these disk-like lamina (Fig. 5B). The mitochondrial derivative with the characteristic disk-like foldings of the outer membrane already exist in immature spermatozoa (Fig. 3B, 4).

For the most part the cytoplasm between the nucleus and the mitochondrial derivative lacks any additional organelles; infrequently very small vesicles are to be found here. Mitochondria of conventional structure and dictyosomes are missing.

In the anterior, tapering part of the head only nuclear material is found. Many small spherical membrane complexes, which are completely isolated from one another, regularly occur in the cytoplasm around this apical part of the nucleus (Fig. 3C, 8B). Sporadically, these membrane complexes are also present in the more distally located cytoplasm areas.

Cortical microtubules underlying the cell membrane are found throughout the head region. These monolayered microtubules form several spirals, running longitudinally parallel to the other helicoidal sperm organelles. The number of these microtubules in cross sections of spermatozoan head regions ranges from about 200 to more than 250. The distances between the single microtubules vary from 700 Å to 1500 Å and there is no evidence of bridging between them.

In the transition zone between the head and the tail region the spermatozoa taper, and the average diameter in the more flattened tail mid-region is 3 μm (Fig. 2C, 7A).

Both the nucleus and the mitochondrial derivative continue from the head region to the tail.

The thickness of the nucleus diminishes considerably at the transition zone (Fig. 6B). Furthermore, the nuclear envelope is not infolded into the interior of the nucleus anymore and the small membrane areas in the nuclear material disappear. On the other hand, the nucleus now forms a prominent ribbon-like process showing extremely electron-dense material enveloped by the nuclear membrane (Fig. 2C, 6B, 7A). In any cross section of the tail mid-region the nucleus shows a slightly bent outline (Fig. 7A). The ribbon-like process rises from the concave side of the nucleus perpendicular to its long axis and both structures form an anchor-shaped figure (Fig. 2C, 7A). The contents of both parts appear homogeneous, but have very different electron densities. The nuclear membrane, usually bilayered, often breaks down.

In the tail region the mitochondrial derivative continues as a solid ribbon with a more dense

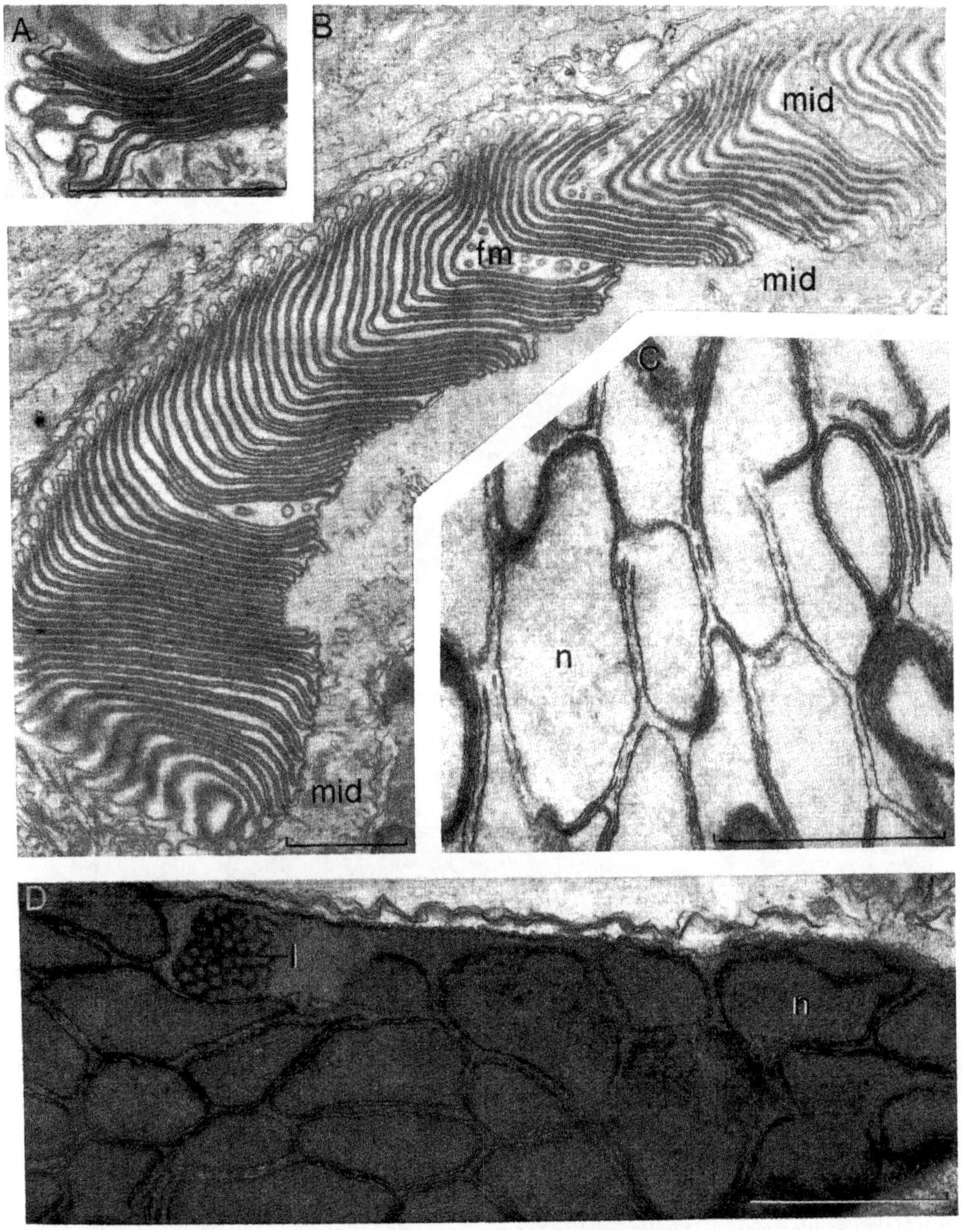

Fig. 5. A & B. Loop-like foldings of the outer membrane of the mitochondrial derivative. C and D. Parts of the nucleus with small membrane bounded nuclear areas (in C less stained than in D) and the electron-dence lattice.

matrix than in the head region. In cross sections the derivative is bean-like, with its concave side faced towards the nuclear process. Prominent cristae are formed by the inner membrane of the derivative; this membrane and the outer one are normally widely separated from each other (Fig. 6B, 7A–C).

In the centre of the concave side, the mitochondrial derivative possesses a longitudinal groove. Cross sections (Fig. 7A) reveal the straitened opening of this groove. The nuclear process is inserted, in the groove, with its slightly thickened distal part.

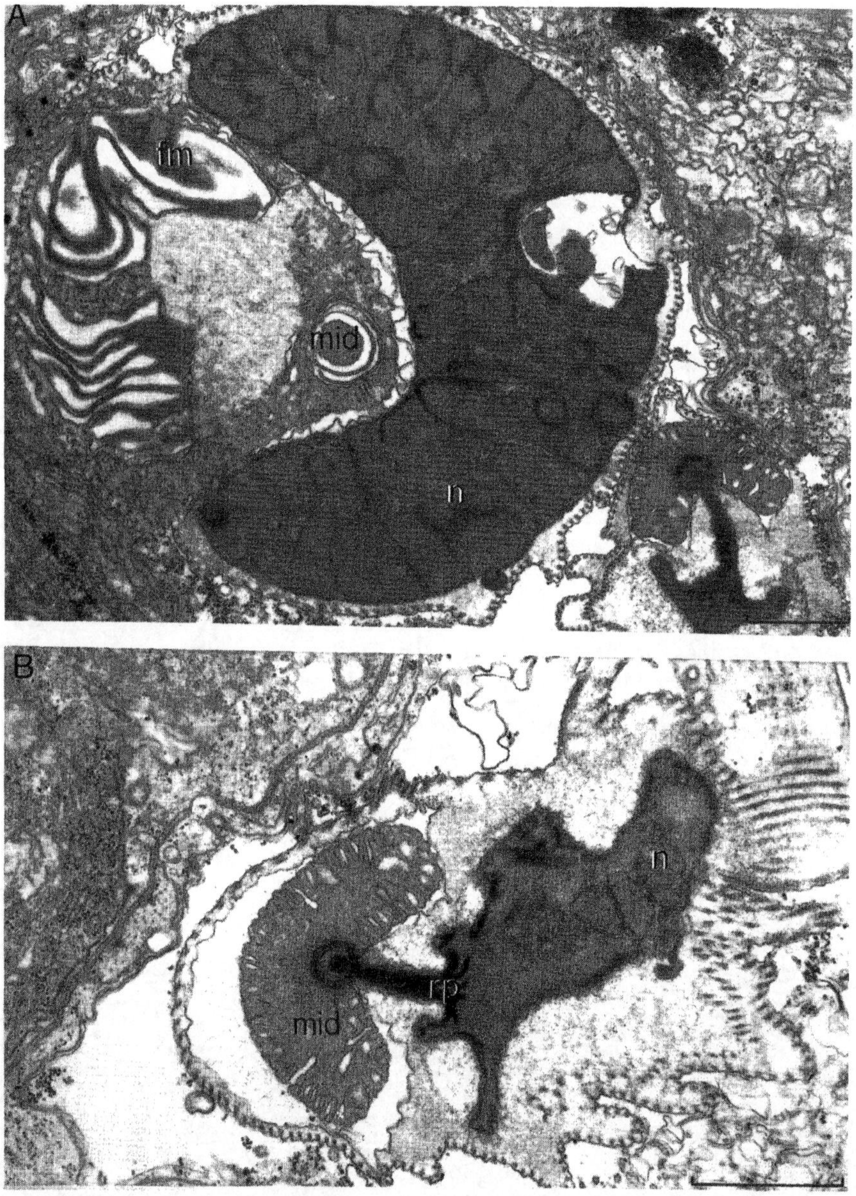

Fig. 6. Mature spermatozoon. A. Cross section of the head region with the condensed nuclear material and cortical microtubules. B. Oblique section of the transition zone with the dense ribbon-like nuclear process and the inner layer of microtubules.

All these organelles – mitochondrial derivative and the nucleus with its ribbon-like process – spiral densely throughout the length of the tail. The process always connects the derivative with the nucleus (Fig. 8A). The thickened part of the process, lying within the groove of the derivative, extends from 0.2 μm in the transition zone (Fig. 6B) to 0.7 μm near the tip of the tail (Fig. 7C).

Within the groove, the bilayered membrane of the thickened process lies on the outer membrane of the mitochondrial derivative (Fig. 7A), whereas this outer membrane and the inner one of the derivative are separated by a distance of 0.05 μm. The space between these two mitochondrial membranes is

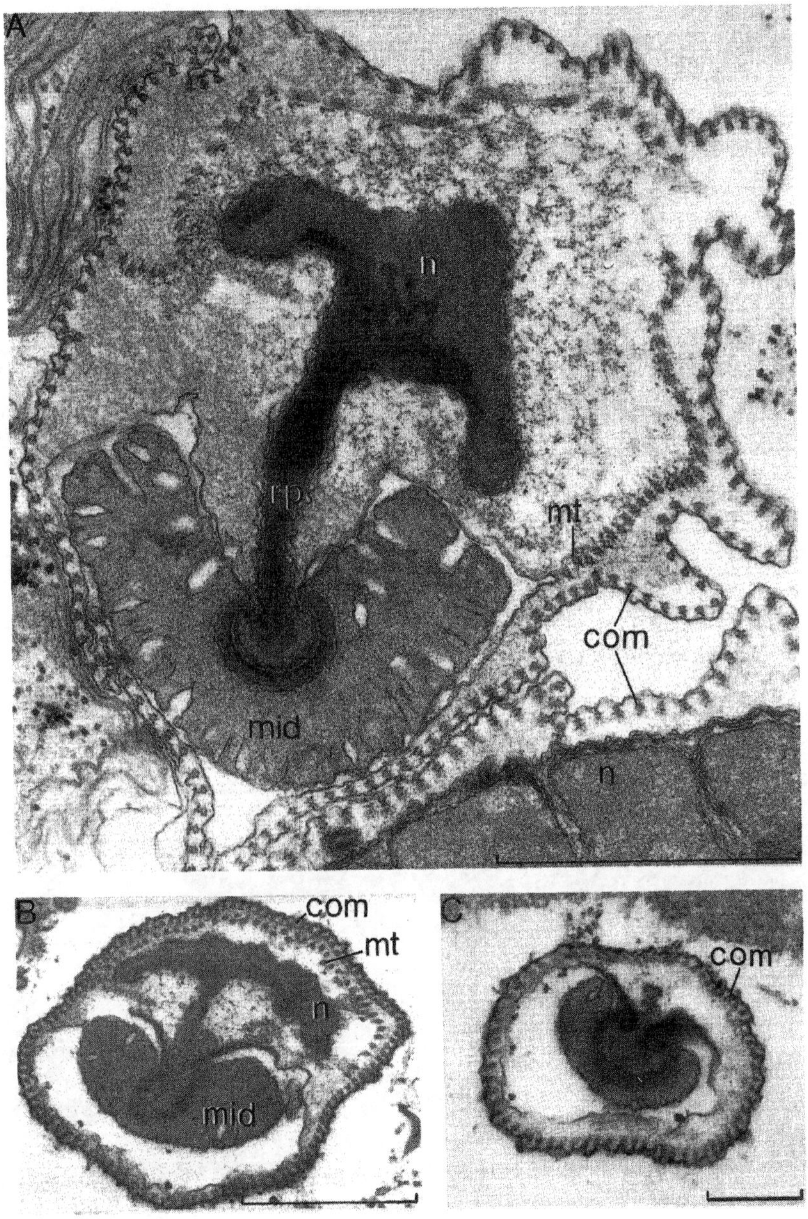

Fig. 7. Cross sections of mature spermatozoa. A. Middle part of the tail region with the two layers of microtubules, nucleus, nuclear process, and mitochondrial derivative. B. More distal part of the tail. C. Part near the tip of the tail without the inner layer of tubules and without the nucleus.

filled with an homogeneous electron-opaque material (Fig. 7A–C), which is especially dense along the outer membrane.

Towards the tip of the tail, the size of the mitochondrial derivative, and especially the nucleus with its process gradually diminishes (Fig. 7B). Thus the electron-dense material in the gap between the two mitochondrial membranes extends from the groove-like cavity along the outer membrane (Fig. 7B, C).

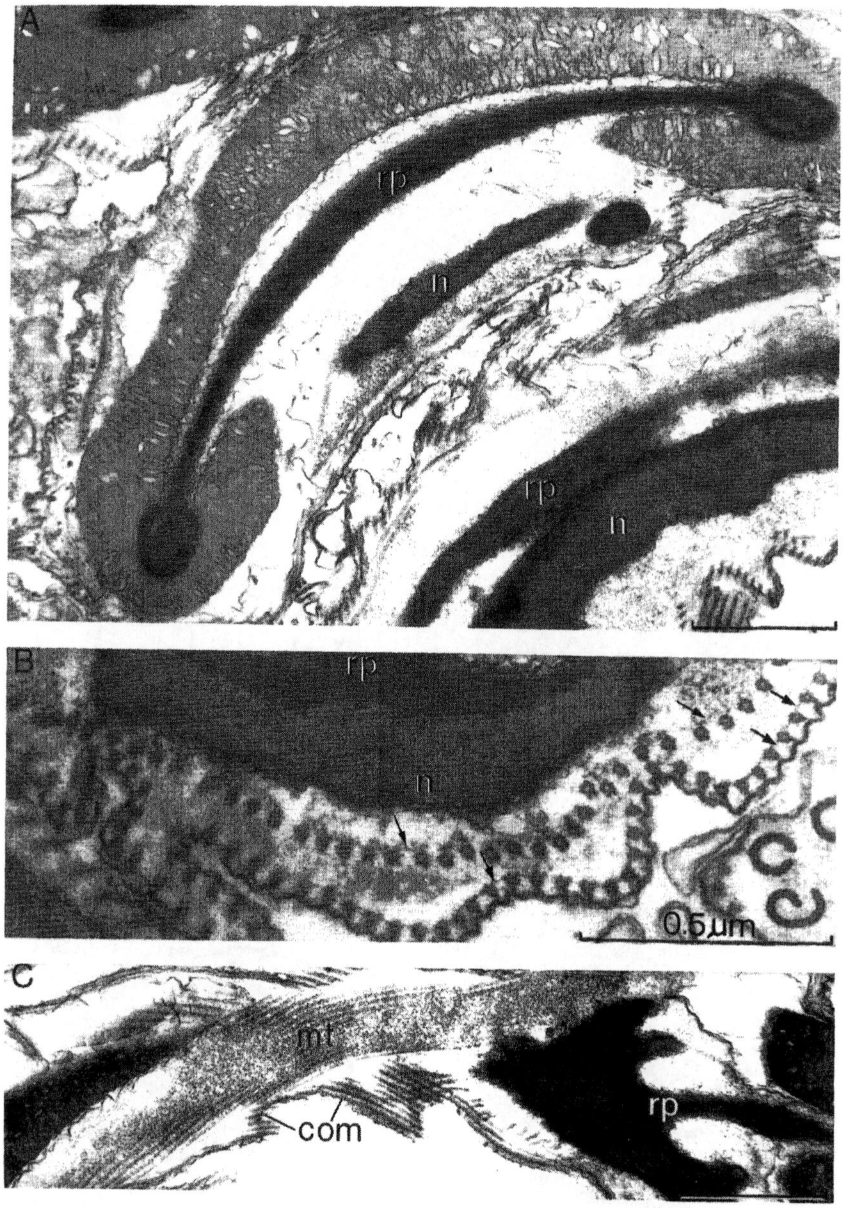

Fig. 8. Tail region of mature spermatozoa. A. Tangential section, showing the long mitochondrial derivative and parts of the nucleus and its process. B. Part of a cross section showing the cross-bridges between the microtubules. On the right side part of a sperm head with isolated membrane complexes. C. Tangential section showing the two layers of microtubules running in an opposite direction.

Near the tip of the tail, the region enveloped by the nuclear membrane lacks any electron-dense nuclear material (lower part of Fig. 4). More distally, the nuclear membrane ends (Fig. 7C), and the ribbon-like process is merely represented by the thickened, lateral part, which is anchored within the groove of the mitochondrial derivative. There exist only some fragmentary remains of the bilayered membrane covering the process.

A layer of cortical microtubules is established underneath the cell membrane of the tail. These peripheral microtubules, located about 420 to 480 Å apart (Fig. 7A, 8B), make several spirals between the transition zone of the spermatozoon and its tail-tip. They can be seen to be attached by short stalklike segments to the cell membrane. Furthermore, there are distinct cross bridges between neighbouring singlet tubules (Fig. 8B).

Within the cytoplasm of the tail there exists a second layer of microtubules (Fig. 2C, 7A). These monolayered tubules run the length of the tail describing serveral spiral courses from the transition zone (Fig. 6B) to the distal region (Fig. 7B). In the most distal part of the tail, which lacks any nuclear material, these tubules are not present (Fig. 7C).

This second layer of tubules only occurs in that half of the tail which contains the ribbon-like nucleus (Fig. 7A). The tubules are separated from one another by distances of 200 Å to more than 500 Å and run in an opposite direction to the tubules of the outer sheath (Fig. 8B, C).

For the most part, the cytoplasm in the tail region between the two layers of tubules and around the mitochondrial derivative exhibits a slightly granular, moderate electron-dense texture. The cytoplasm between the inner layer of tubules and the nucleus with its process is less opaque and contains flocular particles (Fig. 6B, 7A, 8C). At both of the open sides of the inner layer of tubules areas with a more electron-dense cytoplasm can be observed. These areas extend from the tubules up to the adjacent parts of the ribbon-like nucleus (Fig. 7A).

In the most distal part of the tail (upper part of Fig. 3A) all organelles, other than short fragments of bilayered membanes, are lacking.

Discussion

The results obtained at the EM-level confirm the earlier light microscopical observations done by Jensen (1878), v. Graff (1882), and Karling (1940). According to Jensen and v. Graff the spermatozoon of *P. quadrioculatum* reveals a length to 160 μm or 140 μm, respectively; Jensen measured a width of 8.6 μm. V. Graff (1882, p. 154) writes: 'Der letztere (i.e. the tail) ist im Gegensatze zum Kopfe bandförmig, mit hyalinen Seiten und körnigem Mittelstreifen. Der Mittelstreif (i.e. the nucleus) nimmt jedoch nicht das Centrum ein, sondern liegt dem Bande auf einer Seite an, in eine kleine Vertiefung der hyalinen Substanz eingebettet. Dieser granulöse Mittelstreif des stets in lebhafter Spiralbewegung begriffenen Schwanzes setzt sich in der Weise auf den Kopf fort, dass er die Oberfläche desselben in drei Spiralwindungen umkreist'. Jensen (l.c.) also describes and pictures a ribbon forming three spirals within the head, but according to this author the ribbon is placed inside the spermatozoon.

A more detailed description on the spermatozoon is given by Karling (1940, p. 34): 'In dem langen, breiten Kopf bildet ein eosinophiles Band (i.e. the nucleus) die Hauptmasse und es windet sich um den sehr charakteristischen Mittelstreifen (i.e. the mitochondrial derivative). Dieser ist ebenfalls gewunden und liegt gleichsam in einer Rinne des erstgenannten' and 'Nur ist der Mittelstreifen gleich einer Perlschnur gebaut; kleine, gleichmässig grosse Kugeln mit dünner cyanophiler Rinde sind in einer regelmässigen Reihe hintereinander gereiht, die sich aber im Schwanzabschnitt nicht fortzusetzen scheint'. This 'string of pearls' (Karling 1962) corresponds with the loop-like foldings of the outer membrane of the mitochondrial derivative.

According to Westblad (1955, p. 493) the sperm cell only has a length of about 35 μm. Karling (personal communication) also found specimens with relatively very short spermatozoa. Perhaps there exist two closely related species with long and short spermatozoa, respectively.

Several observations on the fine structure of the spermatozoon should be discussed in more detail. I would like to emphasize that no striking morphological differences have been found between the mature spermatozoa in the male genital system and in the bursal organ.

Nucleus. Large nuclei seem to be of widespread occurrence among the Platyhelminthes (Hendelberg 1975). Within the Prolecithophora, complicated nuclear structures are generally found; there exists a spiral nucleus in several species (Böhmig 1890; v. Graff 1904–08; Karling 1940; Kulinitch 1970); the nucleus may also continue in the tail (Karling 1962). The giant size of the nucleus in the mature spermatozoon of *P. quadrioculatum* is striking; generally, sperm nuclei are smaller than those of the somatic cells. Hence several questions must be asked: What can be learned about the infoldings of the nuclear envelope, which forms many small membrane-bounded areas within the nucleus? Does other material besides the chromatin make up part of the electron-dense nucleus?

Sperm cells with an aberrant nucleus are also known from other Platyhelminthes, e.g. the Acoela (Hendelberg 1977a, Fig. 14), Tricladida (Silveira 1970), Rhabdocoela (Hendelberg 1975), and the plagiostomid *Hydrolimax* (Newton 1975). But in none of these Platyhelminthes does the structure of the sperm nucleus resemble that of *P. quadrioculatum*. Further investigations are needed to learn more about this unusual nucleus.

Mitochondrial derivative. This single and extremely long organelle seems to be the result of a fusion of separate mitochondria. Such a confluence of numerous small mitochondria to a giant one is known for many animals (Afzelius 1979; Baccetti & Afzelius 1976; Pladellorens & Subirana 1975), and within the Platyhelminthes long mitochondria situated along the nucleus have also been described (Silveira & Porter 1964; Hendelberg 1975). The kind of functional advantage of such a fusion is not yet clear, but it should be noticed that species with internal fertilization often exhibit modified mitochondria. The many looplike foldings of the outer mitochondrial membrane perhaps strengthen the derivative within the cytoplasm when the sperm cell penetrates the parenchymal tissue. The fusion between the mitochondrial derivative and the nucleus in the tail region via the dense nuclear process may also contribute to strengthen the sperm cell.

Microtubules. It is known, that in aflagellate spermatozoa the monolayered cortical microtubules, also reported from flagellated sperm cells, are not solely rigid structures, but cause the motion of the cells (Hendelberg 1975, 1977b). In *P. quadrioculatum* there is clear evidence of bridging between neighbouring microtubules and between the singlets and the cell membrane, respectively. It is hard to say whether the tubules of the second layer within the tail also take part in the motion of the cell. But this seems possible, because there are also distinct cross bridges between these singlets and because both layers make contact. Perhaps the vigorous motions of the tail are caused by an interaction of the two layers of microtubules, running in opposite directions.

Phylogenetic comments. Hitherto, sperm cells of only four species of the Prolecithophora have been studied at the EM-level. Nevertheless, some general remarks should be attempted.

There is no evidence of any flagella, axonemes, or centrioles in mature spermatozoa. Because primitive Platyhelminthes possess flagella (Hendelberg 1977a; Tyler & Rieger 1977), this absence must be the result of a reduction; therefore the aflagellate condition is an apomorphous character. It is possible that this condition represents a synapomorphy within the Prolecithophora, a group of the Platyhelminthes lacking good synapomorphies (Karling 1974). But the spermatozoa and especially the spermatids of more species should be studied before the conclusion can be drawn that the aflagellated condition has not evolved several times independently.

This is also true for two other possible synapomorphous characters of the prolecithophoran sperm cell: (a) the loop-like foldings of the mitochondrial membrane. Christensen (1961), who studied the spermatozoon of *Plagiostomum morgani*, reports 'whorls of flattened cisternae, resembling endoplasmic reticulum (without ribosomes), which extend out to insert along every or every few pellicular fibrils'. The spermatozoa of *Hydrolimax grisea*, investigated by Newton (1975), possess flattened cisternae (personal communication of R. M. Rieger), and in *Acanthiella chaetonotoides* the spermatozoon also exhibits membranous disk-like foldings (Rieger 1981). (b) the lack of the so-called 'electron-dense bodies' or 'refractile granules', which seem to be a common feature of free-living platyhelminth spermatozoa, even of gnathostomulid spermatozoa (Graebner 1969; Hendelberg 1977b). It should be mentioned here that the aflagellated spermatozoa of the Macrostomida possess electron-dense spheres (Bedini & Papi 1970; Henley 1974).

The other characters of evolutionary progress and specialization in the spermatozoon of *P. quadrioculatum* seem to be autapomorphous characters (e.g. giant size of the nucleus) of this species or at best synapomorphous characters (e.g. nucleus and mitochondrial derivative spirally coiled) between closely related species or species groups within the Prolecithophora (Karling 1940, 1962).

Acknowledgements.

I would like to thank Prof. Dr. Tor G. Karling for

loaning me slides of *P. quadrioculatum* for light microscopic examination, Miss E. Hildenhagen for preparing the material for electron microscopy, and Dipl.-Biol. D. Whybrew for the linguistic correction of the manuscript. Financial support was provided by the Akademie der Wissenschaften und der Literatur, Mainz.

Abbreviations

c – cerebrum	ne – nuclear envelope
com – cortical microtubules	ph – pharynx
cop – copulatory organ	pn – protonephridium
fm – foldings of the mitochondrial membrane	rp – ribbon-like nuclear process
l – lattice	te – testis
mid – mitochondrial derivative	tt – tip of the tail
mt – microtubules	vi – vitellarium
n – nucleus	vs – vesicula seminalis

References

Afzelius, B. A., 1979. Sperm structure in relation to phylogeny in lower metazoa. In: Eds. Fawcett, D. W. & Bedford, J. M., The spermatozoon. Maturation, motility, surface properties and comparative aspects. Baltimore-Munich: Urban and Schwarzenberg, pp. 243–251.

Baccetti, B. & Afzelius, B. A., 1976. The biology of the sperm cell. Monographs in Developmental Biology Vol. 10. Basel: Karger, 254 pp.

Bedini, C. & Papi, F., 1970. Peculiar patterns of microtubular organization in spermatozoa of lower Turbellaria, In: Ed. B. Baccetti, Comparative spermatology. New York – London: Academic Press, pp. 363–366.

Böhmig, L., 1890. Untersuchungen über rhabdocöle Turbellarien. II. Plagiostomina und Cylindrostomina Graff. Z. wiss. Zool. 51: 167–480.

Christensen, A. K., 1961. Fine structure of an unusual spermatozoan in the flatworm *Plagiostomum*. Biol. Bull. 121: 416.

Ehlers, U. & Ehlers, B., 1977. Monociliary receptors in interstitial Proseriata and Neorhabdocoela (Turbellaria Neoophora). Zoomorphologie 86: 197–222.

Franzén, Å., 1956. On spermiogenesis, morphology of the spermatozoon, and biology of fertilization among invertebrates. Zool. Bidr. Uppsala 31: 355–482.

Franzén, Å., 1977. Sperm structure with regard to fertilization biology and phylogenetics. Verh. Deutsch. Zool. Ges. 1977: 123–138.

Graebner, I., 1969. Vergleichende elektronenmikroskopische Untersuchung der Spermienmorphologie und Spermiogenese einiger Gnathostomula-Arten: Gnathostomula paradoxa (Ax 1956), Gnathostomula axi (Kirsteuer 1964), Gnathostomula jenneri (Riedl 1969). Mikroskopie (Wien) 24: 131–160.

Graff, L. von, 1882. Monographie der Turbellaria. I. Rhabdocoelida. Leipzig. 441 pp.

Graff, L. von, 1904–08. Dr. H. G. Bronn's Klassen und Ordnungen des Thier-Reichs, Band IV Vermes, Abtlg. I.c.: Turbellaria, I. Abtlg.: Acoela und Rhabdocoelida. Leipzig: C. F. Winter, pp. 1733–2599.

Graff, L. von, 1913. Turbellaria II. Rhabdocoelida. Tierreich 35: 1–484.

Hendelberg, J., 1969. On the development of different types of spermatozoa from spermatids with two flagella in the Turbellaria with remarks on the ultrastructure of the flagella. Zool. Bidr. Uppsala 38: 1–50.

Hendelberg, J., 1970. On the number and ultrastructure of the flagella of flatworm spermatozoa. In: Ed. Baccetti, B. Comparative spermatology. New York–London: Academic press, pp. 367–374.

Hendelberg, J., 1974. Spermiogenesis, sperm morphology, and biology of fertilization in the Turbellaria. In: Eds. Riser, N. W. & Morse, M. P., Biology of the Turbellaria. New York: McGraw-Hill, pp. 148–164.

Hendelberg, J., 1975. Functional aspects of flatworm sperm morphology. In: Ed. Afzelius, B. A., The functional anatomy of the spermatozoon. Oxford–New York: Pergamon Press, pp. 299–309.

Hendelberg, J., 1977a. Comparative morphology of turbellarian spermatozoa studied by electron microscopy. Acta Zool. fenn. 154: 149–162.

Hendelberg, J., 1977b. Ultrastructure of the cytoplasmic region in spermatozoa of Cryptocelides (Polycladida, Turbellaria). Zoon 5: 107–114.

Henley, C., 1974. Platyhelminthes (Turbellaria), pp. 267–343. In: Eds. Giese, A. C. & Pearse, J. S., Reproduction of marine invertebrates Vol. I. New York–London: Academic Press.

Jensen, O. S., 1878. Turbellaria ad litora Norvegiae occidentalia. Bergen: J. W. Eides, 98 pp.

Karling, T. G., 1940. Zur Morphologie und Systematik der Alloeocoela Cumulata und Rhabdocoela Lecithophora (Turbellaria). Acta Zool. fenn. 26: 1–260.

Karling, T. G., 1962. Marine Turbellaria from the Pacific coast of North America II. Pseudostomidae and Cylindrostomidae. Ark. Zool. 15: 181–209.

Karling, T. G., 1974. On the anatomy and affinities of the turbellarian orders. In: Eds. Riser, N. W. & Morse, M. P., Biology of the Turbellaria. New York: McGraw-Hill, pp. 1–16.

Kulinitch, L. Ja., 1970. Two new plagiostomids (Turbellaria, Prolecithophora) from the Bering Sea. Zool. Sh. 49: 1614–1623.

Newton, W. D., 1975. The ultrastructure of the aflagellate spermatozoon of the freshwater turbellarian Hydrolimax grisea (Platyhelminthes: Plagiostomidae). Biol. Bull. 149: 551–567.

Newton, W. D., 1979. The motile apparatus of the aflagellate spermatozoon of Macrostomum tubum. Am. Zool. 19: 998.

Pladellorens, M. & Subirana, J. A., 1975. Spermiogenesis in the sea cucumber Holothuria tubulosa. J. Ultrastruct. Res. 52: 235–242.

Rieger, R. M., 1978. Multiple ciliary structures in developing spermatozoa of marine Catenulida (Turbellaria). Zoomorphologie 89: 229–236.

Rieger, R. M., 1981. The morphology of Turbellaria at the ultrastructure level. In: Schockaert, E. R. & Ball, I. R. Turbellaria. Proc. Third Int. Symp. Hydrobiologia. This volume.
Silveira, M., 1970. Characterization of an unusual nucleus by electron microscopy. J. Submicr. Cytol. 2: 13-24.
Silveira, M. & Porter, K. R., 1964. The spermatozoids of flatworms and their microtubular systems. Protoplasma 59: 240-265.
Thomas, M. B. & Henley, C., 1971. Substructure of the cortical singlet microtubules in spermatozoa of Macrostomum (Platyhelminthes, Turbellaria) as revealed by negative staining. Biol. Bull. 141: 592-601.
Tyler, S. & Rieger, R. M., 1975. Uniflagellate spermatozoa in Nemertoderma (Turbellaria) and their phylogenetic significance. Science 188: 730-732.
Tyler, S. & Rieger, R. M., 1977. Ultrastructural evidence for the systematic position of the Nemertodermatida (Turbellaria). Acta Zool. fenn. 154: 193-207.
Westblad, E., 1955. Marine 'Alloeocoels' (Turbellaria) from North Atlantic and Mediterranean coasts. I. Ark. Zool. 7: 491-526.

Index auctorum

GPSR Compliance

The European Union's (EU) General Product Safety Regulation (GPSR) is a set of rules that requires consumer products to be safe and our obligations to ensure this.

If you have any concerns about our products, you can contact us on ProductSafety@springernature.com

In case Publisher is established outside the EU, the EU authorized representative is:

Springer Nature Customer Service Center GmbH
Europaplatz 3
69115 Heidelberg, Germany

Batch number: 10800569

Printed by Printforce, the Netherlands